FOOD AND BEVERAGE MYCOLOGY

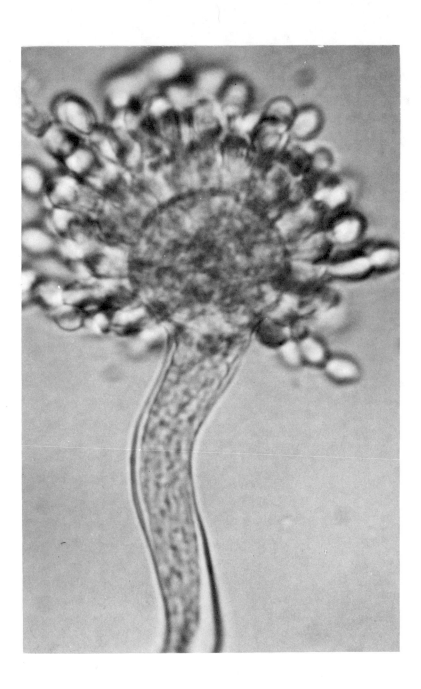

FOOD AND BEVERAGE MYCOLOGY

Larry R. Beuchat

Associate Professor,
Department of Food Science
Agricultural Experiment Station
University of Georgia
Experiment, Georgia

avi

AVI PUBLISHING COMPANY, INC.
Westport, Connecticut

Library of Congress Cataloging in Publication Data

Main entry under title:
　　Food and beverage mycology.

　　Includes index.
　　　1. Food— Microbiology. 2. Mycology.
I. Beuchat, Larry R.
QR115.F63　　　664'.001'5892　　　77-20943
ISBN 0-87055-247-3

Printed in the United States of America

Preface

In the course of studying molds and yeasts and their relationships to food and beverage spoilage and processing, I have found a great need for the compilation of such information into one book. Too often the student of food microbiology, the quality control technician, and the sanitarian must search through several "microbiology" sources in order to obtain specific information on food or beverage mycology. Our increased knowledge of the significant role fungi play with respect to food and beverage deterioration, mycotoxin production, fermentation, and potential sources of dietary protein further emphasizes the need for special consideration of mycology separate from bacteriology.

The plan of the book is, after a chapter on classification and descriptions of fungi associated with foods and beverages, to pass from basic information on one of the most important environmental factors affecting fungal growth, namely water activity, to discussions on specific commodity areas. This is followed by chapters on the use of fungi and fungal by-products as components of the human diet. Public health hazards implicated by the presence of mycotoxin producers in foods are treated in a separate chapter followed by chapters describing the detection of mycotoxins and fungal contaminants in foods and beverages.

It has been necessary to choose what to omit from the various topics presented in the book, since discussions of all aspects in so short a space are not possible. The practical and applied aspects of mycology as it relates to foods and beverages are stressed. Special efforts have been made by the authors to summarize in the form of tables information from a large number of authoritative sources. Also, valuable references are provided for each subject area.

My sincere gratitude is expressed to the authors who have contributed to this book. Their expert knowledge in particular areas of food and beverage mycology has made my job as editor enjoyable. Special thanks are due to Janice Fleming for retyping several of the manuscripts.

LARRY R. BEUCHAT

May 1977
Experiment, Georgia

Foreword

F ungi have contributed to the shaping of man's welfare from the beginning of civilization. However, the acquisition of knowledge regarding this group of microorganisms is too often neglected by students, professionals, and the general public.

Microbiology is too frequently equated with bacteriology, which deals with only one group of those microbes encompassed by the term, and thus the mycological aspects are generally minimized in classes and textbooks. Yet it has been the fungi which have provided much of our information in genetics, biochemistry, and nutrition. In fact the catalysts for biological transformation, enzymes, derive their name from the original isolation source—yeasts.

In the field of food microbiology the role of the fungi is also often under-emphasized. Textbooks concerning themselves with the beneficial and detrimental aspects of fungi as they affect food and drink vary in emphasis and inclusion of current information. *Food and Beverage Mycology* is filling a long-existing hiatus in books available to students, professionals, and lay persons interested in the role fungi play in food availability in both unprocessed and processed forms.

The book, with contributions by many recognized authorities in food microbiology, has concentrated upon fungi involved with food and beverage products. Information is provided on what fungi are, what they do, their involvement in all classes of food products, and methodologies used in detecting and quantifying fungal populations and mycotoxins. For the student, professional and lay person this book will bring together in one place the information they desire on fungi influencing the foods and beverages that are available to the world's population.

MARTIN W. MILLER

May 1977
Davis, California

Contributors

BENEKE, E. S., Professor, Department of Botany and Plant Pathology, Michigan State University, East Lansing, Michigan

BEUCHAT, L. R., Associate Professor, Department of Food Science, Agricultural Experiment Station, University of Georgia, Experiment, Georgia

BOTHAST, R. J., Northern Regional Research Laboratory, Agricultural Research Service, U. S. Department of Agriculture, Peoria, Illinois

BULLERMAN, L. B., Associate Professor, Department of Food Science and Technology, University of Nebraska-Lincoln, Lincoln, Nebraska

CHRISTENSEN, C. M., Regents' Professor Emeritus, Department of Plant Pathology, University of Minnesota, St. Paul, Minnesota

CORRY, JANET E. L., Metropolitan Police Forensic Science Laboratory, 109 Lambeth Road, London SE1 7LP, England

DAVIS, N. D., Professor, Department of Botany and Microbiology, Auburn University Agricultural Experiment Station, Auburn, Alabama

DIENER, U. L., Professor, Department of Botany and Microbiology, Auburn University Agricultural Experiment Station, Auburn, Alabama

HAYES, W. A., Professor, Department of Biological Sciences, The University of Aston in Birmingham, Gosta Green, Birmingham B4 7ET, England

HUMPHREYS, T. W., Central Research and Development Department, John Labatt Limited, London, Ontario N6A 4M3, Canada

JARVIS, B., The British Food Manufacturing Industries Research Association, Randalls Road, Leatherhead, Surrey KT 22 7RY, England

JAY, J. M., Professor, Department of Biology, Wayne State University, Detroit, Michigan

MARTH, E. H., Professor, Department of Food Science, University of Wisconsin, Madison, Wisconsin

MILLER, M. W., Professor, Department of Food Science and Technology, University of California, Davis, California

MUNDT, J. O., Professor, Departments of Food Technology and Science, and Microbiology, University of Tennessee, Knoxville, Tennessee

PONTE, J.G., JR., Professor, Department of Grain Science and Industry, Kansas State University, Manhattan, Kansas

SINSKEY, A. J., Associate Professor, Department of Nutrition and Food Science, Massachusetts Institute of Technology, Cambridge, Massachusetts

SMILEY, K. L., Northern Regional Research Laboratory, Agricultural Research Service, U. S. Department of Agriculture, Peoria, Illinois

SPLITTSTOESSER, D. F., Professor, Department of Food Science and Technology, New York State Agricultural Experiment Station, Geneva, New York

STEVENSON, K. E., Associate Professor, Department of Food Science and Human Nutrition, Michigan State University, East Lansing, Michigan

STEWART, G. G., Beverage Science Department, Labatt Breweries of Canada Limited, London, Ontario N6A 4M3, Canada

TSEN, C. C., Professor, Department of Grain Science and Industry, Kansas State University, Manhattan, Kansas

Contents

Classification of Food and Beverage Fungi

E. S. Beneke and K. E. Stevenson

Fungi comprise a large group of microorganisms which affect our food supply as a result of their infection of living plant tissue and fruit, their ability to invade and decompose harvested and processed foods and beverages, and their capacity to produce toxic secondary metabolites. They are responsible for a major portion of food deterioration in developing countries. On the other hand, the preservative effects brought about by fungal fermentation of foods and beverages are well-known benefits. In addition, various types of mushrooms, referred to as macrofungi, are an important part of human diets in many countries.

Fungi of importance in food and beverage industries can be classified in the Eumycota or true fungi. Traditionally the true fungi have been divided into three classes—Phycomycetes, Ascomycetes, and Basidiomycetes—based on the mode of sexual spore production, with the Deuteromycetes or Fungi Imperfecti reserved for organisms for which no sexual stage was known. However, some taxonomists (e.g., Ainsworth 1966) prefer to divide the Eumycota into five subdivisions which are further divided into classes. The subdivisions of Eumycota can be classified as follows:

(1a)   Motile cells (zoospores) present; perfect-state spores typically oospores   Mastigomycotina

(b)	Motile cells absent	2
(2a)	Perfect state absent	Deuteromycotina
(b)	Perfect state present	3
(3a)	Perfect state spores zygospores	Zygomycotina
(b)	Zygospores absent	4
(4a)	Perfect state spores ascospores	Ascomycotina
(b)	Perfect state spores basidiospores	Basidiomycotina

Since the class Phycomycetes was a miscellaneous assemblage, two of the subdivisions, Mastigomycotina and Zygomycotina, now comprise representatives of this former class.

In most fresh, moist foods, fungi do not grow well due to competition from bacteria. However, in foods which have conditions such as lowered water activity, pH less than 4.5, or refrigerated temperatures, fungi may proliferate. In other words, fungi usually become predominant under conditions which tend to markedly retard bacterial growth. This characteristic will be illustrated in detail in several of the following chapters. Indeed, conditions adverse to bacterial growth are often employed in laboratory media used to detect and identify fungi in foods and beverages (Chap. 16). In the laboratory, microbiologists commonly separate cultures of true fungi into two main groups—yeasts and molds—based on colonial and microscopic appearance. The molds are typically filamentous and are primarily classified according to the morphology of hyphae, sexual spores, and asexual spores. Yeasts, on the other hand, exist mainly as single cells, i.e., they lack many morphological criteria used to classify filamentous fungi, and are normally classified separately with an emphasis on physiological characteristics. Due to the divergent methods used for classification of yeasts and molds, the two groups are discussed separately in the remainder of this chapter.

YEAST CLASSIFICATION

Yeasts as a group are difficult to define. They are fungi which at some stage in their life cycle exist primarily as single cells which reproduce by budding or fission. Typical structural characteristics of a yeast cell are illustrated in Fig. 1.1. By general agreement fungi which are black-pigmented or pluri-nucleate are not considered to be yeasts, although *Aureobasidium (Pullularia) pullulans* has often been referred to as the "black yeast." This is a subjective system and representative yeasts are classified in the Ascomycotina, Basidiomycotina, and Deuteromycotina.

A monumental work, *The Yeasts: A Taxonomic Study* (Lodder 1970) provides the most comprehensive basis for yeast taxonomy at

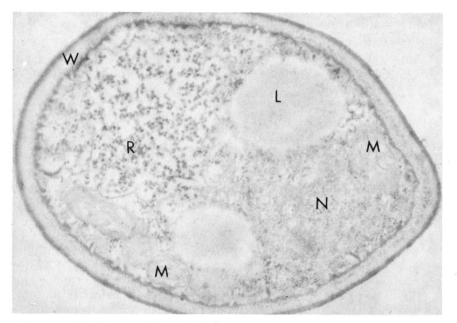

FIG. 1.1. ELECTRON PHOTOMICROGRAPH OF A VEGETATIVE CELL OF *CANDIDA STEATOLYTICA*. Several features common to yeast cells are visible, including cell wall (W), nucleus (N), mitochondria (M), lipid droplets (L), and ribosomes (R); (30,000X).

the present time. Changes in yeast taxa, including the addition of new genera and species, have occurred since 1970, but most of these changes do not involve yeasts which are commonly associated with foods and beverages.

The following key is designed for generic classification of yeasts which are commonly isolated from foods and beverages. In order to facilitate identification of unknown isolates, yeast genera of the Ascomycotina and Deuteromycotina are included in the key:

(1a) Vegetative reproduction exclusively by cross wall formation, i.e., fission *Schizosaccharomyces*

(b) Vegetative reproduction by budding 2

(2a) Vegetative reproduction by bipolar budding, cells usually apiculate or lemon-shaped 3

(b) Vegetative reproduction by multipolar or multilateral budding 5

(3a) Ascospores not formed *Kloeckera*

(b) Ascospores formed 4

(4a) Ascospores spherical *Saccharomycodes*

(b) Ascospores hat- or helmet-shaped, or globose *Hanseniaspora*

(5a) Ballistospores produced   6
(b) Ballistospores not produced   7
(6a) Ballistospores asymmetrical; usually produce
       carotenoid pigments   *Sporobolomyces*
(b) Ballistospores symmetrical; carotenoid pigments not
       produced   *Bullera*
(7a) Cells ogival; strong acetic acid aroma produced; cells
       short-lived on malt agar   8
(b) Cells differ from above description (7a) by one or more
       characteristics   9
(8a) Ascospores formed   *Dekkera*
(b) Ascospores not formed   *Brettanomyces*
(9a) Ascospores formed   10
(b) Ascospores not formed   18
(10a) Nitrate assimilated   11
(b) Nitrate not assimilated   12
(11a) Ascospores spherical with warty walls   *Citeromyces*
(b) Ascospores hat- or saturn-shaped or spherical without
       warty walls   *Hansenula*
(12a) Glucose fermentation weak, slow, or absent   13
(b) Glucose vigorously fermented   16
(13a) Abundant development of true
       mycelium   *Saccharomycopsis*
(b) True mycelium absent or limited if present   14
(14a) Ascospores unusually large and obtuse   *Lodderomyces*
(b) Ascospores differ from above description   15
(15a) Ascospores oval or spherical with warty walls; asci do
       not dehisce readily   *Debaryomyces*
(b) Ascospores spherical, hat- or saturn-shaped with gen-
       erally smooth walls; usually they are easily liberated from
       the ascus   *Pichia*
(16a) Mature asci do not rupture easily; ascospores are
       spheroidal to ellipsoidal   *Saccharomyces*
(b) Mature asci rupture easily   17
(17a) Ascospores reniform to oblong   *Kluyveromyces*
(b) Ascospores spherical, hat- or saturn-shaped   *Pichia*
(18a) Arthrospores and true mycelium are
       produced   *Trichosporon*
(b) Arthrospores absent, true mycelium may be present 19
(19a) Pink or yellow carotenoid pigments formed; no
       fermentation   20
(b) No pigments formed; fermentation may occur   21
(20a) Inositol not assimilated   *Rhodotorula*
(b) Inositol assimilated   *Cryptococcus*

(21a) Pseudomycelium always present; true mycelium may be
       formed   *Candida*
(b) Pseudomycelium absent or rudimentary; no true
       mycelium   22
(22a) Inositol is assimilated; no fermentation   *Cryptococcus*
(b) Inositol is not assimilated; fermentation may
       occur   *Torulopsis*

Speciation of yeasts within genera is accomplished primarily by physiological tests, and currently Lodder (1970) is the best source of information. As an aid to identification of yeasts, Barnett and Pankhurst (1973) have published a key which gives extensive results of physiological tests for 434 yeast species. While Adansonian approaches to yeast classification have been utilized (Campbell 1974), the results do not correspond well with presumed phylogenetic development of yeasts. Recently a simple test was described which reputedly can be used to determine whether an imperfect yeast is related to ascomycetous or heterobasidiomycetous yeasts (van der Walt and Hopsu-Havu 1976). Thus, in the future it may be possible to divide imperfect genera, such as *Candida*, into two phylogenetically distinct groups.

TABLE 1.1

YEAST GENERA WHICH ARE NOT INCLUDED
IN THE IDENTIFICATION KEY

Aessosporon	*Metschnikowia*	*Selenotila*
Ambrosiozyma	*Nadsonia*	*Sporidiobolus*
Arthroascus	*Nematospora*	*Sterigmatomyces*
Coccidiascus	*Oosporidium*	*Sympodiomyces*
Cyniclomyces	*Pachysolen*	*Syringospora*
Filobasidium	*Phaffia*	*Trigonopsis*
Guilliermondella	*Pityrosporum*	*Wickerhamia*
Hormoascus	*Rhodosporidium*	*Wickerhamiella*
Leucosporidium	*Schizoblastosporion*	*Wingea*
Lipomyces	*Schwanniomyces*	

Table 1.1 lists the currently recognized yeast genera which are not included in keys or in discussions in this chapter. Most of these genera are comprised of strains which rarely, if ever, would be encountered in foods and beverages. When isolated from foods and beverages, other genera, such as *Leucosporidium*, *Rhodosporidium*, and *Syringospora*, would only be recognized as their asexual forms (e.g., *Candida*, *Rhodotorula*, and *Candida albicans*, respectively) in most laboratories. Thus, the above key is designed to be utilitarian, and is not comprehensive from a taxonomic standpoint. The following descriptions are given for yeast genera which may be found in foods and beverages. Readers who wish more detailed descriptions, keys, and classification of yeasts into higher taxa should

consult Lodder (1970), Ainsworth (1973), and Kreger-van Rij (1973).

Subdivision Ascomycotina

Yeasts in the Ascomycotina are confined to the class Hemiascomycetes and the order Endomycetales in which zygotes or single cells are transformed directly into asci. For a description and key to the families in this order consult Kreger-van Rij (1973). The following genera of the Ascomycotina are of interest to the food mycologist.

Citeromyces.—Vegetative reproduction is by multilateral budding. Thick-walled asci contain one wrinkled (rugose) ascospore. The single species, *Citeromyces matritensis,* is fermentative, assimilates nitrate, and is osmophilic. It was originally isolated from sweetened, condensed milk and occasionally is isolated from fermenting fruit.

Dekkera.—Forms 1 to 4 hat-shaped or spheroidal ascospores per ascus. The ascospores are usually liberated soon after maturation, and tend to agglutinate. For a general description of the genus see *Brettanomyces* (subdivision Deuteromycotina).

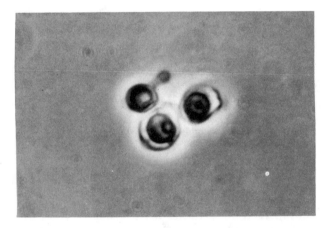

FIG. 1.2. ASCI OF *DEBARYOMYCES HANSENII* CONTAINING ONE ASCOSPORE PER ASCUS. Note the conjugation tube on one ascus and the oil droplet inside each ascospore (phase microscopy, 1740X).

Debaryomyces.—Vegetative reproduction occurs by multilateral budding. Asci usually contain 1 to 2 spores with warty walls; ascospores often have an oil drop inside (Fig. 1.2). Nitrate is not assimilated and sugar fermentation is slow or absent. One species,

Debaryomyces hansenii, is particularly salt-tolerant and can grow in solutions containing 15 to 24% sodium chloride. It is commonly isolated from brined or salt-containing foods. Notably, *D. hansenii* forms slime on certain meat products such as bacon, wieners, and luncheon meats stored under aerobic conditions. In the future these yeasts may be referred to as *Torulaspora* spp.[1]

Endomycopsis.—See *Saccharomycopsis.*

Hanseniaspora.—Asci contain 2 to 4 hat-shaped spores which are released upon maturity, or 1 to 2 spherical spores which are not released at maturity. For a general description, see *Kloeckera* (subdivision Deuteromycotina).

Hansenula.—Vegetative cells reproduce by multilateral budding; pseudomycelia and true mycelia may be formed. Asci contain 1 to 4 hat or saturn-shaped spores which sometimes appear hemispheroidal or spheroidal due to indistinct rings. The ascospores are usually liberated at maturity. Nitrate is assimilated and sugar fermentation is variable. Some strains of *Hansenula* are found as "wild yeasts" contaminating various fermented foods and beverages.

---

[1]Yeast taxonomy is progressing at a rapid pace. As this chapter is being written, there is some acceptance among taxonomists of the contention of van der Walt and Johannsen (1975) that *Torulaspora* is the correct generic name for a large group of ascosporogenous yeasts. They redefined the genus as follows:

Torulaspora: These are predominantly haploid yeasts which form one to four spheroidal to long-oval ascospores without circumfluent ledges. Ascus formation is normally initiated by somatogamous autogamy. Nitrate is not assimilated and fermentation is variable.

The redefined genus *Torulaspora* would be comprised of yeasts which were recently classified (Lodder 1970) into three genera. It would include species of *Debaryomyces*, a generic name which no longer would be recognized. *Torulaspora* would also contain five former *Pichia* species which are haploid, heterothallic, and produce ascospores without a ledge, and the primarily haploid *Saccharomyces* strains (16 species) included in the third and fourth groups discussed under *Saccharomyces* (pages 7-9). This rearrangement of taxa also results in greater homogeneity of the organisms remaining in the genera *Pichia* and *Saccharomyces* (van der Walt and Johannsen 1975). Such a change in taxonomy would naturally necessitate changing the yeast classification key given on pages 3-5. One possible classification scheme would be to substitute the following for sections 14 through 17 in the yeast classification key.

(14a)	Ascospores unusually large and obtuse	*Lodderomyces*
(b)	Ascospores hat- or saturn-shaped	*Pichia*
(c)	Ascospores oval to spheroidal, with or without warty walls	*Torulaspora*
(15a)	Ascospores hat- or saturn-shaped	*Pichia*
(b)	Ascospores spheroidal to oval	16
(16a)	Asci dehisce readily; ascospores reniform to oval	*Kluyveromyces*
(b)	Asci do not dehisce readily	17
(17a)	Ascus formation initiated by somatogamous autogamy; vegetative cells stabilized in haploid state	*Torulaspora*
(b)	Primarily stabilized in diploid state or, if haploid, conjugation of independent cells usually precedes ascus formation	*Saccharomyces*

Many strains produce pellicles (films) on the surfaces of liquids, and some strains are pectolytic.

Kluyveromyces (Fabospora).—Vegetative reproduction is by multilateral budding; pseudomycelia may be formed. Asci contain one to numerous reniform to spheroidal spores which are released at maturity. Nitrate is not assimilated and an exogenous source of vitamins is required. Sugars are vigorously fermented. Two lactose-fermenting species in this genus, *Kluyveromyces fragilis* and *K. lactis,* are important due to their ability to ferment and spoil dairy products. These two species were formerly classified in the genus *Saccharomyces.*

Lodderomyces.—Vegetative reproduction occurs by multilateral budding; pseudomycelium is produced. These yeasts are characterized by formation of 1 to 2 unusually large, obtuse ascospores per ascus. Fermentation of glucose is slow and nitrate is not assimilated. The single species, *Lodderomyces elongisporus,* has been found as a contaminant in soft drinks.

Pichia.—Vegetative reproduction is by multilateral budding; pseudomycelia are usually present. Asci contain 1 to 4 spheroidal, hat- or saturn-shaped spores which are readily liberated from the ascus. Nitrate is not assimilated and fermentation of sugars is variable. Many species produce pellicles in liquid media, i.e., they are film-formers. In the future, strains which form spheroidal ascospores and do not vigorously ferment glucose may be classified in *Torulaspora* (see footnote[1], p. 7).

Saccharomyces.—Vegetative reproduction occurs by multilateral budding; pseudomycelia may be formed. Asci usually contain 1 to 4 smooth-walled spheroidal spores which are not liberated upon reaching maturity (Fig. 1.3.). Nitrate is not assimilated and lactose is

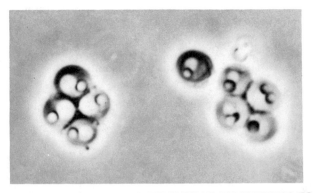

FIG. 1.3. ASCI OF *SACCHAROMYCES CEREVISIAE* CONTAINING FOUR ASCO-SPORES PER ASCUS (PHASE MICROSCOPY, 1740X)

not utilized. Sugars are fermented, usually vigorously.

This genus includes a large number of species which constitute a somewhat heterogenous mixture. There is one group of primarily diploid species which is closely related to *Saccharomyces cerevisiae*. In the current classification system, bakers' yeasts, wine yeasts, and champagne yeasts are all considered to be strains of *S. cerevisiae*. Brewers' yeasts are classified as strains of *Saccharomyces uvarum*; *S. carlsbergensis* is no longer recognized by taxonomists as a distinct species. Some brewers, however, believe that *S. carlsbergensis* exhibits characteristics sufficiently different from some *S. uvarum* strains to justify its existence as a distinct species. (For further discussion, see Chap. 10.) A second group contains primarily haploid species and includes the important osmophilic yeasts *Saccharomyces bailii* var. *osmophilis*, *S. bisporus* var. *mellis*, and *S. rouxii*. Organisms in this group have often been cited as belonging to *Zygosaccharomyces*, a genus which is no longer recognized. A third group contains primarily haploid species related to *Saccharomyces rosei*, which eventually may be classified as *Torulaspora* (see footnote[1]), along with some of the remaining species which represent a heterogenous group. Interested readers should consult van der Walt (1970A) and van der Walt and Johannsen (1975) for a more detailed discussion of these groups of *Saccharomyces*.

Saccharomycodes.—These large apiculate cells reproduce by bipolar budding on a broad base. Asci usually contain four spheroidal spores (in pairs). Nitrate is not assimilated and an external source of vitamins is not required. The single species, *Saccharomycodes ludwigii*, is fermentative and occasionally has been isolated from fermenting materials, particularly fruit musts. This organism is also known to be a symbiont in the "tea fungus" culture used to make an acidic drink from tea infusion and sugar (Phaff *et al.* 1966).

Saccharomycopsis (Endomycopsis).—Budding cells are present and there is abundant formation of septate hyphae. Asci contain up to four hat-shaped or saturnoid spores. Nitrate is not assimilated and fermentation of sugars is weak or absent. This genus contains some species which were formerly classified in *Endomycopsis* (van der Walt and Scott 1971; von Arx 1972). The organism previously classified as *Saccharomycopsis guttulata* has been transferred to a new genus, *Cyniclomyces*.

Schizosaccharomyces.—An ascosporogenous yeast in which vegetative cell reproduction occurs by fission or cross-wall formation (Fig. 1.4). This mode of reproduction distinguishes this genus from all other yeast genera. Asci contain 4 to 8 spheroidal to bean-shaped spores (Fig. 1.5). Nitrate is not assimilated, the organisms are fer-

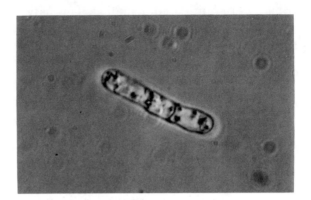

FIG. 1.4.   VEGETATIVE CELLS OF *SCHIZOSACCHAROMYCES* REPRODUCING BY FISSION (CROSS-WALL FORMATION) (PHASE MICROSCOPY, 950X)

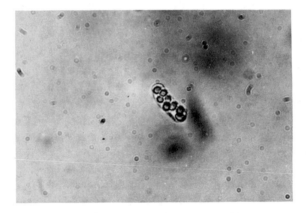

FIG. 1.5.   ASCUS OF *SCHIZOSACCHAROMYCES OCTOSPORUS* CONTAINING EIGHT ASCOSPORES (LIGHT MICROSCOPY, 600X)

mentative and some members of the genus, e.g., strains of *Schizosaccharomyces octosporus*, are osmophilic.

Subdivision Deuteromycotina

The subdivision Deuteromycotina includes a variety of yeasts in which a sexual stage has not been detected. Strains classified as imperfect yeasts are sometimes found to possess a sexual stage and subsequently they are reclassified in the appropriate ascomycetous or basidiomycetous taxon. Some strains of imperfect yeasts are clearly related to particular perfect yeasts, although sexual stages are never observed in those imperfect strains. Other strains of imperfect yeasts, e. g. *Trichosporon*, have no equivalent perfect stage. The Deuteromycotina is normally divided into two families: *Sporobolo-*

mycetaceae, which includes the ballistosporogenous genera *Bullera* and *Sporobolomyces*, and the family Cryptococcaceae which contains the remainder of the imperfect yeast genera. Genera of importance in food and beverage mycology are described below.

Brettanomyces.—These yeasts produce ogival.cells and a large amount of acetic acid under aerobic conditions. They require exogenous vitamins for growth, particularly high concentrations of thiamine. Vegetative reproduction is by multilateral budding; pseudomycelia are produced. Assimilation of nitrate is variable. Oxidation and fermentation of sugars occur. Strains exhibit a characteristic "negative Pasteur effect," i.e., oxygen stimulates fermentation of glucose (Wiken *et al.* 1961). This genus is the asporogenous equivalent of *Dekkera*.

Yeasts belonging to this genus have been isolated from materials associated with fermented beverages. In some instances (e.g., Belgian lambic beer) *Brettanomyces* is used in secondary fermentations (van der Walt 1970B). However, they usually are found as undesirable contaminants in beer, wine, and soft drinks. Yeast isolates belonging to *Brettanomyces* and *Dekkera* are generally short-lived on normal agar media due to their production of acetic acid.

Bullera.—This genus is characterized by the formation of symmetrical ballistospores which are forcibly discharged by a droplet mechanism. Vegetative reproduction occurs by multilateral budding; pseudomycelia are not produced. Nitrate assimilation is variable and sugars are not fermented. These yeasts are usually isolated from plants, but several strains of *Bullera alba* have been isolated from food processing environments. A sensitive test developed by do Carmo-Sousa and Phaff (1962) for the detection of ballistospore discharge is shown in Fig. 1.6.

Candida.—These asporogenous yeasts reproduce by multilateral budding, lack carotenoid pigments, and form pseudomycelia (Fig. 1.7). This genus is the depository for a large heterogeneous group of yeasts which lacks any important criterion for classification except for the production of pseudomycelia. Assimilation of nitrate and sugar fermentation are variable. Over 100 *Candida* spp. have been recognized. Strains which produce teliospores are classified in the perfect genus *Leucosporidium*. The genus *Syringospora* was erected for some strains of *Candida albicans*, the most important pathogenic yeast in which a sexual stage has been observed.

Cryptococcus.—These asporogenous yeasts reproduce by multilateral budding and do not usually form pseudomycelia. Inositol is assimilated, and nitrate assimilation is variable. Sugars are not

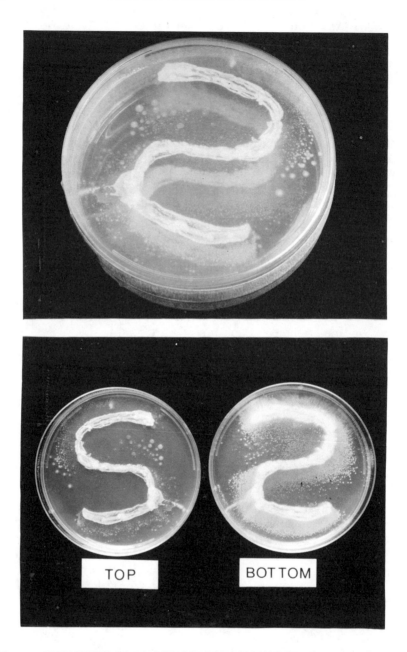

FIG. 1.6.   DETECTION OF BALLISTOSPORE DISCHARGE. The upper photograph
shows a cornmeal agar plate which was streaked with a ballistosporogenous yeast,
inverted over an uninoculated cornmeal agar plate, and incubated at 25°C. After
incubation and ballistospore discharge, the bottom plate contains a mirror image of the
"S" pattern streaked on the top plate (bottom photograph).

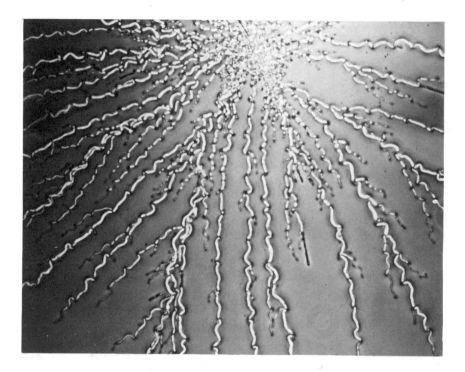

FIG. 1.7. PSEUDOMYCELIUM PRODUCED BY *CANDIDA UTILIS* (PHASE MICROSCOPY, 120X)

fermented. Most strains produce capsules, and small amounts of carotenoid pigments may be produced. These yeasts are commonly found on plants and in soil. For example, cryptococci were shown to be the most common yeasts in an extensive study of the flora of strawberries (Buhagiar and Barnett 1971).

Kloeckera.—Apiculate or ovoidal cells reproduce by bipolar budding (Fig. 1.8) and require an external source of vitamins for growth. Nitrate is not assimilated and all strains ferment sugars. These organisms are most commonly isolated from fresh fruit and materials (including fruit flies) associated with spoiling or fermenting fruit (see Chap. 3). This genus is the asporogenous equivalent of *Hanseniaspora.*

Rhodotorula.—Vegetative reproduction is by multilateral budding; they do not produce pseudomycelia, ascospores, or ballistospores. Red to yellow carotenoid pigments are produced. Nitrate is not assimilated and sugars are not fermented. Inositol is not assimilated. Most strains produce orange or salmon-pink colonies, although

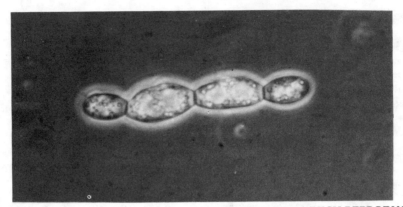

FIG. 1.8. VEGETATIVE CELLS OF A *HANSENIASPORA* SP. WHICH REPRODUCE BY BIPOLAR BUDDING ON A BROAD BASE (PHASE MICROSCOPY, 1410X)

pigmentation can range from yellow to deep red. They may cause "visual" spoilage of foods due to the production of off-colors or simply due to the appearance of pigmented colonies. Many strains are psychrotrophic and may be found in frozen and refrigerated products (Walker and Ayres 1970). Strains which produce teliospores have been classified in the perfect genus *Rhodosporidium*.

Sporobolomyces.—This genus is characterized by the formation of asymmetrical ballistospores which are forcibly discharged. Vegetative reproduction occurs by multilateral budding; pseudomycelia may be produced. Nitrate assimilation is variable. These yeasts are nonfermentative and pigmented (pink), except for one strain which was isolated from the frass of a bark beetle (Phaff and do Carmo-Sousa 1962). They are usually associated with plants or decaying plant material, but several strains have been isolated from foods and food processing environments.

Torulopsis.—These asporogenous yeasts reproduce by multilateral budding and do not form pseudomycelia. They lack carotenoid pigments and do not assimilate inositol. Assimilation of nitrate and fermentation of sugars are variable. This nondescript genus contains a heterogeneous array of yeasts which cannot be conveniently classified elsewhere. Some strains are osmophilic, e.g., *Torulopsis lactis-condensi* which can spoil sweetened, condensed milk, while others are significantly acid-tolerant. Spoilage of fermented foods and salad dressing frequently is caused by *Torulopsis* spp.

Trichosporon—Vegetative reproduction occurs by budding and formation of arthrospores. True mycelium is always present (Fig.

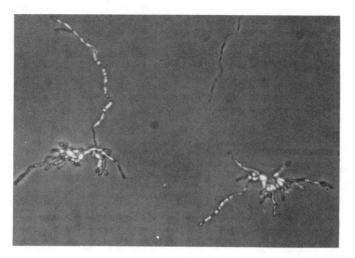

FIG. 1.9. BUDDING CELLS AND TRUE MYCELIUM PRODUCED BY A TRI—CHOSPORON SP. (PHASE MICROSCOPY, 280X)

1.9). Asexual endospores may be formed. Nitrate assimilation is variable and sugar fermentation is absent or weak and latent.

Subdivision Basidiomycotina

Yeasts in the Basidiomycotina are generally considered to belong to the order Ustilaginales. The basidiomycetous genera. *Leucosporidium*, *Rhodosporidium*, *Aessosporon*, and *Filobasidium*, are not described in this chapter. However, they are related to some species of imperfect genera such as *Candida*, *Cryptococcus*, *Rhodotorula*, *Sporidiobolus*, and *Sporobolomyces*. While the ballistosporogenous yeasts, *Bullera*, *Sporidiobolus*, and *Sporobolomyces*, are presumed to be related to the basidiomycetous yeasts, to date sexual reproduction has been clearly demonstrated in only one species, *Sporobolomyces salmonicolor*, whose perfect form is the genus *Aessosporon*.

MOLD CLASSIFICATION

The following genera of filamentous fungi commonly associated with foods and beverages are briefly described under each subdivision along with a key. A comprehensive taxonomic review with keys published by Ainsworth *et al.* in two volumes (1973A, B) provides an up-to-date illustrated taxonomic outline of the major fungal groups down to the generic level. Another useful reference for fungi that sporulate in culture is by von Arx (1974). Additional references and monographs are listed under each subdivision. The taxonomic keys and descriptions of the genera presented here are designed to be useful for identifying molds most frequently encountered in foods and beverages and are not comprehensive. For additional organisms, reference should be made to the bibliography.

Subdivision Mastigomycotina

The three classes in the subdivision Mastigomycotina are the Chytridiomycetes, Hyphochytridiomycetes, and Oomycetes. The order Peronosporales, in the Oomycetes, with two genera, *Phytophthora* and *Pythium,* is of significance in the food industry. The three classes are arbitrarily brought together on the basis of production of uniflagellate or biflagellate zoospore production. The perfect state spores are typically oospores. The class Oomycetes comprises all orders with biflagellate zoospores. The Peronosporales are characterized by a reniform type, biflagellate zoospore, nonseptate hyphae, and single oospores in the oogonia. These organisms grow as parasites or saprophytes. For speciation of *Phytophthora* and *Pythium* reference should be made to Waterhouse (1968, 1970). The key characteristics for separation of the two genera are as follows:

(1a) Sporangia usually ovoid with a papilla; zoospores differentiated before emission into a
vesicle   *Phytophthora*
(b) Sporangia ovoid, filamentous; protoplasm differentiating into zoospores in a vesicle   *Pythium*

Description of selected genera of Mastigomycotina is as follows:

Phytophthora.—The hyphae are nonseptate and well branched, forming sporangia on tips of branches or developing distinct sporangiophores in some species. The ovoid to obpyriform sporangia with an apical pore differentiate zoospores before emission into a vesicle that quickly evanesces. Some species have deciduous sporangia on branched sporangiophores. Oogonia are smooth-walled with one oospore. Certain species cause water-soaked rot of vegetables.

Pythium.—The hyphae are nonseptate and well-branched, forming sporangia on the ends of the branches. The sporangia are filamentous; inflated hyphae are spherical or ovoid in shape, depending upon the species. Sporangia may release protoplasm into a vesicle and differentiate zoospores which are released, or function as a spore and germinate to form hyphae. The oogonium is smooth or spiny-walled, with usually one oospore. Many species of this genus may cause fruit rots and black internal rot of potatoes.

Subdivision Zygomycotina

The subdivision Zygomycotina contains two classes: the Zygomycetes, which include saprobic organisms with mycelium immersed in the host; and the Trichomycetes, which have organisms associated with arthropods and are attached to the digestive tract by holdfasts. The characteristic feature of all these organisms is the production of the zygospore, a perfect-state spore.

The Zygomycotina, represented by three orders, have one order, the Mucorales, of significance as related to food and beverage spoilage and manufacture. Most members of this group are saprophytic, with asexual reproductive structures varying from many-spored sporangia, sporangiola, to one-spored sporangia in the genera. The zygospore forms from fusion of gametangia and may be surrounded by varied shapes of appendages. Two strains must be present in heterothallic strains while only one strain is required if it is homothallic for zygospore formation.

These molds grow rapidly in culture or on various food substrates, producing nonseptate hyphae except where septa are formed in branches developing reproductive structures. Genera in this order are important agents in storage decay, rapid spoilage of bread, and "leak" in small fruits. The following is a key to the most commonly encountered organisms in the Mucorales which are of importance in fermentations and food spoilage. Descriptions of selected genera are presented. For further information on the genera, reference should be made to the extensive coverage of these organisms in Zycha et al. (1969) and Ainsworth et al. (1973B).

(1a)	Rod-shaped merosporangia borne on branched sporangiophores	*Syncephalastrum*
(b)	Not producing rod-shaped merosporangia	2
(2a)	Sporangiospores dark-colored with tufts of bristles at their ends; sporangial wall splitting into halves (Choanephoraceae)	3
(b)	Sporangiospores not with tufts at their ends, usually not dark colored; sporangial wall not splitting into halves	4
(3a)	Sporangia and one-spored sporangiola (conidia) produced	*Choanephora*
(b)	Sporangia and several-celled sporangiola produced	*Blakeslea*
(4a)	Sporangia globose, without a columella	*Mortierella*

(b)   Sporangia globose, with a columella (Mucoraceae)   5
(5a)   Sporangia abortive, many
       chlamydospores   *Chlamydomucor*
(b)   Sporangia not abortive, usually abundant   6
(6a)   A distinct apophysis (swelling) at the base of the
       sporangium; rhizoids at the base of
       sporangiophore   *Rhizopus*
(b)   No distinct apophysis at base of sporangium   7
(7a)   Rhizoids and stolons; not thermophilic; colonies white
       to tan   *Actinomucor*
(b)   No rhizoids or stolons; sporangiophores branched;
       sporangia; sporangiola-like sporangia; spores dark-
       colored   *Mucor*

Description of selected genera of Zygomycotina is as follows:

Actinomucor.—This genus has globose columellate sporangia with hyaline cell walls. The terminal sporangia are large, surrounded by whorls of three to five smaller sporangia. Sporangial walls are roughened by calcium oxalate. Spores are round and smooth-walled. Rhizoids and stolons are formed on the hyphae. No zygospores are known. This genus is used in fermented food processes.

Blakeslea.—This genus has large terminal, columellate sporangia with sporangial walls splitting into two halves. Smaller columellate sporangiola containing usually three spores develop on dichotomous branches. The spores are usually striate and reddish-brown with long stiff spines at both ends. Zygospores form in tong-like suspensors, homothallic. Riboflavin and β-carotene are products of fermentation by this fungus.

Chlamydomucor.—This genus is constantly forming abortive sporangia which resemble sporangia of *Rhizopus*. The globose sporangia are without sporangiola. Many chlamdospores develop in the mycelium. This organism is utilized in fermented Oriental foods.

Choanephora.—Large, terminal, columellate sporangia with walls splitting into two equal halves, and one-spored sporangiola (conidia) are produced by this genus. Spores are not striated in the sporangia; however, sporangiola are striated with long stiff spines at both ends. Zygospores are similar to *Blakeslea*. Riboflavin and β-carotene are products of fermentation by this fungus.

Mortierella.—Branched or unbranched sporangiophores are tapered at the tip with sporangia attached that are globose without a columella. The spores are round, smooth, and colorless. Zygospores, when present, are covered by a thick layer of interwoven hyphae.

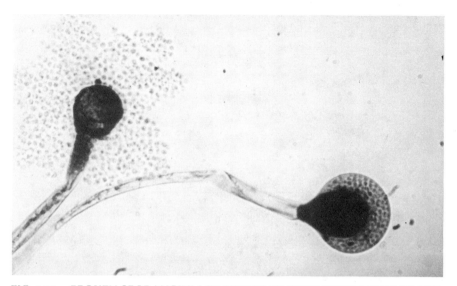

FIG. 1.10. BROKEN SPORANGIUM OF *MUCOR* SP. WITH SPORE RELEASE AND A SPORANGIUM CONTAINING A COLUMELLA AND SPORES (LIGHT MICROSCOPY 350X)

Mucor.—Globose, columellate sporangia are formed with spherical to elliptical, smooth-walled, colorless spores (Fig. 1.10). No rhizoids are present at the base of sporangiophores. This genus develops rapidly, growing cottony colonies. Black, warted zygospores develop in some species. This genus is widely distributed in decaying fruits, vegetables, and fermented foods. A rennet substitute is a commercial product derived from enzymes produced by species of *Mucor*.

Rhizopus.—Sporangia are globose, columellate, white at first, then bluish-black when spores mature. Sporangiophores usually develop in clusters from ends of stolons at the point of origin of rhizoids. Spores vary from round to oval or angular in shape, with smooth, striate or rarely spinulose walls. Colonies are cottony and more vigorous in growth than those of *Mucor*. Zygospores are on suspensors without appendages. This genus is important in fermented food products such as tempé and in the rapid decay of foods.

Syncephalastrum.—Sporangiophores are repeatedly branched, each branch enlarging apically to form a globose head covered with rod-shaped merosporangia, each containing a row of spores inside (Fig. 1.11). Rhizoid-like structures may form. Zygospores are formed between equal gametangia. The organism is found in stored grains and as a contaminant of other foods.

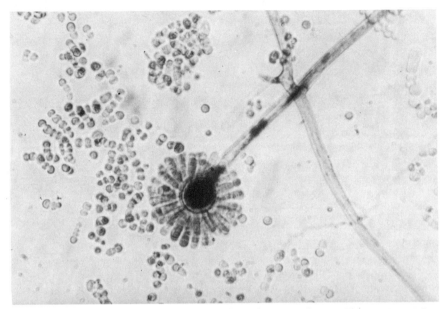

FIG. 1.11. COLUMELLA OF *SYNCEPHALASTRUM* SP. WITH MANY ROD-SHAPED MEROSPORANGIA (LIGHT MICROSCOPY, 500X)

Subdivision Ascomycotina

The organisms in this largest subdivision, the Ascomycotina (Ascomycetes), characteristically produce a sac or ascus with usually eight ascospores by sexual reproduction. These spores frequently are explosively ejected. The vegetative cells are either single cells, in which case they are classified as yeasts as described earlier in this chapter, or they form septate hyphae. They occur in a wide variety of habitats. Some are of interest in food production, food spoilage, synthesis of chemicals, and as pathogens.

The Ascomycotina are divided into six classes: the Hemiascomycetes, Plectomycetes, Pyrenomycetes, Laboulbeniomycetes, Discomycetes, and Loculoascomycetes. The Hemiascomycetes contain mostly ascospore-forming yeasts as described in a previous section. In the Plectomycetes, the fungi have ascocarps or a fruit body developed during sexual reproduction, with a rudimentary loose investment of hyphae or a woven wall around irregularly arranged globose asci known as cleistothecia. Well known genera of importance to food and beverage industries are the imperfect state genera *Penicillium* and *Aspergillus*. The genus *Byssochlamys* is well recognized in the fruit canning industry.

The remaining classes of Ascomycotina have asci regularly arranged in a basal or peripheral layer in the ascocarp. The ascocarps of the Pyrenomycetes are usually round to oval, dark in color, and contain an opening or ostiole. The fruiting bodies are known as perithecia (ascomata) and contain a layer of inoperculate asci with pores or slits. Examples of interest include *Claviceps, Endothia,* and *Gibberella.* The Laboulbeniomycetes, a very specialized group of exoparasites of arthropods, will not be considered. The disc- or cup-shaped ascocarps of the Discomycetes are known as apothecia. The surface of the apothecia has a hymenial layer composed of inoperculate of operculate (lids) asci. Examples are *Monilinia, Morchella,* and *Sclerotinia.* The Loculoascomycetes with ascocarps composed of stroma or tissue containing asci and bitunicate (two walls) ascospores contain fungi of little importance to the food industry. A key to selected genera of Ascomycotina is given below.

(1a) Ascocarp is a cleistothecium with asci scattered inside of loose- or well-formed wall (Plectomycetes)   2

(b) Ascocarp is not a cleistothecium   8

(2a) Wall of cleistothecium composed of scant wefts of thin hyphae around clusters of asci   *Byssochlamys*

(b) Wall of cleistothecium not composed of scant wefts of thin hyphae around individual asci   3

(3a) Wall of cleistothecium composed of interwoven hyphae, white to yellow or brown   *Talaromyces*
(Imperfect state: *Penicillium*)

(b) Wall of cleistothecium not composed of interwoven hyphae   4

(4a) Wall of cleistothecium composed of one layer of pseudo-parenchymatous cells   *Eurotium*
(Imperfect state: Osmophilic *Aspergillus glaucus* group)

(b) Wall of cleistothecium with many layers of pseudo-parenchymatous cells   5

(5a) Cleistothecia thin-walled, invested with sterile hyphae or hülle cells   6

(b) Cleistothecia thick-walled, no sterile hyphae or hülle cells   7

(6a) Cleistothecia reddish to purple, surrounded by hülle cells   *Emericella*
(Imperfect state: *Aspergillus nidulans* group)

(b) Cleistothecia white to yellowish, invested with sterile hyphae, equatorial crests on ascospores   *Sartorya*
(Imperfect state: *Aspergillus fumigatus* group)

(7a) Cleistothecia with thicker outer cell walls, thinner inner cell walls   *Eupenicillium*
(Imperfect state: *Penicillium*)

(b) Cleistothecia with thicker outer cell walls, thinner inner cell walls; ascospores lenticular, hyaline, two equatorial planes   *Hemicarpentales*
(Imperfect state:, *Aspergillus*)

(8a) Ascocarp is a perithecium (ascomata) with an ostiole, and asci usually in a hymenial layer (Sphaeriales)   9

(b) Ascocarp is an apothecium (cup) or modified apothecium; asci operculate or inoperculate   12

(9a) Perithecia not in a stroma; ascospores dark brown, striate, oval, one-celled   *Neurospora*

(b) Perithecia in a stroma (tissue)   10

(10a) Perithecia on a stroma, lilac or violet color; ascospores fusiform, two-celled   *Gibberella*

(b) Perithecia embedded in the stroma   11

(11a) Stroma fleshy, yellow color; ascus with two ascospores, two-celled   *Endothia*

(b) Stroma an apical shaped head on a stalk developed from a sclerotium, dark purple; parasitic on grain; filiform ascospores   *Claviceps*

(12a) Asci thin at the apex, opening by a lid   *Morchella*

(b) Asci thickened at the apex, with an apical pore to discharge ascospores (Helotiales)   13

(13a) Apothecia develop from a globose sclerotia   *Sclerotinia*

(b) Apothecia develop from stromatized host tissue, often mummified fruits   *Monilinia*

Descriptions of genera in the subdivision Ascomycotina which are of interest to the food mycologist are given below. Characteristics of both perfect and imperfect states are important in the identification and utilization of fungi. Cross reference is therefore made for many genera in the Ascomycotina that also exhibit an imperfect state which is described in a later section, subdivision Deuteromycotina.

Byssochlamys.—*Byssochlamys fulva* and *B. nivea* are of special interest as possible causes of spoilage of canned and bottled fruit. They are fairly common in soil of orchards. Asexual reproduction is typical of the form-genus *Paecilomyces* (Deuteromycotina). The cleistothecia are nothing more than clusters of oval asci surrounded by loose network of hyphae. Further discussion of this genus is presented in Brown and Smith (1957).

Claviceps.—*Claviceps purpurea* produces ergot alkaloid, a well known mycotoxin, in the sclerotial stage. In the spring the fungus develops pseudoparenchymatous stroma from the sclerotia that are stalked stromatic heads. Conidia are produced on the surface and later perithecia develop embedded in the tissue with asci containing filiform ascospores. The sclerotium is of concern in some thermally processed food products.

Emericella.—When cleistothecia with ascospores are developed in some *Aspergillus* spp. (subdivision Deuteromycotina), the perfect state name is designated as *Emericella, Eurotium, Hemicarpentales,* or *Sartorya.* Genera in both states grow on a wide range of substrates. In addition to the presence or absence of cleistothecia, genera and species are determined on variations in structure and color of conidiophores, conidia, and colonies.

Endothia.—The causative agent of chestnut blight, *Endothia parasitica* has been a serious plant pathogen. It produces a proteolytic enzyme that is used as a rennet substitute. Perithecia are produced immersed in a yellow fleshy stroma. The asci contain two-celled, fusiform-shaped ascospores.

Eupenicillium.—When cleistothecia with ascospores (Fig. 1.12) are developed in *Penicillium* spp. (subdivision Deuteromycotina), the perfect state is either *Eupenicillium* or *Talaromyces.* Classification is based on variations in structures and colony characteristics.

Eurotium.—See *Emericella* (subdivision Ascomycotina) and *Aspergillus* (subdivision Deuteromycotina).

Gibberella.—*Gibberella fujikuroi* is a well known source of the plant hormone, gibberellin. Blue to violet perithecia develop on the surface of stromata. The asci contain fusiform ascospores. The conidial stage is *Fusarium moniliforme* (Deuteromycotina), a cause of pink-ear rot of corn.

Hemicarpentales.—See *Emericella* (subdivision Ascomycotina) *Aspergillus* (subdivision Deuteromycotina).

Monilina.—Brown rot of peaches and other stone fruits is caused by *Monilinia fructicola.* The lemon-shaped conidia are produced in long chains on branched conidiophores. The conidia are readily dispersed and invade new substrates. Hyphae penetrate the entire

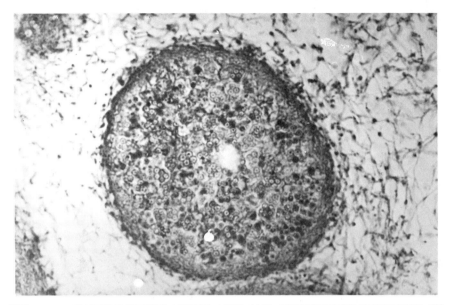

FIG. 1.12. CLEISTOTHECIUM WITH ASCI AND ASCOSPORES PRODUCED BY *EUPENICILLIUM* SP. (LIGHT MICROSCOPY, 450X)

fruit which may drop to the ground and become mummified as a sclerotium. In the following spring apothecia develop from the sclerotia and the cylindrical asci disperse ascospores for reinfection.

Morchella.—The true morel mushrooms belong to this genus. *Morchella esculenta* (Fig. 1.13), *M. angusticeps,* and other species are much sought after as food products. The fleshy portion of the morel is a modified apothecium with pitted areas containing a layer of asci over the surface. The asci open by a lid for spore dispersal. Many reference books are available for identification of morels and mushrooms (see Chap. 11).

Neurospora.—The common red bread mold is known as *Neurospora sitophila.* The organism is also used to ferment peanuts in the preparation of oncom. The conidia disperse readily by wind currents and have evident orange-red color. This is the imperfect state, *Monilia sitophila* (Deuteromycotina). Chains of two sizes of conida are usually produced on branched conidiophores. The smaller conida (microconidia) function in the sexual reproduction of the organism. Mature perithecia are dark colored and beaked with an ostiole. Cylindrical asci contain eight striated ascospores. A great volume of literature has accumulated on the genetics and

FIG. 1.13. ASCOCARP OF A TRUE MOREL, *MORCHELLA ESCULENTA* (HALF ACTUAL SIZE)

biochemistry of *Neurospora*.

Sartorya.—See *Emericella* (subdivision Ascomycotina) and *Aspergillus* (subdivision Deuteromycotina).

Sclerotinia.—Species of this genus produce rots, dry cankers, and blights of vegetables in the field and in storage. One species, *Sclerotinia libertiana*, is used for pectinase production. Conidia of the form-genera *Monilia* or *Botrytis* (Deuteromycotina) develop first on vegetables. Later in the season infected vegetables become covered with sclerotia (Fig. 1.14). In the spring the apothecia develop from the mummified tissue (stromata) and ascospores are released from the asci in the hymenial layer.

Talaromyces.—See *Eupencillium* (subdivision Ascomycotina) and *Penicillium* (subdivision Deuteromycotina).

Subdivision Deuteromycotina

The subdivision Deuteromycotina includes conidial states of organisms whose perfect states are unknown. Some conidial states are cross-referenced with perfect states in the Ascomycotina, Basiodiomycotina, and occasionally Zygomycotina. Many of these fungi are of importance in food processing, food decay, toxin

Courtesy of Dr. S. T. Williams, University of Liverpool
FIG. 1.14. SCLEROTIA OF *SCLEROTINIA* SP. ON CARROT ROOT (SCANNING ELECTRON MICROSCOPY, 1000X)

production, and diseases.

Genera of the Deuteromycotina have been artificially separated by some classification systems into form-genera based primarily on conidial characteristics. Some of these characteristics are the types of fructification, color, shape, and septation of conidia. The hyphae of all of the genera in this subdivision are usually septate. Since there are a great number of organisms in this group in addition to the more important ones listed here that may occur as saprophytes in foods and beverages, a number of useful references are available. Keys and descriptions of the genera by Barnett and Hunter (1972), descriptions of the genera of the Hyphomycetes by Barren (1968), and information on those sporulating in culture by von Arx (1974) are helpful in the identification of the genera of the Deuteromycotina.

The type of fructification is used as a basis to separate the

Deuteromycotina into form-orders.

(1a) Conidia produced in pycnidia   Sphaeropsidales
(b) Conidia not produced in a pycnidium   2
(2a) Conidia usually produced in an acervulus; conidiophores
      single or in compact groups   Melanconiales
(b) Conidia not in an acervulus   3
(3a) Conidia produced directly on mycelium or on sporogenous
      cells or on single or clustered conidiophores   Moniliales
(b) No conidia produced under favorable nutritional
      conditions   Mycelia Sterilia

Each form-order can be keyed to determine the form-genera within it. However, since a relatively few form-genera of importance to food and beverage spoilage and manufacture belong to the form-orders Sphaeropsidales, Melanconiales, and Mycelia Sterilia, these have been combined in a single key.

(1a) Acervuli formed; hyaline, one-celled conidia; dark setae
      present   *Colletotrichum*
(b) Pycnidia formed   2
(c) No conidia formed   6
(2a) Conidia hyaline   3
(b) Conidia dark   5
(3a) Pycnidium with ostiole; one-celled conidia of two types:
      short and ovoid, and large and bent   *Phomopsis*
(b) Pycnidium with ostiole; one-celled conidia of one type 4
(4a) Conidiophores simple; pycnidia superficial on natural
      substrate   *Aposphaeria*
(b) Conidiophores simple: pycnidia submerged in natural
      substrate   *Phoma*
(5a) Conidia large, oval to elongate, one-celled dark;
      pycnidium dark   *Sphaeropsis*
(b) Conidia small, oval to ellipsoid; one-celled dark
      pycnidia   *Coniothyrium*
(c) Conidia dark, two-celled, in pycnidium formed singly,
      dark in color   *Diplodia*
(6) No conidia formed on mycelium; sclerotia variable in form,
      consisting of loosely woven hyphae   *Rhizoctonia*

Form genera in the form-order Sphaeropsidales are characterized by the formation of pycnidia. The following form-genera are important in the food and beverage industries.

Aposphaera.—Pycnidia are dark, on surface substrate, rounded with papillate ostiole. They contain short, tapering, one-celled conidiophores bearing hyaline, one-celled, elongate to globose conidia. This fungus is a psychrophile in frozen foods.

Coniothyrium.—Pycnidia are black, globose and ostiolate. They produce short simple conidiophores bearing small, dark, one-celled, ovoid or ellipsoid conidia. This organism is used for production of pectinases.

Diplodia.—Pycnidia are black, single, globose, ostiolate, and immersed in the substrate. The slender, simple conidiophores produce two-celled, dark, ellipsoid or ovoid conidia. *Diplodia* spp. cause rot of tropical fruits and peaches as well as vegetables in storage.

Phoma.—Pycnidia are dark, ostiolate, lenticular to globose, and immersed in the substrate. They produce short conidiophores bearing one-celled, hyaline, ovoid to elongate conidia. This fungus causes dry rot of beets in storage.

Sphaeropsis.—The pycnidia are globose, black, ostiolate, as individuals or in groups containing short conidiophores. The apex of the conidiophores produces large, dark, one-celled, ovoid conidia. *Sphaeropsis* may occur as a psychrophile in frozen foods.

The form-order Melanconiales contains form-genera which produce acervuli. One genus is especially important to the fruit industry:

Colletotrichum.—The acervuli are disc- or cushion-shaped, typically with dark, septate setae or spines at the end or among the conidiophores. The conidiophores are simple and short, bearing hyaline, one-celled, ovoid or oblong conidia. The setae are not always produced on culture media. *Colletotrichum* causes anthracnose disease on many fruits.

Form-genera in the form-order Mycelia Sterilia do not produce conidia.

Rhizoctonia.—No conidia or asexual fruit bodies develop on the mycelium. The broad hyphae branch more or less at right angles, and anastomose frequently, developing brown or black sclerotia. This fungus is a common cause of vegetable rots. Where in contact with soil particles, large, black sclerotial masses may appear.

The fourth form-order of the subdivision Deuteromycotina, Moniliales, comprises many form-genera which are of significance in food and beverage spoilage and manufacture. A key to selected form-genera of Moniliales followed by descriptions of form-genera are given below.

[1a]   Conidia and conidiophore [if developed] hyaline or
          brightly colored   2

[b]   Either conidia or conidihores or both with dark

	pigments	16
(2a)	Conidia one-celled, globose to cylindrical	3
(b)	Conidia two-celled or three-celled to several-celled	14
(3a)	Apical conidiophore cells distinct from conidia	4
(b)	Conidia not greatly different from apical cells of conidiophore	*Monilia*
(c)	Conidia (arthrospores) formed by segmentation of mycelium	*Geotrichum*
(4a)	Conidiophores simple or sparingly blanched; phialides (sterigmata), if present, not tightly clustered	5
(b)	Conidiophores mostly branched, occasionally simple, phialides may be in heads	6
(5a)	Conidiophores short or indefinite; conidia single or in short chains	*Chrysosporium*
(b)	Conidiophores distinct; conidia globose to ovoid, in heads of slime	*Cephalosporium*
(6a)	Conidia in chains	7
(b)	Conidia not in chains	10
(7a)	Phialides in heads; conidiophores simple	*Aspergillus*
(b)	Phialides not in heads; conidiophores branched	8
(8a)	Conidiophores not in a layer or column	9
(b)	Conidiophores in a layer or column	*Myrothecium*
(9a)	Conidia are phialospores (youngest at base); phialides divergent	*Paecilomyces*
(b)	Conidia are phialospores; phialides close together, brush-like	*Penicillium*
(c)	Conidia are annellospores (ring-like scars built up on apex of conidiophores); lemon-shaped conidia	*Scopulariopsis*
(10a)	Conidia produced in mucilaginous clusters at apex of verticillate conidiophores	*Verticillium*
(b)	Conidiophores not verticillate	11
(11a)	Conidia not aggregated in slime drops	12
(b)	Conidia aggregated together in slime drops	13
(12a)	Conidia abundant on inflated apical cells; dichotomous conidiophores	*Botrytis*
(b)	Conidia not on inflated apical cells; conidiophore branches loose	*Botryotrichum*
(13a)	Conidiophores brush-like, similar to *Penicillium*	*Gliocladium*
(b)	Conidiophores spread out, not brush-like	*Trichoderma*
(14a)	Conidia two-celled, apical, single; conidiophore simple	*Trichothecium*
(b)	Conidia typically three- to many-celled	15

(15a) Macroconidia sickle-shaped; microconidia usually
　　　　present　　　　　　　　　　　　　　　　　　*Fusarium*
(b) Conidia not sickle-shaped, formed sympodially on
　　　　simple conidiophores　　　　　　　　　　　*Pyricularia*
(16a) Conidia one-celled, globose to short cylindrical　　　17
(b) Conidia typically two-celled or three-celled to several-
　　　　celled　　　　　　　　　　　　　　　　　　　　23
(c) Conidia several-celled, muriform　　　　　　　　　　25
(17a) Conidiophores short or absent　　　　　　　　　　18
(b) Conidiophores or phialides well formed　　　　　　19
(18a) Conidia yeast-like, borne on the sides of
　　　　hyphae　　　　　　　　　　　　　*Aureobasidium*
　　　　　　　　　　　　　　　　　　　　　　(Pullularia)
(b) Conidia in upright chains　　　　　　　　　*Wallemia*
　　　　　　　　　　　　　　　　　　　　(Type=*Torula*)
(19a) Conidia dark color in mass; conidia in chains; conidio-
　　　　phore enlarged at the apex　　　　　　　*Aspergillus*
(b) Conidia dark-colored in mass; conidia single or rarely in
　　　　chains　　　　　　　　　　　　　　　　　　　20
(20a) Conidia in clusters　　　　　　　　　　　　　　21
(b) Conidia not in clusters, mostly single　　　　　　　22
(21a) Conidia in slime; conidiophores produce a few thick
　　　　phialides (sterigmata)　　　　　　　　*Stachybotrys*
(b) Conidia in slime; conidiophore a short, single phialide
　　　　with an apical collarette　　　　　　　　*Phialophora*
(22a) Conidiophores short, one to few cells; conidia lenticular
　　　　with subhyaline band　　　　　　　　　　*Arthrinium*
　　　　　　　　　　　　　　　　　　　　　　(Papularia)
(b) Conidiophores short; conidia globose　　　　*Humicola*
(23a) Conidia typically two-celled, in branched
　　　　chains　　　　　　　　　　　　　　*Cladosporium*
(b) Conidia typically three-celled to several-celled　　　24
(24a) Conidiophores single, forming ovoid to cylindrical conid-
　　　　ia from lateral pores; sympodial growth of conidiophores;
　　　　conidia from lateral spores　　　　*Helminthosporium*
(b) No lateral pores; sympodial growth of conidiophore; con-
　　　　idia bent by enlargement of one medial cell　*Curvularia*
(25a) Conidia in chains, muriform　　　　　　　*Alternaria*
(b) Conidia muriform, not in chains; conidiophores poorly
　　　　developed　　　　　　　　　　　　　　　　　26
(26a) Conidia globose; conidiophores clustered　　*Epicoccum*
(b) Conidia ellipsoid; conidiophores single　　　*Pitomyces*
　　　　　　　　　　　　　　　　　　　(Sporodesmium)

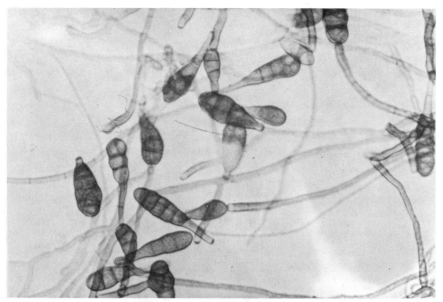

FIG. 1.15. CONIDIA OF *ALTERNARIA* SP. (LIGHT MICROSCOPY, 450X)

Description of the form-genera of Moniliales is as follows.

Alternaria.—Conidiophores are dark, simple, with a chain of dark conidia typically with cross and longitudinal septa (muriform), ovate to obclavate, tapering toward the tip, smooth or rough-walled, borne acropetally (Fig. 1.15). *Alternaria* is a common contaminant in dust and food products such as butter.

Aspergillus.—Conidiophores are erect, usually arising from a foot cell, terminating in a globose or clavate vesicle bearing phialides (sterigmata) over the surface (Fig. 1.16). Flask-shaped phialides produce chains of conidia (phialospores) that are globose and one-celled. Sclerotia or cleistothecia or both are produced by some species. An excellent coverage of the genus is presented in the monograph by Raper and Fennell (1965).

Aspergilli are used commercially to produce enzymes and organic acids. They produce mycotoxins and are a common cause of food decay.

Arthrinium (Papularia).—Conidiophores are more or less cylindrical with a subspherical basal mother cell, forming laterally or terminally on short pegs. Conidia are one-celled, round or flattened, with a hyaline rim, brown in color. *Arthrinium* has been isolated from frozen foods.

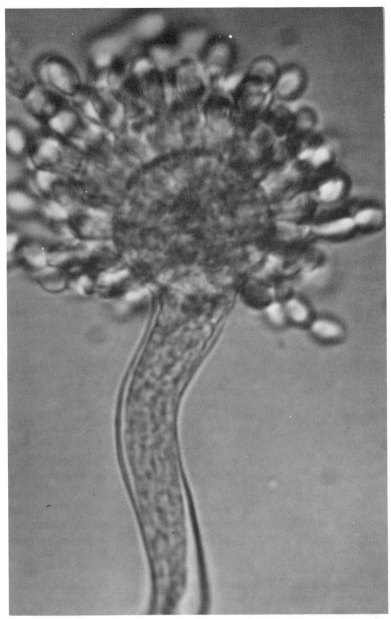

Courtesy of Dr. C. M. Christensen, University of Minnesota
FIG. 1.16. CONIDIOPHORE OF *ASPERGILLUS GLAUCUS* GROUP SHOWING
VESICLE, STERIGMATA, AND CONIDIA (LIGHT MICROSCOPY, 500X)

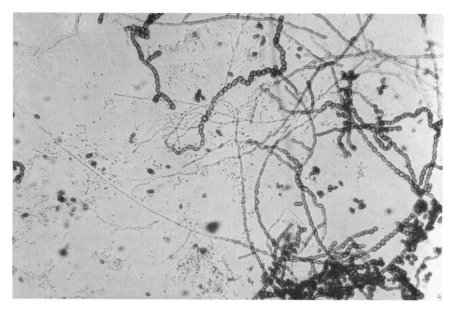

FIG. 1.17. HYALINE AND DARK-COLORED HYPHAE WITH CONIDIA PRODUCED BY *AUREOBASIDIUM* SP. (LIGHT MICROSCOPY, 120X)

Aureobasidium (Pullularia).—Hyphae are hyaline when young, becoming dark on older colonies, black and slimy on surface. Conidia are produced by budding on the sides of the hyphae from short denticles (Fig. 1.17). Colonies are yeast-like in appearance. This fungus grows at low temperatures and can be a problem in the frozen food industry and on paint surfaces.

Botrytis.—Conidiophores are long, hyaline or dark, branched or dichotomous; produce swollen terminal cells with clusters of conidia on sterigmata (ampullae), one-celled, hyaline, or gray in mass. The genus *Botrytis* is found on mature grapes and is known as "gray mold" of many plants.

Botryotrichum.—Setae taper to a point, dark below, and intermixed with short, irregularly branched, hyaline conidiophores. The lateral branches produce clusters of large, spherical, one-celled, brown, thick-walled conidia (aleuriospores). Sometimes small, hyaline, one-celled phialospores are produced on short phialides. This is a psychrophilic fungus in frozen foods.

Cephalosporium.—Conidiophores (or phialides) are usually unbranched, bearing one-celled, hyaline conidia (phialospores) in balls (Fig. 1.18). *Cephalosporium* may occur as a psychrophilic

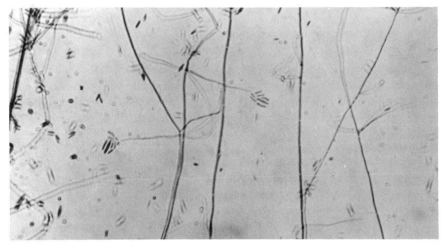

FIG. 1.18. CONIDIOPHORES OF *CEPHALOSPORIUM* SP. WITH HYALINE CLUSTERS OF CONIDIA (LIGHT MICROSCOPY, 500X)

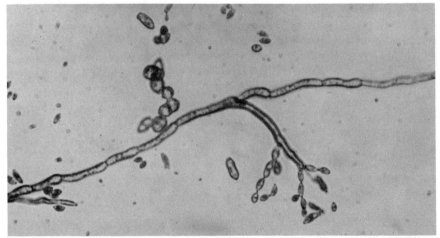

FIG. 1.19. CONIDIOPHORE OF *CLADOSPORIUM* SP. The loose conidia have characteristic attachment scars at each end (light microscopy, 600X)

fungus in frozen and refrigerated foods.

Chrysosporium.—The conidiophores are poorly differentiated from the hyphae. They are hyaline, usually erect, and branch irregularly. The conidia (aleuriospores) are one-celled, hyaline to brightly colored, globose to pyriform, and usually with a broad basal scar. This is a common genus in the soil and occasionally is a psychrophile in refrigerated foods.

Cladosporium.—Conidiophores are upright, dark, branched irregularly near the apex, and tree-like in form. The branches bud off conidia (blastospores), one- or two-celled, variable in shape with attachment scars; spores in chains readily separate (Fig. 1.19). This is a very common fungus on organic substrates.

Curvularia.—Hyphae or stromata give rise to conidiophores, simple and dark in color. The three- to several-celled, obovate to fusiform, smooth or verrucose-walled conidia (porospores) arise through pores in the conidiophore walls. Conidia are typically bent with one of the central cells enlarged. This fungus is a saprophyte and produces a toxin, curvularin.

Epicoccum.—Conidiophores are short, in clusters or on sporodochia, and dark colored. They produce globose, dark colored, muriform (cross walls in two directions) conidia with warty walls. One species, *Epicoccum nigrum,* can frequently be recognized by the yellow or orange hyphae. This fungus is usually found as a saprophyte in organic substrates.

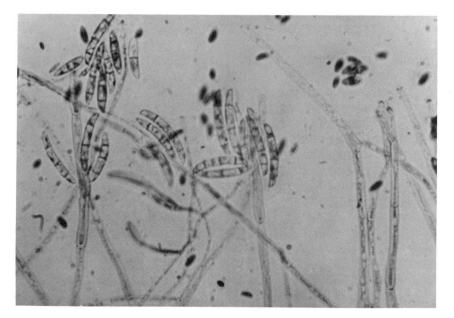

FIG. 1.20. MACROCONIDIA AND MICROCONIDIA OF *FUSARIUM* SP. (LIGHT MICROSCOPY, 500X)

Fusarium.—Colonies are cottony with pink, purple, or yellow coloration in the hyphae. Conidiophores are variable, simple; branched with phialides, or grouped into sporodochia. Conidia (phialospores) are of two sizes: macroconidia are several-celled, curved usually boat-shaped with a foot cell for attachment, and microconidia are one- or two-celled, ovoid or oblong (Fig. 1.20). Both types of conidia may be in spore balls. Species of this genus cause spoilage, produce toxin in grains, and are used in the production of gibberellin.

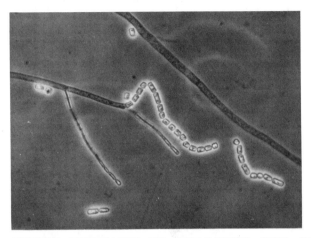

FIG. 1.21. HYPHAE TERMINATE BY SEPARATING INTO INDIVIDUAL ARTHROSPORES IN *GEOTRICHUM* SP. (PHASE MICROSCOPY, 500X). A stained specimen is shown in Fig. 16.3.

Geotrichum.—Hyphae terminate in short cuboid or cylindrical arthrospores with flattened ends (Fig. 1.21). *Geotrichum candidum* is pathogenic to ripe fruits, including tomato, peach, and melons. It is frequently associated with the molding of foods containing lactic acid. Also known as "machinery mold" or "dairy mold," its presence in processing plants has been offered as an index of sanitation (see Chap. 6)

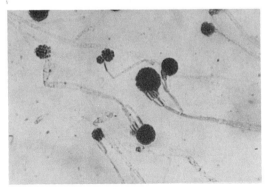

FIG. 1.22. CONIDIOPHORES OF *GLIOCLADIUM* SP. WITH TERMINAL DROPLETS OF CONIDIA (LIGHT MICROSCOPY, 600X)

Gliocladium.—Conidiophores are hyaline, branching in a penicillate manner as *Penicillium*, produce hyaline or brightly colored conidia (phialospores), one-celled in mucilaginous droplets (Fig. 1.22). They are found in organic substrates.

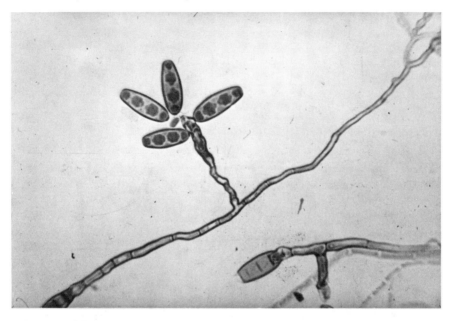

FIG. 1.23. CONIDIOPHORE OF *HELMINTHOSPORIUM* SP. WITH OBCLAVATE CONIDIA (LIGHT MICROSCOPY, 500X)

Helminthosporium.—Hyphae are dark; conidiophores are single or in whorls, and form brown, pseudoseptate, obclavate conidia from pores beneath the septa (Fig. 1.23). A prominent basal scar is at the base of the conidium. This fungus is found in the spoilage of vegetables and as a disease of plants.

Humicola.—The conidiophores are short, cylindrical or slightly inflated, dark, with single, globose, brown, one-celled conidia (aleuriospores). Some species produce phialides with phialospores in chains. *Humicola* spp. are strongly cellulolytic.

Monilia.—The mycelium is similar to the cells of the simple or branched conidiophores. One-celled, short, cylindrical to round, branched chains of budding pink, gray, or tan conidia (blastospores) give a distinctive bead-like appearance. *Monilia sitophila*, the red bread mold, is the conidial state of *Neurospora sitophila*. This contaminant grows rapidly. *Monilia* is the conidial state of *Monilinia fructicola*, the cause of brown rot of stone fruits.

Myrothecium.—Fructification is a sporodochium or a cushion-like mass of conidiophores with a dark green to black covering of one-celled ovoid conidia, viscous and green when young, later becoming hard and black. The conidiophores are repeatedly branched, terminated by verticils of phialides bearing the conidia. This genus is a strong cellulase producer.

Paecilomyces.—Conidiophores are simple or branching, with the sporogenous cell phialides developed singly or in verticils or in penicillate heads, larger at the base and tapering toward the apex, bearing long chains of ovoid to fusoid conidia. *Paecilomyces* is the conidial state of *Byssochlamys*.

Papularia.—See description under *Arthrinium*.

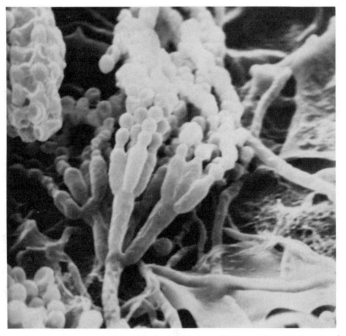

Courtesy of Dr. S. T. Williams, University of Liverpool

FIG. 1.24. CONIDIOPHORE OF *PENICILLIUM* SP. WITH STERIGMATA PRODUCING CHAINS OF CONIDIA (SCANNING ELECTRON MICROSCOPY, 2000X)

Penicillium.—Species are based on variations in structures and colony characteristics, and on the presence or absence of cleistothecia. The culture of these organisms on Czapek's agar is important for identification of species. Conidiophores are erect, arising from hyphae, singly or sometimes in synnemata; they are hyaline, septate, branching at or near the apex in a brush-like head (penicillus) ending in phialides (Fig. 1.24). The one-celled conidia (phialospores) are formed in chains and are hyaline or brightly colored in mass. Sclerotia or cleistothecia or both are produced in some species. When the imperfect state or conidiophores with conidia are the only reproductive structures formed, the genus *Penicillium* is considered under the subdivision Deuteromycotina. When cleistothecia with ascospores are developed in some species, the perfect state is either *Talaromyces* or *Eupenicillium* (subdivision Ascomycotina). A monograph on the penicillia by Raper and Thom (1949) is helpful for species identification.

The penicillia are as common and as important as the aspergilli. Various species cause decay of grains, fruits, vegetables, and cured meats. Others are used in the production of cheese and organic acids.

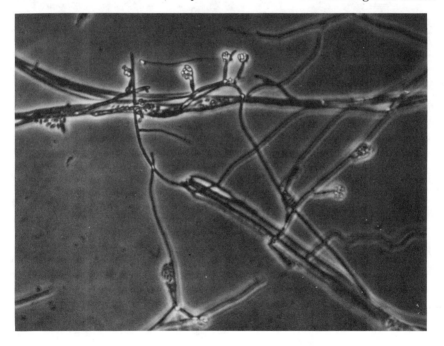

FIG. 1.25. CONIDIOPHORES (PHIALIDES) OF *PHIALOPHORA* SP. ARE FLASK-SHAPED WITH CLUSTERS OF CONIDIA (PHIALOSPORES) (PHASE MICROSCOPY, 1000X)

Phialophora.—Conidiophores are short or lacking; phialides are short, often flask-shaped, with a collarette at the apex. The conidia (phialospores) are hyaline or dark, one-celled, globose to ovoid; they are produced within the collarette and form moist heads (Fig. 1.25).

This genus is isolated from cellulosic materials and from frozen foods.

Pithomyces.—Mycelium is brown in color. Short, lateral peg-like conidiophores are produced which give rise to single conidia (aleuriospores) with up to eight transverse septa, elliptical to oblong, with verrucose or echinulate walls. This is a toxin producer in rye grass, causing facial exema in sheep and cattle.

Pyricularia.—Conidiophores are long and simple; they produce conidia in acropetal succession by sympodial extension of the tip cell. Conidia are ellipsoid to pyriform, attached at the broad end, two- to three-celled, and hyaline or lightly pigmented. This organism occurs in food spoilage.

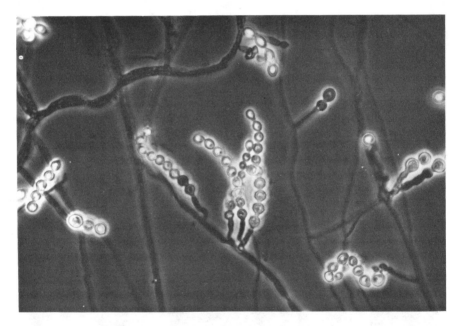

FIG. 1.26. CONIDIOPHORES WITH CHAINS OF CONIDIA PRODUCED BY *SCOPULARIOPSIS* SP. (PHASE MICROSCOPY, 500X)

Scopulariopsis.—Conidiophores are mostly branched or produce at the apex a cluster of sporogenous cells which are somewhat elongated before forming chains of conidia (annellospores) (Fig. 1.26). The conidia are one-celled and globose with a truncate base, smooth or roughened, and hyaline or pigmented (commonly tan). *Scopulariopsis* grows commonly on substrates containing largely protein and in many types of decaying matter. *Scopulariopsis brevicaulis* has been isolated from cheese, eggs, and ham.

Stachybotrys.—Conidiophores are simple, hyaline at first, becoming dark and roughened with age in some cases. The apex of the conidiophores bears verticils of cylindrical phialides, usually three to seven in a cluster, with dark, one-celled, globose to ovoid conidia produced in moist heads. This organism has marked cellulase activity and produces a toxin.

Trichoderma.—Conidiophores are hyaline, branched, sometimes at right angles or verticillate with single phialides or in clusters. The phialides are hyaline, ovate to flask-shaped, producing balls of one-celled, ovoid, hyaline conidia (phialospores). The colonies are easily recognized by the yellow-green or green color with green patches or cushions of conidia. This fungus is common in decaying matter and has marked cellulase activity.

Trichothecium.—The conidiophores are long, slender, septate, and bear hyaline or brightly colored, two-celled, ovoid to ellipsoid conidia in groups or chains held together by mucus. One species, *Trichothecium roseum*, is rose colored and occurs as a saprophyte on fruits and nuts. This organism may produce a trichothecene toxin.

Verticillium.—The conidiophores are hyaline and branched with at least some in verticils (whorls). The conidia (phialospores) are ovoid to ellipsoid, hyaline, and one-celled, appearing singly or in small moist clusters at the apex of the branches (phialides). This organism causes a wilt in higher plants, is a disease of mushrooms, and occurs as a saprophyte.

Subdivision Basidiomycotina

The fungi in the subdivision Basidiomycotina characteristically produce by sexual reproduction a basidium with four basidiospores on projections termed sterigmata. The basidia vary considerably in structure, with the form of the basidium being an important criterion in classification. A single, cylindrical cell without septa, bearing four basidiospores at its apex is a holobasidium (e.g., mushrooms). In other cases the basidium develops from a thick-walled cell (teliospore or chlamydospore) and is divided into four cells (e.g., rusts and smuts). Another variation is the development of transverse or longitudinal septa in the basidium (jelly fungi).

The Basidiomycotina are divided into three classes: Teliomycetes, Hymenomycetes, and Gasteromycetes. The Teliomycetes produce teliospores or chlamydospores in place of basidiocarps in the orders Uredinales and Ustilaginales. The rusts, Uredinales, produce two to four basidiospores, while the smuts, Ustilaginales, have numerous basidiospores. In other Basidiomycotina, basidia (holobasidia) arise from hyphae with the hymenium exposed at maturity in the basidiocarp as in the class Hymenomycetes. This class contains two groups. The first group with basidia not segmented or forked has two orders: the Agariales (agarics and boleti) with fleshy basidiocarps, and the Aphyllophorales (polypores, etc.) with leathery, corky, or woody basidiocarps. The second group with three orders is separated on the basis of segmented or forked basidia. Two orders, the Tremellales and Auriculariales, are known as jelly fungi. In the third class, the Gasteromycetes, the hymenium with basidia and basidiospores remains enclosed. This class contains a number of orders and well known representatives such as puffballs, earthstars, stinkhorns, and bird's nest fungi. Additional information on the Gasteromycetes may be found in references on identification of these organisms and in Ainsworth *et al.* (1973B). A key to selected genera of the Basidiomycotina follows:

(1a) Basidiocarp lacking, replaced by large dark masses of chlamydospores in sori (fruit structure), form indefinite number of basidiospores from a septate basidium
Ustilago

(b) Basidiocarp present, basidia in a hymenium 2

(2a) Basidium is separated into a lower cell, the probasidium, and a cylindrical, straight, or curved one- to three-septate coiled metabasidium; the basidiocarp is fleshy to fibrous *Helicobasidium*
(jelly fungus)

(b) Basidium is not septate 3

(3a) Basidiocarp is an effuse, thin layer of hyphae with a smooth hymenium, resembling a thin coating of gray paint on the substrate surface *Corticium*

(b) Basidiocarp fleshy, hymenium borne on gills or pores 4

(4a) Spore prints purple-brown to chocolate-brown, gills free from stalk, spores gray to pink when young, annulus on stalk *Agaricus*

(b) Spore print white, gills attached to stalk, gills thick or thin, grows on wood *Lentinus*

Description of some genera of Basidiomycotina of commercial interest is as follows:

Agaricus.—Most mushrooms grown commercially in the United States are *Agaricus bisporus*. This species produces two basidiospores on the basidium in contrast to four basidiospores for *Agaricus campestris*, which is a wild species.

Corticium.—One of the primitive polyporoid genera that grows in culture and produces α-galactosidase is *Corticium rolfsii*.

Helicobasidium—One of the jelly fungi (*Helicobasidium mompa*) is associated with stress metabolites in sweet potatoes.

Lentinus.—Known as the shiitake mushroom, *Lentinus edodes* is cultured on a large scale on hardwood logs (mostly oak) in the Orient, especially Japan.

Ustilago.—Commonly known as corn smut, *Ustilago maydis* produces high quantities of lysine in culture.

Many forms of Basidiomycotina known as mushrooms, stinkhorns, puffballs, jelly fungi, bird's nest fungi, rusts, and smuts have been described along with keys in monographs. These monographs and references are listed in Ainsworth *et al.* (1973B) and in the bibliography of Chap. 11. Reference should be made to other publications for discussion of the Basidiomycotina, as there are numerous genera and species which require more space than available for description and identification here. Most of these, however, are of little interest at present to the food mycologist.

REFERENCES

AINSWORTH, G. C. 1966. A general purpose classification for fungi. Bibliogr. Syst. Mycol. Nomen. 1, 1-4.

AINSWORTH, G. C. 1971. Ainsworth and Bisby's Dictionary of the Fungi, 6th Edition. Commonwealth Mycol. Inst., Kew. Surrey, England.

AINSWORTH, G. C. 1973. Introduction and keys to higher taxa. *In* The Fungi, Vol. IVA. A Taxonomic Review with Keys: Ascomycetes and Fungi Imperfecti. G. C. Ainsworth, F. K. Sparrow, and A. S. Sussman (Editors). Academic Press, New York.

AINSWORTH, G. C., SPARROW, F. K., and SUSSMAN, A. S. 1973A. The Fungi, Vol. IVA. A Taxonomic Review with Keys: Ascomycetes and Fungi Imperfecti. Academic Press. New York.

AINSWORTH, G. C., SPARROW, F. K., and SUSSMAN, A. S. 1973B. the Fungi, Vol. IVB. A Taxonomic Review with Keys: Basidiomycetes and Lower Fungi. Academic Press, New York.

ALEXOPOULOS, C. J. 1962. Introductory Mycology, 2nd Edition. John Wiley and Sons, New York.

BARNETT, H.L., and HUNTER, B. B. 1972. Illustrated Genera of Imperfect Fungi, 3rd Edition, Burgess Publishing Co., Minneapolis, Minn.

BARNETT, J. A., and PANKHURST, R. J., 1973. A New Key to the Yeasts. North-Holland Publishing Co., Amsterdam.

BARREN, G. L. 1968. The Genera of the Hyphomycetes from Soil. Williams and Wilkens Co., Baltimore.

BROWN, A. H. S., and SMITH, G. 1957. The genus *Paecilomyces* Bainer and its perfect stage *Byssochlamys* Westling. Trans. Br. Mycol. Soc. *40*, 17-89.

BUHAGIAR, R. W. M. and BARNETT, J. A. 1971. The yeasts of strawberries. J. Appl. Bacteriol. *34*, 727-739.

CAMPBELL, I. 1974. Methods of numerical taxonomy for various genera of yeasts. Adv. Appl. Microbiol. *17*, 135-156.

COOKE, W. B. 1974. Terminology of the fungi imperfecti. Mycopathol. Mycol. Appl. *53*, 45-67.

DO CARMO-SOUSA, L., and PHAFF, H. J. 1962. An improved method for the detection of spore discharge in the Sporobolomycetaceae. J. Bacteriol. *83*, 434-435.

KREGER-VAN RIJ, N. J. W. 1973. Endomycetales, basidiomycetous yeasts, and related fungi. *In* The Fungi, Vol. IVA, A Taxonomic Review with Keyes G. C. Ainsworth, F. K. Sparrow, and A. S. Sussman (Editors). Academic Press, New York.

LODDER, J. 1970. The Yeasts, A Taxonomic Study, 2nd Edition. North-Holland Publishing Co., Amsterdam.

MILLER, O. K., JR. 1972. Mushrooms of North America. E. P. Dutton and Co., New York.

PHAFF, H. J., and DO CARMO-SOUSA, L. 1962. Four new species of yeasts isolated from insect frass in bark of *Tsuga heterophylla* (Raf.) Sargent. Antonie van Leeuwenhoek. J. Microbiol. Serol. *28*, 193-207.

PHAFF, H. J., MILLER, M. W., and MRAK, E. M. 1966. The Life of Yeasts, Harvard Univ. Press, Cambridge, Mass.

RAPER, K. B., and FENNELL, D. I. 1965. The Genus *Aspergillus.* Williams and Wilkens Co., Baltimore.

RAPER, K.B., and THOM, C. 1949. A Manual of the Penicillia. Williams and Wilkens Co., Baltimore.

STEVENS, R. B. 1974. Mycology Guidebook. Univ. of Washington Press, Seattle.

VAN DER WALT, J. P. 1970A. Genus 16. *Saccharomyces* Meyen emend. Reess. *In* The Yeasts, 2nd Edition, J. Lodder (Editor). North-Holland Publishing Co., Amsterdam.

VAN DER WALT, J. P. 1970B. Genus 1. *Brettanomyces* Kufferath et van Laer. *In* The Yeasts, 2nd Edition, J. Lodder (Editor). North-Holland Publishing Co., Amsterdam.

VAN DER WALT, J. P., and HOPSU-HAVU, V. K. 1976. A colour reaction for the differentiation of ascomycetous and hemibasidiomycetous yeasts. Antonie van Leeuwenhoek, J. Microbiol. Serol. *42*, 157-163.

VAN DER WALT, J. P., and JOHANNSEN, E. 1975. The genus *Torulaspora* Lindner. C. S. I. R. Res. Rep. No. *325*, Pretoria, S. Afr.

VAN DER WALT, J. P., and SCOTT, D. B. 1971. The yeast genus *Saccharomycopsis* Schionning, Mycopathol. Mycol. Appl. *43*, 279-288.

VON ARX, J. A. 1972. On *Endomyces, Endomycopsis* and related yeast-like fungi. Antonie van Leeuwenhoek J. Microbiol. Serol. *38*, 289-309.

VON ARX, J. A. 1974. The Genera of Fungi Sporulating in Pure Culture, 2nd Edition. J. Cramer, Lehr, Germany.

WALKER, H. W., and AYRES, J. C. 1970. Yeasts as spoilage organisms. *In* The Yeasts, Vol. 3. Yeast Technology. A. H. Rose and J. S. Harrison (Editors). Academic Press, New York.

WATERHOUSE, G. M. 1968. The genus *Pythium.* Pringsheim. Mycol. Pap. *110*, 1-71.

WATERHOUSE, G. M. 1970. The genus *Phytophthora.* diagnoses and descriptions, Revised Edition. Mycol. Pap. *122*, 1-59.

WEBSTER. J. 1970. Introduction to Fungi. Cambridge Univ. Press, Cambridge, England.

WIKEN, T., SHEFFERS, W. A., and VERHAAR, A. J. M. 1961. On the existence of a negative Pasteur effect in yeasts classified in the genus *Brettanomyces* Kufferath et van Laer. Antonie van Leeuwenhoek J. Microbiol. Serol. *27*, 401-433.

ZYCHA, H. R., SIEPMAN, H. R., and LINNEMANN, G. 1969. Mucorales. J. Cramer, Lehr, Germany.

2

Relationships of Water Activity to Fungal Growth

Janet E. L. Corry

T he oldest method of food preservation is probably drying. Until comparatively recent times, drying and salting, sometimes combined with smoking, were almost the only methods known for preserving foods (excluding fermentation systems where lactic or acetic acid or alcohol acted as chemical preservatives).

Grain was preserved by drying and storage in granaries in ancient Egypt (Genesis 41: 35-36) and, no doubt, for a long time before that. Although methods of drying and storage have been modified in the light of present-day knowledge, the basic technique remains the same today and is no less vital. Drying or salting of fish and meat, and drying of fruit are also ancient arts, although adding sugar as a preservative was not widespread until the development of the sugar industry in the 18th century.

Drying can be achieved either by physically removing water or by adding high concentrations of solutes—traditionally either sodium chloride or sucrose. In the first case, the residual water will contain high concentrations of the solutes already present in the product and in the second case, the added solutes will predominate. Freezing also can be regarded as a method of removing water. As the temperature is lowered, water freezes, leaving an unfrozen portion of progressively more concentrated solutes. The low temperature *per se* also prevents microbial spoilage.

Freeze-preservation of food is today a vast industry and a number of new methods have been developed to produce better dried products—including freeze-drying of meats, vegetables, and instant coffee, spray-drying of egg and milk, reverse osmosis, and the development of new "intermediate moisture foods." Dried products, including sugar confectionery, have the advantages of being relatively resistant to chemical and microbiological spoilage, being light in weight, requiring less exacting conditions of storage than frozen foods, and requiring less packaging material than canned foods. Transportation is consequently easier and less expensive. Another advantage, given the present concern about the toxic effects of chemicals, is that addition of preservative substances is not always necessary, since the dryness of the food can be sufficient for preservation. Molds and yeasts are almost always implicated when spoilage problems occur in dried foods, although halophilic bacteria also can cause difficulties in salt-preserved foods. Losses due to molds in stored food crops have been estimated at over 10% world-wide (Moreau 1974).

ASSESSMENT OF WATER RELATIONS

The percentage (wt/wt) water in a food can readily be measured by various methods, but unfortunately, different foods with the same weight proportion of water may have widely differing stabilities.

TABLE 2.1

WATER CONTENT OF VARIOUS DRIED FOODS AT a_W 0.70 (THE MAXIMUM ALLOWABLE IF THEY ARE TO BE MICROBIOLOGICALLY STABLE)

Food	Water Content (%, wt/wt)
Nuts	4 - 9
Whole milk powder	7
Cocoa	7 - 10
Soybeans	9 - 13
Dried whole egg	10
Skimmed milk powder	10
Dried lean meat and fish	10
Rolled oats	11
Rice	12 - 15
Pulses	12 - 15
Dried vegetables	12 - 22
Wheat flour, noodles	13 - 15
Dried soup mixes	13 - 21
Dried fruits	18 - 25

Source: Mossel (1975A).

Table 2.1 shows a variety of foods and maximum water contents which must not be exceeded if microbial stability is to be assured. The value of "relative humidity" (RH) as an index of the ability of a food to support microbial growth has long been realized (Tomkins 1930; Galloway 1935; Heintzeler 1939; Mossel and Westerdijk 1949; English 1953; Mossel and Ingram 1955). The RH was taken to mean the RH of the atmosphere in equilibrium with the food. An alternative expression, widely used in the food industry, is "equilibrium relative humidity" (ERH). Since the publications of Scott (1953, 1957) concerning the effect of the water content [measured as "water activity" (a_w)] on the growth of microorganisms, a_w has been adopted by food microbiologists as a factor affecting growth to be ranged alongside other parameters such as pH and Eh. However, a_w is a thermodynamic concept derived by physical chemists and adopted by microbiologists and food technologists and scientists to describe systems containing both extremely complex mixtures of chemicals and living organisms, and this has caused considerable confusion. It has been suggested recently that the symbol p_w (relative vapor pressure) should be used to describe this parameter (Mossel 1975B; Tracey 1975). This would help dispel the impression that the state of water in a food can be described solely in terms of its partial pressure. The p_w of a system can be regarded as a factor similar to pH or pO_2 (through perhaps more important) which, like them, interacts with other factors in affecting microbial growth. Nevertheless the symbol a_w will be used here because it is still the generally accepted term, and whatever the parameter is actually called, it will continue to be valuable to food microbiologists.

MEASUREMENT OF WATER ACTIVITY AND EQUILIBRIUM RELATIVE HUMIDITY

Water activity measurements estimate the proportion of "available" water in a system—water available for biological (biochemical) and chemical reactions. For an ideal solution:

$$a_w = \frac{n_1}{n_1 + n_2} = \frac{p}{p_0} \qquad (2.1)$$

where n_1 = moles of solvent; n_2 = moles of solute: p = vapor pressure of solution; p_0 = vapor pressure of solvent (water). Unfortunately most solutions behave in a nonideal fashion. However, it is evident from equation (2.1) that pure water will have an a_w of unity and a completely dry solute will have a_w of zero. Equilibrium relative humidity is numerically equivalent to a_w:

$$a_W = \frac{ERH}{100} \tag{2.2}$$

and since almost all measurements of a_W actually measure the relative humidity of air in equilibrium with the food or the water vapor pressure by direct measurement, many food microbiologists prefer this term.

Calculation of Water Activity Value

Table 2.2 shows the molalities of solutions of various common solutes compared with the molality at equivalent a_W for an ideal solute. (Molality=moles per 1000 g water, whereas molarity=moles per 1000 ml solution.) Molality does not change with temperature, whereas molarity does. For similar reasons, water content in studies involving a_W measurements and solute concentrations are conveniently measured on a wt/wt basis. All solutes deviate from ideal values. The effect is more marked at high concentrations, and ionic compounds generally show greater deviation than nonionic, because of attracting forces between ions and polar water molecules.

TABLE 2.2
MOLALITIES OF SOME SOLUTES FOR VARIOUS VALUES OF a_W AT 25°C

	Solute (Molality)				
a_W	Ideal Value	Sodium Chloride	Calcium Chloride	Sucrose	Glycerol
0.995	0.281	0.150	0.101	0.272	0.277
0.990	0.566	0.300	0.215	0.534	0.554
0.980	1.13	0.607	0.408	1.03	1.11
0.960	2.31	1.20	0.770	1.92	2.21
0.940	3.54	1.77	1.08	2.72	3.32
0.920	4.83	2.31	1.34	3.48	4.44
0.900	6.17	2.83	1.58	4.11	5.57
0.850	9.80	4.03	2.12	5.98	8.47
0.800	13.9	5.15	2.58		11.5
0.750	18.5		3.00		.14.8
0.700	23.8		3.40		18.3
0.650	30.0		3.80		22.0

Source: Adapted from Scott (1957).

There is a relationship between osmotic pressure and a_W. Water activities of individual solutions can be calculated from the molal osmotic coefficient (ϕ):

$$a_W = \exp - \left(\frac{(vm\phi)}{55.51} \right) \tag{2.3}$$

where m = mole concentration of solute; v = number of ions generated by each molecule of solute; 55.51 = number of moles of water in 1 kg.

An alternative method for calculating a_w is via the activity coefficient (γ) and equation (2.1):

$$a_w = \gamma \frac{n_2}{n_1 + n_2} \qquad (2.4)$$

Various equations for the estimation of a_w levels in systems with mixtures of solutes have been suggested—notably (for sugar confections) the methods of Money and Born (1951):

$$a_w = \frac{100}{1 + 0.27n} \qquad (2.5)$$

where n = number of moles of sugar (in solution) per 100 g water. Grover (1947), who used a "sucrose equivalent conversion factor" for other confectionery ingredients, and Norrish (1966), whose equation also utilizes constants for the various components:

$$\log a_w = \log x_1 - [(-K_{1,2})^{1/2} x_2 + (-K_{1,3})^{1/2} x_3 + \dots \text{etc.}] \qquad (2.6)$$

where x_1 = mole fraction of water; x_2, x_3, etc. = mole fractions, of various solutes; $K_{1,2}$, $K_{1,3}$, etc. = constants of each solute in water. Equation (2.6) often gives a very good estimation of actual a_w, but in foods containing other ingredients it is almost impossible to predict a_w levels. There are a number of reasons for this. Although soluble, low molecular weight compounds (e.g. sugars and polyhydric alcohols (polyols) such as glycerol and sorbital, and ionic compounds) affect a_w most, the effect of other high molecular weight materials such as proteins and other compounds not in solution cannot be ignored. In addition, there appears to be a capillary effect from insoluble components, and particular combinations of solutes may interact to give unexpectedly high or low a_w values (Karel 1975). All these considerations make it necessary at least to check c_w of all but the simplest of systems by direct measurement.

Procedures for Measuring Water Activity

Many methods of measuring a_w have been published and detailed reviews have been written (Gal 1967, 1975; Smith 1971). Accurate measurement of a_w is time-consuming and sorption isotherms (percentage [wt/wt] water versus a_w at a particular temperature) are frequently constructed. The a_w values are then deduced from the isotherms. A multilayer adsorption equation has been used to describe reasonably well the water sorption isotherms of a variety of foods and food components (Iglesias et al. 1975). Figure 2.1 shows a typical sorption isotherm. Although, theoretically, the same curve should be obtained whether an a_w is reached by removal (desorption) or addition (adsorption) of water, two different curves are frequently found. This phenomenon is due to hysteresis. The method of

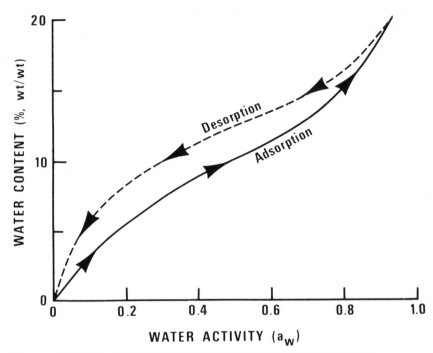

FIG. 2.1.   TYPICAL SORPTION ISOTHERM, SHOWING HYSTERESIS

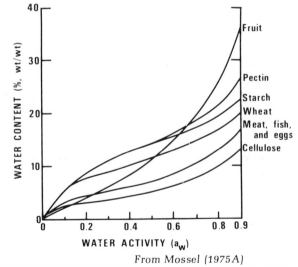

From Mossel (1975A)
FIG. 2.2.   SORPTION ISOTHERMS OF VARIOUS FOODS AT 20°C

preparation, whether by adsorption or desorption of water, can also affect the microbial stability of a food. Figure 2.2 shows sorption isotherms characteristic of various foods. In order to plot a sorption isotherm, both the a_W value and the percentage of water in the food must be measured. Water content can be measured by one of several methods. Conventional methods include oven drying, preferably in a vacuum oven over phosphorus pentoxide, distillation methods, or by use of Karl Fischer reagent (Pande and Pande 1962; Pearson 1970, 1973; Pande 1971; Kirkbright *et al.* 1975). Alternatively, water content may be determined from secondary properties, using nuclear magnetic resonance or infarred spectroscopy after calibration with water content determined by a conventional method. Measurement of these secondary properties can be very rapid, and can be useful for monitoring on an industrial scale (Slight 1971). It should be remembered that different methods of determination will not give exactly the same results (Overbeek and Mossel 1951).

Water activity can be determined in one of the following ways:

(1) *Gravimetric Methods*—These involve either measuring the water vapor pressure of the sample of known water content directly by means of a manometer in an evacuated system, or producing a known water vapor pressure by use of saturated salt solutions or pure water reservoirs at suitable temperatures and then determining the water content of the sample. The simplest methods using the latter system involve leaving a number of samples to equilibrate in closed containers with atmospheres at known RH and at constant temperature. Solutions of sulfuric acid, sodium chloride or sodium hydroxide, or saturated salt solutions of known a_W (ERH) are used (Stokes and Robinson 1949; Landrock and Proctor 1951; Mossel 1951; d'Alton 1966). Table 2.3 shows the a_W of a number of saturated salt solutions. More details can be found elsewhere (Solomon 1951; Robinson and Stokes 1955; Rockland 1960; Young 1967; Smith 1971). Saturated salt solutions have the advantage that they are not affected by loss or gain of small quantities of water during equilibration. However, the number of a_W levels available is limited. When equilibrium (constant weight of sample) is attained, the water content is determined and sorption isotherms can be constructed. The disadvantage of this method is its extreme slowness, which can result in microbial growth at higher a_W, and the necessity of removing the samples from the equilibrating system for weighing. Elaborate and very accurate weighing methods have been devised, using both evacuated manometric systems and systems where water vapor pressures are produced using saturated salt solutions or pure water reservoirs at suitable temperatures. These methods are more rapid, often incorporating electronic balances and continuous recording devices, or even a system to determine sorption isotherms

TABLE 2.3

WATER ACTIVITIES OF VARIOUS SATURATED SALT
SOLUTIONS AT 25°C

Salt	a_w	Salt	a_w
Lithium chloride	0.11	Sodium chloride	0.75
Potassium acetate	0.23	Ammonium sulfate	0.79
Magnesium bromide	0.31	Cadmium chloride	0.82
Zinc nitrate[1]	0.31	Potassium bromide	0.83
Magnesium chloride	0.33	Lithium sulfate	0.85
Lithium nitrate[1]	0.41	Potassium chloride	0.86
Potassium carbonate	0.43	Potassium chromate	0.87
Magnesium nitrate	0.52	Sodium benzoate	0.88
Calcium nitrate[1]	0.54	Zinc sulfate[1]	0.88
Sodium bromide	0.57	Barium chloride	0.90
Cobalt chloride	0.64	Potassium nitrate	0.93
Cupric chloride	0.67	Potassium sulfate	0.97
Lithium acetate	0.68	Disodium phosphate	0.97
Strontium chloride	0.71	Lead nitrate	0.97

Source: Adapted from Rockland (1960).
[1]These salts have large temperature coefficients and therefore temperature should be carefully controlled.

automatically (Bollinger et al. 1972; Best and Spingler 1972). However, these usually have been developed to study simple systems with one or two components only. For complicated heterogeneous mixtures such as foods, less accurate methods are quite sufficient.

Taylor (1961) described a simple manometric method which has been used successfully to give accurate and rapid readings of water vapor pressure (Denizel et al. 1976). The dynamic method of Landrock and Proctor (1951) involves measuring gain or loss of weight by a number of portions of a given sample left for a given time in atmospheres with a range of humidities and then calculating the a_w by interpolation to the ERH which causes no weight change. This procedure requires only the simplest of equipment. The samples must be allowed to equilibrate at an accurately controlled temperature. Methods to determine a_w based on the equilibrium moisture adsorption of standard proteins (Flett 1973) and microcrystalline cellulose (Vos and Labuza 1974) in the presence of foods have been described.

(2) *Hygroscopicity of Salts*—A series of various salt crystals or filter papers impregnated with salt solutions are placed in close proximity to samples in sealed containers. If the a_w of the food is less than the a_w of the saturated salt, the papers or crystals remain dry. If the a_w of the food is greater, the papers or crystals become wet

(Pouncy and Summers 1939; Kvaale and Dalhoff 1963). This is a crude but rapid method to estimate a_W.

(3) *Hygrometric Instruments*—There are a number of instruments available that rely on secondary effects of relative humidity, such as changes of resistivity, conductivity, or conductance of substances. One of the most popular commercial instruments of this type relies on the change of resistance of lithium chloride with changes of relative humidity. Another instrument depends on capacitance changes of an anodized aluminum sensor. A number of instruments are based on similar principles (Mossel and van Kiujk 1955; Spencer-Gregory and Rourke 1957; Jason 1965). All these instruments must be calibrated frequently, are rather slow to use, and must be operated under known and constant temperature.

Dew point methods again measure the ERH of the atmosphere in equilibrium with the sample, as the ratio of the saturated water vapor pressure at the temperature at which dew just forms on a cooled mirror to the saturated vapor pressure of pure water at the temperature of the food sample (Grover and Nicol 1940; Ayerst 1965; d'Alton 1966; Norrish 1969; Anagnostopoulos 1973). The main problem with these instruments is the difficulty in making an accurate assessment of the dew point temperature and the temperature of the sample. Slight inaccuracies in measuring these temperatures can result in large errors in measuring a_W. Electronic methods have been devised for sensing the dew point and recording temperatures. Commercial instruments of this type are also available.

Simple hair hygrometers, although not very accurate, can be useful when an easy and quick method is needed. Rödel and Leistner (1972) developed a polyamide thread hygrometer which is more accurate than the traditional hair type, provided it is calibrated frequently. It has been found useful in the German meat industry.

Control of a_W in Growth Media

Any investigation of the effects of a_W on fungal behavior requires a method whereby a constant known a_W can be maintained. There are several methods of achieving this.

(1) *Equilibration of the Sample with Atmospheres of Known ERH*—The principle here is similar to that used in gravimetric determination of a_W. The sample is equilibrated with atmospheres of known RH by using solutions sealed in the same container. The procedure may be used in growth studies (Snow 1949; Ayerst 1969) or in heat resistance studies (Murrell and Scott 1966). Alternatively, water can be maintained at a constant temperature in one part of the sealed system, with the sample at a different but constant temperature in the

other part (Murrell and Scott 1966).

(2) *From the Sorption Isotherm*—A sorption isotherm for the medium being used can be constructed and the desired a_w obtained by adding water and sealing the system to prevent loss or gain of water (Pitt and Christian 1968).

(3) *Addition of Solutes to the Medium*—A suitable concentration of solute can be incorporated and the system sealed before conducting growth studies (Scott 1953) and heat resistance studies (Härnulv and Snygg 1972). However, the results may be affected by the type of solute as well as the final a_w.

A number of workers have devised methods for observing growth of fungi in systems with controlled a_w that do not necessitate removing the specimen from the controlled RH. Systems used include measurement of optical density in liquid media for yeasts and occasionally molds (Norkrans 1966; Ormerod 1967; Anand and Brown 1968), and a number of systems wherein molds can be observed *in situ* (Pelhate 1968), sometimes by projection of a magnified image onto a screen (Ayerst 1969).

WATER ACTIVITY GROWTH LIMIT OF FUNGI

It is customary to refer to yeasts able to grow at low a_w as "osmophilic" (high osmotic pressure-loving) and molds as "xerophilic" (dry-loving). Neither description is very accurate because, unlike halophilic bacteria, all yeasts and molds grow better at a_w levels considerably higher than their minimum for growth, and many at a_w close to unity. A more accurate description would be xero- or osmo-tolerant and there seems to be no good reason for using different terms for dry-tolerant yeasts and molds. Accordingly, the term "xerotolerant" will be used here for all these organisms.

There is some disagreement, however, about the characteristics possessed by xerotolerant fungi. Heintzeler (1939) and Pelhate (1968) define xerotolerant organisms as those whose spores germinate at less than 0.80 a_w and have an optimum growth level below a_w 0.95. Christian (1963) suggested that the growth limit for yeasts should be 0.85, and Pitt (1975) suggested that all fungi growing below a_w 0.85 should be characterized as xerophilic. The last seems to be a convenient definition, particularly since almost no bacteria except the halophiles have been found that are able to grow below about a_w 0.85.

Habitats of Xerotolerant Fungi

The natural habitats of xerotolerant yeasts include the nectar of flowers, the insects that visit the flowers, honey (Lochhead and Heron

1929; Mossel 1951; Ingram 1957), tree sap exudates (Lodder 1970), and salt water (Norkrans 1966). Xerotolerant molds can be isolated in low numbers from soil, where presumably they are able to take advantage of conditions that periodically prevent the growth of less xerotolerant types. They have also been isolated from salt marshes (Moustafa 1975).

Molds are the predominant spoilage flora of stored seeds (peanuts, wheat, maize, rice, etc.) and dried foods such as milk, flour, egg and meat. In products with high levels of fermentable substrates such as syrups, fruit juices, jams, sugared fruit, condensed sweetened milk and intermediate moisture foods, as well as products of fermentation industries, yeasts also occur, sometimes in high numbers. The ability of the filamentous fungi to grow on materials with low levels of readily fermentable substrates seems to be related to their amylolytic, proteolytic, and lipolytic capabilities. Xerotolerant yeasts are employed during fermentation processes of Oriental condiments such as soy sauce and miso (Onishi 1963). The review of Walker and Ayres (1970) gives much useful information on the occurrence of xerotolerant and other yeasts in foods (see also Ingram 1958, 1960; Moreau 1974).

Water Activity Range for Fungal Growth

Minimum.—The a_W ranges of various foods and the minimum growth limits for the main groups of microorganisms are shown in Fig. 2.3. It is evident that yeasts and molds, apart from halophilic bacteria in salt-preserved foods, are the only organisms likely to grow at a_W levels below about 0.85, and that their ability to tolerate reduced a_W exceeds that of any other group.

Information on a_W minima for growth of a number of species of molds and yeasts is summarized in Table 2.4. Most of the fungi studied have been those isolated from dry situations, so there is little information available on other less tolerant organisms. The minimum a_W at which growth of fungi has been observed is 0.61 to 0.62 (Snow 1949; von Schelhorn 1950; Pitt and Christian 1968) but growth or germination at this a_W is invariably extremely slow. A doubling time of about two months at 25°C has been observed by von Schelhorn (1950) for *Saccharomyces rouxii* at this a_W. Tilbury (1967) noted that visible fermentation occurred about 0.05 a_W units above the limit for growth of yeasts capable of fermenting sugar.

The minimum a_W for growth may appear to be higher than it really is if incubation time is not sufficiently long. This probably explains why the a_W growth limits obtained by Galloway (1935) were consistently higher than those obtained by Snow (1949), for instance. Data concerning germination, however, should always be interpreted

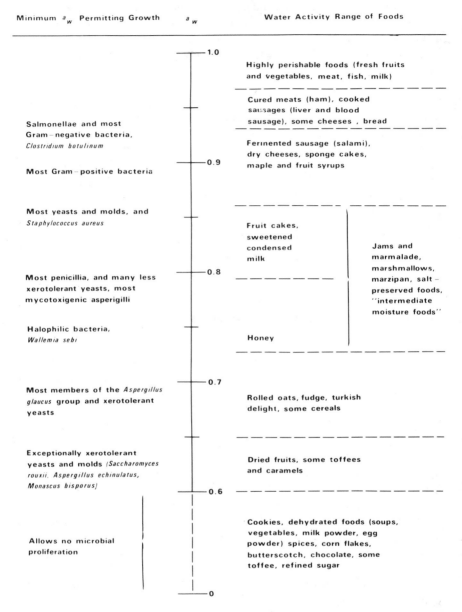

Minimum a_w **Permitting Growth** a_w **Water Activity Range of Foods**

— 1.0

Highly perishable foods (fresh fruits and vegetables, meat, fish, milk)

Cured meats (ham), cooked sausages (liver and blood sausage), some cheeses , bread

Salmonellae and most
Gram-negative bacteria,
Clostridium botulinum

Fermented sausage (salami), dry cheeses, sponge cakes, maple and fruit syrups

— 0.9

Most Gram-positive bacteria

Most yeasts and molds, and
Staphylococcus aureus

Fruit cakes, sweetened condensed milk

Jams and marmalade, marshmallows, marzipan, salt-preserved foods, "intermediate moisture foods"

— 0.8

Most penicillia, and many less
xerotolerant yeasts, most
mycotoxigenic asperigilli

Halophilic bacteria,
Wallemia sebi

Honey

— 0.7

Most members of the *Aspergillus
glaucus* group and xerotolerant
yeasts

Rolled oats, fudge, turkish delight, some cereals

Exceptionally xerotolerant
yeasts and molds *(Saccharomyces
rouxii. Aspergillus echinulatus,
Monascus bisporus)*

Dried fruits, some toffees and caramels

— 0.6

Allows no microbial
proliferation

Cookies, dehydrated foods (soups, vegetables, milk powder, egg powder) spices, corn flakes, butterscotch, chocolate, some toffee, refined sugar

— 0

FIG. 2.3. MINIMUM WATER ACTIVITIES FOR GROWTH OF MICROORGANISMS
AND RANGES OF a_w OF VARIOUS FOODS

TABLE 2.4

MINIMUM a_W FOR GROWTH OF MOLDS AND YEASTS[1]

Fungus	a_W	Temperature (°C)	Reference
Molds			
Rhizoctonia solani	0.96	25	Dube et al. (1971)
Stachybotrys atra	0.94	37	Ayerst (1969)
Botrytis cinerea	0.93	25	Snow (1949)
Mucor spinosus	0.93	25	Snow (1949)
Rhizopus nigricans	0.93	25	Snow (1949)
Epicoccum nigrum	0.90	22	Pelhate (1968)
Mucor circinelloides	0.90	22	Pelhate (1968)
Pythium splendens	0.90	22	Pelhate (1968)
Trichothecium roseum	0.90	22	Pelhate (1968)
	0.90	25	Snow (1949)
Cladosporium herbarum	0.88	25	Snow (1949)
C. cladosporioides	0.85	22	Pelhate (1968)
Alternaria tenuissima	0.85	22	Pelhate (1968)
Penicillium cyclopium	0.85	22	Pelhate (1968)
	0.83	25	Snow (1949)
	0.82	32	Ayerst (1969)
P. rugulosum	0.85	25	Snow (1949)
Paecilomyces varioti	0.84	25	Pitt and Christian (1968)
Penicillium islandicum	0.83	38	Ayerst (1969)
P. expansum	0.83	23	Mislivec and Tuite (1970)
P. palitans	0.83	23	Mislivec and Tuite (1970)
P. patulum	0.81	23	Mislivec and Tuite (1970)
P. puberulum	0.81	23	Mislivec and Tuite (1970)
P. viridicatum	0.81	23	Mislivec and Tuite (1970)
P. citrinum	0.80	25	Galloway (1935)
P. sartoryi	0.80	25	Snow (1949)
P. fellutanum	0.80	25	Snow (1949)
P. martensii	0.79	32	Ayerst (1969)
Aspergillus flavus	0.78	43	Ayerst (1969)
A. niger	0.78	43	Ayerst (1969)
	0.83	25	Snow (1949)
A. tamarii	0.78	43	Ayerst (1969)
A. nidulans	0.78	47	Ayerst (1969)
	0.82	25	Snow (1949)
A. terreus	0.78	47	Ayerst (1969)
A. versicolor	0.78	25	Snow (1949)
A. sydowi	0.78	25	Snow (1949)
A. ochraceus	0.77	25	Pitt and Christian (1968)
A. restrictus	0.75	25	Snow (1949)
	0.75	22	Pelhate (1968)
A. candidus	0.75	25	Snow (1949)
	0.75	44	Ayerst (1969)
Wallemia sebi	0.75	22	Pelhate (1968)
Aspergillus amstelodami	0.75	25	Snow (1949)
	0.71	42	Ayerst (1969)
A. chevalieri	0.71	42	Ayerst (1969)
	0.73	25	Snow (1949)
	0.73	38	Ayerst (1969)
A. repens	0.71	25	Snow (1949)

TABLE 2.4 (Continued)

Fungus	a_W	Temperature (°C)	Reference
Chrysosporium xerophilum	0.71	25	Pitt and Christian (1968)
Aspergillus ruber	0.71	38	Ayerst (1969)
	0.70	25	Snow (1949)
Chrysosporium fastidium	0.69	25	Pitt and Shristian (1968)
Aspergillus echinulatus	0.62-0.64	25	Snow (1949)
Monascus bisporus	0.61	25	Pitt and Christian (1968)

Yeasts

Fungus	a_W	Temperature (°C)	Reference
Schizosaccharomyces octosporus	0.98[2]		
Saccharomyces pastori (Pichia pastoris)	0.98[2]		
Hansenula suaveolens (H. saturnus)	0.97[2]		
H. canadensis	0.97[2]	Not stated	Battley and Bartlett (1966)
Kluyveromyces fragilis	0.96[2]		
Saccharomyces cerevisiae	0.95[2]		
S. microellipsodes	0.95[2]		
S. cerevisiae	0.94[2]		Onishi (1963)
	0.92[3]		Onishi (1963)
Candida utilis	0.94	20	Burcik (1950)
S. cerevisiae	0.90	20	Burcik (1950)
Hansenula anomala	0.88	20	Burcik (1950)
	0.75[4]	27	Tilbury (1967)
Trichosporon pullulans	0.88	20	Burcik (1950)
Saccharomyces rouxii	0.86[2]	Not stated	Onishi (1963)
	0.84[3]	Not stated	Onishi (1963)
Debaryomyces hansenii	0.83[2]	25	Norkrans (1966)
Saccharomyces bailii	0.80	Not stated	Pitt (1975)
S. cerevisiae	0.80[4]	27	Tilbury (1967)
S. bisporus var. mellis	0.70[4]	27	Tilbury (1967)
Torulopsis famata (T. candida) (three strains)	0.65-0.75[4]	27	Tilbury (1967)
S. rouxii	0.65[4]	27	Tilbury (1967)
	0.62[5]	30	von Schelhorn (1950)

[1]Unless otherwise indicated, the medium was equilibrated with a controlled relative humidity.
[2]NaCl-containing medium
[3]Glucose-containing medium
[4]Sucrose/glycerol mixture
[5]Fructose-containing medium

with some caution, since growth may not necessarily follow.

Pitt (1975) states that the minimum a_W for growth of filamentous fungi is characteristic for a particular species. However, minima for different strains of S. rouxii vary widely (von Schelhorn 1950; Onishi

1963; Anand and Brown 1968) and xerotolerance can be gained or lost by this yeast in the presence and absence of high solute levels, respectively (Scarr 1968; Davenport 1976). Yeasts isolated from high-salt environments tend to be more salt tolerant, and those from high-sugar environments more sugar tolerant (Onishi 1963).

All xerotolerant fungi so far isolated belong to a few genera of Ascomycetes or are imperfect forms of those genera. The majority of xerotolerant filamentous fungi belong to the genera *Penicillium* or *Aspergillus*, or are perfect forms of *Aspergillus* (*Eurotium* and *Emericella*). Most xerotolerant aspergilli belong to the *glaucus* group and all members of this group whose water relations have been investigated have been found to be xerotolerant (Pitt 1975). The penicillia are generally less xerotolerant than the aspergilli. Two xerotolerant species each of *Chrysosporium* (*C. fastidium* and *C. xerophilum*) and *Eremascus* (*E. fertilus* and *E. albus*) have been isolated, while *Monascus* (formerly *Xeromyces*), *Paecilomyces*, and *Wallemia* (*Sporendonema*) are reported to contain one xerotolerant species each (*M. bisporus*, *P. varioti*, and *W. sebi* or *epizoum*, respectively). *W. sebi* frequently occurs in high-salt situations (Frank and Hess 1941).

The most commonly isolated and most xerotolerant yeasts are *Saccharomyces rouxii*, *S. bailii* var. *osmophilus*, and *S. bisporus* var. *mellis* (Lodder 1970). These are predominately haploid, and hence were formerly often classified as *Zygosaccharomyces* (Ingram 1950; Mossel 1951). *S. bailii* is considered to be intermediate between *S. bisporus* and *S. rouxii*. It differs from *S. bisporus* in its morphology and from *S. rouxii* because it does not utilize maltose. Strains intermediate between *bisporus* and *bailii*, and *rouxii* and *bailii*, can be found. Isolates are commonly anascosporogenous, especially after being propagated in laboratory media, due to being haploid and heterothallic. *S. bailii* is frequently resistant to antifungal agents and is a particularly important spoilage yeast in sugar-containing soft drinks. Other xerotolerant yeasts include strains of *Hansenula* (Scarr 1954), *Pichia* (Lodder 1970), and *Debaryomyces hansenii*. This last yeast is common in seawater and sea fish (Norkrans 1966; Ross and Morris 1962) and can grow in 24% (wt/vol) sodium chloride. Xerotolerant strains of *Schizosaccharomyces* (*octosporus* and *pombe*), which are considered to be related to the mold *Eremascus* (Lodder 1970), have also been isolated infrequently (Scarr 1951, 1954; Davenport 1976). Imperfect strains of xerotolerant yeasts are generally *Candida* and *Torulopsis* spp. (Ingram 1950; Scarr 1954; Scarr and Rose 1966; Anand and Brown 1968). These yeasts are sometimes imperfect forms of perfect xerotolerant species. For example, *Torulopsis mogii* = *S. rouxii*, and *Torulopsis candida* = *Debaryomyces hansenii*.

TABLE 2.5

PRINCIPAL YEAST SPECIES ISOLATED FROM HIGH-SALT
(10 TO 20% SODIUM CHLORIDE) ENVIRONMENTS, AND THEIR
CHARACTERISTICS FOR DIFFERENTIATION

| Yeast | Assimilation of | | | Fermentation of Glucose |
	Nitrate	Nitrite	Salicin	
Debaryomyces hansenii	—	—	—	strong
Torulopsis famata				
(T. candida)	—	+	+	weak
Pichia ohmeri	—	—	+	strong
Hansenula anomala				
var. anomala	+	—	+	strong

Source: Davenport (1975).

The most common types of xerotolerant yeasts isolated from high-salt or high-sugar environments and the most useful criteria in their differentiation are shown in Tables 2.5 and 2.6. Scarr and Rose (1965) recommend that xerotolerant yeast sugar fermentation/assimilation tests should be carried out at higher sugar concentrations than normally used, although Davenport (1976) considers this unnecessary. Strains of xerotolerant yeasts should be maintained on low a_W media, otherwise their tolerance tends to decrease (Scarr 1968).

Optimum.—Studies by Pelhate (1968) and Ayerst (1969) showed that the optimum a_W for growth of most molds was near a_W 1.0, but that strains of the A. glaucus group had optima near a_W 0.93. Anand and Brown (1968) made similar observations with xerotolerant yeasts. However, the optimum growth rates of xerotolerant strains of molds and yeasts are generally lower than those with less xerotolerance. Observations on the microflora of stored crops at a_W levels above about 0.85 to 0.90 (Moreau 1974; Denizel et al. 1976) show that nonxerotolerant or less xerotolerant strains outgrow the more xerotolerant strains. This is also true for yeasts (Anand and Brown 1968; Brown 1975A).

Maximum.—Although no bacteria except the halophiles seem to have an absolute requirement for a reduced a_W for growth, there is evidence that some molds and yeasts do; Monascus bisporus shows negligible growth above a_W 0.95, A. amstelodami will not grow above a_W 0.99 (Scott 1957), and A. restrictus will not grow above a_W 0.985 (Pelhate 1968). There have been several similar reports on strains of xerotolerant yeasts (Lochhead and Heron 1929; Scarr 1954; Anand and Brown 1968), but some of these reports may have been confused

TABLE 2.6

PRINCIPAL YEAST SPECIES ISOLATED FROM HIGH-SUGAR (60%)
ENVIRONMENTS, AND THEIR CHARACTERISTICS FOR DIFFERENTIATION

Yeast	Type of Vegetative Reproduction	Ascospore Production	Fermentation of		
			Raffinose	Sucrose	Maltose
Saccharomyces rouxii	budding	rarely	–	– or slow	+
S. bailii var. osmophilus	budding	rarely	– or weak	often slow	–
S. bisporus var. mellis	budding	rarely	–	– or weak	– or acquired
Torulopsis lactis-condensi	budding	none	+	+	–
Schizosaccharomyces pombe	fission	readily	+	+	+

Source: Davenport (1975).

by the fact that some xerotolerant yeasts cannot grow at high a_W at elevated temperatures.

EFFECT OF GROWTH CONDITIONS ON LIMITING WATER ACTIVITY

Nutrients, pH, and Temperature

Many studies have indicated that fungi show greatest tolerance to reduced a_W when all other conditions (pH, temperature, Eh, nutrients) are near optimum, and that the growth range (pH range or temperature range for growth) is narrower at low a_W (Tomkins 1930; Heintzeler 1939; Snow 1949; Onishi 1963; Ayerst 1969). This effect can be put to practical advantage in assessing the keeping quality of food products. Foods with low pH will be more stable at a given a_W than those with a higher pH (Leistner and Rödel 1975). Foods with high levels of available nutrients will be more susceptible to spoilage at low a_W than foods with low levels of nutrients (Galloway 1935; Snow et al. 1944; Snow 1949).

An interesting effect is that some molds and yeasts seem to be able to grow at higher temperatures in media of low a_W. Onishi (1959) showed that S. rouxii can grow at 40°C only when 3 to 4% sodium chloride, or more than 40% sugar, is present in the growth medium. Similar effects have been noted with osmophilic yeasts in orange juice (Ingram 1950) and with molds (Ayerst 1969). One strain of a xerotolerant yeast has been observed to grow at 27°C but not at 37°C in the absence of sucrose (Scarr and Rose 1966). Relationships between temperature, a_W, and rate of growth for a xerotolerant and a nonxerotolerant mold are shown in Fig. 2.4 (Ayerst 1969).

Effect of Solutes

When solutes are incorporated into foodstuffs or laboratory media, the minimum a_W for growth of fungi in these substrates is often affected. Scott (1957) reported that growth rates of A. amstelodami were faster in media containing sucrose and glucose than in those with magnesium chloride, sodium chloride, or glycerol at the same a_W. Onishi (1957) observed that osmophilic yeasts, especially those from high sugar sources, grew at lower a_W in media containing sucrose or glucose than in media containing sodium chloride. Charlang and Horowitz (1971) observed that the rate of growth of Neurospora crassa was in the order glycerol>NaCl>glucose>sucrose at a given a_W.

Studies by Horner and Anagnostopoulos (1973) showed that yeasts

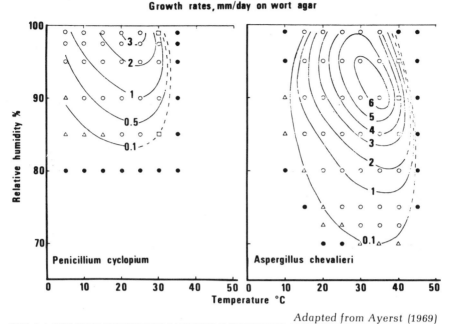

Growth rates, mm/day on wort agar

Adapted from Ayerst (1969)

FIG. 2.4. TEMPERATURE/RELATIVE HUMIDITY RELATIONSHIPS FOR GROWTH OF *ASPERGILLUS CHEVALIERI* AND *PENICILLIUM CYCLOPIUM* ON WORT AGAR. Symbols: o , germination time of 1 to 16 days; △, 16 to 95 days; ●, >95 days; □, germination not followed by growth. Lines join conditions giving the same growth rate (mm/day).

were able to grow faster in glycerol- than in sucrose-adjusted media, while a mold of the *A. glaucus* group, *A. niger,* and *Penicillium* spp. grew faster in media containing sucrose, although the lag was shorter in glycerol media.

Method of Attaining a_W Level

Recent studies (Labuza *et al.* 1972; Acott and Labuza 1975B; Labuza 1975) have indicated that the method of attaining a given a_W in a dried food can affect markedly the response of molds, yeasts, and bacteria. The limiting a_W for growth is higher if the food is prepared by freeze-drying followed by rehumidification (adsorption system) than if the food is prepared by removing water (desorption system). This is a situation where the water content rather than the a_W seems to be more important. The water content of the food is higher in the desorption than the adsorption system (see Fig. 2.1). However, hysteresis is a temporary phenomenon and when equilibrium is finally attained, the a_W levels will have changed. Nevertheless, this effect should be borne in mind, particularly when formulating and processing intermediate moisture foods.

ADAPTATION TO REDUCED WATER ACTIVITY

The reasons why some microorganisms are able to grow in drier situations than others are not fully understood, but there is evidence that internal osmotic pressure is increased in response to increased external osmotic pressure, thus counteracting the drying effect (plasmolysis). In order to achieve this, the concentrations of internal low molecular weight solutes are increased, i.e., water molecules are replaced by solute molecules. The lower a_w limit for growth of a microorganism depends on the characteristics of the solute(s) accumulated (particularly on the ability of solutes to replace water without impairing the normal functioning of the cell system, whether they are "compatible" or not) (Brown 1974) and probably on their solubility also.

Early work indicated that the internal osmotic pressure of microorganisms growing in a medium having high osmotic pressure was similar to that of the medium itself (Christian and Ingram 1959). Levels of potassium ions and free amino acids in bacteria growing in media with high sodium chloride concentrations are elevated (Christian and Waltho 1962; Tempest and Meers 1968; Tempest et al. 1970), although levels of amino acids in S. cerevisiae remain unchanged (Tempest et al. 1970). Recent work has indicated that adaptation to high osmotic pressure is achieved in most bacteria by the accumulation of certain amino acids (Measures 1975), although halophilic bacteria have an absolute requirement for high intracellular ion levels, specifically potassium ions (Brown 1964; Kushner 1968). Nonhalophilic bacteria capable of growing at the lowest a_w levels (e.g., Staphylococcus aureus and Bacillus cereus) accumulate proline, while those with higher minimal a_w limits (e.g., Salmonella and Pseudomonas) accumulate mainly glutamic acid.

Both xerotolerant and nonxerotolerant yeasts growing at reduced a_w accumulate polyalcohols (polyols) such as arabitol or sometimes glycerol and a hexitol (Brown and Simpson 1972; Brown 1974), which are presumably more effective than amino acids as compatible solutes. It is not clear why xerotolerant yeasts are better adapted to conditions of low a_w than other yeasts, since polyols affect enzymes of both yeast types similarly. Brown (1975A) suggests that xerotolerant yeasts, at least those growing in media with high concentrations of glucose, have no catabolite repression.

It remains to be demonstrated that molds also use polyols as compatible solutes, but polyols are one of the principle soluble carbohydrates of the Basidiomycetes, Ascomycetes, and Fungi Imperfecti. Mannitol is the most common polyol, and arabitol is also frequently found (Lewis and Smith 1967). Polyols may function in preventing dehydration of aerial mycelium and as reserves to store reducing power, since they are more reduced than their corresponding sugars. Storage of polyols may also be a method of regulating levels of nicotinamide adenine dinucleotide phosphate (NADP) and regulating the function of the pentose phosphate pathway. However, it would be surprising if an entirely different system of adaptation to high osmotic pressure existed in molds, especially since both molds and yeasts have been shown to excrete large quantities of polyols into the growth medium at high osmotic pressure (Foster 1949; Spencer 1968).

EFFECT OF a_W ON GERMINATION AND SPORULATION

Reduced a_W has the effect of increasing the lag phase (or time before germination), in addition to reducing the rate of growth and final fungal biomass yield (Tomkins 1930; Heintzeler 1939; Snow 1949; Ayerst 1969; Charlang and Horowitz 1971). Growth rate near the minimum a_W for growth after germination is very slow. Snow (1949) found latent periods of almost two years before germination of conidiospores of Aspergillus echinulatus at a_W 0.64 to 0.66, and abnormal germ tubes. Armolik and Dickson (1956) and Mislivec and Tuite (1970) observed abnormal swelling of conidiospores of penicillia at a_W levels approaching the lower limit for germination. Aleuriospores of Monascus bisporus took 120 days to germinate at a_W 0.61 (Pitt and Christian 1968), but the growth rate thereafter was not reported. This very low a_W for germination is particularly interesting, since the pH of the growth medium was also low (pH 3.8) and the minimum a_W levels observed for germination of asexual spores of the A. glaucus group were higher than have been observed by other workers at higher pH levels (Snow 1949; Pelhate 1968; Ayerst 1969).

Germination of fungal spores may not always be followed by mycelial growth, especially under extremes of temperature and with suboptimal nutrient conditions. Ayerst (1969) observed that germination of Aspergillus conidiospores frequently occurred at temperatures near the maximum over a range of a_W levels, but little or no outgrowth occurred (see Fig. 2.4). Pitt (1975) reports germination but no growth of Chrysosporium fastidium aleuriospores in sodium chloride-containing medium between 0.96 and 0.80 a_W. Teitell (1958) observed that conidiospores of A. flavus germinated but did not grow, and eventually lost their viability at 0.75 a_W and 29°C, while at lower a_W levels they remained dormant and viable.

It has been suggested that the minimum a_W for the development of conidia is higher than that for mycelial growth (Panasenko 1967; Mislivec and Tuite 1970), but it is not clear whether this observation might simply be a function of very slow growth rates. Molds exposed to adverse conditions frequently produce asexual spores, presumably as a survival mechanism. There seems to be no doubt, however, that that minimum a_W level for ascospore formation is higher than that required for the formation of asexual spores (Snow 1949; Panasenko 1967; Pitt 1975).

EFFECT OF a_W ON MYCOTOXIN PRODUCTION

There is no evidence that yeasts, xerotolerant or otherwise, produce toxins, but some xerotolerant molds have been reported as toxigenic (Table 2.4; Moreau 1974; Pitt 1975). However, only a small proportion of the total number of mold strains occurring in stored products appear to produce mycotoxins. About 10% of the *Aspergillus* and *Penicillium* strains isolated from ground rice, rice, wheat, and other flours have been shown to be toxic for mice (Kurata and Ichinoe 1967; Kurata *et al.* 1968A,B). About 16% of the *Aspergillus* and *Penicillium* strains isolated from aged cured meats produced substances toxic to chick embryos (Wu *et al.* 1974). The proportion of *A. flavus* and *A. parasiticus* strains from all sources which produce aflatoxin is about 60% (Diener and Davis 1969). However, it is not clear whether those strains which have been shown to produce mycotoxins at high a_W would do so at low a_W levels in food products. Almost all the information available on this subject concerns aflatoxin production by *A. flavus* and related strains. Unfortunately, although a considerable amount of work has been carried out, it is difficult sometimes to compare results because the water relations of the substrate have frequently been quoted in terms of percentage of dry weight, or even percentage of wet weight. The consensus of opinion appears to be, however, that aflatoxin production ceases or becomes very low at a_W below about 0.85, whereas growth can occur at a_W as low as 0.78 to 0.80. This applies to all storage crops so far investigated and includes peanuts, pistachio nuts, Brazil nuts, maize, wheat, rice (Sanders *et al.* 1968; Diener and Davis 1968, 1969; Boller and Schroeder 1974; Denizel *et al.* 1976), raisins (Follstad 1966), and salami (Bullerman *et al.* 1969). Competition from the mixed flora in nonsterile products tends to retard aflatoxin accumulation still further (Boller and Schroeder 1973; Denizel *et al.* 1976).

Bacon *et al.* (1973) studied ochratoxin and penicillic acid production by *A. ochraceus* growing in poultry feed at various a_W levels and temperatures. Their results indicated that ochratoxin production was greatest at a high temperature and high a_W level (30°C, a_W 0.95), while penicillic acid production was favored by a lower a_W level and temperature (optimum at 22°C, a_W 0.90). No ochratoxin was produced at the lowest a_W for growth (a_W 0.76) and sporulation occurred at 30°C but not 22°C at this a_W. Only low levels of ochratoxin were produced at a_W 0.85 at 30°C.

EFFECT OF a_W ON HEAT RESISTANCE

Although low a_W levels reduce or inhibit microbial growth, they have also the unwelcome effect of increasing heat resistance of vegetative cells and spores. Microbiologists have been aware for many years that dry heat is less lethal than wet heat, but quantitative studies were not carried out until comparatively recently and most studies have been made on bacteria (Corry 1973, 1975). When bacteria are heated after equilibration with water vapor to various a_W levels, they show very large increases in heat resistance as a_W is reduced, with maxima at about a_W 0.2 to 0.4 (Murrell and Scott 1966; Alderton and Halbrook 1973). The presence of high levels of solutes affects heat resistance differently, and although heat resistance is increased, the solute present as well as the a_W level affects heat resistance (Härnulv and Snygg 1972; Corry 1974). Observations on the resistance of mold spores heated in the presence of solutes indicate that heat resistance is also enhanced, although not to the extent found for bacterial spores, showing a gradual and continuous increase from a_W 1.0 to 0.0 (Fig. 2.5; Lubienieki-von Schelhorn and Heiss 1975). Studies by Doyle and Marth (1975A,B) showed that heat resistance of conidiospores of *A. flavus* and *A. parasiticus* was increased in solutions containing sucrose, glucose, and sodium chloride, and that below a_W of about 0.93 the order of increased resistance was NaCl sucrose glucose. *Byssochlamys nivea* ascospore heat resistance is greatly enhanced in grape juice containing elevated levels of sucrose (Beuchat and Toledo 1977).

Extensive experiments have not been carried out for a range of a_W values for yeasts, but investigation of the effects of high levels of solutes has again indicated that heat resistance is increased, and that the order of increased heat resistance is sucrose>sorbitol>glucose>fructose>glycerol for strains of *S. rouxii* and *Schizosaccharomyces pombe* at a_W 0.95 (Corry 1976A). Data available on the heat resistance

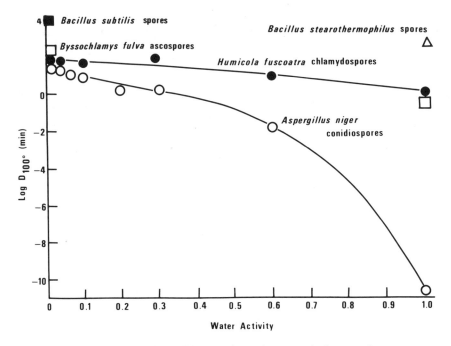

Adapted from Lubienieki-von Schelhorn and Heiss (1975)
FIG. 2.5. EFFECT OF WATER ACTIVITY (RH x 0.01) ON DECIMAL REDUCTION
TIMES AT 100°C FOR MOLD AND BACTERIAL SPORES

of fungi are listed in Table 2.7. It is difficult to compare results for
heat resistance determined at different temperatures and a_w levels,
but comparison of the results from yeast cells (Corry 1976A) and
fungal conidiospores (Doyle and Marth 1975B) indicates similar heat
resistance, assuming z values of about 5°C. There is evidence that z
values under dry conditions approach 10°C (Lubienieki-von Shel-
horn and Heiss 1975). The heat resistance of yeast ascospores seems
to be slightly higher than that of vegetative cells (English 1954; Put
and Sand 1974; Put et al. 1976; Corry 1976A).

There is evidence that the differences in heat resistance observed in
media containing different solutes, both for vegetative bacteria and
xerotolerant yeasts, are due to differences in the ability of solutes to
permeate the cell. Solutes that are unable to penetrate the cytoplasmic
membrane do not enter the cytoplasm. These solutes promote increased heat resistance to a greater extent than those that are able to
permeate the cytoplasm (Corry 1975, 1976B).

TABLE 2.7

HEAT RESISTANCE OF FUNGI AT VARIOUS WATER ACTIVITIES

Fungus	a_W	Temperature (°C)	Solute	D value (min)	Reference
Aspergillus flavus (conidiospores)	0.90	55	NaCl	65[1]	Doyle and Marth (1975B)
			Sucrose	65[1]	
			Glucose	40[1]	
Aspergillus parasiticus (conidiospores)					
1st strain	0.90	55	NaCl	130[1]	
			Sucrose	84	
			Glucose	70[1]	
2nd strain	0.90	55	NaCl	240[1]	
			Sucrose	200	
			Glucose	120[1]	
Aspergillus niger (conidiospores)	1.00	55	None	6	Lubienieki-von Schelhorn
	0.60	70	None	100	and Heiss (1975)
	0.30	80	None	216	
	0.00	100	None	100	
Byssochlamys fulva (ascospores)	0.00	120	None	25	
Byssochlamys nivea (ascospores)	0.98	75	Sucrose	60[1]	Beuchat and Toledo (1977)
	0.92	75	Sucrose	260[1]	
	0.84	75	Sucrose	470[1]	
Humicola fuscoatra (chlamydospores)	1.00	80	None	101[1]	Lubienieki-von Schelhorn
	0.60	80	None	143	and Heiss (1975)
	0.30	100	None	100	
	0.00	120	None	30	
Saccharomyces rouxii (vegetative cells)	0.95	65	Sucrose	2.0	Corry (1976A)
			Sorbitol	0.9	
			Glucose	0.4	
			Fructose	0.4	
			Glycerol	0.3	

Approximate values.

PRACTICAL ASPECTS

Dry Food

"Dry" food includes not only obviously dried materials such as grains, flour, dehydrated products like milk, egg, and soups, dried meat, dried and/or salted fish, and dried fruits and vegetables, but also products that contain high levels of solutes, for example candies, or which may even appear liquid such as sugar or fruit syrups and brines. Many traditional products such as fermented sausages, cheeses, some candies, jam, and crystallized fruits have a_W levels of 0.70 to 0.85 so that some molds and yeasts are able to grow in or on them (see Fig. 2.3). Recently other foods in the same a_w range have been added to these traditional intermediate moisture foods (see reviews of Karel 1973 and Davies *et al.* 1976). The aim is to produce relatively stable foods which can be stored at ambient temperatures and consumed without prior cooking or rehydration. Dryness and suitable textures may be achieved by incorporation of relatively high levels of solutes and humectants, usually glycerol in combination with sucrose, glucose, propylene glycol, 1,3-butanediol, or salt, in conjunction with an antimycotic agent such as sorbic acid. Propylene glycol, 1,3-butanediol, and related compounds also have specific antimicrobial activity (Acott and Labuza 1975A; Sinskey 1976). Although the new intermediate moisture food products have military and space applications, they seem to have been more popular as foods for pets than for humans—perhaps dogs and soldiers have less choice.

The shelf life of the traditional intermediate moisture foods can be extended by use of sulfur dioxide, salts of sorbic or benzoic acids (dried fruits, fruit juices, jams), nitrite (cheese, sausages), or by refrigeration, pasteurization, or sterilization. Partially rehydrated prunes are an example of a traditional dried food that has recently successfully been changed to an intermediate moisture food. The product is sold in sealed plastic pouches and must either be sterilized or preserved with antimycotics (Pitt 1975). Partially rehydrated prunes have an a_w of about 0.86 instead of the traditional a_w of about 0.68.

Conditions Necessary for Stability

Determination of whether a dry or intermediate moisture food will be stable, and for how long, depends on a number of factors, including a_w, solutes and preservatives present, pH, and initial numbers and types of viable organisms. A detailed assessment of each of these parameters should be made for each product. If conditions of high humidity are likely to be encountered, packaging should be such that water cannot permeate the food, and if sudden changes in temperatures are liable to occur, which can result in local condensation of water, allowance should be made. Storage tests should be conducted under conditions of fluctuating temperatures (Mossel and Sand 1968; Leistner and Rödel 1975) before newly formulated dry or intermediate moisture products are marketed.

Mode of Growth in Foods

Besides lowering sugar concentration, producing undesirable flavors, and affecting the appearance of foods, the production of large quantities of carbon dioxide during sugar fermentation causes foaming problems in syrups and even "exploding" candy. One mystery which does not seem to have been solved is that, while many strains of xerotolerant yeasts are unable to assimilate sucrose (Ingram 1960; Scarr 1954; Scarr and Rose 1966), they seem to be capable of fermenting substances consisting almost entirely of sucrose. Small concentrations of invert sugar stimulate yeast growth in sucrose syrups, and one suggested mechanism is that the production of small quantities of acid during the metabolism of invert sugar causes localized acidity and inversion of sucrose, which can then be utilized by the yeast (Scarr 1954). The pH levels characteristically fall during fermentation of sucrose syrups by xerotolerant yeasts (Scarr 1954). Many yeasts, particularly S. rouxii, ferment fructose preferentially and leave high levels of glucose (Tilbury 1967). Fermentation of liquid sugar usually starts in the top layers. This may be because water and/or oxygen levels are higher or because yeast cells tend to float. Centrifugation tests have shown that yeast cells in high concentrations of sugar are less dense than the sugar solution (Scarr 1954; Ingram 1960). After prolonged incubation, the yeast cells sediment. Ingram (1960) suggests that this occurs because of the reduction of sugar concentration, and hence a decrease in the density of the suspending solution results, caused by the metabolic activity of the yeasts. Scarr (1968) considers that sedimentation is associated with loss of viability and plasmolysis. Cells of both xerotolerant and nonxerotolerant yeasts shrink when placed in high osmotic pressure solutions (Scarr 1954, 1968; Corry 1976A,B). The effects of 48% (wt/wt) sucrose compared with water on cells of S. rouxii and Candida utilis are shown in Fig. 2.6. The xerotolerant strain shrinks less, presumably because it accumulates polyols even when grown at low osmotic pressure (Brown and Simpson 1972).

   Molds also may be detected in high-sugar products, especially where water has condensed due to temperature fluctuations (Mossel and Sand 1968). In bulk-stored foods such as grains and oilseeds, xerotolerant mold growth may start in small localized areas of high humidity. "Metabolic" water produced by the mold enables the spoilage to spread and eventually may allow less xerotolerant species to grow. The observation of Denizel et al. (1976) of a succession of mold species on naturally contaminated pistachio nuts incubated in a confined container at 85% relative humidity illustrates this effect. A. glaucus group molds were succeeded by less xerotolerant molds, including A. flavus, which produced aflatoxin. Besides the dangers from mycotoxin production, mold growth can cause considerable heating and even "spontaneous" combustion of stored crops. Even if the product is not rendered totally useless, the organoleptic and

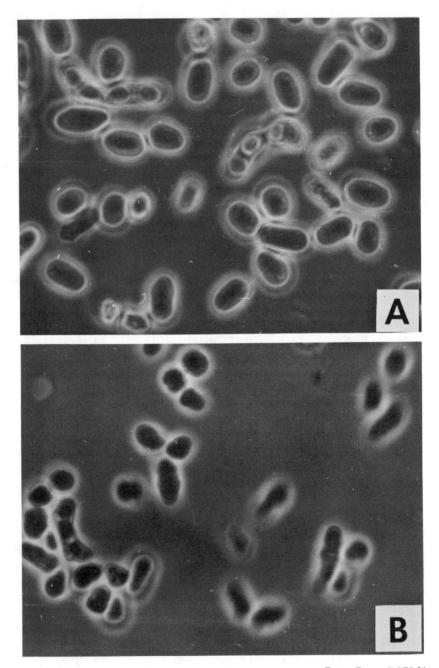

From Corry (1976A)

FIG. 2.6. EFFECT OF CONCENTRATED SUCROSE SOLUTION ON THE MORPHOL-
OGY OF YEASTS (PHASE CONTRAST 4060X). Cells were grown in malt extract
Agar. *Saccharomyces rouxii* was suspended in: A, water; and B, 48% (wt/wt) sucrose in

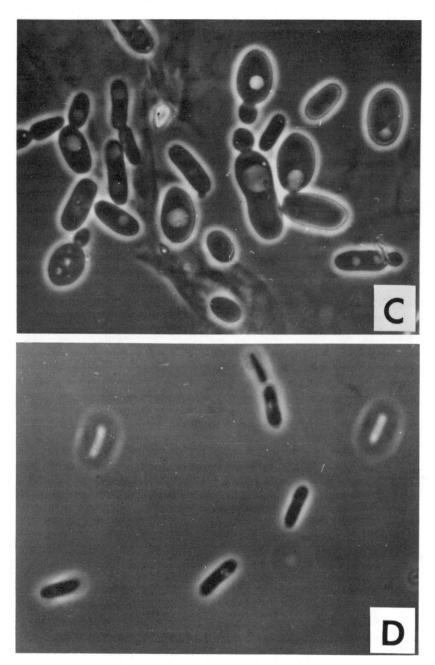

0.1 M phosphate buffer, pH 6.5. *Candida utilis* was suspended in: C, water; and D, 48% (wt/wt) sucrose in 0.1 M phosphate buffer, pH 6.5. Reproduced by permission of the British Food Manufacturing Industries Research Association.

nutritional values, and hence the market value, are reduced.

Problems in Detecting Xerotolerant Fungi

Many different plating media have been described for selecting xerotolerant fungi. All or almost all include high levels (30 to 50% wt/wt) of sugar (usually glucose or fructose), a suitable source of nitrogen, and growth factors such as malt extract, yeast extract, or peptone (Mossel 1951; Scarr 1954, 1959; Ingram 1957, 1960; Mossel and Sand 1968; Pitt 1975; see also Chap. 16). If a selective medium is not used, the xerotolerant fungi may not only be outgrown by non-xerotolerant strains, but they also may be unable to grow at high a_W at all. If dilutions are prepared, the diluent also should contain sugar (20 to 30% glucose or fructose in peptone water) in order to avoid osmotic shock. When difficulty is experienced in detecting xero-tolerant fungi, it may be possible to detect them by microscopic examination, or to enrich them by incubating samples in a suitable liquid medium before plating onto a solid medium. When all else fails, evidence of microbial activity may be obtained by detection of metabolites such as ethanol, carbon dioxide, or organic acids (Mossel and Sand 1968).

Prevention of Spoilage at Various a_W

This problem can be approached from at least two directions:

(1) If the final a_W level is below 0.60 to 0.65 and the product is maintained at this level during storage, problems arising due to microbial spoilage are rare, irrespective of the number of organisms present. However, high levels of contaminating organisms, even if they cannot grow, should be avoided, since they can survive for long periods and may contaminate other foods or cause problems after rehydration. This applies particularly to products such as seed flours, cocoa, dried egg, milk, and products of animal origin which are prone to contamination with food poisoning organisms such as salmonellae and staphylococci (Mossel and Shennan 1976).

(2) At higher a_W levels, particularly over a_W 0.70, there is a possibility of microbial spoilage. The shelf life of a product, provided the a_W level is maintained, depends on a number of factors, including the temperature of storage, pH, availability of oxygen, presence of antimycotic substances, method of attaining a_W (adsorption or desorption), and numbers and types of xerotolerant organisms pres-ent.

Keeping quality of foods can be assessed by storage tests on the product concerned, and monitored by checking a_W level, pH, and numbers of xerotolerant organisms. Frequent checks on the microbiological quality, particularly numbers of xerotolerant organisms and specific organisms associated with spoilage of the composite food, are required (Mossel 1971). Tests should also be made on individual ingredients of the product and on the product during production, coupled with scrupulous hygiene, to prevent most problems with all types of spoilage organisms. If a heating process is involved, the formula may be modified in order to minimize the heat resistance of the contaminating microorganisms (e.g., by replacing sucrose with glycerol) or the ingredients can be treated individually before use.

ACKNOWLEDGEMENT

I should like to thank Professor D.A.A. Mossel, Dr. R.R. Davenport and Dr. B. Jarvis for helpful discussion, information and comment during the preparation of this chapter.

REFERENCES

ACOTT, K., SLOAN, A. E., and LABUZA, T. P. 1976. Evaluation of antimicrobial agents in a microbial challenge study for an intermediate moisture dog food. J. Food Sci. 41, 541-546.

ACOTT, K. M., and LABUZA, T. P. 1975A. Inhibition of Aspergillus niger in an intermediate moisture food system. J. Food Sci. 40, 137-139.

ACOTT, K. M., and LABUZA, T. P. 1975B. Microbial growth response to water sorption preparation. J. Food Technol. 10. 603-611.

ADAMS, G. H., and ORDAL, Z. J. 1976. Effects of thermal stress and reduced water activity on conidia of Aspergillus parasiticus. J. Food Sci. 41, 547-550.

ALDERTON, G., and HALBROOK, W. V. 1973. Effect of water activity on heat resistance of dried vegetative cells (Salmonella typhimurium). Abst. Ann. Meet. Am. Soc. Microbiol. E95, 16.

ANAGNOSTOPOULOS, G. D. 1973. Water activity in biological systems; a dew point method for its determination. J. Gen. Microbiol. 77, 233-235.

ANAND, J. G., and BROWN, A. D. 1968. Growth rate patterns of the so-called osmophilic and non-osmophilic yeasts in solutions of polyethylene glycol. J. Gen. Microbiol. 52, 205-212.

ARMOLIK, N., and DICKSON, J. G. 1956. Minimum humidity requirements for germination of conidia associated with storage of grain. Phytopathology 46, 462-465.

AYERST, G. 1965. Determination of the water activity of some hygroscopic food materials by a dew point method. J. Sci. Food Agric. 16, 71-78.

AYERST, G. 1969. The effects of moisture and temperature on growth and spore germination of some fungi. J. Stored Prod. Res. 5, 127-141.

BACON, C. W., SWEENEY, J. G., ROBBINS, J. D., and BURDICK, D. 1973. Production of penicillic acid and ochratoxin A on poultry feed by Aspergillus ochraceus: temperature and moisture requirements. Appl. Microbiol. 26, 155-160.

BATTLEY, E. H., and BARTLETT, E. J. 1966. A salt-concentration gradient method for the determination of the maximum salt concentration for microbial growth. Antonie van Leeuwenhoek J. Microbiol. Serol. 32, 256-260.

BEST, R., and SPINGLER, E. 1972. Measurement of adsorption and desorption isotherms with a fully automatic apparatus. Chem.-Ing.-Tech. 44, 1222-1226. (German)

BEUCHAT, L. R., and TOLEDO, R. T. 1977. Behavior of Byssochlamys nivea ascospores in fruit syrups. Trans. Br. Mycol. Soc. 68, 65-71.

BOLLER, R. A., and SCHROEDER, H. W. 1973. Influence of Aspergillus chevalieri on production of aflatoxin in fice by Aspergillus parasiticus. Phytopathology 63, 1507-1510.

BOLLER, R. A., and SCHROEDER, H. W. 1974. Influence of relative humidity on production of aflatoxin in rice by Aspergillus parasiticus. Phytopathology 64, 17-21.

BOLLINGER, W., GAL, S., and SIGNER, R. 1972. An apparatus for the automatic determination of water sorption isotherms and isobars. Helv. Chim. Acta 55, 2659-2663. (German)

BROWN, A. D. 1964. Aspects of bacterial responses to the ionic environment. Bacteriol. Rev. 28, 296-329.

BROWN, A. D. 1974. Microbial water relations: features of the intracellular composition of sugar tolerant yeasts. J. Bacteriol. 118, 769-777.

BROWN, A. D. 1975A. Microbial water relations: effects of solute concentration on the respiratory activity of sugar-tolerant and non-tolerant yeasts. J. Gen. Microbiol. 86, 241-249.

BROWN, A. D., and SIMPSON, J. R. 1972. Water relations of sugar-tolerant yeasts: the role of intracellular polyols. J. Gen. Microbiol. 72, 589-591.

BROWN, B. I. 1975B. Further studies on ginger storage in salt brine. J. Food Technol. 10, 393-405.

BULLERMAN, L. B., HARTMAN, P. A., and AYRES, J. C. 1969. Aflatoxin production in meats. II. Aged dry salamis and aged country cured hams. Appl. Microbiol. 18, 718-722.

BURCIK, E. 1950. The relationship between hydration and growth of bacteria and yeasts. Arch. Mikrobiol. 15, 203-235. (German)

CHARLANG, G. W., and HOROWITZ, N. H. 1971. Germination and growth of Neurospora at low water activities. Proc. Natl. Acad. Sci. U.S.A. 68, 260-262.

CHRISTIAN, J. H. B. 1963. Water activity and the growth of microorganisms. In Recent Advances in Food Scienc. Vol. 3. J. Leitch and D. Rhodes (Editors). Butterworths, London.

CHRISTIAN, J. H. B., and INGRAM, M. 1959. The freezing points of bacterial cells in relation to halophilism J. Gen. Microbiol, 20, 27-31.

CHRISTIAN, J. H. B., and WALTHO, J. A. 1962. Solute concentrations within cells of halophilic and non-halophilic bacteria. Biochem. Biophys. Acta 65, 506-508.

CORRY, J. E. L. 1973. The water relations and heat resistance of microorganisms. Prog. Ind. Microbiol. 12, 73-108.

CORRY, J. E. L. 1974. The effect of sugars and polyols on the heat resistance of salmonellae, J. Appl. Bacteriol. 37, 31-43.

CORRY, J. E. L. 1975. The effect of water activity on the heat resistance of bacteria. In Water Relations of Foods. R. B. Duckworth (Editor). Academic Press, London.

CORRY, J. E. L. 1976A. The effect of sugars and polyols on the heat resistance and morphology of osmophilic yeasts. J. Appl. Bacteriol. 40, 269-276.

CORRY, J. E. L. 1976B. Sugar and polyol permeability of Salmonella and osmophilic yeast cell membranes measured by turbidimetry, and its relation to heat resistance. J. Appl. Bacteriol. 40, 277-284.

D'ALTON, G. 1966. Glucose syrups: determination of e.r.h. Confect. Manuf. Mktg. 3, 122-123, 184-186, 233-235.

DAVENPORT, R. R. 1975. The distribution of yeasts an yeast-like organisms in an English vineyard. Ph.D. Thesis. University of Bristol, England.

DAVENPORT, R. R. 1976. Personal communication. Long Ashton, Bristol, England.

DAVIES, R., BIRCH, G. G., and PARKER, K. 1976. Intermediate Moisture Foods. Applied Science Publishers, Barking, England.

DENIZEL, T., ROLF, E. J., and JARVIS, B. 1976. Moisture-equilibrium relative humidity relationship in pistachio nuts with particular reference to the control of aflatoxin production. J. Sci. Food Agric. 27, 1027-1034.

DIENER, U. L., and DAVIS, N. D. 1968. Effect of environment on aflatoxin production in freshly dug peanuts. Trop. Sci. 10, 22-28.

DIENER, U. L., and DAVIS, N. D. 1969. Aflatoxin formation by Aspergillus flavus. In Aflatoxin. L. A. Goldblatt (Editor). Academic Press, New York.

DOYLE, M. P., and MARTH, E. H. 1975A. Thermal inactivation of conidia from Aspergillus flavus and Aspergillus parasiticus. I. Effect of moist heat, age of conidia, and substrate. J. Milk Food Technol. 38, 678-682.

DOYLE, M. P., and MARTH, E. H. 1975B. Thermal inactivation of conidia from Aspergillus flavus and Aspergillus parasiticus. II. Effects of pH and buffers, glucose, sucrose and sodium chloride. J. Milk Food Technol. 38, 750-758.

DUBÉ, A. J., DODMAN, R. L., and FLENTJE, N. T. 1971. The influence of water activity on the growth of Rhizoctonia solani. Aust. J. Biol. Sci. 24, 57-65.

ENGLISH, M. P. 1953. The fermentation of malt extract by an osmophilic yeast. J. Gen. Microbiol. 9, 15-25.

ENGLISH, M. P. 1954. Some observations on the physiology of Saccharomyces rouxii Boutroux. J. Gen. Microbiol. 10, 328-336.

FLETT, H. M. 1973. Water activity determination in foods in the range 0.80 to 0.99. J. Food Sci. 38, 1097-1098.

FOLLSTAD, M. N. 1966. Microflora of nonprocessed raisins. Phytopathology 56, 1413.

FOSTER, J. W. 1949. Chemical Activities of Fungi. Academic Press, New York.

FRANK, M., and HESS, E. 1941. Studies on salt fish. V. Studies on Sporendonema epizoum from "dun" salt fish. J. Fish. Res. Board Can. 5, 276-286.

GAL, S. 1967. Methods for Determining water Sorption Isotherms. Springer-Verlag. Berlin. (German)

GAL, S. 1975. Recent advances in techniques for the determination of sorption isotherms. In Water Relations of Foods. R. B. Duckworth (Editor). Academic Press London.

GALLOWAY, L. D. 1935. The moisture requirements of mould fungi with special reference to mildew in textiles. J. Text. Inst. 26, T123-T129.

GROVER, D. W. 1947. Keeping properties of confectionery as influenced by its water vapour pressure. J. Soc. Chem. Ind. 66, 201-204.

GROVER, D. W., and NICOL, J. M. 1940. The vapour pressures of glycerin solutions at 20°. J. Soc. Chem. Ind. (London) 59, 175-177.

HARNULV, B. G., and SNYGG, B. G. 1972. Heat resistance of Bacillus subtilis spores at various water activities. J. App. Bacteriol. 35, 615-624.

HEINTZELER, I. 1939. The growth of moulds—dependence on hydration and various other limiting factors. Arch. Mikrobiol. 10, 92-132 (German)

HORNER, K. J., and ANAGNOSTOPOULOS, G. D. 1973. Combined effect of water activity, pH and temperature on the growth and spoilage potential of fungi. J. Appl. Bacteriol. 36, 427-436.

IGLESIAS, H. A. CHIRIFE, J., and LOMBARDI, J. L. 1975. An equation for correlating equilibrium moisture content in foods. J. Food Technol. 10, 289-297.

INGRAM, M. 1950. Osmophilic yeasts from concentrated orange juice. J. Gen. Microbiol. 4, ix.

INGRAM, M. 1957. Micro-organisms resisting high concentrations of sugars or salts. In Microbial Ecology, 7th Symp. Soc. Gen. Microbiol., Cambridge Univer. Press. Cambridge, England.

INGRAM, M. 1958. Yeasts in food spoilage. In The Chemistry and Biology of Yeasts. A. H. Cook (Editor). Academic Press, New York.

INGRAM, M. 1960. The physiological properties of osmophilic yeasts. Rev. Ferment. Ind. Aliment, 14, 23-33. (French)

JARVIS, B. 1971. Factors affecting the production of mycotoxins. J. Appl. Bacteriol. 34, 199-213.

JARVIS, B. 1976. The problem of mould growth on intermediate moisture foods with respect to mycotoxins. In Intermediate Moisture Foods. R. Davies, G. G. Birch, and K. Parker (Editors). Applied Science Publishers, Barking, England.

JASON, A. C. 1965. Some properties and limitations of the aluminum oxide hygrometer. In Humidity and Moisture, Vol. I. A. Wexler and R. E. Ruskin (Editors). Reinhold, New York.

KAREL, M. 1973. Recent research and development in the field of low-moisture and intermediate-moisture foods. C.R.C. Crit. Rev. Food Technol. 3, 329-373.

KAREL, M. 1975. Physico-chemical modification of the state of water in foods—a speculative survey. In Water Relations of Food. R. B. Duckworth (Editor). Academic Press, London.

KIRKBRIGHT, G. F., MAYNE, P. J., and WEST, T. S. 1975. Technical note: application of a permutivity method for the rapid determination of water in meat. J. Food Technol. 10, 103-108.

KURATA, H., and ICHINOE, M. 1967. Cited by B. Jarvis. Factors affecting the production of mycotoxins. J. Appl. Bacteriol. 34, 199-213.

KURATA, H., UDAGAWA, S., ICHINOE, M., KAWASAKI, Y., TAKADA, M., TZAWA, M., KOIZUMI, A., and TANABE, H. 1968A. Cited by B. Jarvis. Factors affecting the production of mycotoxins. J. Appl. Bacteriol. 34, 199-213.

KURATA, H., UDAGAWA, S., ICHINOE, M., KAWASAKI, Y., TAZAWA, H., TANAKA, J., TAKADA, M., and TANABE, H. 1968B. Cited by B. Jarvis. Factors affecting the production of mycotoxins. J. Appl. Bacteriol. 34, 199-213.

KUSHNER, D. J. 1968. Halophilic bacteria. Adv. Appl. Microbiol. 10, 73-99.

KVAALE, O., and DALHOFF, E. 1963. Determination of equilibrium relative humidity in foods. Food Technol. 17, 657-661.

LABUZA, T. P. 1975. Interpretation of sorption data in relation to the state of constituent water. In Water Relations of Foods. R. B. Duckworth (Editor). Academic Press, London.

LABUZA, T. P., CASSIL, S., and SINSKEY, A. J. 1972. Stability of intermediate moisture foods. 2. Microbiology. J. Food Sci. 37, 160-162.

LANDROCK, A. H., and PROCTOR, B. E. 1951. A new graphical interpolation method for obtaining humidity equilibria data, with special reference to its role in food packaging studies. Food Technol. 5, 332-337.

LEISTNER, L., and RÖDEL, W. 1975. The significance of water activity for microorganisms in meats. In Water Relations of Foods. R. B. Duckworth (Editor). Academic Press, London.

LEWIS, D. H., and SMITH, D. C. 1967. Sugar alcohols (polyols) in fungi and green plants. New Phytol. 66, 143-184.

LLOYD, A. C. 1975A. Osmophilic yeasts in preserved ginger products. J. Food Technol. 10, 575-581.

LLOYD, A. C. 1975B. Yeasts in ginger brines. J. Food Technol. 10, 407-413.

LOCHHEAD, A. G., and HERON, D.A . 1929. Microbiological studies of honey. Can. Dept. Agric. Bull. 116. (new series)

LODDER, J. 1970. The Yeasts, A Taxonomic Study. North-Holland Publishing Co., Amsterdam.

LUBIENIEKI-VON SCHELHORN, M., and HEISS, R. 1975. The influence of relative humidity on the thermal resistance of mould spores. *In* Water Relations of Foods. R. B. Duckworth (Editor). Academic Press, London.

MEASURES, J. C. 1975. Role of amino acids in osmoregulation of non-halophilic bacteria. Nature (London) *257*, 398-400.

MILLER, M. W., and TANAKA, H. 1963. III. Relation of equilibrium relative humidity to potential spoilage. Hilgardia *34*, 183-190.

MISLIVEC, P. B., and TUITE, J. 1970. Temperature and relative humidity requirements of species of *Penicillium* isolated from yellow dent corn kernels. Mycologia *62*, 75-68.

MONÉY, R. W., and BORN, R. 1951. Equilibrium humidity of sugar solutions. J. Sci. Food Agric. *2*, 180-185.

MOREAU, C. 1974. Toxigenic Fungi in Foods, 2nd Edition. Masson et Cie, Paris. (French)

MORRIS, E. O. 1975. Yeasts from the marine environment. J. Appl. Bacteriol. *38*, 211-233.

MOSSEL, D. A. A. 1951. Investigation of a case of fermentation in fruit products, rich in sugars. Antonie van Leeuvenhoek J. Microbiol. Serol. *17*, 146-152.

MOSSEL, D. A. A. 1971. Physiological and metabolic attricutes of microbial groups associated with foods. J. Appl. Bacteriol. *34*, 95-118.

MOSSEL, D. A. A. 1975A. Occurrence, prevention and monitoring of microbial quality loss of foods and dairy products. C.R.C. Crit. Rev. Environ. Control *5*, 1-139.

MOSSEL, D. A. A. 1975B. Water and micro-organisms in foods—a synthesis. *In* Water Relations of Foods, R. B. Duckworth (Editor). Academic Press, London.

MOSSEL, D. A. A., and INGRAM, M. 1955. Physiology of the microbial spoilage of foods. J. Appl. Bacteriol. *18*, 232-268.

MOSSEL, D. A. A., and SAND, F. E. M. J. 1968. Occurrence and prevention of microbial deterioration of confectionery products. Conserva *17*, 23-32.

MOSSEL, D. A. A., and SHENNAN, J. L. 1976. Micro-organisms in dried foods. Their significance, limitation and enumeration. J. Food Technol. *11*, 205-220.

MOSSEL, D. A. A., and VAN KIUJK, H. J. L. 1955. A new and simple technique for the direct determination of the equilibrium relative humidity of foods. Food Res. *20*, 415-423.

MOSSEL, D. A. A., and WESTERDIJK, J. 1949. The physiology of microbial spoilage in foods. Antonie van Leeuwenhoek J. Microbiol. Serol, *15*, 190-202.

MOUSTAFA, A. F. 1975. Osmophilous fungi in the salt marshes of Kuwait. Can. J. Microbiol. *21*, 1573-1580.

MURRELL, W. G., and SCOTT, W. J. 1966. The heat resistance of bacterial spores at various water activities. J. Gen. Microbiol. *43*, 411-425.

NORKRANS, B. 1966. Studies on marine occurring yeasts: growth related to pH, NaCl concentration and temperature. Arch. Microbiol. *54*, 374-392.

NORRISH, R. S. 1966. An equation for the activity coefficients and equilibrium relative humidities of water in confectionery syrups. J. Food Technol. *1*, 25-39.

NORRISH, R. S. 1969. An apparatus for the measurement of relative humidity in confined static environments. B.F.M.I.R.A. Tech. Circ. *425*. Br. Food Manuf. Ind. Res. Assoc., Leatherhead, Surrey, England.

ONISHI, H. 1957. Studies on osmophilic yeasts. Part III. Classification of osmophilic soy and miso yeasts. Bull. Agric. Chem. Soc. Jpn. *21*, 151-156.

ONISHI, H. 1959. Studies on osmophilic yeasts. Part IV. Change in permeability of cell membranes of osmophilic yeasts and maintenance of their viability in the saline medium. Bull. Agric. Chem. Soc. Jpn. *23*, 332-339.

ONISHI, H. 1963. Osmophilic yeasts. Adv. Food Res. *12*, 53-94.

ORMEROD, J. G. 1967. The nutrition of the halophilic mould *Sporendonema epizoum*. Arch. Microbiol. *56*, 31-39.

OVERBEEK, J. TH. G., and MOSSEL, D. A. A. 1951. Entrainment distillation as a reference method for the determination of the water content of foods. Rec. Trav. Chim. Pays-Bas *70*, 63-70.

PANASENKO, V. T. 1967. Ecology of microfungi. Bot. Rev. *33*, 189-215.

PANDE, A. 1971. Techniques for the measurement of moisture in biological materials. Lab. Pract. *20*, 117-120, 131.

PANDE, A., and PANDE, C. S. 1962. Physical methods of moisture measurement. Instrum. Pract. *16*, 896-903. 988-995.

PEARSON, D. 1970. The Chemical Analysis of Foods. 6th Edition. Churchill, London.

PEARSON, D. 1973. Laboratory Techniques in Food Analysis. Butterworths, London.

PELHATE, J. 1968. A study of water requirements in some storage fungi. Mycopathol. Mycol. Appl. *36*, 117-128. (French)

PITT, J. I. 1975. Xerophilic fungi and the spoilage of foods of plant origin. *In* Water Relations of Foods. R. B. Duckworth (Editor). Academic Press, London.

PITT, J. I., and CHRISTIAN, J. H. B. 1968. Water relations of xerophilic fungi isolated from prunes. Appl. Microbiol. *16*, 1853-1858.

POUNCY, A. E., and SUMMERS, B. C. L. 1939. The micromeasurement of relative humidity for the control of osmophilic yeasts in confectionery products. J. Soc. Chem. Ind. (London) *58*, 162-165.

PUT, H. M. C., DEJONG, J., SAND, F. E. M. J., and VAN GRINSVEN, A. M. 1976. Heat resistance studies on yeast spp. causing spoilage in soft drinks. J. Appl. Bacteriol. *40*, 135-152.

PUT, H. M. C., and SAND, F. E. M. J. 1974. A method for the determination of heat resistance of yeasts causing spoilage in soft drinks. Proc. 4th Int. Symp. on Yeasts. H. Klauschofes and V. B. Sleytr (Editors). Hochschule fur Bodenkultur, Vienna.

ROBINSON, R. A., and STOKES, R. H. 1955. Electrolyte Solutions. Academic Press, London.

ROCKLAND, L. B. 1960. Saturated salt solutions for static control of relative humidity between 5° and 40°. Anal. Chem. *32*, 1375-1376.

RÖDEL, W., and LEISTNER, L. 1972. Measuring the water activity (a_W value) of meat and meat products by means of a dewpoint hygrometer. Fleischwirtschaft *52*, 1461-1462. (German)

ROSS, S. S., and MORRIS, E. O. 1962. Effect of sodium chloride on the growth of certain yeasts of marine origin. J. Sci. Food Agric. *13*, 467-475.

SANDERS, T. H., DAVIS, N. D., and DIENER, U. L. 1968. Effect of carbon dioxide, temperature and relative humidity on production of aflatoxin in peanuts. J. Am. Oil Chem. Soc. *45*, 683-685.

SCARR, M. P. 1951. Osmophilic yeasts in raw beet and cane sugars and intermediate sugar-refining products. J. Gen. Microbiol. *5*, 704-713.

SCARR, M. P. 1954. Studies on the taxonomy and physiology of osmophilic yeasts isolated from sugar cane. Ph.D. Thesis. University of London.

SCARR, M. P. 1959. Selective media used in the microbiological examination of sugar products. J. Sci. Food Agric. *10*, 678-681.

SCARR, M. P. 1968. Studies arising from observations of osmophilic yeasts by phase contrast microscopy. J. Appl. Bacteriol. *31*, 525-529.

SCARR, M. P., and ROSE, D. 1965. Assimilation and fermentation patterns of osmophilic yeasts in sugar broths at two concentrations. Nature (London) 207, 887.

SCARR, M. P. and ROSE, D. 1966. Study of osmophilic yeasts producing invertase. J. Gen. Microbiol. 45,,9-16.

SCOTT, W. J. 1953. Water relations of Staphylococcus aureus at 30°C. Aust. J. Biol. Sci. 6, 549-564.

SCOTT, W. J. 1957. Water relations of food spoilage micro-organisms. Adv. Food Res. 7, 83-127.

SEILER, D. A. L. 1976. The stability of intermediate moisture foods with respect to mould growth. In Intermediate Moisture Foods. R. Davies, G. G. Birch, and K. Parker (Editors). Applied Science Publishers, Barking, England.

SINSKEY, A. J. 1976. New developments in intermediate moisture foods: humectants. In Intermediate Moisture Foods. R. Davies, G. G. Birch, and K. Parker (Editors). Applied Science Publishers, Barking, England.

SLIGHT, H. A. 1971. In line process control report. B.F.M.I.R.A. Res. Rep. 164. Br. Food Manuf. Ind. Res. Assoc., Leatherhead, Surrey, England.

SMITH, P. R. 1971. The determination of equilibrium relative humidity or water activity in foods—a literature review. B.F.M.I.R.A. Scientific and Technical Survey No.70, Br. Food Manuf. Ind. Res. Assoc., Leatherhead, Surrey, England.

SNOW, D. 1949. The germination of mould spores at controlled humidities. Ann. Appl. Biol. 36, 1-13.

SNOW, D., CRICHTON, M. H. G., and WRIGHT, N. C. 1944. Mould deterioration of feeding stuffs in relation to humidity of storage. Ann. Appl. Biol. 31, 102-110.

SOLOMON, M. E. 1951. Control of humidity with potassium hydroxide, sulphuric acid, or other solutions. Bull. Entomol. Res. 42, 543-554.

SOMAVILLA, J. F., TIENDA, P., ARROYO, V., and INIGO, B. 1975. Osmophilic yeasts in "mistelas" (alcoholic grape juice) and in dried fruits or nuts. Rev. Agroquim Tecnol. Aliment. 15, 573-580. (Spanish)

SPENCER, J. F. T. 1968. Production of polyhydric alcohols by yeasts. Prog. Ind. Microbiol. 7, 1-42.

SPENCER-GREGORY, H., and ROURKE, E. 1957. Hygrometry. Crosby Lockwood, London.

STOKES, R. H., and ROBINSON, R. A. 1949. Standard solutions for humidity control at 25°. Ind. Eng. Chem. 41, 2013.

TANAKA, H., and MILLER, M. W. 1963. Microbial spoilage of dried prunes. II. Studies of the osmophilic nature of spoilage organisms. Hilgardia 34, 171-181.

TAYLOR, A. A. 1961. Determination of moisture equilbrium in dehydrated foods. Food Technol. 15, 536-540.

TEITELL, L. 1958. Effects of relative humidity on viability of conidia of aspergilli. Am. J. Bot. 45, 748-753.

TEMPEST, D. W., and MEERS, J. L. 1968. Influence of NaCl concentration of the medium on the potassium content of Aerobacter aerogenes and on the interrelationship between potassium, magnesium and ribonucleic acid in the growing bacteria. J. Gen. Microbiol. 54, 319-325.

TEMPEST, D. W., MEERS, J. L., and BROWN, C. M. 1970. Influence of environment on the content and composition of microbial free amino acid pools. J. Gen. Microbiol. 64, 171-185.

TILBURY, R. H. 1967. Studies on the microbiological deterioration of raw cane sugar, with special reference to osmophilic yeasts and the preferential utilization of laevulose in invert. M. S. Thesis. Univ. of Bristol, England.

TILBURY, R. H. 1976. The microbial stability of intermediate moisture foods with respect to yeasts. *In* Intermediate Moisture Foods. R. Davies, G. G. Birch, and K. Parker (Editors). Applied Science Publishers, Barking, England.

TOMKINS, R. G. 1930. Studies of the growth of moulds—I. Proc. R. Soc. London, Ser. B, *105*, 375-401.

TRACEY, M. V. 1975. "Envoi." *In* Water Relations of Foods. R. B. Duckworth (Editor). Academic Press, London.

VON SCHELHORN, M. 1950. Spoilage of water-proof food by osmophilic micro-organisms. I. Spoilage of foods by osmophilic yeasts. Z. Lebsensm. Unters. Forsch. *91*, 117-124. (German)

VOS, P. T., and LABUZA, T. P. 1974. Technique for measurement of water activity in the high a_W range. J. Agric. Food Chem. *22*, 326-327.

WALKER, H. W., and AYRES, J. C. 1970. Yeasts as spoilage organisms. *In* The Yeasts, Vol. 3. A. H. Rose and J. S. Harrison (Editors). Academic Press, London.

WU, M. T., AYRES, J. C., and KOEHLER, P. E. 1974. Toxigenic aspergilli and penicillia isolated from aged, cured meats. Appl. Microbiol. *28*, 1094-1096.

YOUNG, J. F. 1967. Humidity control in the laboratory using salt solutions—a review. J. Appl. Chem. *17*, 241-245.

3

Fruits and Fruit Products

D. F. Splittstoesser

A characteristic shared by most fruits is their high acidity and, as illustrated in Table 3.1, a pH of under 3.5 is not uncommon. Various organic acids are responsible for this acidity: citric is the principal acid in blueberries, cranberries, strawberries, and citrus fruits; malic acid predominates in apples, apricots, pears, and quinces; and grapes contain almost equal quantities of tartaric and malic acids. Some of the other organic acids that can be found in these fruits are oxalic, quinic, succinic, fumaric, shikimic, isocitric, and acetic. The latter acids usually are present only in trace amounts.

The reason for discussing pH at the start of this chapter is that it is the single most important factor with respect to the type of microorganism that can spoil this class of food. While most species of bacteria are inhibited by the hydrogen ion concentrations cited in Table 3.1, yeasts and molds are more aciduric and many find these pH values to be tolerable if not optimum for growth. It is because of acidity, therefore, that the fungi are the principal spoilage microorganisms of fruits and fruit products.

FRESH FRUIT

The rots and other defects of fresh fruit are termed "market

TABLE 3.1
TYPICAL pH VALUES FOR VARIOUS
FRUITS

Fruit	pH Range
Apple	3.1-3.9
Apricot	3.3-4.4
Blackberry	3.0-4.2
Blueberry	3.2-3.4
Cherry	3.2-4.0
Cranberry	2.5-2.7
Fig	4.8-5.0
Grape	3.0-4.0
Grapefruit	2.9-3.4
Lemon	2.2-2.6
Lime	2.3-2.4
Mango	3.8-4.7
Orange	3.3-4.0
Peach	3.3-4.2
Pear	3.7-4.6
Pineapple	3.4-3.7
Plum	3.2-4.0
Raspberry	2.9-3.5
Strawberry	3.0-3.9

diseases." Losses due to market disease amount to millions of dollars annually in the United States and molds usually are responsible. Although the surfaces of fresh fruits harbor large numbers of both yeasts and molds, yeasts generally lack the mechanisms to invade and infect plant tissue and therefore are secondary rather than primary agents of spoilage.

Infection of fruit and subsequent rotting can occur in the orchard or vineyard as well as during the different marketing steps which include the harvest, grading and packing, transport, storage, and various manipulations in the wholesale and retail markets. Some of the molds responsible for spoilage are true plant pathogens in that they can invade and cause an infection of intact, formerly healthy tissue. Others are saprophytic species which only become established after the fruit has been infected by a pathogenic organism or has been damaged by some physical or physiological cause. Growth of saprophytic molds generally is restricted to dead plant tissue.

Nomenclature

Some of the names given specific diseases are based on the

scientific name of the mold, while others reflect the appearance of the infected fruit. To illustrate—diseases of citrus fruit named after mold genera are *Alternaria, Aspergillus, Botrytis, Fusarium, Phytophthora, Sclerotinia,* and *Trichoderma* rots; citrus disease names derived from the appearance of the infected fruit include black pit, brown rot, stem-end rot, and cottony rot.

As a result of using appearance as a basis for the naming of rots, the market diseases of different fruits may have the same name even though different molds are involved. For example, brown rot of citrus fruits is caused by species of *Phytophthora* while brown rot of cherries and peaches is due to *Monilinia fructicola;* also, *Physalospora obtusa* is responsible for black rot of apples while *Alternaria citri* is the cause of black rot of oranges.

Important Diseases

Some of the more troublesome molds and the diseases they cause are presented in Table 3.2. Obviously, from an economic standpoint, the most important organisms are those responsible for the principal diseases of the major fruit crops. One might single out *Penicillium italicum* which causes blue rot and *Penicillium digitatum* which causes green rot of citrus fruits, and *Penicillium expansum* which is responsible for blue mold rot of apples and cherries. *Botrytis cinerea,* the cause of gray mold rot, is the most important spoilage mold of grapes and strawberries and *Monilinia fructicola* is responsible for brown rot of apricots, nectarines and peaches, the most serious disease of these fruits (Fig 3.1). For a detailed discussion of these and other diseases, the reader is referred to various handbooks published by the United States Department of Agriculture (Harvey and Pentzer 1960; Pierson *et al.* 1971; Smoot *et al.* 1971; Harvey *et al.* 1972).

Control

Orchard and Vineyard.—Pruning, certain cultural and sanitary practices, and the application of fungicides are the principal means for minimizing infection and spoilage prior to harvest.

Plant pathogens may infect a variety of tissues in addition to the fruit, and thus the removal of diseased branches and other plant parts by pruning helps reduce the incidence of spoilage. Cultural practices that lower humidity in the growing area, such as weed control and the proper spacing of plants, often have a beneficial effect because the growth of fungi is favored by moist conditions. Sanitation in the

TABLE 3.2

SOME OF THE MORE IMPORTANT MOLDS RESPONSIBLE FOR
MARKET DISEASES OF FRUITS

Genus	Spoilage Problems
Alternaria	Brown to black spots on apples, stone fruits[1], and figs; stem-end and black rot of citrus fruits; wooly growth on blueberries
Aspergillus	Black rot on peaches, nectarines, apricots, citrus fruits and figs
Botryodiplodia	Cushion or crown rot of bananas; ripe rot of papaws
Botrytis	Gray mold rot of apples, pears, raspberries, strawberries, grapes, figs, blueberries, citrus, and stone fruits
Cladosporium	Restricted rot with a gray-black core on stone fruits, olive-green growth on raspberries, black rot on grapes, and spotting of figs
Colletotrichum	Brown to black spots (anthracnose) on citrus fruits, avocados, mangos, papaws, and papayas
Diaporthe	Stem-end rot of citrus fruits
Diplodia	Stem-end rot of citrus fruit, avocados, mangos, and papayas; watery, tan-brown rot of peaches
Fusarium	Brown rot of citrus fruit and pineapple; soft rot of figs
Geotrichum	Sour rot of citrus fruits and peaches
Gloeodes	Sooty blotch of citrus fruits, apples, pears
Gloeosporium	Anthracnose, black rot, and lesion rot of bananas; bull's eye rot of pome fruits[2]
Guignardia	Black spot of citrus fruit rinds
Monilinia	Brown rot of stone fruits; mummification of blueberries
Nigrospora	Soft, watery (squirter) rot of the pulp of bananas
Penicillium	Blue and green mold rots of citrus fruits; blue mold rot of apples, grapes, pears, stone fruits, and figs; brown rot of pineapple
Phomopsis	Stem-end rot of citrus fruits and avocados
Phytophthora	Brown rot of apples and citrus fruits; leathery rot of strawberries
Rhizopus	Watery, soft rot of apples, pears, stone fruits, grapes, strawberries, avocados, and figs
Sclerotinia	Watery, white rot of strawberries; cottony rot of lemons and other citrus fruits
Septoria	Septoria spot or rot of citrus fruits
Trichoderma	Cocoa-brown to green rot of citrus fruits
Venturia	Black spot of apples, pears

[1]Stone fruits include cherries, peaches, nectarines, apricots and plums.
[2]Pome fruits include apples, pears and quince.

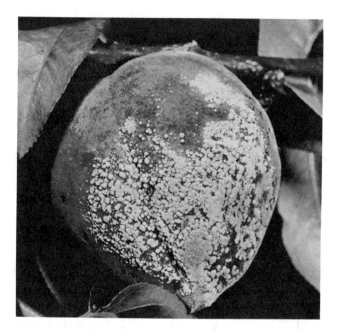

FIG. 3.1. INFECTION OF A PEACH WITH THE BROWN ROT MOLD, *MONILINIA FRUCTICOLA*

orchard and vineyard is important for control of spoilage organisms that overwinter in cankers, dead branches, and fallen fruit. Mummified peaches, for example, are a common reservoir for *Monilinia*, the mold responsible for brown rot.

Spraying with fungicides is perhaps the most effective method for controlling pathogenic fungi. Two of the older fungicides still being used are Bordeaux mixture (basic cupric sulfate) and elemental sulfur. Some of the newer, widely used, organic fungicides are captan, folpet, and benomyl. Vineyard and orchard spray schedules often call for a number of applications of these materials during the growing season.

Harvesting, Grading, and Packing.—Important control practices at these stages are:
(1) Harvest fruit when at optimum maturity.
(2) Handle the fruit gently to prevent bruises and punctures that would permit the entry of saprophytic fungi.
(3) Maintain good sanitation to minimize the buildup of molds on fruit-contact surfaces. Mechanical harvesters, lug boxes, and packinghouse equipment should be cleaned and sanitized regularly. Live steam, formaldehyde, and fumigation with chlo-

rine gas are some of the treatments used to destroy fungal spores.

(4) Moldy fruit as well as that showing skin breaks and bruises should be culled out during the sorting and grading operations.

(5) Wash fruit with fungicides or hot water. Sodium orthophenylphenate, borax, thiabendazole, and biphenyl are compounds used to treat citrus fruit, pineapples, apples and peaches. Troublesome molds controlled by these fungicides are *Phomopsis, Diplodia, Penicillium* and *Botrytis*. The dipping of citrus fruit in hot (45° to 55°C) water is a treatment used to destroy brown rot and stem-end rot infections as well as the spores of other fungi. A hot water dip is also used to control brown rot of peaches.

(6) Fruit should be cooled promptly. Although the lowest temperature just above freezing would be most effective for retarding mold growth on fresh fruit, the physiology of the fruit often dictates a higher storage temperature because many fruits suffer from chilling injury when exposed to refrigeration temperatures.

Storage and Transport.—In addition to refrigeration, methods used to retard spoilage during the storage and shipment of fruits are fumigation, controlled atmospheres, and exposure to certain fungicides.

Fumigation using sulfur dioxide is the procedure of choice of *Vitis vinifera* grapes. A gas concentration of 1% is initially introduced into the storage compartment. This is followed by additions at weekly intervals to maintain levels of approximately 0.25%. Unfortunately, sulfur dioxide cannot be used for the protection of most fruits, including other species of grapes, because their tissues are damaged by the gas. Nitrogen trichloride is a fumigant that is applied to citrus fruits.

Increasing the partial pressure of carbon dioxide in the storage atmosphere is a means of retarding mold growth on cherries and other carbon dioxide-tolerant fruits. Species of *Alternaria, Botrytis,* and *Rhizopus* are important spoilage molds that are inhibited by carbon dioxide concentrations of 10 to 20%.

PROCESSED FRUIT

Field Contamination

Numbers and Kinds.—Fruit as received at the processing plant commonly harbor large populations of fungi. Most are not the cause of the market diseases that were discussed previously but rather are innocuous species that do little harm to sound fruit. The organisms are important in fruit processing, however, in that, unless removed or destroyed, they may be introduced in large numbers into juice and other products.

Contamination of fruit during the harvest as well as growth of fungi on lesions and other damaged areas is largely responsible for the large populations that often are observed. For example, commercially harvested Concord grapes yielded an average count of 680,000 per g (mostly yeasts) compared to a population of only 38,000 per g on fruit that had been collected and transported under relatively aseptic conditions (Moyer et al. 1969); sound oranges gave average counts of 740,000 per orange compared to 21 million on fruit that had skin splits (Murdock and Brokaw 1958); and quality apples yielded maximal yeast populations of 10^3 per apple compared to counts of 10^7 on fruits containing rots and other defects (Marshall and Walkley 1951A).

Many different species of yeasts and molds are present on the surface of fruit. Studies on the yeast flora of citrus fruit (Recca and Mrak 1952), apples (Marshall and Walkley 1952), and grapes (Mrak and McClung 1940) have indicated that the populations of yeasts are about evenly divided between ascosporogenous and imperfect species. *Saccharomyces, Hanseniaspora* and *Pichia* were some of the more numerous ascospore producers while the more common anascosporogenous species were *Torulopsis, Kloeckera, Candida* and *Rhodotorula*. The predominant mold genera were *Penicillium, Aspergillus, Mucor, Alternaria, Cladosporium* and *Botrytis*.

Patulin.—Patulin, a carcinogenic secondary fungal metabolite, may contaminate fruit that are destined for processing. This mycotoxin is produced by species of *Aspergillus, Byssochlamys,* and *Penicillium,* including *P. expansum,* a common mold on many fruits. The probability is high, therefore, that patulin will be present in moldy apples, pears and stone fruits.

Contamination of a food with patulin is prevented by careful inspection of fruit prior to processing. Moldy fruit are discarded or the mold is removed by careful trimming. The latter is effective for larger fruit because, although rotted areas may contain as much as 125 ppm of patulin in infected tissue, adjacent healthy tissue has been found to be free of the toxin (Buchanan et al. 1974).

Processing Effects

Reducing Contamination.—Washing of fruit, usually one of the first processing steps, will remove much of the original microflora. A brush wash followed by a rinse with chlorinated water was found to reduce the microbial population on the surface of oranges by 95% (Murdock and Brokaw 1958), while a reduction in viable count of over 99.9% was observed when apples were washed in running water (Marshall and Walkley 1951A).

Many fruits are subjected to a heat treatment as an early step in processing. Although destruction of contaminating microorganisms generally is not the primary objective of the treatment, it is one of the accomplishments since the temperatures are lethal for yeasts and most molds. To cite several examples: Concord grapes are heated to a temperature of about 60°C prior to pressing to facilitate pigment extraction, and peaches are peeled by exposing them to a hot (102° to 104°C) lye solution or by steaming them in a scalder. Also, fruit juices are commonly pasteurized soon after pressing to inactivate undesirable enzymes; freshly extracted orange juice, for example, is heated to 85° to 98°C to destroy pectinesterases and other enzymes that cause a loss of cloud (turbidity).

Many fruit juices are clarified by precipitation of colloidal substances (fining) followed by filtration or centrifugation. As the studies by Marshall and Walkley (1951B) on apple juice illustrate, these treatments also remove a large number of microorganisms. They found that fining by the use of a gelatin-tannin treatment reduced the yeast population an average of 62% and that subsequent filtration or centrifugation further lowered the count by an additional 96%.

Line Buildup.—Various equipment used in the processing of fruit is difficult to clean and therefore can be a significant source of contamination by yeasts and molds. Unit operations that may be especially troublesome are presses and extractors, finishers, mills, pitters, slicers, and conveyors. Any machine which is operated at a temperature within the physiological growth range of fungi and which permits the accumulation of fruit solubles is a potential source of microbial buildup. Sometimes contamination originates from unexpected sources such as the interior walls of the evaporators used for the concentration of juices; here, certain stages may be operated at temperatures sufficiently low to permit microbial growth.

Bulk Storage.—Certain fruit juices are held in large tanks for extended time periods. Although the juice has been pasteurized, it usually is not completely free of viable organisms because of subsequent recontamination from pumps, pipelines, and the walls of the storage tanks. Concord grape juice, for example, is stored 1 to 6 months at -5.5° to -2.2°C, temperatures just above freezing, in order to precipitate excess crude tartrates. Although mold growth may develop on the surface of the juice during storage, the greater problem is the multiplication of yeasts. Pederson et al. (1959) found yeast populations in excess of 10^6 per ml in some stored juices. The genera responsible for these counts were Saccharomyces, Hanseniaspora, Torulopsis and Candida. Although all of the isolates were able to grow at 0°C, the Candida spp. were true psychrophiles in that they had a low optimal growth temperature of 11°C and were incapable of growth when the temperature exceeded 20°C (Lawrence et al. 1959). The psychrophilic candidas often made up 95% or more of the yeasts present in the stored juices.

The Boehi process has been widely used in Europe for extending the storage time of commercial fruit juices. The principle behind the method is that multiplication of yeasts is arrested by high concentrations of carbon dioxide. In the original process, juice was sparged with carbon dioxide until it contained 1.5% by weight, which is equivalent to a pressure of 7.7 atmospheres. More recent modifications of the process have involved the saturation of juice at 2°C to achieve 0.8% carbon dioxide by weight, a pressure of only 3.5 atmospheres (Luthi 1959). Since fermentation is not inhibited by high levels of carbon dioxide, filtration or some other means must be used to assure that the viable yeast populations in the juice are low at the time the tanks are filled.

Recently a bulk storage technology has been developed in which acid foods are aseptically filled into sterile tanks following sterilization and cooling in tubular heat exchangers (Nelson 1971). The method utilizes pipes, valves, pumps and tanks that can be completely sterilized with respect to potential spoilage organisms. Destruction of contaminants has been accomplished by exposing all surfaces to a 20 ppm acidified iodophor solution for at least 20 min. Although the process is presently being used primarily for chopped tomatoes and tomato paste, it also would be applicable for fruit juices and other fruit products.

Monitoring Contamination.—Various methods have been used to detect high fungal populations in fruit products during processing. The enumeration of viable yeasts and molds is one of the more common means of control. Potato dextrose agar acidified to pH 3.5 has been widely used as a plating medium, although recent studies have indicated that the incorporation of chloramphenicol and chlortetracycline into the medium to inhibit bacteria affords higher recoveries of fungi than does acidification (Koburger and Farhat 1975). A shortcoming of viable count procedures is that a time period as long as five days may be required to obtain results.

The testing for acetylmethylcarbinol and diacetyl using a quantitative Voges-Proskauer test has been suggested as a rapid means for evaluating microbial quality of fruit products (Hill *et al.* 1954; Fields 1962). These metabolic products are produced by many yeasts and molds as well as by lactic acid bacteria (Fields 1964; Murdock 1964). Fields (1962) concluded that apple juice made from sound fruit and processed under sanitary conditions should contain under 1 ppm acetylmethylcarbinol and that concentrations over 1.6 ppm were indicative of moldy apples or excessive contamination during processing. However, it may not always be possible to relate acetylmethylcarbinol concentrations to yeast populations since some fruit products such as orange juice at times contain mostly Voges-Proskauer-negative strains (Murdock 1964).

The enumeration of the mold *Geotrichum candidum* in fruit products has been proposed as a rapid method for assessing the sanitary state of processing lines (Cichowicz and Eisenberg 1974). *Geotrichum* is believed to be the principal mold contaminant of processing equipment (it is referred to as "machinery mold") and therefore a certain incidence in canned fruits and other products will reflect insanitary conditions. The procedure involves the use of sieves to separate quantitatively the mold fragments from the food. After staining, hyphae typical of *Geotrichum* are enumerated on a rot fragment slide using a stereoscopic microscope. See Chap. 16 for further discussion.

Preservation

Heat.—In general, yeasts, molds and other aciduric microorganisms possess little heat resistance in that temperatures as low as 55°C are lethal for many species (Jensen 1954; Stokes 1971). As a result, fruits and fruit products usually require only a relatively mild heat process in order to achieve commercial sterility. A common heat treatment given juices, for example, is merely to fill the containers with juice having a temperature of 82° to 88°C. Immediately after closing, the cans or jars are held for a brief period in an inverted position to destroy microorganisms that may have contaminated the lid. Canned fruit are often heated in boiling water until the center of the container reaches a temperature of 80° to 90°C.

A few molds are able to survive these processes and are occasionally responsible for spoilage outbreaks. The heat resistance of these molds is due to the formation of sclerotia or the production of resistant spores.

Sclerotia, which are hard masses of thick-walled vegetative cells, are produced by many molds. Some of the most heat resistant sclerotia that have been studied were formed by a *Penicillium* sp. that was responsible for the spoilage of canned blueberries (Williams *et al.* 1941). Thermal death time studies on this mold showed that a suspension of 86,000 sclerotia per ml could survive a heat treatment of 30 min at 90.5°C. It was concluded that one means of preventing spoilage by this mold was to remove the sclerotia from blueberries by careful washing (Ruyle et al. 1946).

Asexual fungal spores such as conidia are generally only slightly more resistant than are vegetative structures—it is the ascospores that may possess some degree of heat resistance. Molds reported to produce resistant ascospores include Neurospora, Aspergillus, Penicillium and Byssochlamys spp. Byssochlamys has been the most common cause of spoilage of canned fruits and fruit products, perhaps because its spores have the greatest heat resistance.

Byssochlamys appears to be closely related to Penicillium and its conidiophores are Penicillium-like with the exception that the conidia-bearing tips tend to bend away from the main axis. A distinguishing feature of Byssochlamys is that its eight-spored asci are produced in irregular clusters that are not covered by peridium. Anascosporogenous strains that otherwise are identical to Byssochlamys are placed in the genus Paecilomyces (Brown and Smith 1957). The species of Byssochlamys that are mainly responsible for the spoilage of canned fruit are B. fulva and B. nivea. Major differences between the two species are that the colonies of B. nivea are persistently white while those of B. fulva are yellowish-brown in color; B. nivea produces abundant chlamydospores which are rarely formed by B. fulva; and the ascospores of B. fulva are somewhat larger, 5.2 to 6.5 x 3.2 to 4.0 μm compared to 4.0 to 5.5 x 2.5 to 3.5 μm for B. nivea (Stolk and Samson 1971).

Growth of Byssochlamys spp. in canned fruit often results in a complete breakdown of fruit structure due to the mold's pectinolytic enzymes. Spoiled fruit juice will exhibit off-odors and small quantities of mycelial strands can be detected if the juice is passed through coarse filter paper.

Byssochlamys was first recognized as a spoilage agent of canned fruits in Great Britain (Olliver and Rendle 1934) and for a time it appeared to be a localized problem. However, in recent years spoilage outbreaks due to this mold have been recognized on many continents of the world, including North America (Maunder 1969).

The medium in which Byssochlamys ascospores are heated can have a significant effect on their resistance. Thus, survivals in apple, Concord grape and tomato juices were about 1000-fold higher than those obtained in water or a vinifera grape juice when spores were heated for 3 hours at 85°C (Fig. 3.2). The survival curves also show that the ascospores do not exhibit the logarithmic order of death that is obtained with most bacterial spores. The presence of elevated levels of sucrose in Concord grape juice has been shown to protect

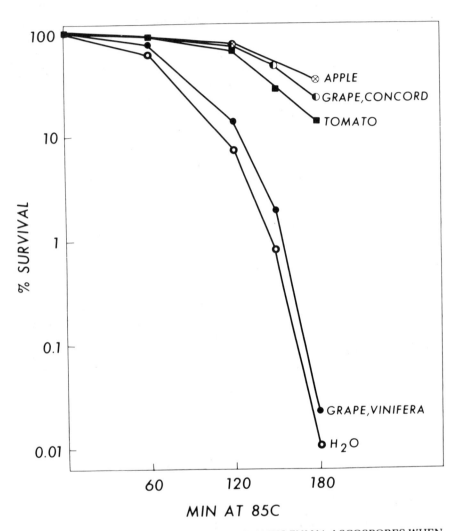

FIG. 3.2. SURVIVOR CURVES OF *BYSSOCHLAMYS FULVA* ASCOSPORES WHEN HEATED IN FRUIT JUICES AND WATER

against heat inactivation of *B. nivea* ascospores (Beuchat and Toledo 1977).

Certain compounds that may be present in canned fruit products have a significant effect on the heat resistance of *Byssochlamys*. As little as 2 ppm sulfur dioxide reduced the thermal death time of 200 asci per ml from 45 to 14 min when heated at 85°C (Gillespy 1940), which may explain the differences in survivals obtained with the two grape juices in Fig. 3.2. Some organic acids also markedly influence heat resistance: pH 3.0 solutions of tartaric and malic acids afforded protection in that they permitted higher survivals than were obtained in distilled water, while fumaric, lactic, and acetic acids caused ascospores to be more sensitive to heat (Fig. 3.3).

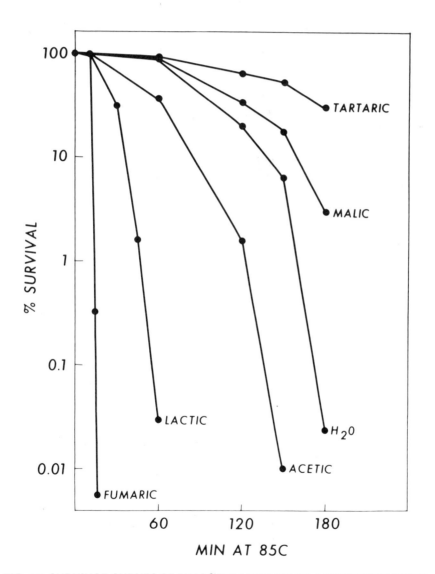

FIG. 3.3. SURVIVOR CURVES OF *BYSSOCHLAMYS FULVA* ASCOSPORES WHEN HEATED IN 0.05 M, pH 3.0 SOLUTIONS OF ORGANIC ACIDS

Although asci of *Byssochlamys* can be found on many different fruits, they usually are present only in low numbers, less than one per gram (Splittstoesser *et al*. 1971). Certain processing operations in addition to the heat treatment may explain why the incidence of spoilage is not higher in view of the mold's wide distribution. It has been shown that a high percentage of the asci, about 99.999%, are removed from grape juice when it is passed through a commercial-type filter (King *et al*. 1969). Also the common practice of transporting cherries from the orchard to the processing plant in water removes asci (Splittstoesser *et al*. 1974), as presumably does washing of fruit.

Cold.—Numerous fruits and fruit products are preserved by freezing. Some are destined for the retail market while others are to be used as food ingredients by bakeries, dairies, and the preserves industry. Important frozen fruit products in the grocery store are citrus, apple, and grape juices that have been concentrated to contain 42 to 45% soluble solids. Orange juice concentrate is particularly popular in the United States with over 450 million liters being frozen annually. Widely used frozen fruits include apple slices, tart cherries, peaches, strawberries and various other berries.

A number of freezing methods are used. Juice concentrates may be filled into retail size containers at a temperature of -6.7° to -3.9°C and then frozen solid in tunnels having a temperature as low as -46°C. Concentrates and purees that will be marketed for remanufacture often are slush-frozen in heat exchangers prior to freezing in 208-liter drums or larger containers. Fruit often are packed in 4.54- or 13.6-kg containers and frozen in blast freezers. They may be frozen without added sugar, in a heavy syrup, or as a dry sucrose pack. The latter may contain 3 or 4 parts of fruit to 1 part of sugar on a weight basis.

Freezing destroys a portion of the microorganisms present on fruit products, and during subsequent storage in the frozen state, microbial death continues, but at a slower rate. The survival of microorganisms during storage is related to temperature. The death rate in orange juice concentrate stored at -62°C was slower than in juice held at -23°C (Wolford 1958). Sucrose has a protective effect against death of *B. nivea* ascospores in fruit juices stored at -30°C (Beuchat and Toledo 1977). While mold spores are usually more freeze-resistant than yeasts, the latter often are more numerous in frozen fruit because they had predominated on the product prior to freezing.

There is surprisingly little published information on the incidence of yeasts and molds in different frozen fruits, perhaps because, in general, they are not a cause of food-borne illnesses. The viable yeast and mold counts as well as total counts on fruits are usually lower than those found on frozen vegetables. Yeasts and molds grow more slowly than bacteria and therefore accumulation on fruit processing lines is easier to control. To cite some viable count data: most samples of 42° Brix orange juice gave counts of under 15,000 per ml of reconstituted juice with the majority of the contaminants being lactic acid bacteria rather than yeasts and molds (Vaughn and Murdock 1956); unpublished work of the author showed that the microflora of frozen apple slices was composed primarily of yeasts and molds, with yeasts making up about 90% of the population. Most samples gave viable counts of under 5000 per g.

Growth of yeasts and molds in frozen fruits does not occur when they are held at a suitably low temperature; storage at -18°C or lower is recommended. A problem may arise when foods are exposed to higher temperatures resulting from defrost cycles in the freezer chest, or mishandling during transport or storage, since the spores of some molds have the ability to germinate and grow at temperatures of -5°C or lower (Elliott and Michener 1965). Multiple exposures to defrost temperatures, therefore, can result in spoilage. Gunderson (1961) isolated 52 species from frozen convenience foods that were able to grow at temperatures under 5°C. The most frequently encountered mold was *Pullularia pullulans* (now *Aureobasidium pullulans*) followed by *Botrytis cinerea*, a *Phoma* sp., and *Geotrichum candidum*. Other psychrophilic species belonged to the genera *Alternaria*, *Anixiopsis*, *Aposphaeria*, *Botryotrichum*, *Cephalosporium*, *Chrysosporium*, *Cladosporium*, *Humicola*, *Monilia*, *Mucor*, *Myxotrichum*, *Papularia* (*Arthrinium*), *Phialophora*, *Plenozythia*, *Rhizopus*, *Sphaeropsis*, and *Penicillium*.

Low Water Activity.—As discussed in Chap. 2, a common feature of dried fruits, certain fruit juice concentrates, jellies, jams and fruit preserves is that a low water activity (a_w) is the basis for their preservation. When spoilage of these foods occurs, the same species of osmophilic yeasts and xerophilic fungi often are responsible.

Important dried fruits, listed in decreasing order of their production in the United States are: raisins, prunes, dates, figs, apples, apricots and peaches. Two basically different methods are used to remove water from these fruits: (1) evaporation, in which the fruit are dried in the sun; and (2) dehydration, which involves the use of various mechanical dryers that permit control of temperature, air flow and humidity. Sun-dried fruit have moisture contents as high as 25%, a level considerably above that of dehydrated fruit, which typically contain less than 5%. Dates are unique in that they usually require no special drying treatment because their moisture content is below 24% at the time of harvest.

Fungal spoilage of dried fruit generally does not occur when the water content is under 25%, because at this level the a_w is too low to support growth of most yeasts and molds. Prunes containing 22% moisture, for example, have a_w of 0.68, which is inhibitory to all but the most xerophilic fungi. High-moisture dried prunes, on the other hand, contain 30 to 35% water and therefore must be treated with a preservative such as potassium sorbate or given a heat treatment to prevent their spoilage (Pitt and Christian 1968; Schade et al. 1973).

The populations of viable yeasts and molds on sound dried fruits are relatively low, with average counts usually under 5000 per g (King et al. 1968). One of the reasons for the low counts is that many fruits are treated with sulfur dioxide prior to drying to prevent browning, and the levels that commonly are absorbed, 1000 to 4000 ppm, destroy a very high percentage of the contaminating microorganisms. Another factor is that drying is a lethal process for many microorganisms, especially when conducted at higher temperatures or when the organisms are being exposed simultaneously to ultraviolet rays from the sun.

Figs may undergo spoilage while still on the tree prior to drying. Yeasts, introduced into the interior of the fruit by insects, ferment the juice of the florets that line the inner walls. This type of spoilage has been referred to as "souring" because the juice that drips from fermented fruit has a vinegar-like odor, presumably because of subsequent oxidation of ethanol by acetic acid bacteria. Studies on the yeasts responsible for the initial fermentation have shown that the predominant organisms were *Saccharomyces* and *Candida* spp. (Mrak et al. 1942). Other yeasts that were isolated were species of *Pichia, Hanseniaspora, Kloeckera, Debaryomyces, Hansenula,* and *Torulopsis.* Most of the isolates were capable of growing in 40° but not 50° Balling fig syrup, which indicated they would not be spoilage agents of dried figs.

Various studies have been concerned with the yeasts and molds that are capable of growing on dried fruits. Mrak et al. (1942) isolated 67 yeast cultures that were responsible for slow spoilage of dates that had been grown in California and Egypt. The most frequently encountered yeasts were species of *Zygosaccharomyces* [now *Saccharomyces rouxii* (Lodder 1970)] followed by *Hanseniaspora* and *Candida:* also isolated were *Saccharomyces, Pichia, Hansenula,* and *Torulopsis* spp. While all of the isolates grew in 40° Balling date syrup, only the *Zygosaccharomyces* grew in 65° syrup.

Tanaka and Miller (1963A, B) investigated the yeasts and molds that caused spoilage of dried prunes. The most common yeasts were *Saccharomyces rouxii, Saccharomyces mellis* (now *S. bisporus*),

Torulopsis magnoliae, and *Torulopsis stellata.* The most numerous molds were members of the *Aspergillus glaucus* group followed by *Aspergillus niger* and *Penicillium* spp. The most osmophilic yeasts were *S. rouxii, S. mellis,* and *T. stellata,* all of which fermented media containing 70% soluble solids but failed to grow in media of 75% soluble solids. Members of the *A. glaucus* group were the most xerophilic of the molds encountered.

Both *S. rouxii* and *A. glaucus* grew on prunes that had been equilibrated at 20°C in an atmosphere of 76% relative humidity. Under these conditions the moisture content of the prunes was about 20% (Miller and Tanaka 1963).

Pitt and Christian (1968) found *Xeromyces bisporus* (now *Monascus bisporus*), which is one of the most xerophilic mold species, to be predominant on spoiled dried prunes. *X. bisporus,* which previously had been thought to be quite rare, may have been overlooked by other investigators because it does not grow at a_w levels above 0.97; Pitt and Christian used isolation media that contained 40 to 65% sucrose to give a_w values of 0.94 and 0.86. Other xerophilic molds that were isolated were species of *Chrysosporium* and members of the *Aspergillus glaucus* group. Their research on *X. bisporus* showed that the mold's aleuriospores germinated in 120 days when incubated in media having an a_w of 0.605. The minimal a_w for the formation of sexual and asexual spores was 0.672 and 0.663. Sporulation was slow under these conditions in that 116 days were required for the production of sexual spores.

A method used for the preservation of certain fruit juices such as apple, prune, grape and citrus is to concentrate them to a soluble solids level where their osmotic pressure is inhibitory to most potential spoilage organisms. When the Brix of these products is 70° or higher, they usually can be stored at ambient temperature without additional treatment.

When spoilage of fruit juice concentrates has occurred, the organisms found are usually the same species that grow on dried fruits. Thus, yeasts that had fermented 63° Brix orange juice were *Zygosaccharomyces (Saccharomyces rouxii)* and research on the isolates showed they could grow in concentrations of sugar as high as 70% (Ingram 1959A). Another yeast found to be capable of growing in orange juice concentrate is *Torulopsis magnoliae* (Kitchel and Miller 1960). It fermented 70° Brix juice within 30 days when held at temperatures of 16° and 27°C. In 60° Brix juice significant growth occurred in 90 days at the low storage temperature of 4.4°C. Juice concentrates stored in sealed drums that have minimal head space are less susceptible to spoilage by xerophilic molds since the organisms are obligate aerobes.

Jellies, jams, preserves and fruit syrups are products whose preservation is achieved by the addition of high concentrations of sugar. The a_w of these foods, which contain a minimum of 65% soluble solids, usually ranges from 0.82 to 0.94 (Kaplow 1970). The osmophilic yeasts and molds that might grow relatively rapidly at these a_w values possess little heat resistance and thus are killed by the extended boiling that is a part of the concentration process, and by the hot fill temperatures of 60° to 82°C that generally are used. Contamination of these foods after the containers have been opened by the consumer generally does not result in spoilage because most yeasts and molds do not grow at this low a_w or they grow very slowly (Mossel and Ingram 1955).

Fermentation.— Wines are semipreserved foods in that many of the yeasts and molds that are able to grow in juice of grapes and other fruits are inhibited by ethanol and, in the case of yeasts, by the lack of a fermentable sugar.

The growth of molds in wine generally can be readily controlled because of anaerobic conditions, the inhibitory effect of ethanol, and because molds are relatively sensitive to sulfur dioxide, a common wine additive. Molds may still be a problem, however, because the damp atmosphere of the typical winery encourages their development on walls and on the surfaces of wooden casks and tanks. Moldy cooperage is particularly serious since a single moldy stave may affect the flavor of an entire tank of wine (Amerine and Joslyn 1970). It is recommended that moldy cooperage first be scraped, then treated with hypochlorite solution and even dilute solutions of hydrogen peroxide or potassium permanganate. Badly molded wood may have to be replaced with new cooperage.

The majority of the yeasts that compose the microflora of grapes, apples and other fruits are either oxidative species or low alcohol producers that would not be capable of carrying out the desired fermentation (Mrak and McClung 1940; Beech and Davenport 1970). The few fermentative strains that are present may also possess undesirable properties such as low tolerance to ethanol or the excess production of organic acids, higher alcohols, hydrogen sulfide, or other compounds that might adversely affect the flavor of wine (Rankine 1968).

To assure that a wine is not spoiled by a bad fermentation, it is customary to eliminate most of the "wild" yeasts prior to inoculation with a selected wine strain. The addition of 50 to 100 ppm of sulfur dioxide will accomplish this as will heating of the fruit to a temperature of 55°C or higher. An inoculum of vigorously fermenting juice of 2 to 3% by volume will produce an initial population of over

10^6 yeasts per ml in the sweet must, a population that will permit the desired strain to predominate.

Bottled wines that have been sweetened or which have not been fermented to complete dryness are subject to refermentation by yeasts. The means for preventing this are sterile filtration, pasteurization, and the use of yeast inhibitors.

At present in the United States, most sterile filtration of wines is conducted with membrane filters. Pore sizes of $1.2\,\mu$m will effectively retain yeasts, although many wineries use filters of $0.65\,\mu$m or smaller pore size in order also to remove bacteria. Because the cells and other particulates are collected on the surface of the membrane, the capacity of this type of filter is relatively low. Wines to be filter sterilized, therefore, must first be highly polished by prefiltration through various types of depth filters.

Various pasteurization procedures are used by the industry. The wine may be given a high-temperature short-time treatment in which it is heated for 1 sec or less at a temperature of 80°C or higher before being cooled and filled into bottles. Another method is to fill cold and then heat the bottled wine in a water bath or with hot water sprays until an internal temperature of about 60°C is achieved. In a third method, hot bottling, the wine is heated to 55° to 70°C and then filled into bottles while still at that temperature.

The alcohol content of a wine has a significant effect on the heat resistance of yeasts and has been found to be more important than other variables such as pH and residual sugar (Splittstoesser et al. 1975). Thus in a wine containing 12.1% ethanol, the D value of Saccharomyces cerevisiae strain GS-1 was 0.5 min at 51°C compared to 3.3 min when the alcohol content was reduced to 8%. The z values at both alcohol concentrations were the same, about 4.4°C, for this yeast. The observation that GS-1 gave higher survivals than 28 other strains is evidence that most yeasts possess little thermal resistance when heated in wines.

The quality of some wines is reduced by pasteurization while that of others may be improved. The response apparently is related to the variety of grape.

Wines that are filled into bottles at ambient temperatures following sterile filtration or pasteurization are subject to chance recontamination from pipelines, filling machines, and bottles; the filling process is not aseptic nor are all bottles sterile. Generally refermentation is not a serious problem because the initiation of growth by a low inoculum (a few yeasts per bottle) is prevented by the free sulfur dioxide, ethanol, and other inhibitory substances that are intrinsic to the wine. Sorbic acid is often added to semidry wines to inhibit growth of yeasts. Diethyl pyrocarbonate (DEPC) was used to destroy contaminating microorganisms in wines until it was discovered that it reacted with ammonia to form urethan, a known carcinogen.

Preservatives.—The most commonly used preservatives for fruit products are benzoic, sorbic, and sulfurous (sulfur dioxide) acids. Foods preserved by these compounds include beverages, pie fillings, fruit salads, and dried fruits. Some of the other applications have been discussed previously. Not all of these preservatives are used in all fruit products; federal regulations, for example, do not authorize the use of benzoates or benzoic acid in wine. The concentrations of preservative that may be added to a food vary but in general 0.1% is the maximum permitted level.

It has long been known that the activity of these compounds increases with a reduction in pH, presumably because it is the undissociated acid that is germicidal (Cruess *et al.* 1931; Rahn and Conn 1944). Thus if the pH of a food is raised by one unit; approximately 10 times more preservative must be added to achieve a given effect (Ingram 1959B). Another factor that may influence the activity of preservatives is the level of contamination; growth may not be arrested if the population of yeasts and molds is too high. One possible explanation for this is that certain preservatives may be broken down by the contaminating microorganisms. *Penicillium* spp. for example, are able to degrade potassium sorbate (Marth *et al.* 1966).

Various hypotheses have been advanced to explain the antimicrobial activity of these acids. One proposes that because they are lipophilic, substrate transport and oxidative phosphorylation are uncoupled. Growth of the organism is inhibited principally because cellular uptake of amino acids. phosphate, and other nutrients is prevented (Freese *et al.* 1973). Other studies have indicated that certain enzyme systems are inactivated: sorbic acid was shown to react with the sulfhydryl-containing enzyme, fumarase (York and Vaughn 1964), and to inhibit catalase (Troller 1965).

The presence of more than one preservative in a fruit product may result in a synergistic effect. While alone neither 480 ppm of sorbic acid nor 129 ppm of sulfur dioxide were able to suppress yeast growth in a sweet table wine, a combination of 80 ppm of each compound was effective (Ough and Ingraham 1960). Also, lower concentrations of sorbic acid were required when the alcohol content was raised.

A number of other antimycotics have been used experimentally. The storage life of strawberries and raspberries was extended by dipping or spraying their surfaces with the macrolide antibiotics, nystatin, rimocidin, and pimaricin (Ayres *et al.* 1964). Vitamin K-5

(4-amino-2-methyl-1-naphthol hydrochloride) was shown to inhibit the growth of various molds when added to berry purees (Merrifield and Yang 1965).

SOFT DRINKS

Composition

Soft drinks are nonalcoholic beverages that contain sugar, flavoring substances, acids, and coloring agents. Most of those consumed in the United States are carbonated (Anon. 1971).

The sugar content of soft drinks ranges from 9.5 to over 13%. Fruit juices prepared from concentrates, natural extracts, and/or synthesized compounds may be responsible for flavor. The pH ranges from 2.6 to 4.0, depending upon type: root beer is one of the less acid drinks. Most soft drinks are acidified with citric acid, although the colas, which have the lowest pH, contain phosphoric acid, and tartaric acid often is added to grape flavored beverages. The amount of carbonation, which is governed by the type of flavor, varies from about 2 to over 4 gas volumes. Of the flavored beverages, ginger ale is most heavily carbonated while the fruit flavored drinks contain the lowest volumes. Sodium benzoate is commonly added as a preservative, its use being permitted by United States Federal Regulations which define the Standard of Identity for nonalcoholic beverages.

Spoilage

Most microbial spoilage of soft drinks is caused by yeasts. Strains of *Saccharomyces cerevisiae* and *Saccharomyces uvarum* have been responsible for "large scale spoilage explosions" while slow spoilage has been caused by members of the genera *Brettanomyces, Candida, Hansenula, Lodderomyces, Pichia, Torulopsis,* and *Saccharomyces* (Put and Sand 1974). Manifestations of yeast growth are the formation of turbidity and sediment, changes in color, and the development of off-flavors.

Contamination of soft drinks with dead mold mycelia has been a problem on rare occasions. It has occurred because bottle washing conditions had been inadequate to completely remove mold growth from the interior of used, returnable bottles.

Preservation

Soft drinks are poor growth media for most yeasts and molds. Mold growth is prevented by carbon dioxide and anaerobic conditions while yeasts are inhibited by the combination of a low pH and carbonation. Witter et al. (1958) observed that a yeast which had been isolated from spoiled orange drink grew over the pH range of 2.0 to 4.5 in noncarbonated beverages but failed to grow at pH values below 4.0 when the drink contained at least two volumes of carbon dioxide. Most rapid death was obtained in a 10° Brix beverage at pH 2.0 having 4.5 volumes of carbon dioxide.

Kelley (1975) studied the effect of pH and benzoate concentration on the growth of S. uvarum and a wild yeast in an experimental grape drink which contained 1.5 volumes of carbon dioxide. At pH of 2.5, 3.0, 3.5 and 4.0, the levels of benzoate needed to prevent growth were, respectively, 0.015, 0.030, 0.040 and 0.050%. Slightly lower concentrations of benzoate were needed when the initial yeast population was reduced from 10^4 to 10^2 cells per ml. Although 0.1% sodium benzoate may be added legally to soft drinks in the United States, the usual amount ranges from 0.03 to 0.05%.

Bottled soft drinks that do not contain a preservative may be protected by pasteurization (Put and Sand 1974). Heating for 20 min at 70° to 80°C is a common process. Studies on yeasts isolated from fermented, pasteurized soft drinks indicated that the ascospores of certain strains of S. cerevisiae were the most heat resistant. Populations of 10^5 per ml survived 10 min at 65°C when heated in thermal death time tubes (Put and Sand 1974). In a later study, Put et al. (1976) investigated the heat resistance of 120 yeast strains representative of the flora in soft drinks and certain acid food products. Asporogenous strains (Brettanomyces, Candida, Kloeckera, Rhodotorula and Torulopsis) were generally less heat resistant than ascomycetous types (Debaryomyces, Hansenula, Kluyveromyces, Lodderomyces, Pichia, Saccharomyces and Saccharomycopsis). Saccharomyces spp., especially S. cerevisiae and S. chevalieri, showed the highest heat resistance.

As with most foods, the successful preservation of soft drinks depends upon controlling contamination by potential spoilage organisms. Areas of concern are ingredients, equipment and containers. The industry has established standards for dry sugar that specify maximum viable counts of 10 yeasts and 10 molds per 10 g (Anon. 1971). Processing equipment that can present a problem are syrup vessels, lines and valves, filters, and filling machines. Control of microbial buildup depends upon sanitary design and effective cleaning, and especially upon adequate washing of returned bottles.

A 5-min wash at 54°C using a 3% alkali solution which is 60% sodium hydroxide is the type of treatment that will destroy spoilage organisms.

REFERENCES

ALLEN, R.N. 1970. Plantation and market diseases of banana fruit. Agric. Gaz. N.S.W. 81, 332-337.

AMERINE, M.A., and JOSLYN, M.A. 1970. Wines: The Technology of Their Production, 2nd Edition. Univ. of California Press, Berkeley.

ANON. 1971. Beverage Production and Plant Operation. Natl. Soft Drink Assoc., Washington, D.C.

AYRES, J. C., KRAFT, A. A., DENISEN, E. L., and PEIRCE, L. C. 1964. The use of macrolide antifungal antibiotics in delaying spoilage of fresh small fruits and tomatoes. In Microbial Inhibitors in Food. N. Molin (Editor). Almqvist and Wiksell, Stockholm.

BALATSOURAS, G. D., and POLYMENACOS, N. G. 1963. Chemical preservatives as inhibitors of yeast growth. J. Food Sci. 28, 267-275.

BARKAI-GOLAN, R. 1974. Species of Penicillium causing decay of stored fruits and vegetables. Mycopathologia 54, 141-145.

BEECH, F. W., and DAVENPORT, R. R. 1970. The role of yeasts in cidermaking. In The Yeasts, Vol. 3. A. H. Rose and J. S. Harrison (Editors). Academic Press, New York.

BEUCHAT, L. R. 1976. Effectiveness of various food preservatives in controlling the outgrowth of Byssachlamys nivea ascospores. Mycopathologia 59, 175-178.

BEUCHAT, L. R., and TOLEDO, R. T. 1977. Behaviour of Byssochlamys nivea ascospores in fruit syrups. Trans. Br. Mycol. Soc. 68, 65-71.

BROWN, A. H. S., and SMITH, G. 1957. The genus Paecilomyces Bainier and its perfect stage Byssochlamys Westling. Trans. Br. Mycol. Soc. 40, 17-89.

BUCHANAN, J. R., SOMMER, N. F., FORTLAGE, R. J., MAXIE, E. C., MITCHELL, F. G. and HSIEH, D. P. H. 1974. Patulin from Penicillium expansum in stone fruits and pears. J. Am. Soc. Hortic. Sci. 99, 262-265.

BUHAGIAR, R. W. M. 1975. Torulopsis bararium, Torulopsis pustula and Torulopsis multi-gemmis sp. nov., three new yeasts from soft fruit. J. Gen. Microbiol. 86, 1-11.

BUHAGIAR, R. W. M., and BARNETT, J. A. 1971. The yeast of strawberries. J. Appl. Bacteriol. 34, 727-739.

CICHOWICZ, S. M., and EISENBERG, W. V. 1974. Collaborative study of the determination of Geotrichum mold in selected canned fruits and vegetables. J. Assoc. Off. Anal. Chem. 57, 957-960.

COLE, G. M. 1955. Microbiology of orange juice and lemonade concentrates. Food Technol. 9, 38-45.

CRUESS, W. V., RICHERT, P. H., and IRISH, J. H. 1931. The effect of hydrogen-ion concentration on the toxicity of several preservatives to microorganisms. Hilgardia 6, 295-314.

DENNIS, C., and COHEN, E. 1976. The effect of temperature on strains of soft fruit spoilage fungi. Ann. Appl. Biol. 82, 51-56.

EDNEY, K. L. 1975. An investigation of persistent infection of stored apples by Gloeosporium spp. Ann. Appl. Biol. 82, 355-360.

ELLIOTT, R. P., and MICHENER, H. D. 1965. Fabtors affecting the growth of psychrophilic microorganisms in food. A review. Agric. Res. Serv., U. S. Dep. Agric. Tech. Bull. *1320*.

FELLERS, C. R., and SMITH, E. G. 1936. Chemical composition and fermentation studies of citron. J. Agric. Res. *53*, 859-867.

FIELDS, M. L. 1962. Voges-Proskauer test as a chemical index to the microbial quality of apple juice. Food Technol. *16*, 98-100.

FIELDS, M. L. 1964. Acetylmethylcarbinol and diacetyl as chemical indexes of microbial quality of apple juice. Food Technol. *18*, 1224-1228.

FREESE, E., SHEU, C. W., and GALLIERS, E. 1973. Function of lipophilic acids as antimicrobial food additives. Nature (London) *241*, 321-325.

GILLESPY, T. G. 1940. Studies on the mould *Byssochlamys fulva* (III). Ann. Rep., Univ. of Bristol Fruit and Veg. Pres. Res. Sta., Campden, England.

GUNDERSON, M. F. 1961. Mold problem in frozen foods. *In* Proc. Low Temperature Microbiology Symp., Campbell Soup Co., Camden, N. J.

HANLIN, R. T., and BLANCHARD, R. O. 1974. Fungi associated with developing pecan fruits. Bull. Ga. Acad. Sci. *32*, 68-75.

HARPER, K. A., BEATTIE, B. B. PITT, J. I., and BEST, D. J. 1972. Texture changes in canned apricots following infection of the fresh fruit with *Rhizopus stolonifer.* J. Sci. Food Agric. *23*, 311-320.

HARVEY, J. M., and PENTZER, W. T. 1960. Market diseases of grapes and other small fruits. Agric. Handb. *189*, U.S. Govt. Printing Office, Washington, D. C.

HARVEY, J. M., SMITH, W. L., JR., and KAUFMAN, J. 1972. Market diseases of stone fruits: cherries, peaches, nectarines, apricots and plums Agric. Handb. *414*, U.S. Govt. Printing Office, Washington, D. C.

HEWITT, W. B. 1974. Rots and bunch rots of grapes. Calif. Afric. Exp. Stn. Bull, *868*.

HILL, E. C., WENZEL, F. W., and BARRETO, A. 1954. Colorimetric method for detection of microbial spoilage in citrus juices. Food Technol. *8*,,168-171.

HSU, E. J. 1975. Factors affecting microflora in processed fruits. *In* Commercial Fruit Processing. J. G. Woodroof and B. S. Luh (Editors). AVI Publishing Co., Westport,

INGRAM, M. 1959A. Comparisons of different media for counting sugar tolerant yeasts in concentrated orange juice. J. Appl. Bacteriol. *22*, 234-247.

INGRAM, M. 1959B. Technical aspects of the commercial use of anti-microbial chemicals as food preservatives. Chem. Ind. 552-557.

INGRAM, M., and LUTHI, H. 1961. Microbiology of fruit juices. *In* Fruit and Vegetable Juice Processing Technology. D. K. Tressler and M. A. Joslyn (Editors). AVI Publishing Co., Westport, Conn.

JENNINGS, D.L., and CARMICHAEL, E. 1975. Resistance to grey mould (*Botrytis cinerea* Fr.) in red raspberry nruits. Hortic. Res. *14*, 109-115.

JENSEN, L. B. 1954. Microbiology of Meats, 3rd Edition. Garrard Press, Urbana, Ill.

KAPLOW, M. 1970. Commercial development of intermediate moisture foods. Food Technol. *24*, 889-893.

KELLEY, S. F. 1975. The effect of sodium benzoate as a preservative for carbonated beverages. Proc. 22nd Annu. Meet. Soc. Soft Drink Technol., Washington, D. C.

KING, A. D., JR., FIELDS, R. K., and BOYLE, R. P. 1968. Dried fruits have low microbial counts. Food Eng. *40*, No. 12, 82-83.

KING, A. D., JR., MICHENER, H. D., and ITO, K. A. 1969. Control of *Byssochlamys* and related heat-resistant fungi in grape products. Appl. Microbiol. *118*, 166-173.

KITCHEL, R. L., and MILLER, M. W. 1960. The viability of a yeast in high density orange concentrates stored at various temperatures. Food Technol. *14*, 547-549.

KOBURGER, J. A., and FARHAT, B. Y. 1975. Fungi in Foods. VI. A comparison of media to enumerate yeasts and molds. J. Milk Food Technol. *38*, 466-468.

LAWRENCE, N. L., WILSON, D. C., and PEDERSON, C. S. 1959. The growth of yeasts in grape juice stored at low temperatures. II. The types of yeast and their growth in pure culture. Appl. Microbiol. *7*, 7-11.

LLOYD, A. C. 1975. Preservation of comminuted orange products. J. Food Technol. *10*, 565-574.

LODDER J. 1970. The Yeasts, A Taxonomic Study. North-Holland Publishing Co., Amsterdam.

LOVETT, J., and PEELER, J. T. 1973. Effect of pH on the thermal destruction kinetics of patulin in aqueous solution. J. Food Sci. *38*, 1094-1095.

LUTHI, H. 1959. Microorganisms in noncitrus juices. Adv. Food Res. *9*, 221-284.

MARSHALL, C. R., and WALKLEY, V. T. 1951A. Some aspects of the microbiology applicable to commercial apple juice production. I. Distribution of microorganisms on fruit. Food Res. *16*, 448-456.

MARSHALL, C. R., and WALKLEY, V. T. 1951B. Some aspects of microbiology applied to commercial apple juice production. II. Microbiological control of processing. Food Res. *16*, 515-521.

MARSHALL, C. R., and WALKLEY, V. T. 1952. Some aspects of microbiology applicable to commercial apple juice production. III. Isolation and identification of apple juice spoilage organisms. Food Res. *17*, 123-131.

MARTH, E. H., CAPP, C. M., HAZENZAHL, L., JACKSON, H. W., and HUSSONG, R. V. 1966. Degradation of potassium sorbate by *Penicillium* species. J. Dairy Sci. *49*, 1197-1205.

MAUNDER, D. T. 1969. Spoilage problems caused by molds of the *Byssochlamys-Paecilomyces* group . In *Byssochlamys*. Seminar Abstracts. Circ. *20*, N. Y. State Agric. Exp. Stn., Geneva.

MERRIFIELD, L. S., and YANG, H. Y. 1965. Vitamin K_5 as a fungistatic agent. Appl. Microbiol. *13*, 660-662.

MILLER, M. W., and PHAFF, H. J. 1962. Successive microbial populations of calimyrna figs. Appl. Microbiol. *10*, 394-400.

MILLER, M. W., and TANAKA, H. 1963. Microbial spoilage of dried prunes. III. Relation of equilibrium relative humidity to potential spoilage. Hilgardia *34*, 183-190.

MOSSEL, D. A. A., and INGRAM, M. 1955. The physiology of the microbial spoilage of foods. J. Appl. Bacteriol. *18*, 233-268.

MOYER, J. C., SPLITTSTOESSER, D. F., and BOURNE, M. C. 1969. The quality of mechanically harvested grapes. In Fruit and Vegetable Harvest Mechanization. B. F. Cargill and G. E. Rossmiller (Editors). Rep. *16*, Michigan State Univ., East Lansing.

MRAK, E. M., and McCLUNG, L. S. 1940. Yeasts occurring on grapes and in grape products in California. J. Bacteriol. *40*, 395-407.

MRAK, E. M., PHAFF, H. J., VAUGHN, R. H., and HANSEN, H. N. 1942. Yeasts occurring in souring figs. J. Bacteriol. *44*, 441-450.

MURDOCK, D. I. 1964. Voges-Proskauer-positive yeasts isolated from frozen orange concentrate. J. Food Sci. *29*, 354-359.

MURDOCK, D. I., and BROKAW, C. H. 1958. Sanitary control in processing citrus concentration. I. Some specific sources of microbial contamination from fruit bins to extractors. Food Technol. *12*, 573-576.

NELSON, P. E. 1971. Technical developments in bulk storage processing. Hortic. Sci. *6*, 222-224.

OGAWA, J. M., RUMSEY, J., MANJI, B. T., TATE, G., TOYODA, J., BOSE, E., and DUGGER, L. 1974. Implication and chemical testing of two *Rhizopus* fungi in softening of canned apricots. Calif. Agric. *28*, No. 7. 6-7

OLLIVER, M., and RENDLE, T. 1934. A new problem in fruit preservation. Studies on *Byssochlamys fulva* and its effect on the tissues of processed fruit. J. Soc. Chem. Ind. (London) 53, 166-172.

OUGH, C. S., and INGRAHAM, J. L. 1960. Use of sorbic acid and sulfur dioxide in sweet table wines. Am. J. Enol. Vitic. 11, 117-122.

PEDERSON, C. S. 1971. Grape juice. In Fruit and Vegetable Juice Processing Technology. D. K. Tressler and M. A. Joslyn (Editors). AVI Publishing Co., Westport, Conn.

PEDERSON, C. S., ALBURY, M. N., WILSON, D. C., and LAWRENCE, N. L. 1959. The growth of yeasts in grape juice stored at low temperature. I. Control of yeast growth in commercial operation. Appl. Microbiol. 7, 1-6.

PERCEBOIS, G., BASILE., A.-M., and SCHWERTZ, A. 1975. The existence on strawberries of ascospores of strains of *Byssochlamys nivea* capable of producing patulin. Mycopathologia 57, 109-111. (French)

PIERSON, C. F., CEPONIS, M. J., and McCOLLOCH, L. P. 1971. Market diseases of apples, pears, and quinces. Agric. Handb. 376, U.S. Govt. Printing Office, Washington, D. C.

PITT, J. I., and CHRISTIAN, J. H. B. 1968. Water relations of xerophilic fungi isolated from prunes. Appl. Microbiol. 16, 1853-1858.

PRICE-DAVIS, W., and McDONALD, D. 1972. Soft drinks. In Quality Control in the Food Industry, Vol. 3. S. M. Herschdoerfer (Editor). Academic Press, London.

PUT, H. M. C., DEJONG, J., SAND, F. E. M. J, and VAN GRINSVEN, A. M. 1976. Heat resistance studies on yeast spp. causing spoilage in soft drinks. J. Appl. Bacteriol. 40, 135-152.

PUT, H. M. C., and SAND, F. E. M. J. 1974. A method for determination of heat resistance of yeasts causing spoilage in soft drinks. Proc. 4th Int. Symp. on Yeasts, Vienna.

RAHN, O., and CONN, J. E. 1944. Effect of increase in acidity on antiseptic efficiency. Ind. Eng. Chem. 36, 185-187.

RANKINE, B. C. 1968. The importance of yeasts in determining the composition and quality of wines. Vitis 7, No. 1, 22-49.

RECCA, J., and MRAK, E. M. 1952. Yeasts occurring in citrus products. Food Technol. 6, 450-454.

RUYLE, E. H., PEARCE, W. E., and HAYS, G. L. 1946. Prevention of mold in kettled blueberries in No. 10 cans. Food Res. 11, 274-279.

SCHADE, J. E., STAFFORD, A. E., and KING, A. D., JR. 1973. Preservation of high-moisture dried prunes with sodium benzoate instead of potassium sorbate. J. Sci. Food Agric. 24, 905-911.

SCOTT, K. J., and ROBERTS, E. A. 1967. Control in bananas of black end rot caused by *Gloeosporium musarum*. Aust. J. Exp. Agric. Anim. Husb. 7, 283-286.

SMOOT, J. J., HOUCK, L. G., and JOHNSON, H. B. 1971. Market diseases of citrus and other subtropical fruits. Agric. Handb. 398, U. S. Govt. Printing Office, Washington, D. C.

SMOOT, J. J., and SEGALL, R. H. 1963. Hot water as a post-harvest control of mango anthracnose. Plant Dis. Rep. 47, 739-742.

SOMMER, N. F., BUCHANAN, J. R., and FORTLAGE, R. J. 1974. Fungus and enzyme activity in fresh apricots as related to softening of canned fruits. Calif. Agric. 28, No. 7, 8.

SOMMER, N. F., and FORTLAGE, R. J. 1966. Ionizing radiation for control of post harvest diseases of fruits and vegetables. Adv. Food Res. 15, 147-193.

SPLITTSTOESSER, D. F., DOWNING, D. L., ROGER, N. F., and MURDOCK, D. I. 1974. Influence of the harvest method on contamination of fruit by *Byssochlamys* ascospores. J. Milk Food Technol. 37, 445-447.

SPLITTSTOESSER, D. F., KUSS, F. R., HARRISON, W., and PREST, D. G. 1971. Incidence of heat-resistant molds in eastern orchards and vineyards. Appl. Microbiol. 21, 335-337.

SPLITTSTOESSER, D. F., LIENK, L. L., WILKISON, M., and STAMER, J. R. 1975. Influence of wine composition on the heat resistance of potential spoilage organisms. Appl. Microbiol. 30, 369-373.

STOKES, J. L. 1971. Influence of temperature on the growth and metabolism of yeasts. In The Yeasts, Vol. 2. A. H. Rose and J. S. Harrison (Editors). Academic Press, New York.

STOLK, A. C., and SAMSON, R. A. 1971. Studies on Talaromyces and related genera. I. Hamigera gen. nov. and Byssochlamys. Persoonia 6, 341-357.

SUBRAMANYAM, H., KRISHNAMURTHY, S., and PARPIA, H. A. B. 1975. Physiology and biochemistry of mango fruit. Adv. Food Res. 21, 223-305.

TANAKA, H., and MILLER, M. W. 1963A. Microbial spoilage of dried prunes. I. Yeasts and molds associated with spoiled dried prunes. Hilgardia 34, 167-170.

TANAKA, H., and MILLER, M. W. 1963B. Microbial spoilage of dried prunes. II. Studies of the osmophilic nature of spoilage organisms. Hilgardia 34, 171-181.

TROLLER, J. A. 1965. Catalase inhibition as a possible mechanism of the fungistatic action of sorbic acid. Can. J. Microbiol. 11, 611-617.

VAUGHN, R. H., and MURDOCK, D. I. 1956. Sanitary significance of microorganisms in frozen citrus products. Am. J. Public Health 46, 886-894.

WEHNER, F. C., and RABIE, C. J. 1970. The microorganisms in nuts and dried fruits. Phytophylactica 2, 165-170.

WILLIAMS, C. C., CAMERSON, E. J., AND WILLIAMS, O. B. 1941. A facultative anaerobic mold of unusual heat resistance. Food Res. 6, 69-73.

WITTER, L. D., BERRY, J. M., and FOLINAZZO, J. F. 1958. The viability of Escherichia coli and a spoilage yeast in carbonated beverages. Food Res. 23, 133-142.

WOLFORD, E. R. 1958. Effect of varied storage temperature on microorganisms in frozen concentrated orange juice. Appl. Microbiol. 6, 139-141.

YACKEL, W. D., NELSON, A. I., WEI, L. S., and STEINBERG, M. P. 1971. Effect of controlled atmosphere on growth of mold on synthetic media and fruit. Appl. Microbiol. 22, 513-516.

YATES, A. R. 1974. The occurrence of Byssochlamys sp. molds in Ontario. Can. Inst. Food Sci. Technol. J. 7, 148-150.

YORK, G. K., and VAUGHN, R. H. 1964. Mechanisms in the inhibition of microorganisms by sorbic acid. J. Bacteriol. 88, 411-417.

4

Fungi in the Spoilage of Vegetables

J. Orvin Mundt

In order that the role of fungal spoilage of vegetables is placed into its proper perspective, it is well to point out immediately that fungi are but one of several major reasons for the loss of harvested vegetables. Bacteria play an approximately coequal role, particularly in the spoilage of succulent and tender greens which have a minimum of hard tissue and are frequently stored under conditions of high relative humidity and reduced oxygen tension in the mass. Significant losses of vegetables are attributable to physical deterioration such as dehydration or wilting, which leads to consumer rejection. Maturation results in flaccidity and changes in color, such as the yellowing of cabbages. Root, tuber and bulb vegetables sprout when withdrawn from cold storage into warmer atmospheres. Vegetables are condemned for the presence of unapproved fungicidal and herbicidal chemicals. Insects and rodents contribute their share to the losses. The internal black spot of beets, the brown stele of parsnips, and the brown and brittle stem of celery are caused by boron deficiency in the soil. Black heart of stored potatoes is a process of darkening which follows freezing injury or bruising.

The primary function of fungi is to digest and oxidize to the simplest constituents the once living, now dead, organic matter in order that the constitutive elements are liberated for use by the next generation of life. Fungal spoilage of vegetables is merely a mani-

FIG. 4.1. BLASTOPHORE (B) OF *AUREOBASIDIUM PULLULANS* ATTACHED TO HYPHA GROWING INTERNALLY IN A CABBAGE LEAF. Other structures are diatoms and amorphous wax. From the author's laboratory (scanning electron microscopy, 3000X).

festation of this function. Many of the fungi attacking harvested vegetables attack living tissue, and in this respect they differ from the vast group of truly saprophytic fungi encountered in nature.

Very few of the numerous species of fungi common to the soil exist in an epiphytic relationship with plant surfaces. *Aureobasidium* (*Hormodendron, Cladosporium*) *pullulans* and *Epicoccum nigrum* appear to colonize green plant surfaces on a worldwide basis. In the author's laboratory, these mold species have been enumerated in populations to 10^5 per g of macerated samples of locally grown asparagus, leaf lettuce, green and pole beans, cucumbers, summer squash, broccoli, okra, cabbage, Chinese cabbage, and kale, chard, and mustard greens. They have been isolated from stem and leaf tissue (Fig. 4.1), and from weeds growing in association with the vegetables. In a review of the literature, Last and Warren (1972) indicate that these fungi grow internally in plant tissues and that they serve a useful function in the repression of plant pathogens. The numbers of propagules are relatively constant throughout the growing season from early spring through late fall, and are influenced little by variations in climatic conditions or elevation above ground.

Species of *Alternaria, Fusarium, Helminthosporium* and *Chaetomium* are usually present on vegetables in low numbers. Although a variety of fungi common to the soil can be isolated from field vegetables, their numbers and infrequency of isolation are indicative of chance contamination.

Yeast populations on vegetables range from less than 10^3 to more than 10^6 per g of macerated tissue, with little consistency in numbers among samples of the same or different vegetables. Pigmented yeasts usually dominate the populations. Last and Warren (1972) indicate that the plant-resident yeasts are chiefly asporogenous and are identifiable as *Cryptococcus, Torulopsis, Rhodotorula, Sporobolomyces* and *Tilletiopsis*.

The majority of all fungi are unable to attack healthy, living plant tissue. Among those which do are the virulent plant pathogens which infect the plant in the field and the opportunistic saprophytes which take advantage of weakened or stressed conditions. The resistance of the vegetable at the time of harvest is characteristic of that of the living, intact plant. Harvested plant parts do not die immediately when excised from the plant or removed from the soil, but events are set into motion which eventually lead to the death of the tissues. The tissue is brought to the threshold of invasiveness of the fungi and ultimately, when the tissue is moribund or dead, many of the nonaggressive fungi are able to grow.

PLANT RESPONSES TO FUNGI

Protective Mechanisms

The stratified epithelial cell barrier of tissues is the primary defense mechanism of the plant. The barrier may be rather thin and fragile, as it is in lettuce, or it may be several layers thick and suberized as it is in the ripened potato. Some plants, among them members of the brassicae, have soft or hard waxes which entrap and immobilize potentially damaging microorganisms, thus enhancing the mechanical barrier. The succulent epithelial surface also harbors vast numbers of saprophytic bacteria whose function it may be to overwhelm microbial pathogens which may be present.

Secondary defense mechanisms come into play when the epithelial barrier is breached by insect stings, breakage, mechanical or windblown abrasions, and various types of damage during harvesting and handling. The soluble solids of exuded cell juices are concentrated at the broken tissue surface through evaporation if the relative humidity is low. This is often seen in the cracked tomato in the field. Invading microorganisms deposited beneath the surface may not tolerate the anerobic conditions. Plants do not possess the equivalent of the animal reticuloendothelial, antibody-producing mechanisms, but their juices do exhibit a generalized low level of inhibitory activity.

Tissue cells immediately adjacent to the site of injury may revert to the embryonic state to bring about healing. In the apple this occurs if the injury takes place before maturation of the seed. The subsequent cornification appears as a swollen, unpalatable and discolored tissue. The exposed cells of the white potato form new periderm successfully if the temperature is below 20°C, while at higher temperatures bacteria outgrow the host cells. Corky tissue forms in the sweet potato if the potato is cured at the proper temperature and relative humidity; otherwise *Rhizopus stolonifer* (formerly *R. nigricans*) outgrows the repair process.

Phytoalexin Accumulation

Plant stems, roots, fruits and seeds have been demonstrated to accumulate specific compounds when infected with a variety of fungi (Table 4.1). These compounds have been labeled as "abnormal" metabolites, "stress" metabolites, or phytoalexins (Baker and Robinson 1975; Kuc and Currier 1976). They may also accumulate in response to chemical or physical alterations other than fungal

TABLE 4.1
FUNGI ASSOCIATED WITH STRESS METABOLITE ACCUMULATION
IN VEGETABLE HOST TISSUE

Host	Fungus	Metabolite
Bean seeds, pods	*Monilinia fructicola, Phytophthora megasperma* var. *sojae, Rhizoctonia solani*	Phaseollin, hydroxyphaseollin
Carrot root	*Ceratocystis fimbriata*	6-methoxymellein
Celery stem	*Sclerotinia sclerotiorum*	Psoralens (5-methoxypsoralen, 8-methoxypsoralen, 4, 5′, 8-trimethylpsoralen)
Pea seeds, pods	*Monilinia fructicola, Rhizoctonia solani*	Pisatin
Rice seeds	*Piricularia oryzae, Cochliobolus miyabeanus, Gibberella fujikuroi*	Coumarin, chlorogenic acid, other phenols
Sweet potato root	*Ceratocystis fimbriata, Fusarium solani, F. oxysporum, Thielaviopsis basicola, Helicobasidium mompa*	Ipomeamarone, ipomeanine, batatic acid, furan-β-carboxylic acid; other phenolics, terpenes, coumarins
White potato tuber	*Phytophthora infestans, Bipolaris zeicola* (formerly *Helminthosporium carbonum*)	Alkaloids, coumarins, phenolics, terpenes; other solanidine derivatives

Source: Adapted from Baker and Robinson (1975); Kuc and Currier (1976).

invasion. The potential for phytoalexin production is present in all healthy plants and may be manifested as a result of synthesis by the plant, by liberation from bound forms by fungal or plant enzymes, or by fungal modifications of host compounds.

Several furanocoumarin compounds collectively called psoralens are examples of phytoalexins or stress metabolites. Xanthotoxin (8-methoxypsoralen) and 4,5′,8-trimethylpsoralen are produced in diseased celery (pink rot) by *Sclerotinia sclerotiorum* (Scheel et al. 1963; Wu et al. 1972). Although *Sclerotinia* grows well on carrots (Fig. 4.2), sweet potatoes, artichokes, cucumbers, and turnips, psoralen production has not been noted on these vegetables. Concurrent metabolism of celery and of *S. sclerotiorum* is required for the production of xanthotoxin and bergapten. Psoralens elicit human skin disorders in the form of blisters or lesions. It is necessary that, following consumption of infected celery, the skin be exposed to sunlight or to ultraviolet radiation in the range of 320 to 380 nm for lesions to be produced (Ayres 1973).

Invading fungi induce the formation or release of antifungal compounds from the sweet potato, Irish potato, carrot, turnip, snap bean, soybean, broad bean, English pea, pepper, barley and rice. Some of these compounds exhibit toxicity to mammalian systems. In addi-

Courtesy of the Agric. Res. Service, USDA
FIG. 4.2. CARROT INFECTED WITH SCLEROTINIA

tion, solanidine and α-solanine which accumulate in Irish potatoes blighted with *Phytophthora infestans* have been associated with gastrointestinal and neurological disorders in humans and farm animals. Sweet potatoes infected with *Fusarium solani* represent a potential health hazard, although no known instances of intoxication have occurred.

STORAGE DISEASES

The term "storage disease" describes the deterioration of vegetables and fruits caused by fungi, bacteria and viruses during storage. The majority of the fungi causing storage diseases are more or less virulent plant pathogens which are also capable of producing serious crop losses in the field. It is generally believed that they invade the vegetable prior to harvest, but have not developed to the point of detectable destruction of tissue. Since the spores and filaments of these fungi are in the soil, any damage brought about during harvesting and subsequent handling facilitates the ability of the organisms to grow. The growth of *R. stolonifer* on cut or damaged surfaces of the sweet potato is an excellent example of destruction following tissue damage. The majority of the fungi listed in Table 4.2 are plant pathogens.

The saprophytic fungi not recognized as plant pathogens are introduced to the vegetable through the adherent soil, by dust, rain, contact with harvesting machinery and lug boxes, bins, truck bodies and other harvesting and transporting equipment, by wash waters, contact with decaying produce in the packing sheds, and from the human handlers. *Fusarium* and *Rhizopus* spp. occupy an intermediate position, since they may infect vegetables during postharvest handling, but they are also "wet weather" fungi in that they infect produce during periods of abundant rain and prevailing high relative humidity. They are also adventitious molds which develop on vegetables late in the storage period.

Some fungi grow on vegetable tissues only during late stages of storage or when the action of other microorganisms has made the vegetable receptive to them. *Scopulariopsis* takes part in the final disintegration of many vegetables. It is seldom encountered during early storage. It is actively proteolytic, producing ammonia abundantly from organic nitrogen compounds and disagreeable odors from vegetables such as turnip and cabbage.

Mucor spp. and *R. stolonifer* are representative of molds which grow only on wounded or damaged vegetables. *Rhizopus* shares with *Ceratocystis* sp. (formerly *Endoconidiophora*) the reputation for being the most serious harvest pathogen of improperly cured and stored sweet potatoes. Like other members of the family Mucoraceae, the growth of *R. stolonifer* is encouraged by high relative humidity, a temperature between 10° and 30°C, and available carbohydrate. It is both pectinolytic and amylolytic. It produces a soft, yet firm, rot of the sweet potato and spreads throughout the container, often binding many potatoes together into a single mass with its intertwining filaments.

Mycosphaerella is a common field pathogen of cucurbits which opens the way for later spoilage by bacteria and yeasts. The initially small, water-soaked spots of infection turn black as pycnidia are formed. *Fusarium* spp., on the other hand, usually are weak, ubiquitous storage pathogens which often follow bacterial rot of vegetables and *Pythium* rot of white potatoes. They grow on wet or undried vegetables such as sweated beans and carrots in storage, forming large, white to pink tufted mycelial masses.

Botrytis cinerea, *Colletotrichum lindemuthianum*, and *Rhizoctonia* spp. attack a wide range of vegetables and fruits. *Botrytis* and *Rhizoctonia* cause slimy rots which progress rapidly through stored

TABLE 4.2
COMMON STORAGE DISEASE FUNGI OF VEGETABLES

Fungus	Commonly Observed Type of Spoilage and Representative Vegetable
Alternaria	Leaf spot or nail head spot of cabbage, cauliflower, turnip, melon, cucumber, tomato; initially small, inconspicuous black spots enlarging and coalescing; superficial growth on cortical parenchyma of potato; calyx rot of peppers
Aspergillus	A. alliaceus causes dry "charcoal" rots; black rot of onion, garlic
Botrytis	B. cinerea causes soft, dark rots of most vegetables; B. allii causes neck rot of onion
Ceratocystis	Black rots of vegetables; with Rhizopus, most serious cause of loss of stored sweet potatoes: initial brown spots enlarge, become black with perithecial formation
Choanephora	Blossom end rot of squash
Cladosporium	Scab on cucumber, pumpkin: black and green-black rots on aging produce
Colletotrichum	Anthracnose or bitter rot: brown surfaces; wide host range among vegetables and fruits
Diaporthe	D. batatis causes firm dry decay of sweet potato; roots turn dark brown, becoming black as pycnidia form; D. vexans causes fruit rot of eggplant
Diplodia	Browning and shriveled areas of cucurbits, sweet potatoes; stem-end rots of melons
Fusarium	Wet weather infection of wide variety of vegetables; adventitious on aging produce; field-infecting agent, manifest through growth in storage; bulb rot of Allium spp.; often follows bacterial soft rots: wound infection of sweet potato; penetrating black dry rot of white potato
Geotrichum	Sour rot of vegetables; also known as "machinery mold" or "dairy mold" in processing plants
Mucor	Wound infections; prominent growth; filamentous; nonfermenting
Mycosphaerella	Cucurbits; small, water-soaked spots turning black with pycnidial formation; followed by yeasts, bacteria
Penicillium	Charcoal rot of sweet potato; common on wet and aging produce; P. hirsutum is a major cause of loss of stored horseradish; blue mold rots
Phoma	Dry brown rot of beet
Phomopsis	Phomopsis blight or market rot of eggplant
Phytophthora	Buckeye rot of tomato: leathery, watersoaked rot of vegetables; downy mildews
Pythium	Major spoilage agent of potatoes in early storage; watery, cream-gray to black rot known as "leak" is caused by P. debaryanum; frequently precedes Fusarium
Rhizoctonia	With Botrytis, most common cause of mold loss of vegetables; slimy brown rots, often where vegetable is in contact with soil; forms large, black sclerotial masses
Rhizopus	Adventitious on wet vegetables; wound-infecting on sweet potato; rapid decay; late storage rots; often characterized by fermentative odor
Trichoderma	Green and black rots of brassicae
Trichothecium	Soft rots of cucurbits; T. roseum causes pink rot

vegetables. *Botrytis* is the common gray mold of vegetable rots, resembling somewhat the color of *Rhizopus*, but with a much restricted aerial growth. *Colletotrichum* causes anthracnose in plants in the field and rather hard, nodular rots. Although it often occurs in the inedible portions of the vegetable such as the leaf and seed pod, it makes the vegetables unsightly and unappealing.

Diaporthe batatis is an example of a fungus with a very restricted host range. It produces a firm, brown rot of the sweet potato which ultimately becomes dry, hard and mummified. Small, dome-like protuberances just visible to the eye cover the surface. The imperfect stages of *Diaporthe* spp. belong the form-genus *Phomopsis*.

Aspergilli and penicillia are representative molds which are primarily saprophytic, but which contain members capable of attacking living plant tissue. *Aspergillus alliaceus* is a specific spoilage agent of stored onion and garlic, producing dry, black "charcoal" rots. *Penicillium hirsutum* is the major cause of loss of stored horseradish. Species of penicillia grow adventitiously on onion and garlic, and their pectinolytic enzymes have been indirectly incriminated in the softening of some types of cucumber pickles.

FRESH VEGETABLES

All vegetables have a natural postharvest storage life which is limited by many of the factors mentioned earlier. Deterioration by fungi is common among hard, firm vegetables and is not commonly encountered with greens which are packed and stored in ice. Prevention of fungal deterioration involves cleanliness, sanitation and a knowledge of the proper conditions for storage of specific vegetables.

All harvesting equipment should be maintained in a clean condition. Knives, belts and other surfaces should be cleaned daily to remove accumulated grime. The grime is a mixture of vegetable tissue, cell juices, field dirt and microorganisms. Initially bacteria constitute the dominant populations; however, yeasts and molds develop over a period of several days and thereafter contaminate the vegetables passing over or through the equipment.

Lug boxes, crates, sacking, wagon beds and other receptacles for the harvested vegetables should be cleaned daily to reduce the numbers of mold spores which may come into contact with the harvested tissues. Lug boxes and other receptacles often are constructed of wood which has a rasping effect on vegetables and which is difficult to sterilize. In some critical areas metal receptacles which can be more easily cleansed and sanitized are used.

Care should be taken at all stages of vegetable harvesting and handling to prevent mechanical damage such as breakage, piercing with forks, abrasion and crushing. Transfer of vegetables from one point to another should be done gently, without throwing. The height of packing in hauling vehicles and in storage compartments should be limited by the amount of resistance the vegetable exhibits to mechanical crushing or bruising.

Vegetables should be kept as cool as possible during transport from the field to the packing sheds. Handling should be such that the heat of respiration does not accumulate in the mass of vegetables. During hot harvest months, particularly, the heat of respiration in addition to the field heat hastens the susceptibility of the vegetable to microbial attack.

Field heat should be removed from the produce as soon as possible after harvest. Cooling is accomplished by methods which are most economical and feasible with the least harm to the vegetable. Practically all hot-weather vegetables and many fall vegetables are cooled before entering shipment or storage to relieve the strain on permanent refrigerating equipment and also to bring about rapid cooling. Vegetables are cooled by vacuum, dehydrocooling, packing in ice, and by cold air. After cooling and packing, crates or cartons should be placed in trucks or warehouses so as to prevent warm spots which could be generated by respiration.

Field soil should be removed from root and tuber vegetables wherever possible, either by washing or by vibratory cleaning. Soils have high populations of molds and their spores. Adhering soils place mold filaments and spores into intimate contact with the vegetable, and also maintain a microclimate of relative humidity which favors the growth of molds. Water used in washing should be potable and with low microbial content.

Only sound, healthy vegetables should be stored. Bruised vegetables and those suffering abrasions, mechanical breakage, or crushing deteriorate rapidly and should not be stored. Vegetables which show evidence of disease or overheating should not enter storage.

Some vegetables must be cured after harvest to assure successful storage. Curing involves drying at a suitable temperature and a low relative humidity. The curing period permits the healing of injuries received during harvest. Sweet potatoes are held at 27° to 35°C with good ventilation for 10 to 21 days to reduce the moisture content of the potato. The relative humidity should be below 70% during curing and from 75 to 85% in storage. The temperature during storage should be between 10° and 13°C. Garlic, onions and chili peppers are stored at 60 to 70% RH. Garlic and onions should be well cured to prevent neck rot caused by *Botrytis allii*. Onions with thick necks are slow to cure and they succumb quickly to disease.

Cucumbers, eggplants, okra, sweet peppers, pumpkins and winter squashes store well at 7° to 10°C in an atmosphere of high relative humidity. They develop pitting, water-soaked spots, discoloration and decay caused by *Alternaria* spp. at lower temperatures. Cucumbers and rutabagas often are dipped in wax, not so much to retard decay but to retard shriveling from loss of moisture and to enhance the appearance by imparting a glossy appearance.

Controlled atmospheres, chemical treatments, and irradiation are neither practical nor advisable in the storage of fresh vegetables. Ozonated atmospheres are detrimental to some vegetables and ineffective in the treatment of subsurface mold. Antibiotics are effective in prolonging the storage life of greens and packaged salad mixes, but the use of antibiotics in foods in the United States is illegal. With the exception of acetic and lactic acids which are used as pickling agents, neither these nor benzoic, propionic or sorbic acids are used in vegetable preservation. Ultraviolet irradiations reduce the populations of airborne microorganisms, but the rays neither penetrate protective surfaces nor bend beneath them. The effectiveness of ultraviolet rays decreases with the square of the distance so that much of the effectiveness of the ultraviolet light is lost at a distance of several feet from the light source.

Shippers of fresh vegetables use chlorine at 150 ppm in a water spray to control fungi and bacteria on cauliflower, celery, lettuce and other vegetable crops. Chlorine is used commonly in the washing of salad greens to reduce microbial activity and to prolong the fresh, crisp appearance. A mixture of citric and sorbic acids delays browning and softening of sliced celery for several days (Johnson et al. 1974).

DRIED VEGETABLES

Vegetables were commonly preserved by drying in the home before the development of the Mason jar late in the last century. Presently white potatoes, sweet potatoes, sweet peppers to be ground into paprika, hot peppers for cayenne pepper, garlic and onions are the major dried vegetable products. Carrots, turnips and peas are dried in small quantities for dried soup mixes. Finely shredded celery, including both the stalk and the leaf, is thoroughly dried and then ground to a fine powder to produce celery salt.

With the exception of peppers, garlic, onion and horseradish, most vegetables are blanched to inactivate peroxidase and other enzymes, or they are cooked to expedite rehydration. Because vegetables generally are more sensitive than fruits to high temperatures during drying, they are dried at lower temperatures, in the range of 54° to 71°C (Fabian 1943). Although molds and yeasts recontaminate the blanched or cooked vegetable, they rarely grow during the process of dehydration.

After vegetables are dried, they exhibit the customary hygroscopic behavior of all dried products when exposed to the atmosphere. To protect from fungal growth when stored in areas of high relative humidity, dried vegetables are packed in vapor barrier containers.

PRESERVATION WITH SALT

A number of vegetables can be preserved in brines with low or high salt content. Fermentation occurs in vegetables packed in low-salt brines. Sodium chloride is the only salt used, since all other salts either are bitter, toxic or cathartic. Household and commercial brining practices are well established (LeFevre 1927; Etchells and Jones 1943; Fabian 1943; Fabian and Blum 1943).

Strong brines contain between 10.5 to 15% salt. At high concentrations, neither bacterial spoilage nor lactic acid fermentation occurs. Vegetables commonly preserved in strong salt brines are those which do not shrivel when placed directly into strong brines. These include cauliflower, green beans, onions, green tomatoes and small peppers.

There appears to be no literature concerning yeast or mold spoilage of these brined vegetables. Presumably, unless the products are placed under anaerobic conditions, spoilage similar to that occurring with brined cucumbers will occur (vide infra).

Brine Fermentations

In dilute salt brines the sugars withdrawn from the vegetables are fermented by lactic acid-producing bacteria. In most instances the final acidity is measured at pH 3.6. The salt and the acid, coupled with anaerobiosis or storage near freezing, are preserving agents.

Film-forming, scum or oxidative yeasts grow on the surfaces of fermenting vegetables and utilize the lactic acid. Pichia and Candida vini (C. mycoderma) are common in brines of lower salt concentration, whereas species of Debaryomyces become prevalent at higher concentrations. As a result of the lowered acidity, pectinolytic bacteria, most likely members of the genus Erwinia, grow. The rotting effect moves downward through the container and may result in the loss of a major portion of the fermenting vegetable in home fermentations. In the case of commercial sauerkraut fermentation, the loss of as much as a metric ton or more of cabbage may occur. In several instances, the elevated pH has permitted the growth of *Clostridium botulinum* with the subsequent intoxication of consumers of homemade dill pickles and peppers.

Commercial fermentations of cabbage, cucumbers and olives are carried out in large wooden or concrete vats. Cabbage vats are sheltered from the weather, but vats for the fermentation of cucumbers and olives are placed in open fields. The ultraviolet rays in the sunlight destroy film-forming yeasts and molds at the surface of the brine. The sun's ray may not kill molds growing in the shadow of the vat wall of the olive tank, since olives are harvested and fermented in the fall of the year. Location in the open field has been generally effective in controlling the growth of surface yeasts, but a disadvantage is the exposure of the vat contents to debris from the atmosphere and the dilution of the brines with rain and snow. Growth of the film yeasts in fermenting vegetables is now prevented by covering the vat with a tough plastic sheet which is weighted with several centimeters of water (Fig. 4.3).

The detrimental effects of the film-forming yeasts and their aerobic nature was recognized by the Chinese at least as early as the time of the building of the Great Wall of China. They developed a crock with a moat to receive the deep flange of the cover, and to provide a water seal to prevent access of oxygen to the fermenting vegetable (Fig. 4.4).

Sauerkraut.—Wilted, trimmed, washed, and shredded cabbage is packed firmly into crocks or vats with 2.5% salt. The differential in osmotic pressure resulting from the dissolved salt causes a withdrawal of the cell juices to form the brine. The salt prevents the growth of the majority of the soil and surface-resident plant microflora, and imparts flavor to the finished product. Usually, three successive species of bacteria are involved in the lactic acid fermentation of the sugars: *Leuconostoc mesenteroides, Lactobacillus brevis* and *Lactobacillus plantarum*. At the end of the fermention the brine should contain a titratable acidity of 1.5 to 1.6%, calculated as lactic acid. Occasionally fermentative yeasts are active during the early stages of the cabbage fermentation, when sugars are abundant in the brine. The sauerkraut then has a high ethanol content and the titratable acidity is reduced. If the salt is unequally distributed, pockets of cabbage with high and low salt contents occur. *Rhodotorula* spp. grow in pockets with high salt content and the cabbage develops an unsightly pink color. Where the salt content is

FIG. 4.3. REDWOOD VEGETABLE FERMENTATION VAT COVERED WITH A PLASTIC SHEET AND ABOUT 15 CM WATER. Vats are generally 2.0 to 2.5 m in depth and about 5 m in diameter.

Courtesy of Tennessee Farm and Home Science
FIG. 4.4. CHINESE FERMENTATION JAR. Note the moat for receiving water and the flange of the cover to fit into the moat. Plate in foreground is used to weight the fermenting vegetable in brine.

low, cabbage softens or rots as a result of the growth of pectinolytic bacteria. According to Steinbuch (1965), *Saccharomyces exiguus*, often present in discolored European sauerkraut, secretes enzymes which, in the presence of oxygen, presumably convert cabbage substances into pink-purple, dirty pink or gray materials.

Cucumbers.—During the short harvest season, cucumbers flow from the fields too rapidly to enable packers to convert them into finished pickle products. Cucumbers not directly packed as pickles are fermented in brines containing 7 to 10% salt for conversion into salt stock. When the lactic acid fermentation is completed, the brine concentration is raised slowly by weekly increments to a final concentration of 15% salt. Cucumbers thus fermented can be kept for several years. During the process of fermentation and curing the interior of the cucumber changes in appearance from white and opaque to olive green and translucent. Fresh fermented dill pickles undergo a fermentation similar to that of sauerkraut, but in smaller containers. Pectinolytic and cellulolytic softening of cucumbers may occur during brine fermentation, storage, and after cucumbers are freshened and converted into finished pickle products. Both filamentous molds and yeasts are involved.

Molds grow freely in the bell-shaped flower of the cucumber during the blossom season. The most frequently occurring molds identified by Etchells *et al.* (1958) were *Penicillium oxalicum, P. janthinellum, Ascochyta cucumis, Fusarium roseum, F. oxysporum, F. solani, Cladosporium cladosporioides, Alternaria tenuis, Trichoderma viride,* and *Mucor silvaticus.* With the exception of *M. silvaticus* and *F. roseum,* all were cellulolytic and caused a reduction in firmness when culture filtrates were added to jars of commercially prepared, pasteurized pickles. Both pectinolytic and cellulolytic enzymes are present on cucumber flowers in the field. At the time of brining, the enzymes are extracted from remnants of the flowers remaining attached to the cucumbers and become adsorbed on the surface of the cucumber. One means of combatting enzymatic softening has been to drain and replace the brine, particularly of the vats containing the smaller cucumbers, about 48 hours after filling the vat. Whether softening occurs during the fermentation, much later in storage, or when the finished pickle has been prepared is dependent upon the amount of the enzyme extracted from the flower.

Debaryomyces membranaefaciens (now *D. hansenii*), *Candida krusei, Brettanomyces versatilis* (now *Torulopsis versatilis*), *Endomycopsis ohmeri* (now *Pichia ohmeri*), *Hansenula subpelliculosa, Torulaspora rosei* (now *Saccharomyces rosei*), *Torulopsis caroliniana* (now *T. lactis-condensi*), *T. holmii, Saccharomyces globosus,*

Zygosaccharomyces halomembranosus, Z. globiformis, and *Z. pastori*
(*Zygosaccharomyces* spp. are now classified as *Saccharomyces* and
Pichia spp.) have been isolated from films and the interior of pickle
brines (Etchells *et al.* 1953A,B). These yeasts exhibit varying tolerances to salt. *D. membranaefaciens* and *Z. halomembranosus* grow
luxuriantly and rapidly on the surface of 20% salt broths, while *C.
krusei* tolerates a maximum of 10% sodium chloride in broth. The
authors found it of interest that although *Rhodotorula, Candida* spp.
other than *C. krusei,* and *Kloeckera* sp. occur commonly on flowers,
they are not found in salt brines. Many of the yeasts isolated from
cucumber brines are capable of deesterifying pectin. The fermentative yeasts are numerous during the first few days of brining when the
brines have a relatively high content of sugars which have been
leached from the cucumber and the competing lactic acid-producing
bacteria have not become predominant. The carbon dioxide produced
during the fermentation is entrapped in the cucumbers to form
"bloaters." These are cucumbers with either numerous small, lens-
shaped pockets of gas or a large, gas-filled cavity with the contents of
the cucumber pressed against its skin. Among the measures
recommended to suppress yeast action in brines and to rapidly
remove the carbon dioxide are inoculation with the homofermentative bacteria *Pediococcus cerevisiae* and *L. plantarum,* and purging of
the brine tanks periodically with gaseous nitrogen.

Fresh-pack dill pickles are pasteurized for preservation. The jars
are heated slowly to 71° to 77°C and held at this temperature for 15
min (Monroe *et al.* 1969). Sour and sweet-sour cucumber pickles are
preserved by taking advantage of the inhibitory effects of acetic acid
alone, or of acetic acid coupled with sugar. *Saccharomyces rosei*
(formerly *Zygosaccharomyces globiformis*) is the most common
spoilage agent of unpasteurized sweet pickles. The "spoilage
prediction curve" of Bell and Etchells (1952) specifies the minimum
concentration of each component with respect to the other to ensure
preservation of the product.

Olives.—Olives, like cucumbers, are fermented in salt brines. The
majority of the domestic crop is treated with dilute lye solution to
hydrolyze the bitter oleorupein before the olives are fermented. Lye
treatment is stopped and leaching is begun when the lye has
penetrated nearly to the pit. During the leaching process as much as
65% of the sugar is leached from the olive and bacterial malfermen-
tations may occur unless the sugar is replaced. Desirable fermenting
bacteria are *L. mesenteroides, L. brevis* and *L. plantarum.*

Fermentation of olives is carried out in vats or barrels. Molds and
pectinolytic yeasts are troublesome in open fermentations. *Fusarium,*

Penicillium, Aspergillus, Geotrichum, Paecilomyces and *Verticillium* have been isolated from the surfaces of outdoor brines. They grow in areas of the vat protected from the sun's rays. Spoilage fungi are either pectinolytic or cellulolytic, or both. The yeasts *Saccharomyces kluyveri* and *S. oleaginosus* are also pectinolytic. *Rhodotorula* spp. cause stem-end softening.

Olives are fermented in brines with 6 to 7% salt. Recent attempts to reduce the salt content to 3 to 4% because of ecological factors have resulted in gas pocket formation and softening. Three species of yeasts have been isolated. *S. kluyveri* and *S. oleaginosus* are, as stated above, pectinolytic, while *Hansenula anomala* is strongly fermentative (Vaughn *et al.* 1972).

Oriental Vegetable Fermentations.—These are lactic acid fermentations, usually of mixed vegetables, in which the same bacteria active in the making of sauerkraut are found. The fermentations differ somewhat according to the locale of the fermentation, whether in China (Chao 1949), Korea (Kim and Whang 1959) or the Philippines (Orillo *et al.* 1969). Such products are known as paw-tsay, yen tsai, kimchi, sajur asin, mostasa and nukomiso pickles. The ingredients also are determined by local availability and local preference and practice. Basically the vegetables to be fermented are mixtures of cabbage, sweet turnips, radishes, carrots, peppers, asparagus and mustard greens. Cowpeas, soybeans, leeks, onions, garlic, ginger, watercress, seaweed, spices, mushrooms, oysters, sugar and still other materials may be added. All fermentations are subject to spoilage caused by a variety of scum-forming yeasts and filamentous molds which reduce the acidity and result in pectinolytic spoilage.

REFERENCES

AYRES, J. C. 1973. Antibiotic, inhibitory and toxic metabolites elaborated by microorganisms in foods. Acta Aliment. Acad. Sci. Hung. *2,* 285-302.

BAKER, R. C., and ROBINSON, W. B. 1975. Potential increases in food supply through research in agriculture. Food science research needs for improving utilization, processing and nutritive value of food products. (Panel report to the National Science Foundation). Agric. Exp. Stn., Cornell Univ., Ithaca, N. Y.

BELL, T. A., and ETCHELLS, J. L. 1952. Sugar and acid tolerance of spoilage yeasts from sweet cucumber pickles. Food Technol. *6,* 468-472.

BELL, T. A., ETCHELLS, J. L., and COSTILOW, R. N. 1958. Softening enzyme activity of cucumber flowers from northern production areas. Food Res. *23,* 198-204.

BURKA, L. T., and WILSON. B. J. 1976. Toxic furanosesquiterpenoids from molddamaged sweet potatoes *(Ipomoea batatas).* In Mycotoxins and Other Fungal Related Food Problems. J. V. Rodricks (Editor). Am. Chem. Soc., Washington, D. C.

CHAO, H. H. 1949. Microbiology of paw-tsay. 1. Lactobacilli and lactic acid fermentation. Food Res. *14*, 405-412.

CICHOWICZ, S. M., and EISENBERG, W. V. 1974. Collaborative study of the determination of *Geotrichum* mold in selected canned fruits and vegetables. J. Assoc. Off. Anal. Chem. *57*, 957-960.

CODNER, R. C. 1971. Pectinolytic and cellulolytic enzymes in the microbial modification of plant tissues. J. Appl. Bacteriol. *34*, 147-160.

DUCKWORTH, R. B. 1966. Fruits and Vegetables. Pergamon Press, New York.

ETCHELLS, J. L., BELL, T. A., and JONES, I. D. 1953A. Morphology and pigmentation of certain yeasts from brines and the cucumber plant. Farlowia *4*, 265-304.

ETCHELLS, J. L., BELL, T. A., MONROE, R. J., MASLEY, P. M., and DEMAIN, A. L. 1958. Populations and softening enzyme activity of filamentous fungi on flowers, ovaries, and fruit of pickling cucumbers. Appl. Microbiol. *6*, 427-440.

ETCHELLS, J. L., COSTILOW, R. N., and BELL, T. A. 1953B. Identification of yeasts from commercial cucumber fermentations in northern areas. Farlowia *4*, 249-264.

ETCHELLS, J. L., and JONES, I. D. 1943. Commercial brine preservation of vegetables. Fruit Prod. J. *22*, 242-245.

FABIAN, F. W. 1943. Home Food Preservation. AVI Publishing Co., Westport, Conn.

FABIAN, F. W., and BLUM, H. B. 1943. Preserving vegetables by salting. Fruit Prod. J. *22*, 228-236.

FIELDS, M. L., and DeGUZMAN, A. 1968. Fungal metabolites as indicators of quality of tomato products. Appl. Microbiol. *16*, 948-949.

JARVIS, B. 1972. Mould spoilage of foods. Process Biochem. *7*, No. 5, 11-14.

JOHNSON, C. E., VON ELBE, J. H., and LINDSAY, R. C. 1974. Extension of postharvest storage life of sliced celery. J. Food Sci. *39*, 678-680.

KIM, H. S., and WHANG, K. C. 1959. Cited by T. W. Kwon. Fermented Foods in Korea, Annotated Bibliography (1917-1971). Korea Inst. Sci. Technol., Seoul, 1972.

KUC, J., and CURRIER, W. 1976. Phytoalexins, plants, and human health. In Mycotoxins and Other Fungal Related Food Problems. J. V. Rodricks (Editor). Am. Chem. Soc., Washington, D. C.

LAST, F. T., and WARREN, R. C. 1972. Non-parasitic microbes colonizing green leaves: their form and functions. Endeavor *31*, 143-150.

LeFEVRE, E. 1927. Making fermented pickles. U.S. Dep. Agric. Farmer's Bull, *1438*.

LUTZ, J. M., and HARDENBURG, R. E. 1968. The commercial storage of fruits, vegetables, and florist nursery stocks. U. S. Dep. Agric. Handb. *66*.

MARTIN, W. J., HASLING, V. C., and CATALANO, E. A. 1976. Ipomeamarone content in diseased and nondiseased tissues of sweet potatoes infected with different pathogens. Phytopathology *66*, 678-679.

McCOLLOCH, L. P., COOK, H. T., and WRIGHT, W. R. 1968. Market diseases of tomatoes, peppers, and eggplants. U. S. Dep. Agric. Handb. *28*.

MONROE, R. J., ETCHELLS, J.L., PACILIO, J. C., BORG, A. F., WALLACE, D. H., ROGERS, M. P., TURNEY, L. J., and SCHOENE, E. S. 1969. Influence of various acidities and pasteurizing temperatures on the keeping quality of fresh-pack dill pickles. Food Technol. *23*, 71-77.

ORILLO, C. A., SISON, E. C., LUIS, M., and PEDERSON, C. S. 1969. Fermentation of Philippine vegetable blends. Appl. Microbiol. *17*, 10-13.

PEDERSON, C. S. 1971. Microbiology of Food Fermentations. AVI Publishing Co. Westport, Conn.

SCHEEL, L. D., PERONE, V. B., LARKIN, R. L., and KUPAL, R. E. 1963. The isolation and characterization of two phototoxic furocoumarins (psoralens) from diseased celery. Biochemistry *2*, 1127-1131.

SCOTT, P. M., HARWIG, J., CHEN, Y.-K., and KENNEDY, B. P. C. 1975. Cytochalasins A and B from strains of *Phoma exigua* var. *exigua* and formation of cytochalasin B in potato gangrene. J. Gen. Microbiol. *87*, 177-180.

SOMMER, N. F., and FORTLAGE, R. J. 1966. Ionizing radiation for control of postharvest diseases of fruits and vegetables. Adv. Food Res. *15*, 147-193.

STEINBUCH, E. 1965. Preparation of sauerkraut. Ann. Rep. Sprenger Inst., Wageningen, Netherlands.

THORNE, S. 1975. Studies of the invasion of carrot root parenchyma by *Rhizopus stolonifer* Lind. J. Sci. Food Agric. *26*, 933-940.

VAUGHN, R. H. 1954. Lactic acid fermentation of cucumbers, sauerkraut and olives. *In* Industrial Fermentations, Vol. 1. L. A. Underkofler and R. J. Hickey (Editors). Chemical Publishing Co., New York.

VAUGHN, R. H., STEVENSON, K. E., DAVE, B. A., and PARK, H. C. 1972. Fermenting yeasts associated with softening and gas-pocket formation in olives. Appl. Microbiol. *23*, 316-319.

WARD, E. W. B., UNWIN, C. H., HILL, J., and STOESSL, A. 1975. Sesquiterpenoid phytoalexins from fruits of eggplants. Phytopathology *65*, 859-863.

WOOD, G. E. 1976. Stress metabolites of white potatoes. *In* Mycotoxins and Other Fungal Related Food Problems. J. V. Rodricks (Editor). Am. Chem. Soc., Washington, D. C.

WU, C. M., KOEHLER, P. E., and AYRES, J. C. 1972. Isolation and identification of xanthotoxin (8-methoxypsoralen) and bergapten (5-methoxypsoralen) from celery infected with *Sclerotinia sclerotiorum*. Appl. Microbiol. *23*, 852-856.

5

Meats, Poultry, and Seafoods

James M. Jay

T he ubiquity of fungi in the environments in which meats are handled and stored is such that these organisms should always be found in and on these products. While this is probably true, molds and yeasts are rarely involved in the spoilage of fresh meats except under special circumstances. The reasons for this become clear when the ecological parameters of growth of these organisms are considered.

It is well established that molds can grow over a wide range of pH— from around 2 to 11; over an a_W range of about 0.62 to 0.995; over a temperature range from -10°C to around 60°C; and over a wide range of nutrient limitations. Some mold conidia, chlamydospores, and vegetative cells are known to retain viability for more than 22 years while spores may remain viable for more than 30 years (Sussman 1968). Yeasts are more similar to molds than to bacteria relative to growth parameters. Molds are not as versatile, however, in their capacity to grow over wide ranges of oxidation-reduction potentials (Eh). Most food-borne isolates require aerobic conditions, unlike many bacteria which can proliferate under -Eh conditions. When the above parameters are assessed for meats, poultry, and seafoods, the growth of molds should occur on all of these products where +Eh values exist. The fact that this does not happen under normal circumstances suggests the operation of at least one other factor: the relative rates of growth of fungi and competing bacteria and their relative success in acquiring growth needs. Under natural conditions that permit the growth of a variety of bacteria and molds, the former invariably outgrow the latter and consume the available oxygen in the process. Unless conditions are made unfavorable for bacterial growth, the growth of molds is not observed. This accounts for the general absence of visible fungal growth on fresh and processed meats, poultry, and seafood products. In order to demonstrate the presence of molds and yeasts on products of these types, it is necessary that one or more of the growth parameters be altered so that bacterial growth is not favored while at the same time the development of the slower growing molds is favored. Refrigerated ground beef normally undergoes bacterial spoilage without any signs of visible mold growth. When a bacterial inhibitor such as chlortetracycline is added, however, visible mold growth will invariably result over a 5- to 7-day period at refrigerator temperatures. Similar results can be achieved by lowering the pH or a_w below the growth range for bacteria. Because of the capacity of most meats, poultry, and seafoods to support the growth of bacteria, the mycology of these products has not received much attention until recently.

MOLD CONTAMINATION AND GROWTH

Meats

While Brewer (1925) was apparently the first United States investigator to record the presence of molds in fresh meats, most of the early studies on the incidence and types of these organisms in meats were carried out by British workers (Brooks and Kidd 1921; Brooks and Hansford 1923; Wright 1923). During the long-term storage and transportation of beef quarters in the bottom of sailing vessels, the occurrence of such surface conditions as "black spot," "whiskers" and "white spot" was noted and later shown to be caused by molds such as Cladosporium, Sporotrichum, and Thamnidium spp. The growth of these organisms on the surfaces of beef occurs because of a lack of enough moisture to permit bacterial growth.

A relatively large number of mold genera and species that grow on and cause the spoilage of meats is now known. Most of the twenty genera listed in Table 5.1 were isolated from fresh, stored, cured, or processed meats. Aspergilli and penicillia are most frequently recorded. It is not clear as to why these two groups are so much more

TABLE 5.1

MOLDS ISOLATED FROM MEATS, POULTRY, AND SEAFOODS

Mold	Product	Reference
Alternaria spp.	Beef	Ayres (1960); See Ayres (1951)
	Bacon	Jensen (1954)
	Untreated poultry	Njoku-Obi et al. (1957)
		Ayres et al. (1950)
	Country-cured hams	Sutic et al. (1972)
Aspergillus amstelodami	Moldy hams, salami	Ayres et al. (1967)
	Country-cured hams	Sutic et al. (1972)
A. aurantiobrunneus	Country-cured hams	Sutic et al. (1972)
A. candidus	Country-cured hams	Sutic et al. (1972)
A. chevalieri	Moldy hams, salami	Ayres et al. (1967)
A. clavatus	Meats	Yesair (1928)
A. conicus	Country-cured hams	Sutic et al. (1972)
A. flavus	Italian type salami	Bullerman and Ayres (1968)
	Country-cured hams	Strzelecki et al. (1969);
		Sutic et al. (1972)
		Ayres et al. (1967)
A. glaucus	Meats	Yesair (1928)
	Katsuobushi (fermented fish)	Graikoski (1973)
	Country-cured hams	Sutic et al. (1972)
A. gracilis	Country-cured hams	Sutic et al. (1972)
A. niger	Meats	Yesair (1928)
	Moldy hams, salami	Ayres et al. (1967)
	Country-cured hams, fermented sausage	Ayres et al. (1967)
	Chilled beef carcass	Stringer et al. (1969)
A. ochraceus	Moldy hams, salami	Ayres et al. (1967)
A. penicilloides	Country-cured hams	Sutic et al. (1972)
A. pseudoglaucus		
	Country-cured hams	Sutic et al. (1972)
A. repens	Moldy hams, salami	Ayres et al. (1967)
	Country-cured hams	Sutic et al. (1972)
A. restrictus	Country-cured hams	Sutic et al. (1972)
A. ruber	Country-cured hams, fermented sausage	Ayres et al. (1967)
A. sydowi	Country-cured hams	Sutic et al. (1972)
A. tamarii	Moldy hams, salami	Ayres et al. (1967)
A. versicolor	Country-cured hams	Sutic et al. (1972)
A. viride-nutans	Country-cured hams	Sutic et al. (1972)
A. wentii	Moldy hams, salami, fermented sausage	Ayres et al. (1967)
Aspergillus spp.	Bacon	Jensen (1954)
	Refrigerated beef	Ayres (1960)
	Moldy hams, salami	Ayres et al. (1967)
	Dry European sausage	Bullerman et al. (1969B)

TABLE 5.1 (Continued)

Mold	Product	Reference
Aspergillus spp.	Poultry meat	Njoku-Obi et al. (1957)
	Fish protein concentrate	Fox (1973)
	Smoked fish	Graikoski (1973)
Botrytis spp.	Bacon	Jensen (1954)
Cladosporium herbarum	"Black spot" of beef	Brooks and Kidd (1921)
	"Black spot" of frozen mutton	Wright (1923)
Cladosporium spp.	Refrigerated beef	Ayres (1960)
	Chlortetracycline-treated poultry	Njoku-Obi et al. (1957)
	Country-cured hams	Sutic et al. (1972)
Fusarium spp.	Meats	Yesair (1928)
	Bacon	Jensen (1954)
Geotrichum candidum	Poultry	Njoku-Obi et al. (1957)
Geotrichum spp.	"White spot" of beef	Brooks and Hansford (1923)
	Chicken	Ayres et al. (1950)
	Fresh sausage	Dowdell and Board (1968; 1971)
Monilia spp.	Bacon	Jensen (1954)
	Refrigerated beef	Ayres (1960)
Monascus purpureus	Meats	Yesair (1928)
Mortierella spp.	Meats	Yesair (1928)
Mucor lusitanicus	"Whiskers" of beef	Frazier (1967)
M. mucedo	"Black spot" of frozen mutton	Wright (1923)
M. racemosus	Meats	Yesair (1928)
	"Whiskers" of beef	Frazier (1967)
Mucor spp.	Cured meats	See Ayres (1951)
	Refrigerated beef	Ayres (1960)
	Bacon	Jensen (1954)
	"Black spot" of beef	Brooks and Hansford (1923)
	"Whiskers" of beef	Frazier (1967)
	Untreated poultry	Njoku-Obi et al. (1957)
Neurospora sitophila	Meats	Yesair (1928)
Oidium carnis (Geotrichum sp.)	"Black spot" of beef	Brooks and Kidd (1921)
O. lactic (Geotrichum candidum)	Meats	Yesair (1928)
Oidium spp.	Bacon	Jensen (1954)
Oospora nikitinskii (Geotrichum sp.)	Salted fish	Malevich (1936)
Oospora spp.	Salted fish "dun"	Frank and Hess (1941A)
Penicillium asperulum	Green patches on beef	Frazier (1967)
P. frequentans	Moldy hams, salami, fermented sausage	Ayres et al. (1967)
	Beef	Yesair (1928)
P. expansum	Beef	Yesair (1928)
	Green patches on beef	Frazier (1967)

Table 5.1 (Continued)

Mold	Product	Reference
P. glaucum	"Black spot" of frozen mutton	Wright (1923)
P. melzynski	Moldy hams, salami	Ayres et al. (1967)
P. oxalicum	Green patches of beef	Frazier (1967)
P. puberulum	Moldy hams, salami, fermented sausage	Ayres et al. (1967)
P. spinulosum	Dried beef	Frazier (1967)
P. variabile	Moldy hams, salami, fermented sausage	Ayres et al. (1967)
Penicillium spp.	Refrigerated beef	Ayres (1960)
	Chicken	Ayres et al. (1950)
	Bacon	Jensen (1954)
	Country-cured hams	Ayres et al. (1967)
	"Black spot" of beef	Brooks and Kidd (1921)
	Chlortetracycline-treated poultry	Njoku-Obi et al. (1957)
	Smoked fish	Graikoski (1973)
Pulluleria pullulans (Aureobasidium pullulans)	Fresh shrimp	Phaff et al. (1952) Koburger et al. (1975)
Rhizopus nigricans (R. stolonifer)	"Black spot" of beef	Brooks and Kidd (1921)
	Chlortetracycline-treated and untreated poultry	Njoku-Obi et al. (1957)
Rhizopus spp.	Bacon	Jensen (1954)
	"Black spot" of frozen mutton	Wright (1923)
	"Whiskers" of beef	Brooks and Hansford (1923)
	Refrigerated beef	Ayres (1960)
Scopulariopsis spp.	Dry European type salami	Bullerman et al. (1969B)
Sporendonema epizoum	Salted fish	Van Klavern and Lequendre (1965)
Sporendonema spp. Sporotrichum carnis	"Dun" of dry-salted fish	Frank and Hess (1941A)
(Aleurisma carnis)	"White spot" of beef	Haines (1930)
Sporotrichum spp. Thamnidium chaetocladioides	Refrigerated beef	Ayres (1960)
(T. elegans)	'Whiskers" of beef	Frazier (1967)
	Salted, minced pork	Tanner (1944)
	Beef	Brooks and Hansford (1923)
T. elegans	"Black spot" of beef	Brooks and Kidd (1921)
	"Whiskers" of beef	Frazier (1967)
Thamnidium spp.	Refrigerated beef	Ayres (1960)
	"Black spot" of beef	Brooks and Hansford (1923)
Zygorrhynchus spp.	Frankfurters	See Ayres (1951)

prominent than other mold groups but this may be a reflection of their generally higher incidence in meat environments. It may also be a reflection of the greater ease of their identification, of their ability to grow at low a_W or of some other property of these two groups to adapt to meats with greater ease than others. Some approximate minimal a_W for multiplication of fungi associated with meat and meat products are listed in Table 5.2. It doesn't seem feasible at this time to assign any rank priorities to mold genera and species and specific types of meats. This degree of precision should await further research by more investigators on a wider variety of meats. Also, the origin of molds in some meats such as cured meats seems unclear. Hadlok (1969) found that spices were good sources of penicillia and aspergilli when added to meat products. A better understanding of ecological niches is highly desirable before a better appreciation of an association between mold types and specific products can be attained.

TABLE 5.2

MINIMAL a_W FOR MULTIPLICATION OF SOME FUNGI
ASSOCIATED WITH MEAT AND MEAT PRODUCTS

Fungus	Minimal a_W
Rhizopus, Mucor	0.93
Rhodotorula, Pichia	0.92
Hansenula, Saccharomyces	0.90
Candida, Torulopsis, Cladosporium	0.88
Debaryomyces	0.87
Penicillium	0.85
Aspergillus	0.65

Source: Adapted from Leistner and Rödel (1975).

In the case of country-cured hams, the work of Sutic et al. (1972) indicates that penicillia are by far the most predominant molds, followed by the aspergilli. These investigators recovered 562 molds from these products and found that 403 were penicillia while 121 were aspergilli. The latter represented 8 groups and 15 species. In a study of the mold flora of country-cured hams, Bullerman et al. (1969B) found that while the initial flora consisted of penicillia, the aspergilli became dominant later in the aging process. This was apparently due to the higher initial moisture level in the hams which tended to favor the penicillia. These investigators also reported that xerotolerant types such as A. ruber emerged when the a_W was lowered to around

0.65. In a somewhat similar manner, dry European type salami developed heavy mold growth during its aging process for up to 60 days at room temperature and the predominant molds were penicillia, aspergilli, and *Scopulariopsis* spp. (Bullerman et al. 1969B). Ninety percent of country-cured hams yielded aspergilli while one-third of the mold-ripened sausages harbored these types.

Meats such as smoked and salted bacon, sausages, frankfurters, and the like can and do undergo mold spoilage, and the aspergilli and penicillia appear to be the most abundant types. Vacuum-packed meat products tend not to undergo moldiness because of reduced Eh conditions.

Fresh meats such as ground beef may be expected to contain mold fragments and spores but under normal spoilage conditions the molds do not proliferate. The true incidence and types of molds in products of this type are not known due to the fact that microbiologists in the past have not sought this information with any consistency. Numerous investigators have recorded the presence of molds in a wide variety of meat products over the years but most failed to report the identity of these organisms.

Mycotoxins.—Since the discovery that some *Aspergillus flavus* and *A. parasiticus* strains produce aflatoxins, a large amount of research has been devoted to determining whether these organisms produce toxins in all products that support their growth. With respect to meats, Bullerman et al. (1969A) inoculated both *A. flavus* and *A. parasiticus* into ground beef, fresh ham, and bacon and demonstrated the production of aflatoxins when the products were stored at 15°C and above but not at 10°C. These investigators prevented the growth of bacteria by use of a combination of antibiotics. It appeared that around 14 days were necessary to detect aflatoxins B_1 and G_1 with an inoculum of 10^2 *A. parasiticus* conidia per 100 g of beef. In another study these authors found that Italian type salami contaminated with *A. flavus* was more likely to favor aflatoxin production than was smoked Hungarian type salami under the same conditions. Aflatoxin production did not occur on these products when the holding temperature was below 15°C and the relative humidity was below 75% (Bullerman et al. 1969B).

Incze and Frank (1976) recently investigated the capacity of eight aflatoxin- and eleven sterigmatocystin-producing aspergilli to produce these carcinogenic mycotoxins on Hungarian salami. It was found that the sum of ecological parameters tested (mixed cultures of molds, 13°C, and an a_w of about 0.94) is sufficient in each case to prevent mycotoxin production.

In summarizing the work done in his laboratory on the aflatoxin production potential of 76 strains of molds from country-cured hams and fermented sausages, Ayres (1973) noted that only the *A. flavus* isolates were positive. Two strains of *A. ochraceus* tested for ochratoxin production were found to be negative (Strzelecki *et al.* 1969). However, in a later study Escher *et al.* (1973) were successful in demonstrating ochratoxin A and B production by strains of *A. ochraceus* on country-cured ham. These investigators noted that one-third of the toxin was found in the mycelial mat on the surface of ham 21 days after incubation while two-thirds had penetrated into the meat to a distance of 0.5 cm. Wu *et al.* (1974A,B) reported that several aspergilli and penicillia isolated from aged, cured meats were toxic to chicken embryos. Aflatoxins and other mycotoxins are discussed in more detail in Chap. 14.

Poultry

Refrigerated poultry meats almost always undergo bacterial surface spoilage and the paucity of reports of fungi from these products is understandable. Several investigators have, however, sought to isolate and identify molds from poultry, and six of the recorded genera are indicated in Table 5.1. Some of these genera were identified only after treatment of poultry with chlortetracycline to retard the usually present Gram-negative bacterial flora.

Fish

Of the mold genera indicated in Table 5.1 only four are indicated as having been recorded on fish or fish products. The paucity of reports on the incidence and types of molds in products of this type cannot be due to a lack of molds in waters, as Johnson (1968) has reported finding taxa of Phycomycetes, Ascomycetes, and Fungi Imperfecti throughout the marine environment. Wood (1965) has reviewed the subject of fungi in the oceans and indicates that they are important parts of the phytoplankton and relatively abundant. The mold flora of fresh water is generally thought to be similar to that of the surrounding land areas, the primary habitat of molds. This points up the lack of importance of these organisms in the spoilage of fresh fish. Shewan (1961) has reported that molds are seldom if ever recorded on newly caught fish. They are known, however, to exist and grow on smoked fish and Shewan believes that the sawdust used in smoking is their primary source of entry. The most common genera found on smoked fish are penicillia and aspergilli and both groups are believed to be the cause of spoilage of this product (Graikoski 1973).

In dry-salted fish, a spoilage condition known as "dun" or "mite" sometimes occurs. This condition manifests itself as small brown, black, or fawn-colored spots or tufts on the surface of the fish and represents the growth of halophilic or halotolerant molds of the genera *Sporendonema* or *Oospora* (*Geotrichum*) (Malevich 1936; Frank and Hess 1941A, B; Shewan 1961). These organisms have been reported to have a pH range for growth of 3.3 to 7.5, an optimum relative humidity for growth of around 75%, to require at least 5 to 10% sodium chloride for growth, and to fail to grow below 5°C (Frank and Hess 1941A). The spoilage condition can be prevented either by storage below 5°C or by use of sorbic acid (Boyd and Tarr 1955). The dun condition of salted cod has been reported to be caused by *Sporendonema epizoum* (Van Klavern and Lequendre 1965).

Fish protein concentrate may be spoiled by the growth of aspergilli if the product contains more than 17% moisture (Fox 1973).

Mycotoxins.—Love (1968) reported that when the moisture content of fishmeal was 10% or less, *A. flavus* failed to produce aflatoxins. On the other hand, fishmeal containing 18% or more moisture supported toxin production by massively inoculated *A. flavus* when stored for several weeks at 24°C or above. Toxin production occurred in five days when a massively inoculated product was stored at 30°C.

YEAST CONTAMINATION AND GROWTH

Meat

Like the lactic acid bacteria, with which they are often associated in processed meats, yeasts tend to grow better at low pH values. Unlike most bacteria, some can grow at extremely low a_w levels. Those that are encountered in foods are favored by +Eh conditions.

Table 5.3 presents yeasts isolated and identified from meats, poultry, and seafoods. Of the genera listed, five were recovered from frankfurters, sausages, and beef, with *Debaryomyces*, *Candida*, and *Torulopsis* spp. having been recorded most often. Yeasts are common in products such as luncheon meats and certain sausages and their presence in products of this type has been known since Cary's studies on sausages in 1916. Many investigators have noted and recorded the numbers of yeasts in processed meats without reporting identifications. Among them are Rogers and McCleskey (1957), who noted that along with micrococci and bacilli, yeasts constituted one of the three largest groups of microorganisms at the beginning of a 14-day storage

TABLE 5.3

YEASTS ISOLATED FROM MEATS, POULTRY, AND SEAFOODS

Yeast	Product	Reference
Candida catenulata	Frankfurters	Drake *et al.* (1959)
C. guilliermondii	Fresh shrimp	Phaff *et al.* (1952)
	Chlortetracycline-treated poultry	Njoku-Obi *et al.* (1957)
C. intermedia	Eviscerated poultry	Walker and Ayres (1959)
C. krusei	Eviscerated poultry	Walker and Ayres (1959)
C. lipolytica	Fresh fish	Ross and Morris (1965)
	Frankfurters	Drake *et al.* (1959)
C. parapsilosis	Fresh shrimp	Phaff *et al.* (1952)
	Chlortetracycline-treated poultry	Njoku-Obi *et al.* (1957)
	Fresh fish	Ross and Morris (1965)
C. pelliculosa	Eviscerated poultry	Walker and Ayres (1959)
C. rugosa	Eviscerated poultry	Walker and Ayres (1959)
C. scottii	Eviscerated poultry	Walker and Ayres (1959)
C. zeylanoides	Frankfurters	Drake *et al.* (1959)
	Fresh fish	Ross and Morris (1965)
Candida spp.	Frankfurters	Drake *et al.* (1959)
	Fresh fish	Ross and Morris (1965)
	Refrigerated beef	Ayres (1960); Haines (1933)
Cryptococcus diffluens (C. albidus var. diffluens)	Fresh fish	Ross and Morris (1965)
C. laurentii	Fresh shrimp	Koburger *et al.* (1975)
Debaryomyces hansenii	Slimy wiener sausage	Mrak and Bonar (1938)
(D. guilliermondii	Slimy sausage	Cesari and Guilliermond (1920)
var. *nova-*	Meat brines	Drake *et al.* (1959)
zeelandicus)	Frankfurters	Drake *et al.* (1959)
D. kloeckeri	Meat brines	Costilow *et al.* (1954)
	Fresh fish	Ross and Morris (1965)
	Frankfurters	Drake *et al.* (1959)
	Slimy sausage	Cesari and Guilliermond (1920)
D. nicotianae	Frankfurters	Drake *et al.* (1959)
D. subglobosus	Frankfurters	Drake *et al.* (1959)
	Fresh nish	Ross and Morris (1965)
Debaryomyces spp.	Lunch meat	Wickerham (1957)
Hansenula californica	Fresh shrimp	Phaff *et al.* (1952)
Pichia membranaefaciens	Fresh fish	Ross and Morris (1965)
	Fresh shrimp	Koburger *et al.* (1975)
Rhodotorula aurantiaca	Eviscerated poultry	Walker and Ayres (1959)
R. glutinis	Fresh shrimp	Phaff *et al.* (1952); Koburger *et al.* (1975)
	Fresh fish	Ross and Morris (1965)
R. marina	Fresh shrimp	Phaff *et al.* (1952) Koburger *et al.* (1975)

TABLE 5.3 (Continued)

Yeast	Product	Reference
R. minuta	Poultry	Njoku-Obi et al. (1957)
	Fresh fish	Ross and Morris (1965)
	Fresh and stored shrimp	Koburger et al. (1975)
R. mucilaginosa	Eviscerated poultry	Walker and Ayres (1959)
(R. rubra)	Fresh fish	Ross and Morris (1965)
	Fresh shrimp	Phaff et al. (1952)
R. peneaus	Fresh shrimp	Phaff et al. (1952)
(Cryptococcus		
laurentii var. flavescens)		
R. rubra	Fresh fish	Ross and Morris (1965)
	Fresh and stored shrimp	Koburger et al. (1975)
R. texensis	Fresh shrimp	Phaff et al. (1952)
(R. minuta var.		
texensis)		
Rhodotorula spp.	Refrigerated beef	Haines (1933); Ayres (1960)
	Haddock	See Ayres (1951)
	Chicken	Ayres et al. (1950)
	Dried beef	Frazier (1967)
	Iced shrimp	Koburger et al. (1975)
Saccharomyces	Chlortetracycline-treated	
cerevisiae	poultry	Njoku-Obi et al. (1957)
S. dairensis	Chlortetracycline-treated	
	poultry	Njoku-Obi et al. (1957)
Saccharomycetaceae	Chlortetracycline-treated	
	poultry	Ziegler and Stadelman (1955)
Sporobolomyces albo-		
rubescens	Fresh shrimp	Koburger et al. (1975)
Torula spp.	Frozen oysters	Fieger and Novak (1961)
Torulopsis albida		
(Cryptococcus	Poultry	Njoku-Obi et al. (1957)
albidus var. albidus)	Fresh shrimp	Phaff et al. (1952)
T. aeria		
(Cryptococcus		
albidus var. aerius)	Fresh shrimp	Phaff et al. (1952)
T. candida	Frankfurters	Drake et al. (1959)
	Fresh shrimp	Koburger et al. (1975)
	Fresh fish	Ross and Morris (1965)
T. famata	Eviscerated poultry	Walker and Ayres (1959)
(T. candida)	Fresh fish	Ross and Morris (1965)
T. glabrata	Fresh shrimp	Phaff et al. (1952)
	Eviscerated poultry	Walker and Ayres (1959)
T. gropengiesseri	Frankfurters	Drake et al. (1959)
T. holmii	Chlortetracycline-treated	
	poultry	Njoku-Obi et al. (1957)
T. inconspicua	Fresh fish	Ross and Morris (1965)
T. pseudaria	Fresh fish	Ross and Morris (1965)
(Cryptococcus		
albidus var. albidus)		

TABLE 5.3 (Continued)

Yeast	Product	Reference
Torulopsis spp.	Chicken	Ayres et al. (1950)
	Haddock	See Ayres (1951)
	Frankfurters	See Ayres (1951)
	Refrigerated beef	Ayres (1960)
Trichosporon cutaneum	Fresh shrimp	Phaff et al. (1952)
		Koburger et al. (1975)
T. diddensii	Fresh shrimp	Phaff et al. (1952)
(Candida diddensii)		
T. lodderi	Shrimp	Phaff et al. (1952)
(Candida tropicalis)		
T. pullulans	Eviscerated poultry	Walker and Ayres (1959)
	Frankfurters	Drake et al. (1959)

period of ground beef; Cavett (1962) who found large numbers on vacuum-packed bacon; Tonge et al. (1964) who did not find yeasts on vacuum-packed bacon but did find them reaching a maximum of around 8×10^5 per g on high-salt bacon after storage for 15 days; Miller (1967) who found that yeast counts usually outnumbered bacterial counts on sliced and cured bacon; and Dowdell and Board (1968) who found yeasts to range from 10 to 240,000 per g in sausage.

The absence of more reports on the type of yeasts in meats seems clearly to be a case of the lack of interest on the part of most investigators, coupled with the fact that bacteria tend to be more prominent when the two groups appear together. This may be due in part also to the general lack of familiarity of many microbiologists with the techniques of identifying yeasts. It may be due also to a lack of success in isolating these organisms by use of the traditional means of employing acidified media with a pH of around 3.5. Koburger (1970) found that antibiotics in potato dextrose agar allowed for a better recovery of yeasts and molds than did acidified media alone. He also found slightly higher counts when plates were incubated at 22°C as compared to 32°C. These organisms are apparently able to proliferate on processed meats because of their capacity to grow at lower a_w values than Gram-negative bacteria. Bacteria generally out-compete fungi and obscure their growth on the surfaces of fresh meats at refrigerator temperatures.

Poultry

In Table 5.3 five genera of yeasts are represented among those isolated from poultry, Candida spp. being the most commonly isolated group. The occurrence of yeasts on poultry and other meats has been reviewed by Ayres (1951).

Fish and Shellfish

Phaff *et al.* (1952), Ross and Morris (1965), and Koburger *et al.* (1975) made rather extensive studies of yeasts in fish and seafoods, and all of the genera noted in Table 5.3 except *Saccharomyces* spp. have been recorded for this group of foods. The apparent ease of finding yeasts on seafoods is not surprising in view of the relative abundance of these organisms in the oceans (Van Uden and Fell 1968). In their review of marine yeasts, these authors noted the isolation of yeasts from a large number of marine animals, including little-neck clam livers, shrimp eggs, and fishes. While some marine yeasts are thought to be pathogenic to their hosts, it appears that most are saprophytes. It may be assumed that they play roles in the decomposition of dead animals.

Molluscan shellfish undergo a fermentative type of spoilage due to their higher content of glycogen. The presence of yeasts in this instance is understandable. Crustacean shellfish, on the other hand, tend to undergo microbial spoilage much as do bony fish and what role, if any, yeasts play in this instance is not clear. Most research on the microbial spoilage of seafoods has been focused toward bacteria, being of greatest importance, but the yeasts by virtue of their numbers may be more important than is currently believed.

Courtesy of Dr. C. W. Hesseltine, USDA
FIG. 5.1. KATSUOBUSHI (DRIED FERMENTED FISH)

FERMENTED FISH PRODUCTS

The growth of molds and yeasts on fish is not always undesirable. In several Oriental countries the fermentative capacities of bacteria (usually lactic acid of halophilic) and fungi are exploited in the preparation of sauces, pastes, and dried condiments from fish. A fermented fish product of Japan known as katsuobushi becomes moldy during its production. The ripening of katsuobushi can be hastened by artificially inoculating the fish with *Aspergillus glaucus* (Graikoski 1973). The product is made from skipjack tuna or bonito which is repeatedly dried and allowed to mold over a several-week period. The finished product is dark in color, dry, and very hard, resembling petrified wood (Fig. 5.1).

Hesseltine (1965) briefly described bagoong, a fermented paste made from salted saltwater fish in the Philippines. Fermented fish sauces are known as nuoc-mam in Cambodia and Vietnam, ngapi in Burma, patis in the Philippines, nam-pla in Thailand and Laos, and ketjap-ikan in Indonesia. Bacteria are the predominant micro-organisms involved in these fermentations, but osmophilic yeasts may also play a role.

REFERENCES

AYRES, J. C. 1951. Some bacteriological aspects of spoilage of self-service meats. Iowa St. Coll. J. Sci. *26*, 31-48.

AYRES, J. C. 1960. Temperature relationships and some other characteristics of the microbial flora developing on refrigerated beef. Food Res. *25*, 1-18.

AYRES, J. C. 1973. Aflatoxins as contaminants of feeds, fish, and foods. *In* Microbial Safety of Fishery Products. C. O. Chichester and H. D. Graham (Editors). Academic Press, New York.

AYRES, J. C., LILLARD, D. A., and LEISTNER, L. E. 1967. Mold-ripened meat products. Proc. 20th Annu. Reciprocal Meats Conf., Am. Meat Sci. Assoc., Natl. Livestock and Meat Board, Chicago, Ill.

AYRES, J. C., OGILVY, W. S., and STEWART, G. F. 1950. Postmortem changes in stored meat. I. Microorganisms associated with development of slime on eviscerated cut-up poultry. Food Technol. *4*, 199-205.

BOYD, J. W., and TARR, H. L. A. 1955. Inhibition of mold and yeast development in fish products. Food Technol. *9*, 411-412.

BREWER, C. M. 1925. The bacteriological content of market meats. J. Bacteriol. *10*, 543-560.

BROOKS, F. T., and HANSFORD, C. G. 1923. Mould growth upon cold-store meat. Trans. Br. Mycol. Soc. *8*, 113-141.

BROOKS, F. T., and KIDD, M. 1921. Black spot of chilled and frozen meat. Spec. Rep. No. 6, F. Inv. Board. London.

BRUCE, J., and MORRIS, E. O. 1973. Psychrophilic yeasts from marine fish. Antonie van Leeuwenhoek J. Microbiol. Serol. *39*, 331-335.

BULLERMAN, L. B., and AYRES, J. C. 1968. Aflatoxin-producing potential of fungi isolated from cured and aged meats. Appl. Microbiol. *16*, 1945-1946.

BULLERMAN, L. B., HARTMAN, P. A., and AYRES, J. C. 1969A. Aflatoxin production in meats. I. Stored meats, Appl. Microbiol. *18*, 714-717.

BULLERMAN, L. B., HARTMAN, P. A., and AYRES, J. C. 1969B. Aflatoxin production in meats. II. Aged dry salamis and aged country cured hams. Appl. Microbiol. *18*, 718-722.

CARY, W. E. 1916. The bacterial examination of sausages and its sanitary significance. Am. J. Public Health *6*, 124-135.

CAVETT, J. J. 1962. The microbiology of vacuum packed sliced bacon. J. Appl. Bacteriol. 25, 282-289.

CESARI, E. P., and GUILLIERMOND, A. 1920. The yeasts of sausages. Ann. Rev. Inst. Pasteur 34, 229-233. (French)

CIEGLER, A., MINTZLAFF, H. J., WEISLEDER, D., and LEISTNER, L. 1972. Potential production and detoxification of penicillic acid in mold-fermented sausage (salami). Appl. Microbiol. 24, 114-119.

COSTILOW, R., ETCHELLS, J. L., and BLUMER, T. N. 1954. Yeasts from commercial meat brines. Appl. Microbiol. 2, 300-305.

DOWDELL, M. J., and BOARD, R. G. 1968. A microbiological survey of British fresh sausage. J. Appl. Bacteriol. 31, 378-396.

DOWDELL, M. J. and BOARD, R. G. 1971. The microbial associations in British fresh sausages. J. Appl. Bacteriol. 34, 317-337.

DRAKE, S. D., EVANS, J. B., and NIVEN, C. F., JR. 1959. The identity of yeasts in the surface flora of packaged frankfurters. Food Res. 24, 243-246.

ESCHER, F. E., KOEHLER, P. E., and AYRES, J. C. 1973. Production of ochratoxins A and B on country cured ham. Appl. Microbiol. 26, 27-30.

FIEGER, E. A., and NOVAK, A. F. 1961. Microbiology of shellfish deterioration. In Fish as Food, Vol. 1. G. Borgstrom (Editor). Academic Press, New York.

FOX, W. E. 1973. Sanitary considerations in the production of fish protein concentrate. In Microbial Safety of Fishery Products. C. O. Chichester and H. D. Graham (Editors). Academic Press, New York.

FRANK, M., and HESS, E. 1941A. Studies on salt fish. V. Studies on Sporendonema epizoum from "dun" salt fish. J. Fish Res. Board Can. 5, 276-286.

FRANK, M. and HESS, E. 1941B. Studies on salt fish. VI. Halophilic brown molds of the genus Sporendonema emend (Ciferri et Redacili) J. Fish. Res. Board Can. 5, 287-292.

FRAZIER, W. C. 1967. Food Microbiology, McGraw-Hill Book Co., New York.

GRAIKOSKI, J. T. 1973. Microbiology of cured and fermented fish. In Microbial Safety of Fishery Products. C. O. Chichester and H. D. Graham (Editors). Academic Press, New York.

HADLOK, R. 1969. Mould contamination of meat products caused by spices. Proc. 15th Eur. Meat Res. Conf., Helsinki.

HAINES, R. B. 1930. The influence of temperature on the rate of growth of Sporotrichum carnis, from -10°C to +30°C. J. Exp. Biol. 8, 379-388.

HAINES, R. B. 1933. Observations on the bacterial flora of some slaughterhouses. J. Hyg. 33, 165-170.

HALLS, N. A., and AYRES, J. C. 1973. Potential production of sterigmatocystin on country cured ham. Appl. Microbiol. 26, 636-637.

HESSELTINE, C. W. 1965. A millennium of fungi, food, and fermentation. Mycologia 57, 149-197.

INCZE, K., and FRANK, H. K. 1976. Is there danger of mycotoxins in Hungarian salami? I. The influence of substrate, a_w, value, and temperature on toxin production in mixed cultures. Fleischwirtschaft 56, 219-225 (German)

JENSEN, L. B. 1954. Microbiology of Meats. Garrard Press, Champaign, Ill.

JOHNSON, T. W., JR. 1968. Saprobic marine fungi. In The Fungi, Vol. III. G. C. Ainsworth and A. S. Sussman (Editors). Academic Press, New York.

KOBURGER, J. A. 1970. Fungi in foods. I. Effect of inhibitor and incubation temperature on enumeration. J. Milk Food Technol. 33, 433-434.

KOBURGER, J. A., NORDEN, A. R., and KEMPLER, G. M. 1975. The microbial flora of rock shrimp — Sicyonia brevirostris. J. Milk Food Technol. 38, 747-749.

LEISTNER, L., and AYRES, J. C. 1968. Molds and meats. Fleischwirtschaft 48, 62-65.

LEISTNER, L., and RODEL, W. 1975. The significance of water acivity for microorganisms in meats. In Water Relations of Foods R. B. Duckworth (Editor). Academic Press, London.

LOVE, T. D. 1968. Relation of temperature, time, and moisture to the production of aflatoxin in fish meal. U. S. Fish Wildl. Serv., Fish. Ind. Res. 4, 139-142.

MALEVICH, O. A. 1936. A new species of halophilic mold isolated from salted fish (Oospora nikitinskii n. sp.). Mikrobiology 5, 713-718.

MILLER, W. A. 1967. Bacteria and yeast counts of prepackaged, sliced, cured bacon, and sliced, fresh side pork. J. Milk Food Technol. 30, 36-38.

MORRIS, E. O. 1975. Yeasts from the marine environment. J. Appl. Bacteriol. 38, 211-223.

MRAK, E. M., and BONAR, L. 1938. A note on yeast obtained from slimy sausage. Food Red. 3, 615-618.

NJOKU-OBI, A. N., SPENCER, J. V., SAUTER, E. A., and ECKLUND, M. W. 1957. A study of the fungal flora of spoiled chlortetracycline treated chicken meat. Appl. Microbiol. 5, 319-321.

PHAFF, H. J., MRAK, E. M., and WILLIAMS, O. B. 1952. Yeasts isolated from shrimp. Mycologia 44, 431-451.

ROGERS, R. E., and McCLESKEY, C. S. 1957. Bacteriological quality of ground beef in retail markets. Food Technol. 11, 318-320.

ROSS, S. S., and MORRIS, E. O. 1965. An investigation of the yeast flora of marine fish from Scottish coastal waters and a fishing ground off Iceland. J. Appl. Bacteriol. 28, 224-234.

SHEWAN, J. M. 1961. The microbiology of sea-water fish. In Fish as Food, Vol. 1. G. Borgstrom (Editor). Academic Press, New York.

STRINGER, W. C., BILSKIE, M. E., and NAUMANN, H. D. 1969. Microbial profiles of fresh beef, Food Technol. 23, 97-102.

STRZELECKI, E., LILLARD, H. S., and AYRES, J. C. 1969. Country cured ham as a possible source of aflatoxin. Appl. Microbiol. 18, 938-939.

SUSMAN, A. S. 1968. Longevity and survivability of fungi. In The Fungi, Vol. III. G. C. Ainsworth and A. S. Sussman (Editors). Academic Press, New York.

SUTIC, M., AYRES, J. C., and KOEHLER, P. E. 1972. Identification and aflatoxin production of molds isolated from country cured hams. Appl. Microbiol. 23, 656-658.

TANNER, F. W. 1944. Microbiology of Foods, Garrard Press, Champaign, Ill.

TONGE, R. J., BAIRD-PARKER, A. C., and CAVETT, J. J. 1964. Chemical and microbiological changes during storage of vacuum packed sliced bacon. J. Appl. Bacteriol. 27, 252-264.

VAN KLAVERN, F. W., and LEQUENDRE, R. 1965. Salted cod. In Fish as Food, Vol. 3. G. Borgstrom (Editor). Academic Press, New York.

VAN UDEN, N., and FELL, J. W. 1968. Marine yeasts. In Advances in Microbiology of the Sea, Vol. 1. M. R. Droop and E. J. F. Wood (Editors). Academic Press, New York.

WALKER, H. W., and AYRES, J. C. 1959. Characteristics of yeasts isolated from processed poultry and the influence of tetracyclines on their growth. Appl. Microbiol. 7, 251-255.

WICKERHAM, L. J. 1957. Presence of nitrite assimilating species of Debaryomyces in lunch meats. J. Bacteriol. 74, 832-833.

WOOD, E. J. F. 1965. Marine Microbial Ecology. Chapman and Hall, London.

WRIGHT, A. M. 1923. Moulds on frozen meats. N. Z. J. Sci. Technol. Sect. A, 6, 208-211.

WU, M. T., AYRES, J. C., and KOEHLER, P. E. 1974A. Production of citrinin by Penicillium viridicatum on country cured ham. Appl. Microbiol 27, 427-428.

WU, M. T., AYRES, J. C., and KOEHLER, P. E. 1974B. Toxigenic aspergilli and penicillia isolated from aged, cured meats. Appl. Microbiol. 28, 1094-1096.

YESAIR, J. 1928. The action of disinfectants on molds. Ph.D. Dissertation. Univ. of Chicago.

ZIEGLER, F., and STADELMAN, W. J. 1955. The effect of aureomycin treatment on the shelf life of fresh poultry meat. Food Technol. 9, 107-108.

6

Dairy Products

Elmer H. Marth

Several species of molds are used to manufacture certain varieties of cheese, whereas some of these same molds together with numerous others can spoil ripened cheeses, fresh cheeses, yogurt, butter, and some other dairy products. Additionally, some molds can produce toxic metabolites which can get into milk (and hence dairy products) via the cow that has consumed moldy feed, or which can get into dairy products directly when molds grow on these foods.

Yeasts are important for their role in ripening certain cheeses and in the production of some fermented milks. In addition, yeasts can be used to ferment whey, a major by-product from cheesemaking. Growth of yeasts also can be responsible for spoilage of dairy foods such as cheese and cottage cheese.

This chapter will first consider the role of molds in the manufacture of some varieties of cheese. This will be followed by a discussion on how yeasts contribute to production of surface-ripened cheeses and how yeasts can be used to ferment whey. Next, spoilage of dairy products by yeasts and molds will be considered. Finally, some comments will be made about the possible occurrence of mycotoxins in dairy products.

ROLE OF MOLDS IN THE MANUFACTURE OF CHEESE
Blue Cheese
Blue is an example of cheese that is ripened primarily by growth

and activity of mold throughout the cheese mass rather than on the surface only, as is true of Camembert and related varieties. Blue cheese was not made successfully in the United States until about 1918; information on appropriate procedures for making this cheese was not available earlier (Walter and Hargrove 1969). A blue cheese is about 19 cm in diameter and weighs about 2 to 2.3 kg; it is round with a flat top and bottom.

Manufacturing Process.—The following steps (Kosikowski 1966) are involved in producing blue cheese.

(1) Whole milk from cows is separated into cream and skim milk fractions, and the skim milk is pasteurized.

(2) Cream is bleached by adding benzoyl peroxide (maximum 0.002% of the weight of milk), pasteurized and homogenized. Bleaching is done so that the finished cheese is white (except for mold growth) in color and homogenization increases the surface area of milkfat globules and thus facilitates lipolytic action that occurs during ripening of the cheese.

(3) Cream and skim milk are combined and 0.5% of an active lactic (Streptococcus lactis and/or Streptococcus cremoris) starter culture is added.

(4) Inoculated milk is held at 30°C for 1 hr to allow some acid production.

(5) A suitable coagulant is added, milk is allowed to coagulate, and the resultant curd is cut into cubes with 1.6-cm wire knives.

(6) Curds are allowed to remain in whey for about 1 hr while additional acid develops. Curds and whey are then heated to 33°C, held briefly, and whey is drained from the curds.

(7) Curd is trenched and inoculated with Penicillium roqueforti spores; salt is also added and then the curd is stirred. Spores can be obtained in powdered form from commercial culture firms.

(8) Curd is placed into stainless steel blue cheese hoops.

(9) Hoops of curd are turned once every 15 min for 2 hr and then are allowed to drain overnight at 22°C.

(10) The next day curd is removed from the hoop and salt is applied to the surfaces of the cheese. Cheese is then stored at 16°C and 85% relative humidity and is salted once daily for 4 more days.

(11) After salting is completed, each flat surface of the cheese is pierced about 50 times with a suitable needle-like steel rod. This facilitates escape of carbon dioxide from the cheese and entrance of air so that growth of the mold is encouraged.

(12) Pierced cheese is then stored at 10° to 13°C and 95% relative humidity (Fig. 6.1).

(13) After 1 month, surfaces of cheese are cleaned; cheeses are

Courtesy of Treasure Cave Blue Cheese

FIG. 6.1. CURING DOMESTIC BLUE CHEESE

wrapped in foil and then stored at about 2°C for 3 to 4 months to allow additional ripening.

Changes During Ripening.—The high acidity and the increasing amount of salt in the cheese cause a rapid demise of the lactic starter bacteria so that only a few viable cells remain in cheese that is 2 to 3 weeks old. Growth of *P. roqueforti* inside cheese becomes evident about 8 to 10 days after the cheese is pierced (Foster *et al.* 1957). Development of mold is maximal in 30 to 90 days; by then the mold has grown throughout the spaces between curd particles and along holes made when the cheese was pierced.

Penicillium roqueforti is primarily responsible for ripening blue cheese. Proteolytic enzymes from the mold act to soften the curd and thus to produce the desired body in the cheese (Marth 1974). Some components of blue cheese flavor may result from this proteolytic action. Perhaps more important, the mold produces water-soluble lipases which hydrolyze milkfat to free fatty acids. Included are caproic, caprylic, and capric acids which together with their salts are responsible for the sharp peppery flavor of blue cheese (Ernstrom and Wong 1974). *Penicillium roqueforti* forms heptanone-2 from caprylic acid and this ketone is an important component of blue cheese flavor. Other ketones (pentanone-2 and nonanone-2) have been recovered from blue cheese and probably contribute to its flavor (Ernstrom and Wong 1974). Additionally, *P. roqueforti* reduces methyl ketones to form secondary alcohols (pentanol-2, heptanol-2, and nonanol-2) which also contribute to the flavor of the cheese (Jackson and

Hussong 1958). The pH of blue cheese initially is at 4.5 to 4.7 and increases to 6.0 to 6.25 after 2 to 3 months of ripening.

Growth of microorganisms can occur on blue cheese after 2 to 3 weeks of ripening. This growth consists of yeasts, micrococci, and *Brevibacterium* spp. and contributes to the final flavor of the cheese. If cheese is waxed, microorganisms will not grow to produce the surface slime.

Defects.—Improper development of *P. roqueforti* can cause an assortment of defects. Too much growth can result in a musty, unclean flavor or in loss of the typical flavor, whereas too little growth is accompanied by defects in color, a body that is too firm, and insufficient flavor. Growth of unwanted molds can cause defects on the surface of blue cheese.

Cheese Similar to Blue.—Roquefort is the original blue cheese. The designation "Roquefort" is applicable only to cheese made from ewe's milk in the Roquefort area of France. A similar product made elsewhere in France is called bleu cheese. A peculiarity of Roquefort cheese is the fact that it ripens in a network of caves and grottoes where cool moist air moves briskly, temperature never exceeds 10°C, and relative humidity remains at about 95% throughout the year (Walter and Hargrove 1969).

Gorgonzola is the principal blue-mold cheese of Italy where it is claimed to have been made in the Po Valley since 879 A.D. The English blue-mold cheese is Stilton which has been made since about 1750. Stilton is milder than Roquefort or Gorgonzola. The texture of Stilton is sufficiently open so that the cheese usually does not need piercing to facilitate mold growth.

Camembert Cheese

Camembert is an example of cheese that is made with a mold developing only on the surface rather than throughout the mass of cheese as happens with blue cheese. Apparently cheese ripened through the action of mold growth on the surface has been produced in France for centuries. According to Kosikowski (1966) it was in 1791 that Marie Harel, who lived in the village of Camembert in Normandy, developed a product similar to the present Camembert cheese.

The typical Camembert cheese is about 11 cm in diameter, 2.5 to 3.8 cm thick, and weighs 225 to 250 g. The interior is light yellow and waxy, and creamy or almost fluid in consistency, depending on the degree of ripening (Walter and Hargrove 1969). The rind is a thin felt-like layer of mold mycelium and dried cheese. The mold is gray-white in color; sometimes bacterial growth occurs on the surface of the cheese and results in development of areas that are reddish-yellow in color.

Manufacturing Process.—The basic procedure for the manufacture of Camembert cheese employs the following steps.

(1) Pasteurized whole milk with about 3.5% milkfat is adjusted to about 32°C and is inoculated with 2% of an active lactic starter culture (*S. lactis* and/or *S. cremoris*) plus a sporulated culture of *Penicillium camemberti* (alternatively, spores of the mold can be applied to the surface of the cheese later in the manufacturing process).

(2) Annatto (yellow coloring) may be added to the milk.

(3) Inoculated milk is allowed to "ripen" for 15 to 30 min so that a titratable acid of 0.22% develops. Rennet extract (or other suitable coagulant) is added and the milk is stirred and then held quiescently until a firm curd develops.

(4) Curd is cut into cubes with 1.6-cm knives. Alternatively, uncut curd can be ladled into hoops.

(5) Curd is not cooked but is placed into open-ended, round, perforated, stainless steel molds or hoops. Filled hoops are allowed to drain for about 3 hr at about 22°C; no pressure is applied to the cheese during draining.

(6) Hoops of cheese are turned and draining continues. The turning process is repeated 3 to 4 times at 30-min intervals.

(7) Both flat sides of curd in hoops may now be inoculated by spraying the surface with a fine mist of *P. camemberti* spores suspended in water.

(8) After an hour, cheese is removed from hoops, placed on a drain table, and held at 22°C for 5 to 6 hr. Weights are generally not placed on the cheese.

(9) Dry salt is applied to the surface of cheese which is then held overnight at about 22°C.

(10) Cheese is held for 1 or 2 weeks at 10° to 15°C and 95 to 98% relative humidity. It may be turned once during storage to facilitate uniform development of mold on the surface.

(11) Cheese is moved to storage at 4° to 10°C after being wrapped in foil. Storage under these conditions may be for several weeks before the cheese is packaged and moved into distribution channels. Final ripening occurs during distribution.

Camembert cheese should be consumed within 6 to 7 weeks after it is made. The process used to make Camembert cheese has been mechanized in the United States and Europe.

Changes During Ripening.—Foster *et al.* (1957) have outlined the major changes that occur when Camembert cheese ripens. Normally film yeasts and *Geotrichum* appear on the surface of the cheese within 3 to 4 days after it is placed in the "warm" room. Growth of *P. camemberti* is evident a few days later and becomes maximal when the cheese is 10 to 12 days old. Development of mold is followed by the appearance of a reddish growth comprised of *Brevibacterium linens* and related pigmented, rod-shaped bacteria.

Film yeasts and *Geotrichum* are believed to ferment residual lactose at the surface of the cheese and also are thought to reduce acidity, thereby facilitating later growth of other organisms. Yeasts and *Geotrichum* do contribute to the flavor of Camembert cheese; however, excessive growth of these fungi leads to excessive softening of the rind and undesirably strong flavors in the cheese.

Development of *P. camemberti* is essential for production of the normal body and flavor of Camembert cheese. Pigmented bacteria (*Brevibacterium*) develop after fungi have reduced the acidity of the cheese at its surface. These bacteria also contribute to the flavor of the ripened cheese.

Defects.—Common defects associated with Camembert cheese include (a) early gas production during drainage of the cheese, and (b) growth of undesirable "wild" molds on the surface. The first defect can be minimized or eliminated by adequate sanitation and use of high quality milk. The second can be controlled by maintaining adequate humidity in the room where cheese is ripening; wild molds tend to develop on cheese when its surface becomes too dry for normal development of *P. camemberti*.

Late in 1971 at least 227 persons in eight states of the United States became ill with acute gastroenteritis about 24 hr after consuming French Camembert or Brie cheese (Barnard and Callahan 1971; Schnurrenberger *et al.* 1971). The illness was attributed to the presence in cheese of enteropathogenic *Escherichia coli*. Park *et al.* (1973) investigated the fate of enteropathogenic *E. coli* during the manufacture and ripening of Camembert cheese. They added *E. coli* to cheese milk and observed that growth of the bacterium was minimal until after the curd was cut and hooped, when a population in excess of 10^4 per g developed. Overnight storage of cheese in hoops was accompanied by a decline in the number of viable *E. coli* and this trend continued during the ripening period. When production of acid was inadequate during cheesemaking, approximately 10^9 *E. coli* developed per gram of 24-hr-old cheese and 10^7 per g persisted after 9 weeks of ripening.

Cheeses Similar to Camembert.— Walter and Hargrove (1969) and Weigmann (1933) describe several kinds of cheese other than Camembert that are ripened largely through the activity of surface mold. Brie is probably the best known of these Camembert-like cheeses.

Brie is made in three sizes: (a) large—about 40 cm in diameter and 3.8 to 4.2 cm thick, about 2.7 kg; (b) medium—about 30 cm in diameter and somewhat thinner than the large size, about 1.6 kg; and (c) small —14 to 20 cm in diameter and 3.2 cm thick, 0.45 kg. The difference in size between Brie and Camembert causes differences in the ripening process of the two cheeses. This, together with variations in the manufacturing process, causes the flavor and aroma of Brie to differ from Camembert.

Coulommiers is a cheese similar to the small Brie but is ripened for less time. Another cheese similar to Brie is Monthery, which is made in two sizes roughly equivalent to the large-sized and medium-sized Brie. Monthery can be made from whole or partially skimmed milk.

Melum is similar to the small Brie but has a firmer body and sharper flavor. This cheese also is designated as Brie de Melum. Other cheeses similar to Camembert and primarily produced in France include Olivet and Vendome. The latter is sometimes buried in ashes in a cool, moist cellar during ripening.

ROLE OF YEASTS IN THE MANUFACTURE OF CHEESE AND FERMENTED MILKS

Surface-ripened Cheeses

Elsewhere in this chapter it has been indicated that yeasts develop on the surface of some mold-ripened cheeses. Additionally, yeasts are an important component of the microflora that develop on the surface of some cheeses that ripen without the aid of molds. Principal examples of such cheeses include brick and Limburger. Brick cheese will be discussed first and then other kinds of surface-ripened cheese will be mentioned.

Manufacturing Process.—Olson (1969) has described two methods that are generally used to make brick cheese. The following steps are involved in the first method.

(1) Pasteurized whole milk at 32°C is inoculated with *Streptococcus thermophilus*. Alternatively, a combination of *S. thermophilus* and *S. cremoris* or *Lactobacillus bulgaricus* may be used.

(2) After brief incubation, rennet or another suitable coagulant is added to the milk; the resultant curd is cut into 0.64-cm or 0.95-cm cubes and cooked at 38° to 45°C.

(3) After the curd is sufficiently firm, enough whey is drained so that about 2.5 cm remains above the curd surface.

(4) Curd and whey are dipped or pumped into rectangular hoops held on perforated screens.

(5) Hoops of curd are allowed to drain for 6 to 18 hr and are turned at intervals. Weights can be placed on the cheese during the draining process.

(6) Blocks of cheese are removed from hoops and immersed into brine containing 22% sodium chloride. Alternatively, salt can be applied to the exterior of cheese.

(7) After 24 to 36 hr in brine, cheese is removed and placed in a room at 15°C for 4 to 10 days. During this time the "smear" develops on the surface of the cheese.

(8) The smear is washed off, the cheese is waxed or packaged in plastic, and is ripened for 4 to 8 weeks at about 4°C. More flavor can be obtained by leaving the smear intact for the entire ripening period.

The second or "sweet curd" method for making brick cheese employs *S. lactis* and/or *S. cremoris* as the starter culture and addition of water to the curd-whey slurry to control development of acid in the cheese. Some other modifications in manufacturing must be made to ensure that the minimum pH is 5.1 to 5.2 in 3-day-old cheese. Salting and ripening proceed as outlined above.

Brick cheese contains not more than 44% moisture and at least 50% of the total dry matter must be milkfat. A typical brick cheese is about 12 cm wide, 25 cm long and 7.5 cm thick; it weighs about 2.25 kg.

Changes During Ripening.—Yeasts predominate in the surface microflora of brick cheese during the initial stages of ripening (Olson 1969). This is because of their ability to grow at the temperature and relative humidity (approximately 95%) used for ripening as well as the low pH and high concentration of salt at the surface of the cheese. Depending on the water activity of the cheese, yeasts in one or more of the following genera may be present: *Debaryomyces, Rhodotorula, Trichosporon, Candida* and *Torulopsis.*

Growth of yeasts serves to modify the surface of cheese so that *B. linens* and micrococci can grow. This is accomplished by metabolizing lactic acid and thus raising the pH of cheese at its surface above the minimum for growth of the bacteria. Additionally, yeasts produce vitamins which may enhance growth of the bacteria. Growth of yeasts also may contribute to the final flavor of brick cheese.

Brevibacterium linens and micrococci (*Micrococcus varians, M. caseolyticus* and *M. freudenreichii*) develop after sufficient growth of yeasts has taken place. These bacteria release proteolytic enzymes that are largely responsible for producing the characteristic flavor of brick cheese that has had a surface smear develop during ripening.

Defects.—Olson (1969) described the major defects that can appear in brick and other surface-ripened cheeses. Included are defects in flavor, body and texture, and surface microflora.

Flavor defects include:
(1) Sour or acid—caused by excessive fermentation of lactose or

inadequate washing of curd, or both. Too much acid retards development of surface microflora and too little acid results in fruity and gassy cheese.

(2) Bitterness—caused by abnormal protein degradation when the starter culture contains undesirable lactic acid bacteria such as *Streptococcus faecalis* var. *liquifaciens.*

(3) Flat—caused by insufficient growth of surface microflora.

(4) Fruity and fermented—caused when the pH of cheese is high and the salt content is low so that anaerobic spore-forming bacteria can grow.

Defects in body and texture include:

(1) Corky—caused by inadequate acid development or excessive washing of curd, or both.

(2) Weak or pasty—caused by a combination of excessive moisture, too much or too little acid, and inadequate salt.

(3) Mealy—too much acid.

(4) Openness— caused by whey being trapped between firm cubes of curd; openings remain when whey drains from cheese.

(5) Gassiness—caused by growth in the cheese of coliform bacteria, yeasts, certain strains of lactic acid bacteria, *Bacillus polymyxa,* or anaerobic spore-forming bacteria.

(6) Split cheese—caused by gas from anaerobic sporeformers.

Defects in the surface microflora include:

(1) Lack of growth—caused by low temperatures during ripening, too much salt in cheese, or drying of the surface of cheese.

(2) Mold growth—results when the surface smear fails to develop because the surface of cheese is too dry.

Other Varieties of Surface-Ripened Cheese.—Limburger cheese has up to 50% moisture and is made by the procedures used for brick cheese. The initial ripening at 16°C and at a high relative humidity is longer than for brick cheese, thus allowing extensive growth of B. *linens.* Surface growth is not removed when cheese is wrapped and moved to storage at 4° to 10°C. Extensive growth of B. *linens* on relatively small pieces of cheese accounts for the strong, pungent flavor and aroma of Limburger.

Port du Salut, Trappist, and Oka are wheel-shaped cheeses developed by Trappist monks and made by procedures somewhat similar to those for brick cheese. *Geotrichum* may appear in the surface microflora and contribute a distinctive flavor to the cheese.

Other varieties of surface-ripened cheese that are more common in Europe than in the United States include Saint Paulin, Bel Paese, Konigkase, Bella Alpina, Vittoria, Fleur des Alpes, Butter, and Tilsit.

Liederkranz is a trade name for a surface-ripened cheese made in the United States by procedures similar to those for Limburger.

Fermented Milks

Kefir.— This cultured milk originated in the Caucasus mountains and should not be confused with kaffir, a beer fermented from kaffir corn (see Chap. 10). Kefir can be made from the milk of goats, sheep or cows. Although not common in the United States, it is distributed widely in Europe. The finished kefir contains about 0.8% lactic acid, 1% ethanol and carbon dioxide (Kosikowski 1966).

Kefir is made from whole milk after it has been held at 85°C for 30 min. The heated milk is cooled to 22°C, inoculated with kefir grains, and incubated overnight at 22°C. The resultant smooth curd is strained through a wire sieve to recover the kefir grains. The product is then cooled, after which it is ready for consumption.

Kefir grains are whitish or yellowish irregular granules that contain the mixture of microorganisms needed to culture the milk. Included are yeasts (*Candida*, *Saccharomyces* and *Torulopsis* spp.), streptococci, micrococci and bacilli; all are not necessarily useful or required for making kefir. Kefir grains recovered from a lot of kefir can be washed in clean water and stored moist at 4°C. Alternatively they can be dried at room temperature for 36 to 48 hr. Dried kefir grains retain activity for 12 to 18 months whereas moist grains retain activity for only 8 to 10 days (Kosikowski 1966).

Kumiss (Koumiss).—Kumiss is a fermented milk product found in Russia where it is commonly made from the milk of mares. Because of the composition of mare's milk, finished kumiss has a uniform consistency with no tendency to whey-off (Kosikowski 1966). Russians believe kumiss to be of value in treating tuberculosis and a typical medical dose is 1.5 liters of kumiss daily for 2 months. Finished kumiss contains 0.6 to 1.8% lactic acid and 1.0 to 2.5% ethanol.

Kumiss is made from fresh mare's milk warmed to 28°C. The milk is then inoculated with 30% of a starter consisting of lactic acid bacteria and *Torula kefyr* (now classified as *Kluyveromyces* sp. [Lodder 1970]). The culture is prepared by growing each organism separately in cow's milk, then adding both to mare's milk and after suitable incubation adding more mare's milk so that after 4 days the volume of culture is sufficient to serve as inoculum.

The inoculated mare's milk is agitated to facilitate growth of yeast. Approximately 2 hr of incubation will result in a product with sufficient acidity. Kumiss is bottled warm, capped and incubated for an additional 2 hr at 20°C. It is then stored at 4°C and distributed promptly.

PRODUCTION OF YEAST FROM WHEY

Whey can serve as the substrate for production of food-grade and feed-grade yeast and yeast-whey products. These products, if properly produced, are highly nutritious, nontoxic sources of protein and vitamins that, under the right circumstances, can find application in human and animal nutrition. Worldwide protein deficits make production of yeast from whey particularly attractive since a much needed food supplement could be made from a substrate which is often used inefficiently or wasted. Unfortunately, available whey is seldom located in those areas of the world that suffer from acute protein shortages. In addition to transportation problems, difficulties associated with acceptance and palatability of yeast products have not been resolved. Furthermore, consumption of excessive amounts of purines and pyrimidines from yeast can lead to high concentrations of uric acid in blood and thus to gout. These problems are covered in more detail in Chap. 12.

The Organism

Undoubtedly *Kluyveromyces fragilis* is most widely recognized as the organism of choice for producing yeast from whey, although others have been suggested. Myers and Weisberg (1938) in their early work found *K. fragilis* suitable to make delactosed whey. Other investigators who have found *K. fragilis* suitable for growth in whey include Amundson (1966, 1967), Naiditch and Dikansky (1960), Simek *et al.* (1964), Porges *et al.* (1951), Stimpson and Young (1957), Wasserman (1960A) and Stuiber (1966).

Graham *et al.* (1953A, B) compared *Candida krusei, Candida utilis, C. utilis* var. *thermophilus,* and *Torula cremoris* (now classified as *Candida pseudotropicalis*) for their ability to grow in whey and concluded that *T. cremoris* was suitable for use to increase the value of whey as an animal feed. Successful growth in whey was attributed to *Torulopsis utilis (Candida utilis), Torula casei (Candida pseudotropicalis)* and *T. cremoris* by Tomisek and Gregr (1961). Other yeasts suggested as suitable for growth in whey include *Candida tropicalis* (Davidov *et al.* 1963), *Candida utilis* (Demmler 1950), *Torulopsis sphaerica (Kluyveromyces lactis)* and *Torula lactosa (Candida kefyr)* (Enebo *et al.* 1941) and *Torula* spp. (Naiditch 1965). (The genus *Torula* is no longer recognized [Lodder 1970].)

The Process

Numerous procedures have been suggested for growth of yeast in whey. The most detailed studies on methods for maximum yield of yeast cells (and protein) in whey are those of Wasserman and his associates (Wasserman 1960A,B,C, 1961; Wasserman and Hampson

1960; Wasserman *et al.* 1958, 1959, 1961). The process which was developed through these studies will be described first and other options will be considered later.

Medium.—Wheys resulting from cottage, Cheddar, ricotta, Italian or other cheeses made by similar procedures are satisfactory, although some (especially those resulting from the manufacture of Italian hard cheeses) contain more milkfat which may make drying of the finished product difficult.

Supplementation of whey with added nutrients is necessary to obtain maximal yields. Phosphorus must be added in the amount of 0.225% and, since only 25% of whey nitrogen is available to yeasts, additional nitrogen must be supplied. Ammonium sulfate can be added in the amount of 0.85% to compensate for the nitrogen deficiency. Yeast extract or dried brewers' yeast is another necessary additive.

The optimum pH for maximal yeast production is in the range of 5.0 to 5.7, but it may rise during incubation. If it is allowed to reach 8.5, growth of yeast is impaired and the pH must be reduced by addition of acid or by an interruption in aeration so that the yeast produces its own acid under anaerobic conditions.

Heating of whey alters the medium to make it more suitable for yeast growth. If a short growth cycle is used (as described below), sterile conditions are not necessary since the yeast outgrows contaminating microorganisms.

Temperature.—The optimum temperature for growth of *K. fragilis* in whey ranges from 31° to 33°C, although good yields have been obtained up to 43°C. The wide temperature range in which growth is possible reduces the need for cooling since proliferation of yeast is accompanied by liberation of heat.

Inoculum.—The time required to attain the yield limit depends to a great extent on the size of the inoculum. For example, when 3.5 g of yeast are added per liter, the maximum yield can be expected in 8 hr, but when 23.8 g of yeast are used, the maximum yield is reached in 3 to 4 hr. The former condition involves a nine-fold increase in number of cells, whereas the latter requires only a two-fold increase. Inoculation with yeast equivalent in dry weight to approximately 30% of the weight of lactose in the medium results in a 4-hr incubation for maximum yield.

Aeration.—Sufficient oxygen must be provided to enable yeast to oxidize approximately 35% of the lactose in whey to carbon dioxide and water. The remainder of the carbohydrate is used to produce new cell materials or nonoxidizable metabolic products. From 13 to 15 liters of oxygen are needed per liter of whey, although the calculated requirement for whey with 4% lactose is only 10.5 liters. The excess oxygen is required for endogenous cell respiration and for oxidizing small amounts of protein and lactic acid.

The demand for oxygen rises to a peak midway during the 4-hr incubation period, hence information on the total amount of oxygen required is not enough. If the total amount of oxygen is supplied in uniform increments there will be both excessive and inadequate amounts during the fermentation. Anaerobic conditions occur quickly and growth becomes limited if not enough oxygen is supplied. The peak requirement, occurring after approximately 2 hr of incubation in a 4-hr fermentation, is 100 to 120 ml oxygen per liter per min. The increase in oxygen demand rises virtually in a straight line from approximately 30 ml per liter per min initially to approximately 110 ml per liter per min after 2 hr and then declines sharply, and nearly in a straight line, to less than 20 ml per liter per min at the end of 4 hr. The work of Stuiber (1966) further emphasizes the importance of adequate aeration in this fermentation. He observed a marked decrease in yield of yeast (33.1 versus 46.9 g of cells per 100 g of lactose) under conditions of oxygen starvation. A higher proportion of carbon in the medium was converted to carbon dioxide when oxygen was limited than when the supply was ample (63 versus 47.5%).

Oxygen absorption by the medium is dependent on: (a) quantity of air passing through the fermentor, (b) type of impeller used in agitation (turbine impeller preferable to Rheinhutte or propeller types), and (c) speed of the agitator (slow speed is associated with lower absorption). Use of the Waldhof propagator, as opposed to the conventional stirred fermentor, results in good yields of yeast with lower oxygen input. The Waldhof fermentor employs both an agitator (turbine type) and a draft tube. Demmler (1950) also reported successful use of the Waldhof fermentor for yeast propagation and Stimpson and Trebler (1956) suggested modifying this type of fermentor by allowing air to enter from the bottom to minimize foaming.

Harvesting.—The yeast suspension is concentrated by centrifugation so that a slurry with 15 to 18% solids is produced. Yeast may be washed at this point so that the final product is bland in flavor. The yeast cream can be spray-dried or fed to a drum dryer operating with 85 psi (about 8300 g/cm^2) steam pressure and rotating at a speed of 12 rpm. Dried yeast may be pulverized with a hammer mill before bagging.

Yield.—The yield will be variable, depending on fermentation conditions. Reports suggest that approximately 0.42 kg of yeast (dry) can be obtained per kg of lactose. This crop represents a yield of 75% of that which could theoretically be expected, a considerable improvement over the 24 to 45% yield reported by Porges et al. (1951) or the 50% yield obtained by Müller (1949).

Suggested Modifications in the Process

Numerous investigators other than Wasserman and his associates have studied production of yeast from whey and some have suggested procedures which differ from those outlined above. These procedures will be described in the following paragraphs.

Medium.—Several suggestions have been made regarding the heat treatment given to whey before fermentation. Myers and Weisberg (1938) suggest pasteurization (flash at 74° to 85°C or 30 min at 63°C), whereas Stuiber (1966) and Amundson (1966, 1967) recommend heating to 93°C and holding for 5 min. Other recommendations include heating to 60°C and holding at that temperature, followed by heating to 77°C (Stimpson and Young 1957), and heating to 85° to 90°C (Müller 1949).

Some researchers have claimed improved yields from whey fortified with compounds other than those listed earlier. For example, Siman and Mergl (1962) suggested that nitrogen and phosphorus be supplied in the form of ammonium nitrate, potassium nitrate, sodium nitrate, ammonium chloride, ammonium sulfate, ammonium (mono- and di-) phosphate or disodium phosphate. Another combination of ingredients suggested by Müller (1949) consists of 9 g ammonium sulfate, 1 g diammonium phosphate, 0.4 g potassium sulfate, 0.2 g magnesium sulfate and several drops of a 5% ferric chloride solution (all on a per liter basis). Eichelbaum (1902) recommended the acid or enzyme hydrolysis of whey protein before fermentation. Other additives which have been found useful are corn steep liquor (Amundson 1967), ammonia (Stimpson and Young 1957), and by-products of the manufacture of glutamic acid and sodium glutamate (Société des Alcools du Vexin 1963). The suggested pH for this fermentation ranges from 3.5 obtained with added phosphoric acid (Stuiber 1966; Amundson 1966, 1967) to 8.0 (Hanson et al. 1949), although most reports recommend a pH of 5.0 or below.

Temperature.—Although various temperatures of incubation have been suggested, most of them are close to 30°C. Specifically, the following temperatures have been found suitable for use: 26°C (Anon. 1961), 28° to 31°C (Anon. 1946), 25° to 30°C (Graham et al. 1953B; Myers and Weisberg 1938), and 32° to 35°C (Müller 1949).

Inoculum.—The volume of inoculum suggested by Wasserman and coworkers is quite high but is necessary for rapid fermentation. Lower rates of inoculation and correspondingly slower fermentations have been found satisfactory by some workers. For example, Graham et al. (1953B) reported the use of a 5 to 10% inoculum of T. cremoris and Stimpson and Young (1957) suggested that inoculated whey should contain 800 million to 1 billion yeast cells per ml.

Harvesting.—Several procedures different from those described earlier have been reported and may be useful in certain applications. Myers and Weisberg (1938) suggested that yeast and whey solids be concentrated in a vacuum pan before drying. An alternative procedure also was offered in which the fermented whey was first filtered or centrifuged to recover yeast cells and precipitated protein. This was followed by adjusting the pH of the liquor to 7.0, heating to precipitate soluble albumin, and filtering to recover the protein. The filtrate was then condensed in a vacuum pan, calcium lactate was allowed to crystallize, and it was removed by centrifugation. The resultant liquor contained vitamins and further concentration was suggested. Stuiber (1966) and Amundson (1966, 1967) proposed harvesting of yeast and whey protein by means of a self-cleaning clarifier which produced a sludge with 30% solids. The sludge was then pasteurized and spray-dried.

Composition of Dried Products

The composition of various yeast products derived from whey by different investigators is given in Tables 6.1 and 6.2. An examination of the reported data reveals that the dried products are high in protein, although there is considerable variation ranging from a low of 19% to a high of 54.4%. Manufacturing procedures, including the degree of washing given the product, influence the final protein content as well as the content of other constituents (Table 6.2).

It is also readily apparent that the mineral (ash) content of dried products is high, ranging from 14 to 30.42% (Table 6.1). This condition may limit their use in diets of some animals, although washing (Table 6.2) can reduce the ash content to reasonable levels. A comment is in order concerning the relatively high fat content of the product described by Graham et al. (1953C) and listed in Table 6.1. The fat in the dried product undoubtedly is attributable to either or both of the following: (a) the whey contained a substantial amount of fat, or (b) the yeast (T. cremoris) used was able to produce this material.

The amino acid composition of protein produced by K. fragilis is given in Table 6.3. Examination of these data reveals the presence of a substantial amount of lysine (an amino acid usually deficient in cereals) and a shortage or absence of sulfur-containing amino acids.

TABLE 6.1

COMPOSITION OF VARIOUS DRIED YEAST PRODUCTS DERIVED FROM
WHEY AS REPORTED BY VARIOUS INVESTIGATORS

Component (%)	Myers and Weisberg (1938)	Simek et al. (1964)	Stimpson and Young (1957)[1]	Davidov et al. (1963)	Naiditch and Dikansky (1960)	Graham et al. (1953C)	Mergl and Siman (1963)
Protein	31.70	54.0	≥40.0	46.0	25-36	39	19
Ash	30.42	22.0	—	—	16-20	30	14
Lactic acid	13.63	—	—	—	—	—	—
Moisture	7.28	—	—	—	—	—	5
Lactose	0	0	—	—	17-30	—	—
Reducing substances	0.59	—	—	—	—	19	—
Fat	—[2]	4.0	—	—	30-40	12	60
Nitrogen-free extract	—	19.0	—	—	—	—	—
Riboflavin	—	—	P[3]	—	—	P	—
Thiamine	—	—	P	—	—	P	—

[1]Niacin, pyridoxine, pantothenic acid and folic acid also reported as present.
[2]Not reported.
[3]Reported as present.

TABLE 6.2

COMPOSITION OF DRIED UNWASHED AND WASHED
YEAST-WHEY AND DRIED WHEY

Component (%)	Unwashed Yeast-whey[1]	Washed Yeast-whey[1]	Dried Whey[2]
Protein	25.9	54.4	12-14
Fat	8.6	10.3	0.7-1.3
Lactose	5.8	2.4	78-81
Moisture	2.0	2.6	—
Ash	18.6	5.7	7.7-8.7
Lactic acid	0.8	—[3]	0.5-2.0

[1]From Amundson (1967).
[2]From Olling (1963).
[3]Not reported.

TABLE 6.3

AMINO ACID COMPOSITION (G PER 16 G N) OF
KLUYVEROMYCES FRAGILIS

Essential		Nonessential	
Lysine	10.20	Arginine	7.08
Threonine	6.46	Histidine	1.87
Valine	7.78	Asparagine	11.16
Methionine	1.25	Serine	6.96
Isoleucine	6.00	Glutamic acid	13.26
Leucine	9.60	Proline	4.31
Aromatic amino acids		Glycine	4.63
Tyrosine	3.42	Alanine	8.17
Phenylalanine	5.39		

Source: Adapted from Wasserman (1960A).

Table 6.4 summarizes information on the vitamin content of dried yeast-whey products, whey, and *K. fragilis*. Uniformly good sources of the B vitamins are found in the dried yeast-whey.

PRODUCTION OF ETHANOL FROM WHEY

Ethyl alcohol can be produced from whey by one of the lactose-fermenting yeasts. Although ethanol was a by-product of the whey fermentation proposed by Myers and Weisberg (1938), these investigators did not concern themselves with this product in their

TABLE 6.4

VITAMIN CONTENT OF UNWASHED AND WASHED YEAST-WHEY, WHEY AND
KLUYVEROMYCES FRAGILIS (DRY WEIGHT BASIS)

Vitamin (μg/g)	Unwashed Yeast-whey[1]	Washed Yeast-whey	Whey[2]	Kluyveromyces fragilis[3]
Pantothenic acid	94.8	65.2	22-90	67.2
Riboflavin	57.5	42.5	5-68	36.0
Pyridoxine	12	15	0.5-15	13.6
Niacin	61	104	3-22	280
Thiamine	32.8	29.7	4-15	24.1
Folic acid	3.1	2.06	0-1	6.83
p-Aminobenzoic acid	—[4]	—	—	24.2
Biotin	—	—	—	1.96
Choline	—	—	—	6670
Inositol	—	—	—	3000

[1]Adapted from Amundson (1967).
[2]Adapted from Wasserman (1961).
[3]Adapted from Wasserman (1960A).
[4]Not reported.

patent issued in 1938. One year later a patent describing a procedure for production of ethanol from whey was issued to Kauffmann and Van der Lee (1939). They employed pasteurized whey, a lactose-fermenting yeast, and a 3-day incubation. During this time 65% of the lactose was converted to ethanol.

Some years later Rogosa *et al.* (1947) compared different lactose-fermenting yeasts for their ability to produce ethanol in whey and concluded that *T. cremoris (Candida pseudotropicalis)* was the most efficient of those yeasts tested. Their experiments led them to suggest the following procedure for production of ethanol from whey.

(1) Residual milkfat is removed from whey.

(2) Whey is heated to 100°C and the pH is adjusted to 4.7 to 5.0 using sulfuric acid.

(3) Hot whey is filtered to recover the precipitated protein and is then cooled to 34°C.

(4) An inoculum of 0.45 kg per 454 liters of whey is added and the fermentation is allowed to proceed for 48 to 72 hr.

(5) Fermented whey is subjected to centrifugation to recover the yeast cells.

(6) The liquid is distilled to recover the ethanol.

According to Rogosa *et al.* (1947), 84 to 91% of the theoretical ethanol yield was obtained with *T. cremoris.*

Wilharm and Sack (1947) increased the lactose content of whey to

10, 20 and 30% and subjected the fortified wheys to an alcoholic fermentation. Their results indicated that up to 90%, 75 to 80%, 60 to 75%, and 37 to 40% of the lactose was fermented in ordinary whey, and in wheys containing 10, 20, and 30% lactose, respectively. These data suggest that some fortification (or concentration) of single-strength whey might be possible to increase the yield of alcohol, and that excessive supplementation with lactose apparently reduces the efficiency of the fermentation. Ethanol derived from whey by fermentation may be used for conversion to vinegar or in other food applications.

PRODUCTION OF LACTASE FROM WHEY

Lactase is an enzyme which catalyzes hydrolysis of the galactosidic linkage of lactose. Since the linkage between the hexoses of lactose is of the β-D configuration, the enzyme is also termed β-D-galactosidase (Pomeranz 1964A). Lactase occurs naturally in some plants and in the intestines of various animals. Additionally, it is produced by a variety of molds, yeasts and bacteria (Pomeranz 1964A) (see Chap. 13).

Although a widespread application has not been found for lactase in the food industry, some limited uses accompanied by beneficial results have been reported. Pomeranz (1964B) summarized this information and he listed the following possibilities: (a) minimize age-thickening of frozen, concentrated milk (patented by Stimpson 1954C), (b) permit an increase in nonfat milk solids content of ice cream by hydrolyzing lactose to prevent it from crystallizing, (c) hydrolysis of lactose in animal feeds, since many animals cannot tolerate high concentrations of this sugar in their diets, (d) improve the color of some fried foods by making hexoses available, and (e) produce fermentable carbohydrates from nonfat dry milk in bread dough. The last suggested use has not proved to be very successful (Pomeranz 1964B). Recently lactase has been found useful for treating milk so it can be consumed by persons suffering from lactose intolerance.

The use of a lactose-fermenting yeast as a source of lactase and growth of the yeast in whey has been suggested. Most of the early work in this area was done in the research laboratories of National Dairy Products Corp. (now Kraft, Inc.) and results are recorded in patents issued to Stimpson (1954A, B), Morgan (1955), Myers (1956), Myers and Stimpson (1956), Young and Healey (1957), Connors and Sfortunato (1956), Sfortunato and Connors (1958), Stimpson and Stamberg (1956), and Stimpson and Whitaker (1956). Wendorff

(1969) and Wendorff et al. (1965) at the University of Wisconsin also have conducted studies on lactase production.

According to Morgan (1955), lactase may be derived from Kluyveromyces fragilis, Torulopsis sphaerica (now Kluyveromyces lactis), Zygosaccharomyces lactis (K. lactis), Candida utilis, and Candida pseudotropicalis, although K. fragilis appears to be the organism of choice for large-scale production of the enzyme.

The process for lactase production, as recorded by Morgan (1955), Stimpson (1954A), and Myers and Stimpson (1956) involves the following.

(1) Whey containing 2 to 8% solids is adjusted to pH 4.5 and heated to 85°C for 30 min to precipitate the whey proteins.

(2) The liquid fraction is fortified with 0.1% corn steep liquor and with a source of nitrogen (0.2% urea, 0.14% ammonia, or 0.4% dibasic ammonium phosphate).

(3) The fortified liquid is cooled to 30°C and inoculated with 10% of an actively growing culture of K. fragilis so that the freshly inoculated whey will contain 10 to 60 million cells per ml.

(4) The fermentation is completed in 2 to 8 hr with aeration, or in 30 hr without. According to Young and Healey (1957), optimum lactase production was obtained when the medium contained 0.06% available nitrogen and was aerated at the rate of 1 vol of air per vol of medium per min.

(5) Yeast cells are separated from whey and washed if a product with good flavor is desired. Washing is not necessary if the lactase preparation is to be used in animal feeds.

(6) The concentrated yeast is dried and stored at 4°C until used.

Drying of the yeast accomplishes two purposes. The first is the obvious one of removing water and the second is that of destroying "zymase" activity without markedly affecting the lactase activity. Zymase is a collective term referring to the group of enzymes in yeast which catalyze reactions involved in converting glucose and galactose into ethanol and carbon dioxide. Procedures which must be followed in drying have been outlined by Stimpson (1954B), Myers (1956), and Myers and Stimpson (1956). The following methods have been found to yield a satisfactory product.

(1) A yeast cream (10 to 18% yeast solids) is fed into a spray drier with an inlet air temperature of 93° to 154°C and an outlet air temperature of 54° to 96°C. The optima are 121°C for inlet air and 77°C for outlet air.

(2) Tray dry at 66°C or below for no more than 4 hr unless a vacuum is applied, in which case the period may be extended to 8 hr. Alternatively, drying at 46°C for 8 hr without vacuum is satisfactory.

(3) Freezing yeast at a temperature of -43° to -18°C followed by

freeze-drying, with the yeast remaining in the frozen state during drying. The temperature during freezing is not critical, although rapid freezing results in retention of more enzyme activity. Temperature during drying is not critical so long as the yeast remains frozen.

Although roller drying inactivates lactase as well as zymase (Stimpson 1954B), yeast can be treated with toluene, chloroform, or ethyl ether, or heated to 52°C at pH 7 to destroy zymase activity without untoward damage to lactase (Myers and Stimpson 1956).

An additional treatment to kill yeast cells as well as chance contaminants without inactivating lactase was proposed by Morgan (1955). To accomplish the stated purpose he suspended wet yeast cake in an 8 to 17.5% aqueous ethanol solution and allowed the mixture to stand for 1 to 5 hr. After this treatment yeasts were removed from the alcoholic solution and dried by procedures described earlier.

Wendorff et al. (1965) reported the following conditions were needed for maximum lactase production by K. fragilis: (a) whey containing 10 to 15% lactose, (b) addition to whey of corn steep liquor or a casein digest, (c) pH of 4.0 to 4.5, and (d) incubation at 28°C. Yeast grown under optimal conditions yielded 175 units of lactase per g or 1300 units per liter of whey. A unit of lactase was defined by Young and Healey (1957) as the grams of lactose hydrolyzed by 1 g of lactase preparation when done in 30% skim milk solids at 51°C and pH 6.8 for 4 hr.

Additional studies were conducted by Wendorff (1969) to characterize the lactase enzyme produced by K. fragilis. He found the enzyme to be stable at pH 6.0 to 7.0 and that optimum lactose hydrolysis occurred at 37°C and a pH of 6.5. Activation of the enzyme was attributed to potassium ions, whereas manganese ions served as cofactors for the lactase. Inactivation of lactase was associated with urea and inhibition with heavy metals, p-chloromercuribenzoate, iodoacetate, cysteine, galactose, glucose and various amines. Milk products were found to contain a naturally occurring lactase inhibitor which could be inactivated by heating the products at 74°C for 30 min.

SPOILAGE OF DAIRY PRODUCTS BY YEASTS AND MOLDS

Spoilage of dairy products by yeasts has been discussed by Walker and Ayres (1970). Problems caused by yeasts include: (a) production of gas and off-flavors in improperly handled cream, (b) production of gas in sweetened condensed milk, (c) production of gas and off-flavors in cheese, (d) growth, often on the surface, in cottage cheese with accompanying off-flavors and reduction in shelf-life, and (e) production of rancidity and other flavor defects in butter, particularly if the yeast population is large. Control of these problems generally involves one or more of the following: (a) use of accepted hygienic practices before, during, and after manufacture of the product, (b) use of proper heat treatments and refrigerated storage, and (c) use of an acceptable antimycotic agent such as sorbic acid or its salts.

Of all the dairy products, cheese is probably most often spoiled by growth of molds. This can occur under some conditions during ripening of cheese as well as later when cheese is cut and packaged for the consumer. Process cheese, if contaminated after heating, can also be spoiled by molds provided the cheese is not packaged to exclude virtually all oxygen.

As might be expected, a variety of molds can grow on and spoil cheese. In one study Gaddi (1973) isolated 144 mold cultures from 14 varieties of cheese made or distributed by three United States food processors. Of the isolates, 69% were *Penicillium* spp., 9% *Aspergillus* spp., 8% *Scopulariopsis* spp., 3% *Mucor* spp. and 2% *Cephalosporium* spp. Some cultures of *Cladosporium* spp. and *Syncephalastrum* spp. also were encountered.

Molds also have been found responsible for spoilage of farm-separated cream (*Geotrichum candidum*), butter (several genera), cottage cheese, yogurt and sweetened condensed milk.

Control of spoilage by molds is much like control of spoilage by yeasts. Careful attention must be given to hygienic practices so that contamination of products with spores of molds is minimized. Fungal spores are readily inactivated by heat processes common to the dairy industry (Doyle and Marth 1975A, B) and hence it is especially important for heated products to be handled in a sanitary manner. Packaging to essentially eliminate oxygen from the atmosphere surrounding a product is also useful to control mold growth. Finally, when appropriate, use of sorbates and propionates will inhibit mold growth and thus increase the shelf-life of dairy products. It must be remembered, however, that some penicillia can grow in the presence of sorbate and can metabolize it to produce 1,3-pentadiene, a compound which imparts a hydrocarbon-like odor and flavor to cheese (Marth *et al.* 1966).

MYCOTOXINS IN DAIRY PRODUCTS

Mycotoxins are toxic metabolites that are produced by certain molds during growth on foods or feeds. Dairy products can become contaminated with these toxins indirectly if the cow consumes toxic feed and the toxins are transferred to milk, or directly if the toxigenic fungus grows on the food.

Mycotoxins in Cow's Milk

Aflatoxins, a group of toxic metabolites of *Aspergillus flavus* and *A. parasiticus*, have been found in cow's milk. In 1963 Allcroft and Carnaghan first reported that a factor toxic for ducklings was contained in milk from cows whose rations contained peanut meals contaminated with aflatoxin. They also demonstrated that the toxic factor remained with the casein fraction that precipitated when milk was treated with rennet. According to Kiermier (1973) various workers fed controlled amounts of aflatoxin to cows and found that milk generally contained less than 1% of the amount of toxin fed to the cows. Allcroft *et al.* (1966) designated the milk toxin as aflatoxin M and Holzapfel *et al.* (1966) found aflatoxins M_1 and M_2. The latter researchers also determined chemical structures and concluded that M_1 was hydroxyaflatoxin B_1 (see Fig. 14.1) and M_2 was dihydroxy-aflatoxin B_2.

Kiermier (1974) fed cows with peanut meal that contained 10.9 mg of aflatoxin per kg. In one experiment the cow received a single dose of 830 g of feed, in another 415 g were given to a cow on each of two successive days, and in a third, two cows each received 914 g daily for 5 days. Results indicated that: (a) aflatoxin appeared in milk within 12 hr after toxin was consumed by the cow, (b) the amount of aflatoxin in milk varied with the individual cow, (c) excretion of aflatoxin in milk decreased markedly about 1 day after feeding of toxin had stopped, although small amounts appeared in milk for 2 to 3 additional days, and (d) from 0.18 to 0.39% of the amount consumed was excreted in milk.

Several surveys have shown that aflatoxin sometimes occurs in milk produced under normal farm conditions. During 1972 Kiermier (1974) sampled raw milk produced in the vicinity of Freising, Federal Republic of Germany. Twelve of 36 samples taken from individual farms during February to April contained from 0.03 to 0.25 μg aflatoxin M_1 per liter. In May, 9 of 12 samples taken from tank trucks (6000 to 9000 liters per truck) contained 0.01 to 0.08 μg per liter. Small amounts of aflatoxin M_1 also appeared in milks taken during February from the holding tank in the factory, but none was found when milks were tested in March and June, suggesting that the diet of cows was less contaminated with aflatoxin in spring than winter. In 1973, the United States Food and Drug Administration surveyed milk products for presence of aflatoxin M_1. The toxin (0.05 to 0.5 μg per liter) was detected in all 16 samples from one milk shed where use of contaminated feed was suspected (Stoloff 1976).

Processing of milk can reduce the amount of detectable residual aflatoxin M_1. Purchase *et al.* (1972) found a reduction (from 385 μg/kg) of 32% when milk was pasteurized at 62°C for 30 min; 45%

when pasteurized at 72°C for 45 sec; 64% when pasteurized at 80°C for 45 sec; 64% when processed into condensed milk; 61% when roller dried at low pressure; 76% when roller dried at 4.9 kg/cm^2; 81% when sterilized at 115°C; and 86% when spray-dried. In spite of this, aflatoxin has been found in commercial dried milks, cottage cheese curd, and evaporated milk (Hanssen and Jung 1972; Stoloff 1976).

Mycotoxins in Moldy Dairy Products

Kiermier and Böhm (1971) examined 19 varieties of German cheese and found aflatoxin B$_1$ in samples of Tilsit, Edelpilzkäse, butter cheese, smoked cheese, Parmesan and Romadur. Additionally, aflatoxin G$_1$ was found in samples of Emmentaler, Gouda, Tilsit and Romadur. A combination of B$_1$ and G$_1$ occurred in some samples of still other varieties of cheese. That toxigenic aspergilli can produce aflatoxin during growth on Cheddar and brick cheese was demonstrated by Lie and Marth (1967) and Shih and Marth (1972). Furthermore, these authors demonstrated that aflatoxin can penetrate into cheese to a depth of up to 4 cm from the surface. Production of toxic metabolites by some strains of P. roqueforti and P. camemberti has been reported (Kiermier 1974).

It is evident from these brief comments that dairy products will be free of aflatoxin only if cows receive high quality feed free of the toxin and if growth of molds on finished products is prevented. A more extensive discussion of aflatoxins and other mycotoxins is given in Chap. 14.

REFERENCES

ALLCROFT, R., and CARNAGHAN, R. B. A. 1963. Groundnut toxicity: An examination for toxin in human food products from animals fed toxic groundnut meal. Vet. Rec. 75, 259-263.

ALLCROFT, R., ROGERS, H., LEWIS, G., and NABNEY, J. 1966. Metabolism of aflatoxin in sheep: Excretion of the "milk toxin." Nature (London) 209, 154-155.

AMUNDSON, C. H. 1966. Increasing the protein content of whey through fermentation. Proc. 33rd Wash. State Univ. Inst. Dairy., 23-30.

AMUNDSON, C. H. 1967. Increasing protein content of whey. Am. Dairy Rev. 29, No. 7, 22-23, 96-99.

ANON. 1946. Baker's yeast from whey. Südd. Molkereiztg. 67, 114-115. (German)

ANON. 1961. On the utilization of whey from cheese. Technicien Lait 13, 13-14. (French)

BARNARD, R., and CALLAHAN, W. 1971. Follow-up on gastroenteritis attributed to French cheese. Morbidity Mortality Rep. 25, 445.

CONNORS, W. M., and SFORTUNATO, T. 1956. Purification of lactase enzyme and spray-drying with sucrose. U. S. Pat. 2,773,002. Dec. 4.

DAVIDOV, R. B., GUL'KO, L. E., and FAINGER, B. I. 1963. Enrichment of whey with protein and vitamins. Izv. Timiryazevsk. Skh. Akad. *1963*, No. 5, 166-171. (Russian)

DEMMLER, G. 1950. Production of yeast from whey using the Waldof method. Milchwissenschaft *5*, 11-17. (German)

DOYLE, M. P., and MARTH, E. H. 1975A. Thermal inactivation of conidia from *Aspergillus flavus* and *Aspergillus parasiticus*. I. Effects of moist heat, age of conidia, and sporulation medium. J. Milk Food Technol. *38*, 678-682.

DOYLE, M. P., and MARTH, E. H. 1975B. Thermal inactivation of conidia from *Aspergillus flavus* and *Aspergillus parasiticus*. II. Effects of pH and buffers, glucose, sucrose, and sodium chloride. J. Milk Food Technol. *38*, 750-758.

EICHELBAUM, G. 1902. Process of obtaining food extracts. U.S. Pat. 708,330. Sept. 2.

ENEBO, L., LUNDIN, H., and MYRBÄCK, K. 1941. Yeasts from whey. Sven. Kem. Tidskr. *53*, 137-147. (Swedish)

ERNSTROM, C. A., and WONG, N. P. 1974. Milk-clotting enzymes and cheese chemistry. *In* Fundamentals of Dairy Chemistry, 2nd Edition. B. H. Webb, A. H. Johnson, and J. A. Alford (Editors). AVI Publishing Co., Westport, Conn.

FOSTER, E. M., NELSON, F. E., SPECK, M. L., DOETSCH, R. N., and OLSON, J. C., JR. 1957. Dairy Microbiology. Prentice-Hall, Englewood Cliffs, New Jersey.

GADDI, B. L. 1973. Mycotoxin-producing potential of fungi isolated from cheese. Ph.D. Thesis. Univ. of Wisconsin, Madison.

GRAHAM, V. E., GIBSON, D. L., and KLEMMER, H. W. 1953A. Increasing the food value of whey by yeast fermentation. II. Investigations with small scale fermenters. Can. J. Technol. *31*, 92-97.

GRAHAM, V. E., GIBSON, D. L., KLEMMER, H. W., and NAYLOR, J. M. 1953B. Increasing the food value of whey by yeast fermentation. I. Preliminary studies on the suitability of various yeasts. Can. J. Technol. *31*, 85-91.

GRAHAM, V. E., GIBSON, D. L., and LAWTON, W. C. 1953C. Increasing the food value of whey by yeast fermentation. III. Pilot plant studies. Can. J. Technol. *31*, 109-113.

HANSON, A. M., RODGERS, N. E., and MEADE, R. E. 1949. Method of enhancing the the yield of yeast in a whey medium. U.S. Pat. 2,465,870. Mar. 29.

HANSSEN, E., and JUNG, M. 1972. On the occurrence of aflatoxins in foods that are not moldy and suggestions for sampling. Z. Lebensm. Unters.-Forsch. *150*, 141-145. (German)

HOLZAPFEL, C. W., STEYN, P. S., and PURCHASE, I. F. H. 1966. Isolation and structure of aflatoxins M_1 and M_2. Tetrahedron Lett. *25*, 2799-2803.

HUANG, H. T., and DOOLEY, J. G. 1976. Enhancement of cheese flavors with microbial esterases. Biotechnol. Bioeng. *18*, 909-919.

JACKSON, H. W., and HUSSONG, R. V. 1958. Secondary alcohols in blue cheese and their relation to methyl ketones. J. Dairy Sci. *41*, 920-924.

JOLLY, R. C., and KOSIKOWSKI, F. V. 1975. Quantitation of lactones in ripening pasteurized milk blue cheese containing added microbial lipases. J. Agric. Food Chem. *23*, 1175-1176.

KAUFFMAN, W., and VAN DER LEE, P. J. 1939. Method of fermenting whey to produce alcohol. U. S. Pat. 2,183,141. Dec. 12.

KIERMIER, F. 1973. Aflatoxin M excretion in cow's milk in relation to the quantity of aflatoxin B_1 ingested. Milchwissenschaft *28*, 683-685. (German)

KIERMIER, F. 1974. The significance of aflatoxins in the dairy industry. Int. Dairy Fed. Doc. No. 30.

KIERMIER, F., and BÖHM, S. 1971. On aflatoxin formation in milk and milk products. V. Application of the chick embryo test for the affirmation of thin-layer chromatographic determination of aflatoxin in cheese. Z. Lebensm. Unters.-Forsch. *147*, 61-64. (German)

KINSELLA, J. E., and HWANG, D. H. 1976A. Biosynthesis of flavors by *Penicillium roqueforti*. Biotechnol. Bioeng. *18*, 927-938.

KINSELLA, J. E., and HWANG, D. H. 1976B. Enzymes of *Penicillium roqueforti* involved in the biosynthesis of cheese flavor. CRC Crit. Rev. Food Sci. Nutr. *8*, 191-228.

KOSIKOWSKI, F. 1966. Cheese and Fermented Milk Foods. Edwards Brothers, Ann Arbor, Mich.

LIE, J. L., and MARTH, E. H. 1967. Formation of aflatoxin in Cheddar cheese by *Aspergillus flavus* and *Aspergillus parasiticus*. J. Dairy Sci. *50*, 1708-1710.

LODDER, J. 1970. The Yeasts: A Taxonomic Study. North-Holland Publishing Co., Amsterdam.

LOMBARD, S. H., and BESTER, B. H. 1975. Camembert cheese: certain aspects of manufacture. S. Afr. J. Dairy Technol. *7*, 141-146.

MARTH, E. H. 1974. Fermentations. *In* Fundamentals of Dairy Chemistry, 2nd Edition. B. H. Webb, A. H. Johnson, and J. A. Alford (Editors). AVI Publishing Co., Westport, Conn.

MARTH, E. H., CAPP, C. M., HASENZAHL, L., JACKSON, H.W., and HUSSONG, R. V. 1966. Degradation of potassium sorbate by *Penicillium* species. J. Dairy Sci. *49*, 1197-1205.

MERGL, M., and SIMAN, J. 1963. Dried yeast feeding stuffs from fermented whey. Drubeznictvi *11*, No. 9, 136. (Czechoslovakian)

MORGAN, E. R. 1955. Lactase enzyme preparation. U.S. Pat. 2,715,601. Aug. 16.

MÜLLER, W. R. 1949. Growth of yeast in whey. Milchwissenschaft *4*, 147-153. (German)

MYERS, R. P. 1956. Drying of yeast to inactivate zymase and preserve lactase. U.S. Pat. 2,762,748. Sept. 11.

MYERS, R. P., and STIMPSON, E. G. 1956. Production of lactase. U.S. Pat. 2,762,749. Sept. 11.

MYERS, R. P., and WEISBERG, S. M. 1938. Treatment of milk products. U.S. Pat. 2,128,845. Aug. 30.

NAIDITCH, V. 1965. Method and plant for treating whey. French Pat. 86,177. Sept.5.

NAIDITCH, V., and DIKANSKY, S. 1960. Method and equipment for the treatment of whey. French Pat. 1,235,978. June 9.

OLLING, C. H. J. 1963. Composition of Friesian whey. Neth. Milk Dairy J. *17*, 176.

OLSON, N. F. 1969. Ripened semisoft cheese. Chas. Pfizer & Co., New York.

PARK, H. S., MARTH, E. H., and OLSON, N. F. 1973. Fate of enteropathogenic strains of *Escherichia coli* during the manufacture and ripening of Camembert cheese. J. Milk Food Technol. *36*, 543-546.

POMERANZ, Y. 1964A. Lactase (beta-D-galactosidase). I. Occurrence and properties. Food Technol. *18*, 682-687.

POMERANZ, Y. 1964B. Lactase (beta-D-galactosidase). II. Possibilities in the food industry. Food Technol. *18*, 690-697.

PORGES, N., PEPINSKI, J. B., and JASEWICZ, L. 1951. Feed yeast from dairy by-products. J. Dairy Sci. *34*, 615-621.

PURCHASE, I. F. H., STEYN, M., RINSMA, R., and TUSTIN, R. C. 1972. Reduction of the aflatoxin M content of milk by processing. Food Cosmet. Toxicol. *10*, 383-387.

ROGOSA, M., BROWNE, H. H., and WHITTIER, E. O. 1947. Ethyl alcohol from whey. J. Dairy Sci. *30*, 263-269.

SCHNURRENBERGER, L. W., BECK, R., and PATE, J. 1971. Gastroenteritis attributed to imported French cheese. Morbidity Mortality Rep. *20*, 427-428.

SCOTT, P. M., and KENNEDY, B. P. C. 1976. Analysis of Blue cheese for roquefortine and other alkaloids from *Penicillium roqueforti*. J. Agric. Food Chem. *24*, 865-868.

SFORTUNATO, T., and CONNORS, W. M. 1958. Conversion of lactose to glucose and galactose with a minimum production of oligosaccharides. U.S. Pat. 2,826,502. Mar. 11.

SHAHANI, K. M., ARNOLD, R. G., KILARA A., and DWIVEDI, B. K. 1976. Role of microbial enzymes in flavor development in foods. Biotechnol. Bioeng. *18*, 891-907.

SHIH, C. N., and MARTH, E. H. 1972. Experimental production of aflatoxin on brick cheese, J. Milk Food Technol. *35*, 585-587.

SIMAN, J., and MERGL, M. 1962. Growth of strains of *Torulopsis* and *Candida* in whey with added sources of inorganic N and P. Sb. Vys. Sk. Chem. Technol. v Praze Potraviny Tech. *6*, 127-138. (Czechoslovakian)

SIMEK, F., KOVACS, J., and SARKANY, I. 1964. Production of fodder yeast using dairy by-products. Tejipar *13*, 75-78. (Hungarian)

SIMOVA, J., and RUZICKOVA, J. 1975. Nutritional value of kefir milk. Vyz. Lidu. *30*, 147-149. (Czechoslovakian)

SOCIÉTÉ DES ALCOOLS DU VEXIN. 1963. Method and equipment for the treatment of whey. French Pat. 80,198. Mar. 22.

STIMPSON, E. G. 1954A. Conversion of lactose to glucose and galactose. U.S. Pat. 2,681,858. June 22.

STIMPSON, E. G. 1954B. Drying of yeast to inactivate zymase and preserve lactase. U.S. Pat. 2,693,440. Nov. 2.

STIMPSON, E. G. 1954C. Frozen concentrated milk products. U.S. Pat. 2,668,765. Feb. 9.

STIMPSON, E. G., and STAMBERG, O.E. 1956. Conversion of lactose to glucose, galactose, and other sugars in the presence of lactase activators. U.S. Pat. 2,749,242. June 5.

STIMPSON, E. G., and TREBLER, H. A. 1956. Aerating method and apparatus. U.S. Pat. 2,750,328. June 12.

STIMPSON, E. G., and WHITAKER, R. 1956. Ice cream concentrate. U.S. Pat. 2,738,279. Mar. 13.

STIMPSON, E. G., and YOUNG, H. 1957. Increasing the protein content of milk products. U.S. Pat. 2,809,113. Oct. 8.

STOLOFF, L. 1976. Occurrence of mycotoxins in foods and feeds. *In* Mycotoxins and Other Fungal Related Problems. J. V. Rodricks (Editor). Adv. Chem. Ser. No. *149*, 23-50. Am. Chem. Soc., Washington, D.C.

STUIBER, D. A. 1966. Whey Fermentation by *Saccharomyces fragilis*. M.S. Thesis. Univ. of Wisconsin, Madison.

TOMISEK, J., and GREGR, V. 1961. Yeast protein manufacture from whey. Kvasny Prum. *7*, 130-133. (Czechoslovakian)

VAN VEEN, A. G., GRAHAM, D. C. W., and STEINKRAUS, K. H. 1969. Fermented milk wheat combination. Trop. Geogr. Med. *21*, 47-52.

WALKER, H.W., and AYRES, J. C. 1970. Yeasts as spoilage organisms. *In* The Yeasts, Vol. 3, Yeast Technology. A. H. Rose and J. S. Harrison (Editors). Academic Press, London.

WALTER, H. E., and HARGROVE, R. C. 1969. Cheese varieties and descriptions. U.S. Dep. Agric. Handb. 54.

WASSERMAN, A. E. 1960A. The rapid conversion of whey to yeast. Dairy Eng. *77*, 374-379.

WASSERMAN, A. E. 1960B. Whey utilization. II. Oxygen requirements of *Saccharomyces fragilis* growing in whey medium. Appl. Microbiol. *8*, 291-293.

WASSERMAN, A. E. 1960C. Whey utilization. IV. Availability of whey nitrogen for the growth of *Saccharomyces fragilis*. J. Dairy Sci. *43*, 1231-1234.

WASSERMAN, A. E. 1961. Amino acid and vitamin composition of *Saccharomyces fragilis* grown in whey. J. Dairy Sci. *44*, 379-386.

WASSERMAN, A. E., and HAMPSON, J.W. 1960. Whey utilization. III. Oxygen absorption rates and the growth of *Saccharomyces fragilis* in several propagators. Appl. Microbiol. *8*, 293-297.

WASSERMAN, A. E., HAMPSON, J., ALVARE, N.F., and ALVARE, N. J. 1961. Whey utilization. V. Growth of *Saccharomyces fragilis* in whey in a pilot plant. J. Dairy Sci. *44*, 387-392.

WASSERMAN, A. E., HOPKINS, W. J., and PORGES, N. 1958. Whey utilization—growth conditions for *Saccharomyces fragilis*. Sewage Ind. Wastes *30*, 913-920.

WASSERMAN, A.E., HOPKINS, W. J., and PORGES, N. 1959. Rapid conversion of whey to yeast. Int. Dairy Congr. 1959, *2*, 1241-1247.

WEIGMANN, H. 1933. Handbook of practical cheesemaking, 4th Edition. Verlag Paul Parey, Berlin. (German)

WENDORFF, W. L. 1969. Studies on the β-galactosidase activity of *Saccharomyces fragilis* and effect of substrate preparation. Ph.D. Thesis. Univ. of Wisconsin, Madison.

WENDORFF, W. L., AMUNDSON, C. H., and OLSON, N. F. 1965. Production of lactase by yeast fermentation of whey. J. Dairy Sci. *48*, 769-770.

WILHARM, G., and SACK, U. 1947. The properties of several lactose-fermenting yeasts. Milchwissenschaft *2*, 382-389. (German)

YOUNG, H., and HEALEY, R. P. 1957. Production of *Saccharomyces fragilis* with an optimum yield of lactase. U.S. Pat. 2,776,928. Jan. 8.

7

Storage Fungi

C. M. Christensen

T he term "storage fungi" was coined in the 1940s to describe a
group of fungi adapted to growth in an environment of relatively low moisture content and high osmotic pressure, and
that were involved in and responsible for deterioration of stored
grains. They comprise a diverse taxonomic group, but are ecologically
similar in their ability to grow in an environment with limited free
water; some of them, in fact, require a low water activity (a_w) or a high
osmotic pressure to grow. Storage fungi, as the term ordinarily is used,
include a few "group" species of *Aspergillus*, a few species of
Penicillium, *Wallemia (Sporendonema) sebi*, and a few still unidenti-
fied sqecies of Fungi Imperfecti. They are of major significance in the
deterioration of grains and seeds, but they can grow on and in many
other kinds of materials also, including most of the ingredients of
foods and feeds. Some of them can grow in very low concentrations of
oxygen and very high concentrations of carbon dioxide.

WHEN STORAGE FUNGI INVADE SEEDS

With few exceptions, storage fungi do not invade seeds to any
significant degree or extent before harvest (Tuite and Christensen
1957; Qasem and Christensen 1958; Tuite 1959, 1961). However, if ears
of corn on plants in the field are damaged by insects such as weevils,
corn ear worms, or corn borers, as commonly occurs in the south-

eastern United States and in many other regions of relatively mild climate, the injured ears may be invaded by *Aspergillus flavus.* If the conditions are right, aflatoxin may then be produced in the corn before harvest. Inoculum of storage fungi, like that of many other kinds of fungi, is more or less universally present, so that if grains or other materials subject to invasion by these fungi are kept under conditions favorable to the growth of fungi, they may be heavily invaded within a few days.

CONDITIONS REQUIRED FOR THE GROWTH OF STORAGE FUNGI

Like other living things, storage fungi require nutrients, water, a favorable temperature, a stable atmosphere, and time. The rate at which they develop and cause spoilage in a given lot of grain or a given substrate depends to some extent also on the degree to which that lot or substrate already has been invaded by these fungi when it reaches a given point in its storage or processing life. Each of these factors will be discussed.

Nutrients

Just about all the materials that serve as food for man or his domestic animals furnish suitable substrates for the growth of storage fungi. Many of the materials that do not serve as food for man or other animals also provide nourishing substrates for fungi. To quote from Christensen (1951), "*Penicillium,* for example, can subsist on the remains of thousands of different kinds of plants, on cloth, leather, paper, wood, tree bark, cork, animal dung, animal and insect carcasses, ink, syrup, seeds of all kinds, manufactured cereal products and the boxes in which they are packed, including the wax and ink on the outside, on stored fruits and vegetables, soil, glue, paint, liquid drugs, hair and wool of all kinds, on the wax in our ears, and on literally thousands of other common products." Probably the same could be said of a number of other common storage fungi. Some of them, however, have special requirements. *Aspergillus halophilicus* has been isolated only from seeds stored for 6 months or longer at moisture contents in equilibrium with relative humidities of 65 to 70% (a_w of 0.65 to 0.70). It has never been isolated from soil or from freshly harvested or newly stored seeds, or from any other source or substrate taken directly from nature. *A. halophilicus* evidently has a very narrow ecological niche, but inocula of it must be generally present, since it almost invariably can be recovered from seeds stored as indicated in Fig. 7.1. The source of this inoculum remains a mystery. Similarly *Chrysosporium inops* has been encountered only on samples of corn stored for a year or more in tightly closed containers at

FIG. 7.1. ABUNDANT CLEISTOTHECIA OF *ASPERGILLUS HALOPHILICUS* THAT DEVELOPED ON THE SURFACE OF A BEAN SEED *(PHASEOLUS VULGARIS)*. Seed had been stored for several months at a moisture content of about 12% and then placed on an agar medium of low a_W.

moisture contents of 15 to 16%. In that environment *C. inops* may predominate to the almost total exclusion of other fungi. There may be other and still undiscovered storage fungi that inhabit unusual and limited environments, but certainly most of the common storage fungi, if their requirements for moisture and temperature are met, are more or less generally present. All of the multitude of grains and other food ingredients and finished food products that are subject to invasion by fungi will, if stored at moisture contents giving a_W values of 0.73 to 0.78 and at temperatures between about 5° and 40°C, be invaded by *Aspergillus glaucus*. *A. glaucus* is a "group" species, and includes some 15 to 20 more or less distinct species described by various researchers. In a given material or substrate at a given time and place one or another of these species may predominate, but materials stored under the conditions given, anywhere in the world, will be invaded primarily and predominately by *A. glaucus*.

Moisture Content

By definition, storage fungi are those capable of growing on substrates having moisture contents in equilibrium with relative humidities of 65 to 90% (a_W of 0.65 to 0.90). Some of them probably grow with less water and at a higher osmotic pressure than any other kind of living organism. *A. halophilicus* will grow in an agar medium containing concentrations of sodium chloride or sucrose high enough to result in salt or sugar crystal formation above the surface of the agar. It will not grow in an agar medium containing less than about 10% sodium chloride or less than 45% sucrose (Christensen *et al.* 1959). *Aspergillus restrictus* and *Wallemia sebi* predominate almost only in products with a_W values of 0.68 to 0.75. At moisture contents higher than this other storage fungi take over. So far as is known none of the storage fungi can grow at a_W less than 0.65. The lower limits of moisture that permit the growth of many common species of storage fungi are summarized by Ayerst (1969).

Obviously if moisture is a factor in determining the growth of storage fungi, in protecting foods and food ingredients from damage by these fungi it is of utmost importance to know fairly precisely the moisture contents of the products involved. Grain and grain products sometimes are invaded by storage fungi to such an extent as to render them unfit for human consumption, and sometimes to such an extent as to render them unfit for consumption by domestic animals. Usually this is because personnel in charge of the stored grains or other materials do not know the moisture content of the products. They assume that they know, but actually they do not. There are several reasons for this, as outlined below.

Methods of Measuring Moisture Content.—For grains, the Official United States Standards for Grain (USDA 1974) specify that "Moisture shall be ascertained by the air-oven method prescribed by the United States Department of Agriculture, as described in Service and Regulatory Announcement No. 147, issued by the Agricultural Marketing Service, or ascertained by any method which gives equivalent results." Hunt and Pixton (1974) list 27 "official" drying schedules or methods for determining moisture content of grains and seeds, including six different ones for whole grain corn. These procedures do not all give the same results by any means. Hart *et al.* (1959) enumerate drying procedures for many kinds of seeds that give results in close agreement with the Karl Fischer titration method. In practice, moisture contents of grains are determined by means of electric moisture meters. Until 1962 the Tag-Heppenstal or Weston meter was generally used in grain inspection offices. Since 1964 the Motomco Meter, Model 919 (Motomco, Inc., Electronics Division, Clark, N.J.) has been the officially approved meter for use in USDA grain inspection offices. Other makes may give equally good results. It is desirable to occasionally check the accuracy of the meter by oven drying samples, using a schedule and procedure known to give accurate results.

The Sample.—Usually a single sample is used to determine the moisture content of a given truckload or carload lot of grain. A considerable amount of work must be done to make sure that the sample taken is truly representative of the lot from which it came. Such a sample does not indicate the *range* in moisture content within the lot, and this range determines the risk of spoilage. To determine storability of grains or other materials stored in bulk, it is necessary to take a number of samples and to determine the moisture content of each one separately. In large quantities of grains and seeds stored in bulk it is not at all unusual to find portions with moisture contents from 2 to 3% higher or lower than the average.

Moisture Transfer.—If different temperatures prevail in different portions of bulk-stored materials, moisture will be transferred from the warmer to the cooler portions. The rapidity of the moisture transfer depends on the moisture content of the material and on the magnitude of temperature differential. This is why grain that has been stored for months in an aerated bin, with a uniform moisture content and temperature throughout the mass, may, after it has been loaded into a barge or ship, undergo serious fungal spoilage within a month or two. This moisture transfer may be surprisingly rapid. In the laboratory, with grain volumes of much less than a cubic meter, and with moisture contents of 15 to 18%, a difference of temperature of only 10°C between the warm and cool sides of the container may result in an increase of 1 to 2% in moisture content of the grain on the cool side within 24 hr. As a general rule, unless the temperature throughout the mass is kept uniform, as with forced aeration, moisture migration or transfer from the warmer to the cooler portions of the mass is inevitable. Also respiration by insects, mites, and fungi produces water, so that once spoilage gets under way it is self-perpetuating, and usually self-accelerating.

At times all of the above several factors may combine to bring about a considerably higher moisture content in one or another portion of the bulk than is indicated on the warehouse records. This is the simple explanation for the many cases of fungal spoilage in bins of grain whose moisture contents, according to the warehouse records, were too low to permit storage fungi to grow.

Table 7.1 gives the moisture contents of various grains in equilibrium with different relative humidities. Table 7.2 summarizes minimum a_W values that permit the growth of common storage fungi. At any given relative humidity between 65 and 90%, the equilibrium moisture content of grains and grain products may vary as much as ±1.0% or even more. For this reason it is not possible to establish an absolute and unvarying moisture content-deterioration risk relation-

TABLE 7.1

MOISTURE CONTENTS OF GRAINS AND SEEDS IN EQUILIBRIUM WITH
VARIOUS RELATIVE HUMIDITIES AT 25° to 30°C

Relative Humidity (%)	Wheat, Corn Sorghum	Moisture Content (%)			Sunflower	
		Rice		Soybeans	Seeds	Meats
		Rough	Polished			
65	12.5–13.5	12.5	14.0	11.5	8.5	5.0
70	13.5–14.5	13.5	15.0	12.5	9.5	6.0
75	14.5–15.5	14.5	15.5	13.5	10.5	7.0
80	15.5–16.5	15.0	16.5	16.0	11.5	8.0
85	18.0–18.5	16.5	17.5	18.0	13.5	9.0

Source: Christensen (1972).

TABLE 7.2

MINIMUM a_w FOR THE GROWTH OF COMMON STORAGE FUNGI AT THEIR
OPTIMUM TEMPERATURE FOR GROWTH (26° to 30°C)

Fungus	Minimum a_w for Growth
Aspergillus halophilicus	0.68
A. restrictus, Wallemia (Sporendonema) sebi	0.70
A. glaucus	0.73
A. candidus, A. ochraceus	0.80
A. flavus	0.85
Penicillium, depending on species	0.80-0.90

Source: Christensen (1972).

ship for a given kind of grain or grain product. The general limits of various categories of moisture content—spoilage risk—have been established, but some variation around the norms must be accepted and recognized.

Temperature

Most storage fungi have a minimum temperature for growth of 0° to 5°C, an optimum of 25° to 30°C, and a maximum of 40° to 45°C. Both Aspergillus candidus and A. flavus, however, can grow vigorously at 50° to 55°C, and can raise the temperature of materials in which they are growing to that figure. Several species of Penicillium (P. brevicompactum, P. cyclopium, P. puberulum, and P. expansum) can grow slowly at temperatures down to -2°C (Mislivec and Tuite 1970), but require a high moisture content to do so. Cladosporium herbarum, which is not usually included among the storage fungi, may grow at -5°C. It has been found colonizing the filling of meat and turkey pies kept at approximately that temperature.

Under some circumstances low temperature can be just as effective as low moisture content in preventing damage by storage fungi and in preserving quality. However, as stated by Christensen and Kaufmann (1969), "The effect of temperature on the growth of storage fungi and on the damage they do has some complexities too. Corn or wheat that has not been invaded by storage fungi to any serious degree or extent, and which is otherwise sound and in good condition, can be stored safely at a moisture content of 15.0% for nine months to a year at a temperature of 7.2° to 10°C. If, on the other hand, the grain already has been moderately to heavily invaded by storage fungi, and is stored at a moisture content of 15.0% and a temperature of 7.2° to 10°C, the fungi may continue to grow, and within six months may cause extensive damage."

Oxygen and Carbon Dioxide

Many of the fungi that cause deterioration of stored grains can grow in an atmosphere containing only 0.1 to 0.2% oxygen or in an atmosphere containing more than 80% carbon dioxide, and a few of them are at least facultative anaerobes. Thus while airtight storage is used to some extent to preserve grains from spoilage (Hyde 1974), if the moisture content-temperature-time combination permits these microflora to grow, the grain may develop off-flavors that make it unsuitable for food.

Previous Invasion by Storage Fungi

Grain that already is invaded to some extent by storage fungi is partly deteriorated, whether or not this is evident in naked eye inspection. If such grain is stored at a combination of temperature and moisture content that permits these fungi to grow, spoilage will be more rapid than if the grain had been relatively free of fungi when exposed to adverse storage conditions.

EFFECTS OF STORAGE FUNGI

Invasion of stored grains and food- and feedstuffs can result in various kinds of damage. In grains and seeds this damage includes: (1) decrease in germinability; (2) discoloration; (3) production of mycotoxins; (4) heating; (5) mustiness; (6) caking; and (7) total decay. Each of these will be discussed.

Decrease in Germinability

Loss of germinability resulting from invasion of seeds by storage fungi is illustrated in Fig. 7.2 and 7.3. Note that in the tests by Fields and King (1962) with field peas (Pisum sativum) the samples free of storage fungi retained a germination of 98% throughout the 8-month test, whereas all samples that had been inoculated with various storage fungi were reduced to zero germination. Many other agents and processes can result in loss of germinability of seeds also, but there is no question that storage fungi, under conditions favorable to their growth, can be primary killers of many kinds of seeds. Extensive evidence on this is summarized by Christensen (1972).

Both field fungi (those infecting seeds while they are still on the plants in the field) and storage fungi can cause discoloration of whole seeds or of portions of them. Often this fungus-induced discoloration reduces the grade and therefore the monetary value of the affected grains and seeds. According to the Official United States Standards for Grain (USDA 1974), Grade No. 2 corn may contain no more than 5.0% "damaged" kernels. Damaged kernels are defined as "...kernels and pieces of kernels of corn which are heat-damaged, sprouted,

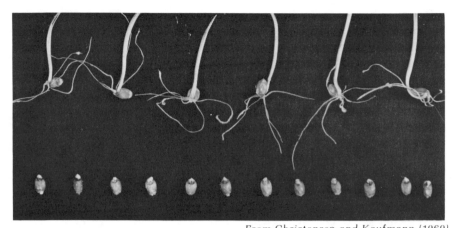

From Christensen and Kaufmann (1969)

FIG. 7.2. REDUCTION IN GERMINABILITY OF WHEAT SEED CAUSED BY ASPERGILLUS CANDIDUS. Top row—controls, free of fungi, germination 100%; bottom row—seeds inoculated with A. candidus, germination 0%. Both stored for 25 days with 15.4% moisture and at 25°C.

frosted, badly ground-damaged, badly weather-damaged, moldy, diseased, or otherwise materially damaged." The term "heat-damaged" in the above suggests that the damage actually is caused by heat, but in many cases this is not so. Schroeder and Sorenson (1961) stated that what is called heat damage in rough rice develops without any detectable rise in temperature—presumably as a result of the growth of storage fungi. Lutey (1961) isolated a bacterium from stained barley that, when inoculated onto barley kernels with 40 to 45% moisture, stained the kernels jet black in a few days. After the stained kernels were dried they became very friable, so that in both color and texture they resembled kernels partially carbonized by heat. Grains stored in the laboratory under conditions that permit heavy invasion by storage fungi, accompanied or followed by invasion by bacteria, may become jet black without ever having been above 25°C. Wheat that develops dark brown to black germs in storage is designated "germ-damaged" or "sick" wheat (Fig. 7.4). Such discolored germs are high in fatty acids and also are very friable. When the grain is milled, fragments end up in the flour, which is highly undesirable. This germ damage is almost solely the product of invasion by storage fungi (Christensen and Kaufmann 1969).

Mycotoxins

Mycotoxins are discussed at length in Chap. 14 and will not be mentioned further here.

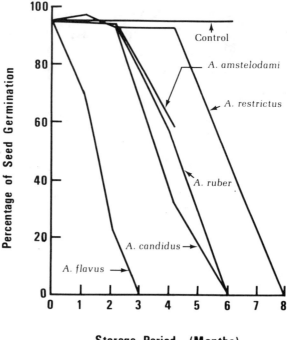

Storage Period   (Months)

Adapted from Fields and King (1962)

FIG. 7.3. EFFECTS OF VARIOUS SPECIES OF *ASPERGILLUS* UPON GERMINATION OF INOCULATED PEA SEEDS STORED AT 30°C AND 85% RELATIVE HUMIDITY.

Heating

At one time it was thought that the respiration of moist grain itself was responsible for the heating that sometimes accompanies spoilage, but none of the tests on which this opinion was based separated respiration of the seeds themselves from respiration of the microflora. Bailey and Gurjar (1918), for example, stored moist wheat in sealed containers, exposed these to different temperatures and, after 3 days, removed samples of air from the containers and measured the amount of carbon dioxide present. They found the highest respiratory rate to be at 55°C. They did not determine viability of the grain at the beginning or end of the tests. Christensen and Kaufman (1974) reported that wheat of 16.0% moisture exposed to 55°C for 2 days was reduced to zero percent germination. Dead seeds do not respire. It seems probable that Bailey and Gurjar measured the respiration of microflora on the wheat, not wheat itself. Darsie *et al.* (1914) found that germinating seeds in Dewar bottles, with moisture contents and

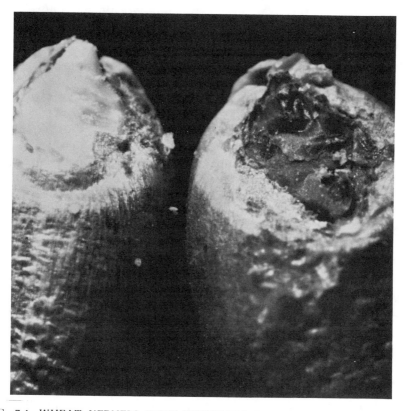

FIG. 7.4. WHEAT KERNELS WITH PERICARPS REMOVED TO EXPOSE THE EMBRYOS. Left—sound; right—"germ-damaged" or "sick" kernel.

respiration rates much higher than those in storage, have raised the temperature within the bottles from 1° to 3°C, whereas temperature in the lots of seeds overgrown with fungi increased 10°C in 4 to 5 days, at which time the tests were ended. Hummel *et al.* (1954) found that respiration of wheat free of storage fungi, with moisture contents of 14 to 18%, and kept at 35°C, was so low that it was not detectable. The evidence by Ramstad and Geddes (1943), Milner and Geddes (1945, 1946), Carlyle and Norman (1941), Carter (1950), and Christensen and Gordon (1948) is consistent in showing that microflora are primarily, and often solely, responsible for the heating of moist grains and other moist organic materials.

Mustiness, Caking and Total Decay

These are the final stages of spoilage caused by fungi, when the organisms involved become detectable by sight and smell. To those unfamiliar with fungi these may be thought to be the first stages of spoilage, but that is of course erroneous. We do not know what chemical compounds are mainly responsible for the musty odors, but we do know that the odors persist through whatever processing that grains or grain products may be subjected to in the preparation of foods and beverages. Flour, bread and even whiskey made from musty grains have a musty flavor.

CHARACTERISTICS OF THE MAJOR STORAGE FUNGI

The salient characteristics of the major storage fungi are summarized below, with the fungi grouped according to the minimum moisture contents at which they will grow, from the lowest to the highest. The moisture contents are on a wet-weight basis.

Aspergillus halophilicus, A. restrictus, and *Wallemia sebi*
> Lower limit of moisture for growth: %
> Cereal grains (corn, wheat, sorghum, rice)   13.5-14.5
> Soybeans   12.0-12.5
> Peanuts, sunflower seeds, copra   9.0-10.0

Effects.—*A. halophilicus* and *W. sebi* cause little damage to seeds and other materials in which they grow. *W. sebi* has been encountered as the major and sometimes the only organism in high-sugar-content, long-shelf-life cakes, but even when the cakes were heavily invaded they had no musty odor or flavor. *W. sebi* commonly occurs in high-protein products such as soybean meal that has been exposed for a time to relative humidities of about 70 to 75%. *A. restrictus,* when growing in stored cereal grains at moisture contents of 14 to 15%, can kill and discolor the germs or embryos, but it grows too slowly to cause detectable heating. This is one example of fungal spoilage that is not accompanied by a temperature rise.

Possible Toxicity.—None of these fungi are known to produce any compounds that are toxic when consumed by man or domestic animals.

Aspergillus glaucus
> Lower limit of moisture for growth: %
> Cereal grains (corn, wheat, sorghum, rice)   14.0-14.5
> Soybeans   12.5-13.0
> Peanuts, sunflower seeds, copra   8.0-9.0

Effects.—*A. glaucus* kills and discolors seed embryos slowly at moisture contents near the lower limit for growth. The process is more rapid at higher moisture contents. It causes "blue-eye" in corn containing 14.5 to 15.0% moisture and may cause mustiness and caking. Normally it does not grow rapidly enough to cause an appreciable rise in temperature of the grain, but an increase in surface-

FIG. 7.5. SEED OF A COMMON BEAN *(PHASEOLUS VULGARIS)* FROM WHICH SPOROPHORES OF *ASPERGILLUS GLAUCUS* ARE GROWING. The bean was heavily invaded internally by the fungus, but this was not outwardly evident until the investigator (Luis Cesar Lopez) scratched his initials in the seed coat and so allowed the fungus to grow out and sporulate.

disinfected kernels yielding *A. glaucus* between one sampling period and the next is a warning that spoilage may be under way (Fig. 7.5).

Possible Toxicity.—Some isolates of individual species within the *A. glaucus* group, when grown in pure culture in the laboratory, produce compounds toxic to animals. There is circumstantial evidence to implicate feed ingredients invaded by one or more species of the group in cases of toxicity in the field. We have not observed any ill effects in rats and chicks fed rations containing grain heavily invaded by a mixture of species in the *A. glaucus* group.

Aspergillus candidus

Lower limit of moisture for growth:	%
Cereal grains (corn, wheat, sorghum, rice)	15.0-15.5
Soybeans	14.5-15.0
Peanuts, sunflower seeds, copra	10.0-11.0

Effects.—*A. candidus* kills and discolors germs of seeds very rapidly. It causes heating up to 55°C, discoloration of entire kernels, and total decay. In commercial storage of all kinds of seeds, *A. candidus* and *A. flavus* are the major causes of heating. Its presence in surface-disinfected kernels is evidence either of poor storage conditions in the past or of present deterioration.

Possible Toxicity.—As with *A. glaucus,* some isolates of *A. candidus,* when grown under the right conditions in the laboratory, produce compounds toxic to animals. However, samples of severely damaged corn decayed by a variety of fungi, including *A. candidus,* when added as 10% of the total to an otherwise balanced ration and fed to rats and chicks, produced no observable ill effects.

Aspergillus ochraceus

Lower limits of moisture for growth: Same as for *A. candidus.*

Possible Toxicity.—Some isolates of *A. ochraceus* produce ochratoxin, which is similar to and nearly as toxic as aflatoxin. Ochratoxin is also produced by *Penicillium viridicatum,* perhaps more commonly than it is by *A. ochraceus.* For a further discussion of ochratoxin see Chap. 14.

Aspergillus flavus

Lower limit of moisture for growth:	%
Cereal grains (corn, wheat, sorghum, rice)	18.0–18.5
Soybeans	17.0–17.5
Peanuts, sunflower seeds, copra	11.0–12.0

Effects.—*A. flavus* kills and discolors germs and decays whole kernels. It can cause heating up to 50° to 55°C.

Possible Toxicity.—Some isolates of the group, under some conditions of growth, produce aflatoxins. For further discussion of these see Chap. 14.

Other Comments.—The presence of *A. flavus* in surface-disinfected kernels usually is evidence of poor storage in the past or of spoilage under way at present in the bin from which the samples were taken. In some regions *A. flavus* may invade corn in the field before harvest.

Penicillium Species

Lower limit of moisture for growth:	%
Cereal grains (corn, wheat, sorghum, rice)	16.5–19.0
Soybeans	16.0–18.5
Peanuts, sunflower seeds, copra	11.0–13.0

Effects.—*Penicillium* kills and discolors germs and whole kernels and causes mustiness and caking. It causes blue-eye in corn stored at low temperature and with a moisture content above about 18 to 19%.

Possible Toxicity.—Isolates of several species of *Penicillium* are capable of producing toxins, for a further discussion of which see Chap. 14.

CONTROL OF STORAGE FUNGI

Low Moisture Content and Temperature

By far the most generally used method of preserving quality in stored grains and seeds and their products is storage at moisture contents too low for fungi to grow. A low and uniform moisture content and a moderately low and uniform temperature throughout the bulk combine to greatly reduce the possibility of moisture transfer within the bulk. This is the function of aeration systems. Temperatures throughout the bulk usually are monitored by means of thermocouples attached to cables that extend from the top to the bottom of the storage bin or tank. Most processes that result in spoilage of stored grains and similar materials are accompanied by a rise in temperature. Any rise in temperature, even a few degrees, is evidence that spoilage is under way.

Fungicides

It often is assumed that because many plant diseases, as well as decay and rot of many kinds of materials, can be prevented with fungicides, the same should apply to storage fungi on seeds. However, the storage fungi grow on substrates having a_w values of 0.65 to 0.90. Fungicides whose effectiveness depends on their being dissolved in water may not, under those conditions, be at all fungicidal. If seeds are treated with some dust fungicides that protect seedlings against damping off and then are stored at 75% relative humidity, heads of A. glaucus may grow out with a visible dab of fungicide on each of them. Milner et al. (1947) tested more than 100 supposedly fungicidal compounds for control of storage fungi on wheat. None of them greatly inhibited the fungi without also killing the seed. Various compounds submitted for testing as inhibitors of storage fungi on seeds have been found to have serious limitations of one sort or another, such as toxicity to animals, excessive cost, difficulty of application, undesirable effects on the processing quality of grain, and lack of toxicity to storage fungi. Some of them were effective at one combination of moisture content-temperature-time, but not at another. In practice, a whole range of moisture content-temperature combinations may occur within a given bin at a given time and, unless protection can be assured under all of these conditions throughout the storage life of the grain, the fungicide is of little value.

Propionic acid, combinations of propionic and acetic acids, and propionic acid-formaldehyde are effective preservatives of high-moisture grains to be used for feeds. The high-moisture, acid-treated grain has a higher feed efficiency than dry grain for beef and dairy cattle (Lane 1972; Perry 1972). The odor and flavor of such acid-treated grain probably would make foods processed from it unacceptable to most people.

Evaluation of Condition and Storability

A program of regular sampling and testing enables those in charge of stored grains or similar materials to determine the condition of the grain at the time of sampling, to predict its future storage risk, and to detect the incipient stages of spoilage long before they become of practical importance. Samples of approximately 0.5 kg are taken from different portions of the bulk, tested for moisture content, examined with a stereoscopic microscope for any evidence of fungus growth, and plated on a suitable agar medium to detect any internal invasion by storage fungi. Any increase in storage fungi between one sampling period and the next is evidence that deterioration is under way. The grain can then be processed before the deterioration becomes serious, or can be removed and dried. As stated by Burrell (1974), "In cool or temperate climates a well-planned and well-managed aeration system can greatly reduce the damage by insects, mites, and fungi, the chief hazards to quality in stored grains and seeds. Even in warm climates, selective ventilation can effectively limit damage caused by these agents, and so can be of tremendous value in preserving quality. The prevention of moisture transfer, through maintenance of a uniform temperature throughout the bulk, greatly reduces the chance of development of hidden or unexpected local areas of deterioration within the bulk. Unquestionably, the development of effective aeration systems has been a major contribution of modern engineering technology to grain storage. Combined with the use of temperature detection systems and a program of regular sampling and testing, it has revolutionized grain storage practices and has brought these far along the road from a somewhat chancy art to at least a moderately exact science."

REFERENCES

AYERST, G. 1969. The effects of moisture and temperature on growth and spore germination in some fungi. J. Stored Prod. Res. 5, 127-141.

BAILEY, C. H., and GURJAR, A. J. 1918. Respiration of stored wheat. J. Agric. Res. 12, 685-713.

BOTHAST, R. J., ADAMS, G. H., HATFIELD, E. E., and LANCASTER, E. B. 1975. Preservation of high-moisture corn: A microbiological evaluation. J. Dairy Sci. 58, 386-391.

BOTHAST, R. J., ROGERS, R. F., and HESSELTINE, C. W. 1974. Microbiology of corn and dry milled corn products. Cereal Chem. 51, 829-838.

BURRELL, N. J. 1974. Aeration. In Storage of Cereal Grains and Their Products. C. M. Christensen (Editor). Am. Assoc. Cer. Chem., St. Paul, Minn.

CARLYLE, R. E., and NORMAN, A. G. 1941. Microbial thermogenesis in the decomposition of plant materials. J. Bacteriol. 41, 699-724.

CARTER, E. P. 1950. Role of fungi in the heating of moist wheat. U.S. Dep. Agric. Circ. 838.

CHRISTENSEN, C. M. 1951. The Molds and Man. Univ. of Minnesota Press, Minneapolis, Minn.

CHRISTENSEN, C. M. 1972. Microflora and seed deterioration. In Viability of Seeds. E. H. Roberts (Editor). Chapman and Hall, London.

CHRISTENSEN, C. M., and GORDON, D. R. 1948. The mold flora of stored wheat and corn and its relation to the heating of moist grain. Cereal Chem. 25, 42-51.

CHRISTENSEN, C. M., and KAUFMANN, H. H. 1969. Grain Storage—The Role of Fungi in Quality Loss. Univ. of Minnesota Press, Minneapolis, Minn.

CHRISTENSEN, C. M., and KAUFMANN, H. H. 1974. Microflora. In Storage of Cereal Grains and Their Products. C. M. Christensen (Editor). Am. Assoc. Cer. Chem., St. Paul, Minn.

CHRISTENSEN, C. M., PAPAVIZAS, G. C., and BENJAMIN, C. R. 1959. A new halophilic species of Eurotium. Mycologia 51, 636-640.

DARSIE, M. L., ELLIOTT, C., and PEIRCE, G. J. 1914. A study of the germinating power of seeds. Bot. Gaz. (Chicago) 58, 101-136.

FIELDS, R. W., and KING, T. H. 1962. Influence of storage fungi on deterioration of stored pea seed. Phytopathology 52, 336-339.

FLANNIGAN, B., and DICKIE, N. A. 1972. Distribution of micro-organisms in fractions produced during pearling of barley. Trans. Br. Mycol. Soc. 59, 377-391.

HANLIN, R. T. 1973. The distribution of peanut fungi in the southeastern United States. Mycopathol. Mycol. Appl. 49, 227-241.

HANSEN, A. P., WELTY, R. E., and SHEN, R. 1973. Free fatty acid content of cacao beans infected with storage fungi. J. Agric. Food Chem. 21, 665-670.

HART, J. R., FEINSTEIN, L., and GOLUMBIC, C. 1959. Oven methods for precise measurement of moisture content of seeds. U.S. Dep. Agric., Agric. Mark. Serv., Mark. Res. Div., Mark. Rep. 304.

HESSELTINE, C. W., BOTHAST, R. J., and SHOTWELL, O. L. 1975. Aflatoxin occurrence in some white corn under loan, 1971: IV. Mold flora. Mycologia 67, 392-408.

HESSELTINE, C. W., SHOTWELL, O. L., KWOLEK, W. F., LILLIHOJ, E. B., JACKSON, W. K., and BOTHAST, R. J. 1976. Aflatoxin occurrence in 1973 corn at harvest. II. Mycological studies. Mycologia 68, 341-353.

HUMMEL, B. C. W., CUENDET, L. S., CHRISTENSEN, C. M., and GEDDES, W. F. 1954. Grain storage studies. XIII. Comparative changes in respiration, viability and chemical composition of mold-free and mold-contaminated wheat upon storage. Cereal Chem. 31, 143-150.

HUNT, W. H., and PIXTON, S. W. 1974. Moisture: its signficance, behavior, and measurement. In Storage of Cereal Grains and Their Products. C. M. Christensen (Editor). Am. Assoc. Cer. Chem., St. Paul, Minn.

HYDE, M. B. 1974. Airtight storage. In Storage of Cereal Grains and Their Products. C. M. Christensen (Editor). Am. Assoc. Cer. Chem., St. Paul, Minn.

KAVANAGH, T. E., REINECCIUS, G. A., KEENEY, P. G., and WEISSBERGER, W. 1970. Mold induced changes in cocoa lipids. J. Am. Oil Chem. Soc. 47, 344-346.

LANE, G. T. 1972. Preventing mold growth in high moisture grain. In Master Manual on Molds and Mycotoxins. G. L. Berg (Editor). Farm Technol. Agri-Fieldman 28, No. 5, 34a-41a.

LILLEHOJ, E. B., FENNELL, D. I., and HARA, S. 1975. Fungi and aflatoxin in a bin of stored white maize. J. Stored Prod. Res. 11, 47-51.

LUTEY, R. W. 1961. Staining of barley kernels by bacteria. Proc. Minn. Acad. Sci. 29, 174-179.

MILNER, M., CHRISTENSEN, C. M., and GEDDES, W. F. 1947. Grain storage studies. VII. Influence of certain mold inhibitors on respiration of moist wheat. Cereal Chem. 24, 507-517.

MILNER, M., and GEDDES, W. F. 1945. Grain storage studies. II. The effect of aeration, temperature, and time on the respiration of soybeans containing excessive moisture. Cereal Chem. 22, 484-501.

MILNER, M., and GEDDES, W. F. 1946. Grain storage studies. III. The relation between moisture content, mold growth, and respiration of soybeans. Cereal Chem. 23, 225-247.

MISLIVEC, P. B., DIETER, C. T., and BRUCE, V. R. 1975. Effect of temperature and relative humidity on spore germination of mycotoxic species of Aspergillus and Penicillium. Mycologia 67, 1187-1189.

MISLIVEC, P. B., and TUITE, J. 1970. Temperature and relative humidity requirements of species of Penicillium isolated from yellow dent corn kernels. Mycologia 62, 75-88.

PERRY, T. W. 1972. Improving feed efficiency with organic acids. In Master Manual on Molds and Mycotoxins. G. L. Berg (Editor). Farm Technol. Agri-Fieldman 28, No. 5, 42a-45a.

QASEM, S. A., and CHRISTENSEN, C. M. 1958. Influence of moisture content, temperature, and time on the deterioration of stored corn by fungi. Phytopathology 48, 544-549.

RAMSTAD, P. E., and GEDDES, W. F. 1943. The respiration and storage behavior of soybeans. Minn. Agric. Exp. Stn. Tech. Bull. 156.

SCHROEDER, H. W., and SORENSON, J. W., JR. 1961. Mold development of rough rice as affected by aeration during storage. Rice J. 64, 8-10.

SEMENIUK, G. 1954. Microflora. In Storage of Cereal Grains and Their Products. J. A. Anderson and A. W. Alcock (Editors). Am. Assoc. Cereal Chem., St. Paul, Minn.

SHOTWELL, O. L., GOULDEN, M. L., BOTHAST, R. J., and HESSELTINE, C. W. 1975. Mycotoxins in hot spots in grains. I. Aflatoxin and zearalenone occurrence in stored corn. Cereal Chem. 52, 687-697.

TUITE, J. F. 1959. Low incidence of storage molds in freshly harvested seed of soft red winter wheat. Plant Dis. Rep. 43, 470.

TUITE, J. F. 1961. Fungi isolated from unstored corn seed in Indiana in 1956-1958. Plant Dis. Rep. 45, 212-215.

TUITE, J. F., and CHRISTENSEN, C. M. 1957. Grain storage studies. XXII. Time of invasion of wheat seed by various species of Aspergillus responsible for deterioration of stored grain, and source of inoculum of these fungi. Phytopathology 47, 265-268.

U.S. DEP. AGRIC. 1974. The official United States standards for grain. U.S Gov. Printing Office, Washington, D.C.

VANDERGRAFT, E. E., HESSELTINE, C. W., and SHOTWELL, O. L. 1975. Grain preservatives: Effects on aflatoxin and ochratoxin production. Cereal Chem. 52, 79-84.

WALLACE, H. A. H. 1973. Fungi and other organisms associated with stored grain. In Grain Storage—Part of a System. R. N. Sinha and W. E. Muir (Editors). AVI Publishing Co., Westport, Conn.

WALLACE, H. A. H., and SINHA, R. N. 1975. Microflora of stored grain in international trade. Mycopathologia 57, 171-176.

WALLACE, H. A. H., SINHA, R. N., and MILLS, J. T. 1976. Fungi associated with small wheat bulks during prolonged storage in Manitoba. Can. J. Bot. 54, 1332-1343.

WEIDNER, H. 1975. Literature on stored product (mainly grain) production, 1974. Z. Pflanzenkr. Pflanzenschutz. 82, 236-254. (German)

WILSON, D. M., HUANG, L. H., and JAY, E. 1975. Survival of Aspergillus flavus and Fusarium moniliforme in high-moisture corn stored under modified conditions. Appl. Microbiol. 30, 592-595.

8

Bakery Products

J. G. Ponte, Jr. and C. C. Tsen

T he aim of this chapter is to discuss some of the ways in which yeasts and molds influence products manufactured by the baking industry. The effects of fungi can be either beneficial to the industry or quite deleterious. On the one hand, bakers' yeast is indispensable to the production of white bread, the staff of life, and on the other hand, mold growth can render bread and other baked foods unpalatable.

Cereals have been the basic food of man since prehistoric times, and were consumed long before breadmaking was developed. Gruels made from grains were probably the first way cereals were prepared for human consumption. Subsequently, simple pastes of ground grain and water were cooked, either directly in hot ashes or on flat stones, into a kind of unleavened bread. Leavened bread appears to have been first prepared in Egypt, where a combination of factors proved favorable for the development of breadmaking: (1) the availability of wheat, (2) the discovery of dough fermentation, and (3) the building of ovens (Reed and Peppler 1973). Information on the history of baking is obtainable from several sources (Storch and Teague 1952; McNeil 1967; Jacob 1970).

Today most of the wheat consumed by humans is eaten in the form of leavened baked foods. Cakes, cookies, and the like are leavened by chemical means (e.g., baking powder), but breads and rolls are primarily leavened by yeasts.

BAKERS' YEAST

Yeasts have been used for brewing and baking for thousands of years. Only in recent times, however, has yeast been grown in relatively pure culture for food use. In the past century, bakers obtained yeast as a by-product of the brewing and distilling industries. After about 1860, certain forces came into play which led to the production of yeast specifically for baking purposes (White 1954). These were the increased manufacture of bread in commercial bakeries because of the growth of industrial centers, and an emerging yeast technology which showed that large yeast crops could be grown when great volumes of compressed air were utilized during yeast growth. The fundamental work of Pasteur and Hansen paved the way for basic yeast physiology studies and for the essential concepts of pure culture techniques.

Bakers' yeast, *Saccharomyces cerevisiae,* is a unicellular fungus measuring 6 to 8 microns in diameter at maturity. About 3500 billion cells are contained in 0.454 kg of compressed bakers' yeast (Buckheit 1971). Encasing the yeast cell is a thin cell wall composed of glucan (30 to 35%), mannan (30%), protein (6 to 8%), chitin (1 to 2%), lipid (8.5 to 13.5%), and several percent of inorganic materials, especially phosphate (Phaff et al. 1966). The interior of the cell contains cytoplasm, a colorless matrix in which are suspended the nucleus, ribosomes, mitochondria, vacuoles, and lipid globules. Figure 1.1 shows structural components common to many yeasts, including *S. cerevisiae.*

Saccharomyces cerevisiae reproduces by forming ascospores or by budding. Ascospores may germinate into small haploid vegetative cells; two of these cells may fuse to result in a diploid generation of somewhat larger cells. An ungerminated ascospore may also fuse with a haploid vegetative cell to result in a diploid generation. The more common mode of *S. cerevisiae* reproduction is by budding in which a bulge or bud appears, usually at the end of the longitudinal axis or on the shoulder of the diploid vegetative cell. The bud expands to a size similar to the mother cell and then separates to yield an autonomous diploid unit. Two stages of vegetative reproduction by *S. cerevisiae,* showing the budding phenomenon, are illustrated in Fig. 8.1.

Saccharomyces cerevisiae is a remarkable organism. It can metabolize either aerobically or anaerobically. The bakers' yeast industry is a reality because *S. cerevisiae* can be economically grown under aerobic conditions, and the baking industry is able to produce bread, rolls, crackers, and the like because the organism can evolve

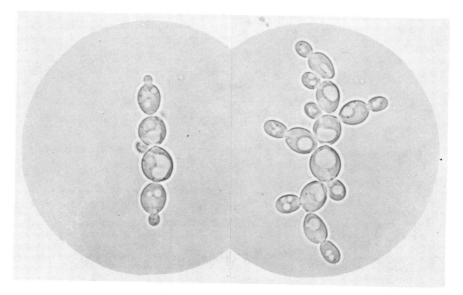

Courtesy of Fleischmann Co.
FIG. 8.1. TWO STAGES OF REPRODUCTION OF SACCHAROMYCES CEREVISIAE BY BUDDING (PHASE MICROSCOPY, 1200X). From Buckheit (1971).

carbon dioxide in dough under a variety of conditions (Pomper 1969).

Alcoholic fermentation is described by the classical Gay-Lussac equation:

$$C_6H_{12}O_6 \longrightarrow 2\ C_2H_5OH + 2\ CO_2$$

$$\text{(glucose)} \qquad\qquad \text{(ethanol)}\quad \text{(carbon dioxide)}$$

Fig. 8.2 summarizes the Embden-Meyerhof-Parnas scheme for alcoholic fermentation of simple sugars. The reactions accounting for the various compounds formed include phosphorylations, dephosphorylations, oxidations, reductions, and isomerations. *Saccharomyces cerevisiae* metabolizes carbohydrates aerobically using a combination of the Embden-Meyerhof-Parnas scheme and the Krebs tricarboxylic acid cycle (Fig. 8.3). It is well recognized that other end products are formed during metabolism of glucose and other substrates in fermenting doughs. These components are responsible in part for the flavor and aroma development in yeast-leavened bakery products.

Requirements for Production

The characteristics of bakers' yeast produced today are largely a

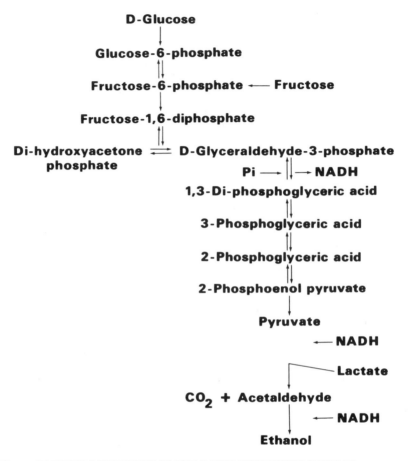

FIG. 8.2. EMBDEN-MEYERHOF-PARNAS FERMENTATION SCHEME

summation of requirements imposed by both the baker and the yeast producer. The baker needs a yeast that is reasonably tolerant to a variety of storage conditions and shows good fermentative activity in various types of doughs under different processing conditions. The producer needs a yeast that, for economic reasons, can be grown rapidly and reproduces itself without any alteration in essential characteristics. Over many years of research strains of *S. cerevisiae* have been evolved that meet these criteria.

The last century has seen several major changes in the way bakers' yeast is manufactured (Reed and Peppler 1973). These can be enumerated as follows.

Use of Aeration.—As indicated previously, the use of vigorous aeration during yeast respiration minimizes alcoholic fermentation and greatly enhances yeast growth.

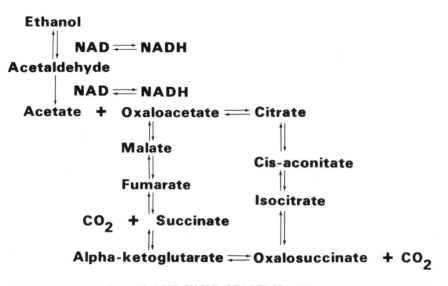

FIG. 8.3. TRICARBOXYLIC ACID CYCLE (STARTING WITH ETHANOL DERIVED FROM AEROBIC FERMENTATION)

Incremental Feeding.—The incremental feeding of yeast was introduced about 1915. The presence of glucose tends to inhibit respiration and encourage fermentation to ethanol; hence, it is important to provide the carbon and energy source (glucose or fructose) at a very low concentration so as to encourage respiration and the efficient reproduction of yeast cells.

Replacement of Grain with Molasses.—At the turn of the century the traditional source of carbon and energy in the mash (yeast growth medium) was corn, malt, and malt sprouts. During the 1920s and 1930s molasses, a by-product of the sugar beet and sugar cane refining industry, began to be utilized because of its lower cost. Other carbon and energy sources which have been used to a much lesser extent to support the growth of bakers' yeast include hydrolyzates of corn grits and wood, sugar-containing wastes from the confectionery industry, and organic chemicals such as ethanol, acetic acid, and lactic acid.

The Manufacturing Process

A flow diagram showing a typical bakers' yeast manufacturing process is illustrated in Fig. 8.4. For purposes of discussion, the commercial production of yeast can be divided into the following basic steps.

Yeast Propagation.—The commercial production of bakers' yeast is initiated in the laboratory. From pure stock cultures maintained in the laboratory, a loop of yeast cells is transferred to a small flask con-

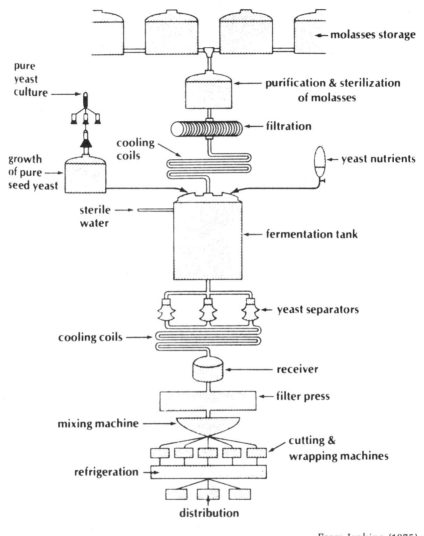

From Jenkins (1975)

FIG. 8.4. DIAGRAM OF BAKERS' YEAST MANUFACTURE

taining a sterile grain or molasses medium fortified with necessary growth factors. After a 2-day incubation period, the flask contents are transferred to pure culture tanks; these tanks may comprise a series of fermentors increasing in capacity from about 76 to 378 liters. The growth medium is sterilized directly in the culture tanks. Other precautions taken during these stages include the sterile transfer of yeast from one culture stage to the next, and the filtration of air.

Contamination of the culture by other microorganisms would ruin the subsequent scale-up to full-size fermentation.

Little aeration is employed in the early culture stages and incremental feeding is not practiced. The level of yeast growth is therefore relatively low, but this is not important, as the principal concern in these early propagation stages is to retain purity.

After the early stages, several more culturing stages are practiced in which incremental feeding and aeration are utilized. Finally, sufficient yeast cells have been grown to support a full-scale fermentation.

Fermentation.—The full-scale or "trade" fermentation of yeast is carried out in large vessels (75 to 225 m³) typically constructed of stainless steel. As previously indicated, molasses is the chief sugar source used in yeast production. Prior to its use, the molasses is made into a concentrated "wort" that has been sterilized and clarified. Certain minerals, vitamins, and a source of nitrogen are incorporated into the growth medium, depending on the deficiencies of these in the raw materials. Phosphorus is supplied as ammonium phosphate or alkaline phosphate salts, and nitrogen is added as ammonia, urea, or ammonium sulfate or phosphate. Biotin and thiamine are also added to the fermentation mixture.

The seed yeast is suspended in a start-volume of water and introduced into the fermentor. The water will contain the minerals needed, as noted above. The sugar-containing wort and the nitrogen source are then added to the yeast suspension.

Fermentation is conducted over a period of about 10 to 13 hr at 30°C. During this time nutrients are incrementally fed so that the sugar concentration is always very low, and the yeast suspension is vigorously aerated. Under these conditions, the nutrients are efficiently assimilated into yeast biomass, anaerobic fermentation is discouraged, and yields are favorable.

The presence of lactic acid-producing bacteria in yeast ferments was at one time thought to be desirable from the standpoint of ultimately imparting improved flavor to bread. Today heterofermentative lactics such as *Leuconostoc* spp., homofermentative lactobacilli, and other bacterial and fungal contaminants are viewed as undesirable.

Separation.—At the completion of fermentation the concentration of yeast in the suspension is about 3.5 to 4.5%. The first step in the recovery of the yeast from the spent fermented wort involves a concentration of the cells into a yeast cream of approximately 18 to 21% solids. This concentration is achieved by means of two passages through a centrifugal separator; the second passage is preceded by a water wash of the yeast cells. The yeast cream is relatively stable and can be kept for several days under refrigeration without loss of quality.

Filtration.—The yeast cream may be further concentrated by either passing through filter presses or rotating vacuum filters.

With filter presses, the yeast cream is pumped into the press which is divided into a number of compartments separated by frames fitted with tightly woven cloth filters. A pressure of 100 to 150 psi (about 9800 to 14,700 g/cm^2) is applied to reduce the water content of the yeast cream.

Vacuum filters are comprised of rotating drums fitted with filters and a vacuum pump. The yeast cream is introduced in the drum usually by directly rotating the drum in a vat of cream, or sometimes by spraying the cream onto the filter surface. As a thin layer of yeast cake forms on the filter surface, it is scraped off by knives and recovered.

Packaging and Distribution.—The yeast cake is prepared for packaging by first mixing with small amounts of water (to adjust the solids level to a target value of about 30%), emulsifiers, and oils. The emulsifiers function to improve yeast appearance and the oils aid in yeast extrusion and cutting operations.

The blended yeast and additives are extruded into continuous ribbons that are then sliced into blocks of desired weight. The blocks of yeast are wrapped with waxed paper and sealed by heat.

A temperature of 4°C or less is utilized to store the yeast. After manufacture the yeast is rapidly moved by refrigerated trucks to distribution centers, and from there to the consumer.

Greater detail on bakers' yeast manufacture is obtainable from several sources (White 1954; Peppler 1960; Reed and Peppler 1973).

Forms of Availability of Bakers' Yeast

Three basic forms of yeast are available to the baking industry. These forms and certain handling characteristics are discussed below.

Compressed Yeast.—Produced by the methods outlined in the preceding discussion, compressed yeast is available in 0.454- and 2.27-kg blocks. In the bakery these blocks should remain wrapped and stored under refrigeration (1° to 5°C) until needed. Even in the compressed form, yeast cells are alive and require energy to survive. This energy is derived from the yeast's reserve carbohydrate, glycogen. The utilization of glycogen by cells is called autofermentation, an exothermic process. Storage of the yeast at cool temperatures retards autofermentation, minimizes heat buildup, and prolongs yeast activity. Cool storage temperatures also inhibit autolysis, another yeast deteriorative process. Autolysis is a self-destructive process whereby proteases within the yeast begin to digest cellular protein. These proteases become activated when the yeast is not kept sufficiently cool.

When fresh, the yeast blocks have a creamy color, and characteristically break apart cleanly. Older yeast has a more brownish color, and tends to crumble when broken apart.

Bulk Yeast.—In recent years a new form of yeast has become available to bakers, namely bulk yeast (Schuldt and Seeley 1966). This yeast is in the form of a free-flowing, readily handled granular material packed in 11.4- to 22.7-kg bags. Bulk yeast is produced from the same strains of yeast as regular compressed yeast and has a similar composition. Bags used to package bulk yeast are of a multiwall design that allows the exchange of yeast respiratory gases with the outside atmosphere, while excluding oxygen. The exclusion of oxygen inhibits yeast respiration and thus inhibits heat build-up and loss of yeast strength. Good handling practices for bulk yeast include proper cool storage and the prompt closing of partly utilized bags to exclude as much air as possible.

Active Dry Yeast (ADY).—Bakers' yeast is available in a dried form that possesses a relatively long shelf-life. ADY is manufactured by a process similar to that for regular fresh bakers' yeast except that the moisture content of the filtered or extruded yeast cake is substantially reduced. Strains of *S. cerevisiae* normally used for fresh yeast manufacture are not suitable for ADY production; special strains better able to withstand the rigors of drying are used for this purpose.

Several methods are available for preparing ADY. Early procedures were directed toward combining compressed yeast with edible dry materials such as starch, calcium salts, or flour. Extruded yeast cake can be either tumble or tunnel dried at controlled forced-air temperatures ranging from about 28° to 40°C. Tunnel drying can be a continuous or batch process. Rotolouver drying is a batch process which involves feeding extruded yeast strands into a revolving metal cylinder equipped with baffles. Warm air is forced through louvers into the cylinder to remove moisture from the yeast. Air-lift or fluid-bed drying of yeast on a metal screen or perforated plate is accomplished by blowing warm air from the bottom at a velocity which suspends the strands in a fluid bed.

The moisture content of commercially marketed ADY ranges from 7.5 to 8.3%. The addition of emulsifiers such as sorbitan esters at 1% and an antioxidant such as butylated hydroxyanisole at 0.1% imparts prolonged stability to low-moisture ADY.

At the same solids level, ADY does not exhibit the baking activity of fresh yeast. The use of ADY by commercial bakers has therefore been limited. ADY is extensively used in home baking and in foreign and domestic bakeries that are outside of distribution regions able to provide refrigeration.

ADY packed in vacuum or under nitrogen gas loses about 1% of its activity per month at ambient temperature, or about 10% per year (Reed and Peppler 1973). On storage in air, ADY loses about 7% of its activity during the first month.

While ADY has a much longer shelf-life compared to compressed yeast, it is still a living organism and is best stored under cool, dry conditions. Rehydration should be carried out using at least four parts of water per part of ADY, and the water temperature should be 40° to 45°C. At lower temperatures components are leached from the yeast cells and baking activity declines (Ponte et al. 1960); at higher temperatures cells may be thermally inactivated.

ROLE OF YEAST IN BAKERY PRODUCTS

Yeast, as previously indicated, is indispensable to the production of many familiar baked foods. Bread, rolls, sweet goods (e.g., Danish pastry), crackers, some doughnuts, bagels, and pretzels are among the products in which S. cerevisiae is utilized as a leavening agent. Space does not permit a discussion of all yeast-leavened baked products; hence, primary attention here will be focused on the predominant product, white pan bread. The importance of the bread market and of the baking industry itself may be attested by statistics. In 1972 sales of bread and rolls amounted to almost $5 billion, while consumers spent over $11 billion for all bakery products (Anon. 1974).

An understanding of the role of yeast in bread production should be undertaken with a consideration of the processes used to make bread. A description of these processes is provided below. More information on breadmaking technology can be obtained from other sources (Fance 1966; Pyler 1973; Cotton and Ponte 1974; Jenkins 1975).

Bread Processing

White pan bread is produced today in the United States by a number of procedures, as well as hybrids of these procedures. For purposes of this chapter, four principal methods can be somewhat arbitrarily singled out and described: (1) sponge dough, (2) straight dough, (3) continuous mix, and (4) liquid ferment.

Sponge Dough.—The sponge dough process began to be commercially utilized in the 1920s and today remains the most widely used breadmaking method. Fig. 8.5 schematically depicts the sponge dough process, while Table 8.1 presents a formula representative of that used to make sponge dough bread.

The "sponge," comprising about 65% of the total flour plus a portion of the total dough water, yeast, and "yeast food," is first mixed. This

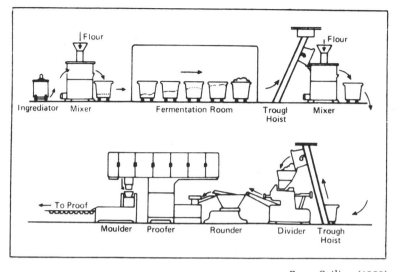

From Seiling (1969)
FIG. 8.5. PRINCIPAL STAGES OF THE SPONGE DOUGH PROCESS FOR BREADMAKING

mixing period is relatively brief and merely aims at uniformly combining the sponge ingredients. The sponge is then discharged into a trough, where it will undergo a fermentation period of some 4.5 hr in a controlled environment. From a starting temperature of about 25°C, the final temperature will increase by approximately 6°C due to the exothermic reactions brought about by yeast activity. Sponge volume will also increase by a factor of 4 or 5 as a consequence of carbon dioxide production during fermentation. Complex biochemical and physical events occurring during the fermentation will be considered in a later section.

At the termination of sponge fermentation, the sponge is transferred into a dough mixer. Also loaded into the mixing machine are the balance of the flour, water, and remaining ingredients. The mixer is operated first slowly to incorporate and blend these components, then it is speeded up (typically the mixer arms rotate at about 72 rpm) until the dough is completely mixed and properly developed. At this point the dough has been transformed from a sticky, wet-appearing mixture into a smooth, cohesive dough, characterized by a glossy sheen. This change occurs because of the unique properties of wheat flour. Upon the addition of water and the input of energy, wheat proteins and lipids form gluten. Gluten comprises the continuous phase of dough and possesses film-forming and gas-retaining properties. As S. cerevisiae evolves carbon dioxide, the dough is able to retain the gas and is thereby leavened.

TABLE 8.1

WHITE PAN BREAD FORMULATIONS FOR
SPONGE DOUGH AND STRAIGHT DOUGH[1]

Ingredient	Sponge Dough		Straight Dough
	Sponge	Dough	
Flour	65.0	35.0	100
Water (variable)	40.0	25.0	65
Yeast	2.5	—	3.0
Yeast food	0.2-0.5	—	0.2-0.5
Salt	—	2.25	2.25
Sweetener (solids basis)	—	8-10	8-10
Fat	—	3.0	3.0
Dairy product	—	0.5-3.0	0.5-3.0
Crumb softener	—	0.2-0.5	0.2-0.5
Rope and mold inhibitor	—	0.125	0.125
Dough improver	—	0-0.5	0-0.5
Enrichment	—	as needed	as needed

Source: Cotton and Ponte (1974).
[1]Ingredients based on 100 parts flour.

The mixed dough is transferred and allowed to rest for 20 to 30 min. During this period the dough recovers from mechanical stress, relaxes, and is better able to undergo subsequent processing.

Dividing is the next stage. As the term implies, the dough is cut into pieces of desired weight by a machine that volumetrically divides the pieces and discharges them onto a moving belt. The dough pieces are conveyed to a rounder, where the rough-appearing pieces are forced along a metal sleeve so that the pieces become rounded and have a smooth, dry skin. In this condition the dough pieces retain more carbon dioxide and are less sticky.

The steps of dividing and rounding involve a certain amount of abuse to the dough pieces, with the result that they are somewhat degassed and unpliable. To overcome this condition, the dough pieces upon leaving the rounder are accorded another rest period, or intermediate "proof," of some 8 to 12 min. This takes place in tray-type conveyors enclosed within cabinets of varying design.

From the intermediate proofer the dough pieces are conveyed to molding machines, which transform the more or less round pieces of dough into cylinders. Molders perform their functions with a series of rollers which sequentially squeeze the dough piece into a sheet, curl the sheet into a cylinder, and finally roll and seal the cylinder. Automatic molders feed the dough cylinders into bread pans.

The pans containing the dough pieces are put into fermentation enclosures called proof boxes for the last fermentation period prior to baking. The environment in these units is typically maintained at 35° to 43°C at a relative humidity of 80 to 95%. The dough pieces expand in the pans to a desired volume, a process usually requiring about 60 min. The proofed loaves are then placed in a humidity-controlled oven for baking. Gas within the dough fabric expands and the "oven spring" is produced. Steam and alcohol vapors also contribute to this expansion. Enzymes are active until the bread reaches about 75°C. At this temperature the gluten matrix coagulates and the dough structure is set. When the bread surface temperature reaches 130° to 140°C, sugars and soluble proteins react chemically to produce an attractive crust color. The center of the loaf may not exceed 100°C.

Remaining stages in the breadmaking process include cooling of the baked bread, slicing, wrapping, and distribution to stores for sale to the consumer.

Straight Dough.—In this method, all of the ingredients (see Table 8.1) are combined and mixed in one stage. After mixing, the straight dough is accorded a bulk fermentation period of 2 to 4 hr and is then processed in a manner similar to that described for the doughs in the sponge dough process.

Straight doughs are now made primarily by smaller bakeries or for production of some specialty breads. Straight doughs require less labor, time, and equipment compared to sponge doughs, but exhibit less tolerance to processing variations. This factor, plus the characteristically blander flavor of straight dough bread, has led over the years to the commercial preference of sponge doughs over straight doughs.

Continuous Mix Doughs.—During the 1950s, continuous mix methods of producing bread were introduced. As with other major industries, the objectives of continuous processing were to lower manufacturing costs and to increase product uniformity. The Do-Maker process (Baker 1954) and the Amflow method (Anon. 1958) are the principal continuous mix methods practiced in the United States today.

Fig. 8.6 and Table 8.2 show, respectively, a flow diagram and formula for the Amflow procedure. This method and the Do-Maker method share essential features.

First, a ferment or brew (stage 1) is prepared (with or without some flour in the case of the Do-Maker), and fermentation is allowed to proceed in a holding tank for about 1 hr. A second stage is added to the system, and fermentation continues for a total of about 2.5 hr. At the end of fermentation the brew is pumped to a premixer; cooling of the brew is achieved by means of a heat exchanger. Also fed into the

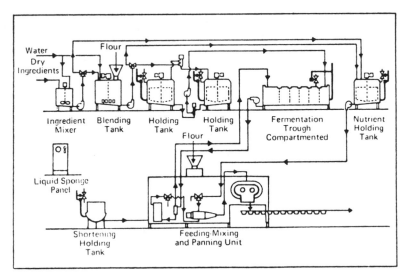

From Seiling (1969)

FIG. 8.6. SCHEMATIC DEPICTION OF CONTINUOUS MIX BREADMAKING (AMFLOW METHOD)

TABLE 8.2

CONTINUOUS-MIX PROCESS FORMULA FOR FLOUR BREW[1]

Ingredient	Formula	Stage 1	Stage 2	Dough Stage
Flour	100.0	30	—	70
Water	67.0	56	4.0	7.0
Yeast	3.0	3.0	—	—
Yeast food	0.5	0.5	—	—
Salt	2.0	—	2.0	—
Sugar	6.0	—	1.0	5.0
Milk	3.0	—	—	3.0
MCP	0.1	0.1	—	—
Mold inhibitor	0.1	—	—	0.1
Shortening	3.0	—	—	3.0
Oxidation/variable				

Source: Trum (1964).

[1]Ingredients based on 100 parts flour.

premixer are the remaining flour and water, sugar, melted fat, oxidant, and other optional ingredients. After the premixer combines all of these ingredients, the loosely mixed mass is pumped to a developing chamber. This chamber contains two counter rotating arc impellers. As the dough passes through the chamber it is under pressure and is subjected to intense mechanical energy. Under these conditions the dough is quickly developed and is then extruded into pans from a slit in the bottom of the chamber by a system of intermittently actuated knives. The dough from the continuous mix developer is much softer and warmer (about 38°C) compared to sponge doughs. The dough pieces are proofed and baked in a manner similar to that of sponge doughs.

Breads made by continuous mix processes characteristically have a much finer, more fragile crumb (inner structure). These features, plus a flavor that some feel is less desirable than that of sponge dough bread, have limited the acceptance of continuous mix bread in some markets.

Liquid Ferment.—Improvements in equipment design brought about by continuous mix technology have, in part, led to liquid ferment breadmaking methods. These methods combine traditional equipment with liquid brews that are easily metered and pumped, and thus offer advantages of space savings, labor savings, more processing flexibility, and better sanitation.

A flow diagram and formula for a liquid ferment process are shown in Fig. 8.7 and Table 8.3, respectively. In this system, a liquid ferment or brew is prepared and allowed to undergo fermentation much as described for continuous mix operations. After fermentation the brew is pumped to a dough mixer (same type as used in conventional sponge dough processing), along with remaining ingredients. The dough is mixed and then subjected to the same processing steps as outlined for the dough in the sponge dough process.

Functions of Yeast in Breadmaking

Yeast, as noted in the preceding discussions, plays an indispensable part in breadmaking. An examination of breadmaking technology reveals at least three major functions of yeast: (1) leavening, (2) flavor development, and (3) dough maturing.

Leavening.—Yeast, as observed earlier, evolves carbon dioxide gas as a consequence of its metabolism of simple sugars. In breadmaking, S. cerevisiae utilizes sugars derived from flour, and/or sugars added as an ingredient.

During fermentation in the sponge of the traditional sponge dough process, the only source of fermentable carbohydrate for the yeast is that obtained from the flour. Early in the fermentation carbon dioxide

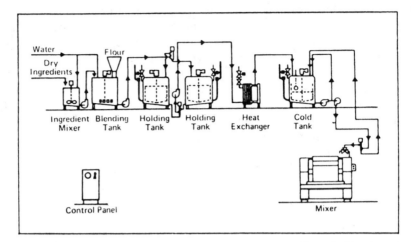

From Seiling (1969)
FIG. 8.7. SCHEMATIC OUTLINE OF LIQUID FERMENT PROCESS FOR BREAD-MAKING

TABLE 8.3

BREAD FORMULA MADE WITH FLOUR-CONTAINING FERMENT[1]

Ingredient	Formula	Liquid Sponge Formula	Materials Added at Mix
Flour	100	50.02	49.98
Water	66	55.77	10.23
Yeast	3	3	—
Yeast food	0.625	0.625	—
Salt	2.25	0.5	1.75
Sugar	8	1.5	6.5
Milk	3	—	—
Mold inhibitor	0.125	0.125	—
Fat	3	—	3
Emulsifier	0.5	—	0.5
Total	186.500	111.540	74.960

Source: Euverard (1967).
[1]Ingredients based on 100 parts flour.

is evolved rapidly as *S. cerevisiae* utilizes the available free sugars from the flour (Cooper and Reed 1968). A sharp drop-off in the rate of carbon dioxide production occurs after about an hour or so when the free sugars are depleted. At this point the yeast adapts to the fermentation of maltose, and gas production again rises. Maltose becomes available as a result of amylase hydrolysis of a part of the flour starch that has been mechanically damaged during milling. At the end of about 3 hr this source of fermentable sugar is exhausted and gas production drops off.

Added Sucrose.—During the dough stage of sponge dough processing and in straight doughs or other breadmaking systems all where all ingredients are present, *S. cerevisiae* will utilize sugar added to the dough as an ingredient. Added sucrose is almost immediately hydrolyzed into the constituent monosaccharides glucose and fructose due to the action of invertase. The yeast will then ferment both simple sugars, but at differing rates; glucose is preferred and is fermented at a faster rate than is fructose. Bread made with added sucrose typically will contain more fructose as a residual sugar than glucose. The relative rates of utilization of fructose, glucose and maltose in dough are shown in Fig. 8.8.

Added Amylases.—Amylases, collectively called diastatic enzymes, randomly split starch molecules to form smaller units of varying size (α-amylase) and progressively release maltose from the terminal portion of starch molecules (β-amylase). These enzymes are important in providing fermentable sugars for yeast growth and gas

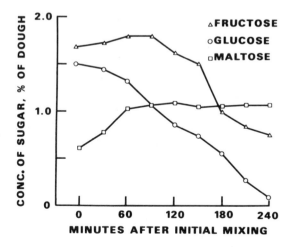

From Tang et al. (1972)
FIG. 8.8. CHANGES IN SUGAR CONCENTRATIONS IN FERMENTING DOUGH

production, and in providing machining and other functional properties in doughs. Amylase activity also affects the external surface color, volume, and keeping quality of bakery products.

Prior to mechanical harvesting, uncontrolled germination of some crops of wheat and barley led to the accumulation of high levels of α-amylase. Ungerminated wheat and barley contain low levels of α-amylase. In contrast, β-amylase activity is substantial and changes little upon seed germination. In order to elevate enzyme activity in flours milled from mechanically harvested cereal grains, malted wheat or barley have been traditionally added. More recently the addition of fungal α-amylase to flours at the mill or bakery has been practiced. The enzyme is derived from *Aspergillus oryzae, A. niger*, or *A. awamori* and is available in the form of diluted powders or water-dispersible tablets.

The influence and effects of fungal α-amylase supplementation on the development of desired quality of bread, rolls, buns, and crackers are numerous (Barrett 1975). Hydrolysis of dextrins yields maltose and glucose which sustains the fermentation rate and also produces small amounts of sugar in the finished product. Fungal enzyme mixtures contain higher levels of glucoamylase than do cereal and bacterial amylase preparations. This is important where sugars are not added to the formula and where the yeast has low maltase activity. α-Amylase activity lowers dough viscosity and affects dough softness. Grain, a term used to describe the fineness of the cell structure of the crumb of yeast-leavened bakery products, is affected by the amount and type of α-amylase used in dough formulations.

Consequences of Carbon Dioxide Production.—Evolution of carbon dioxide into the dough by the action of S. *cerevisiae* leads to the characteristic porosity of bread that enhances its palatability. This porosity is due to the carbon dioxide produced by the yeast and to the ability of the dough to retain the gas. The latter ability is related to the film-forming properties of dough made with wheat flour. Gluten, the protein complex of dough, forms a continuous structure in which are embedded starch granules; the gluten can be stretched into viscoelastic films that form the walls of gas cells. Carbon dioxide diffuses into these cells during the course of fermentation. The gas cells undergo subdivision during various dough-processing steps and eventually form the basic, cellular or porous structure of bread.

The classic investigations of Baker and Mize (1941) indicate that carbon dioxide generated during yeast fermentation does not in itself produce gas cells. Yeast-generated carbon dioxide diffuses into cells already formed from two sources, viz., air bubbles occluded in the dough during mixing and air adsorbed on flour particles (one-fifth of the volume of flour is entrapped air).

Flavor Development.—Fresh bread has a pleasing and appealing flavor that in part is responsible for its universal acceptance as a human food. Yet the flavor is subtle and difficult to characterize in spite of considerable effort to do so. A comprehensive review of this subject has recently been compiled (Maga 1974).

Bread flavor is thought to emanate from two main sources: yeast fermentation and crust browning.

Yeast Fermentation.—The precise role of yeast fermentation has been difficult to establish. Many compounds—organic esters and acids, alcohols, carbonyls—are formed as by-products of yeast fermentation. Some of these compounds are volatile and are baked out of the bread. Of those compounds remaining in the bread, it has not been possible to determine that any of these are specifically related to bread flavor. Attempts to synthesize bread flavor by combining compounds found in dough, liquid ferments, or bread have not been successful.

Many bakers feel that traditional methods of breadmaking involving substantial periods of bulk fermentation yield bread with the most desirable flavor. A number of studies have been done showing that certain organic compounds do increase in concentration as a function of fermentation time (Maga 1974). Countering this view, however, is the work of other investigators (Collyer 1966; Kilborn and Tipples 1968) which has shown that taste panel members could not distinguish flavor differences in bread made with or without bulk fermentation. These findings would indicate that flavor compounds formed during baking and during proofing were sufficient to yield satisfactory bread flavor.

It should be noted here that some of the organic compounds formed during fermentation may arise from bacterial action (Robinson et al. 1958). Lactic acid bacteria found in dough presumably are associated with yeast; commercial bakers' yeast normally contains a low number of contaminating microorganisms. In addition to S. cerevisiae, other yeasts may be responsible for characteristic flavors of certain breads (Ng 1976).

The above discussion indicates that the exact role of yeast fermentation in bread flavor development is not well understood. Yet, it would seem clear that yeast fermentation by-products must play an important role in bread flavor. Even if traditional bulk fermentation is not used in breadmaking, the loaves must be proofed. During this period S. cerevisiae proliferates and produces compounds which undoubtedly contribute to flavor, either directly or as flavor precursors. Breakdown products of flour protein undoubtedly play an important role in flavor and color development. Yeast proteolytic enzymes modify peptones and polypeptides for growth; however, a portion of these products of proteolysis react with sugars to impart desirable flavors upon baking.

Proteolytic enzymes are in fact added to doughs on a large commercial scale, and it is estimated that two-thirds of the white bread baked in the United States is treated with enzymes derived from *Aspergillus oryzae* (Barrett 1975). Proteolysis during the fermentation stage results in shortened protein chains which realign into sheets of protein film. As a result, less time is required before the point of maximum extensibility is reached. Proper levels of fungal proteinase improve handling and machining properties of dough and yield bread loaves having increased volume and better symmetry.

Fungal and bacterial proteases mellow the gluten during fermentation to yield the proper balance of extensibility and strength in cracker dough. These doughs can be rolled out very thinly without tearing and lie flat in the oven without bubbling or curling at the edges.

Crust Browning.—The extent of crust browning is influenced substantially by previous activities of *S. cerevisiae* in the dough, as discussed above. The importance of crust browning in flavor development can be shown by removing the crust of freshly baked bread and storing the bread until it has cooled, or by baking bread in a microwave oven where crust browning does not occur. Under these conditions the product will lack the flavor usually associated with bread. Presumably, therefore, a part of bread flavor is formed in the crust during baking and then diffuses into the crumb where it becomes absorbed. Over 100 flavor compounds have been found in bread (Maga 1974). Table 8.4 lists some of the compounds reportedly produced during fermentation and/or baking (Magoffin and Hoseney 1974).

Dough Maturing.— A third function of yeast is to contribute to many changes in dough that are collectively termed dough "maturing" or "ripening." A properly matured dough is one that exhibits optimum rheological properties (optimum balance of extensibility and elasticity), such that it may be machined well and will lead to bread with desirable volume and crumb characteristics. Some of the reactions leading to dough maturing are as follows.

Yeast fermentation yields alcohol and carbon dioxide, among other products. Alcohol is water miscible and, since appreciable amounts are formed, it therefore influences the colloidal nature of the flour proteins and alters the interfacial tension within dough. Some of the carbon dioxide dissolves in the aqueous phase of the dough and forms weakly ionizable carbonic acid, which lowers the pH of the system. Another effect of the carbon dioxide is to distend the dough as the gas expands, thus imparting some physical energy to the dough.

Saccharomyces cerevisiae liberates ammonia from ammonium sulfate and ammonium chloride added as yeast foods, thus liberating

TABLE 8.4

COMPOUNDS REPORTEDLY PRODUCED DURING FERMENTATION
AND/OR BAKING

Organic Acids		Alcohols	Aldehydes and Ketones	Carbonyl Compounds
Butyric	Acetic	Ethanol	Acetaldehyde	Furfural
Succinic	Lactic	n-Propanol	Formaldehyde	Methional
Propionic	Formic	Isobutanol	Isovaleraldehyde	Glyoxal
n-Butyric	Valeric	Amyl alcohol	n-Valeraldehyde	3-Methyl butanal
Isobutyric	Caproic	Isoamyl alcohol	2-Methyl butanol	2-Methyl butanal
Isovaleric	Caprylic	2,3-Butanediol	n-Hexaldehyde	Hydroxymethyl
Heptanoic	Isocaproic	β-Phenylethyl	Acetone	furfural
Pelargonic	Capric	alcohol	Propionaldehyde	
Pyruvic	Lauric		Isobutyraldehyde	
Palmitic	Myristic		Methyl ethyl ketone	
Crotonic	Hydrocinnamic		2-Butanone	
Itaconic	Benzylic		Diacetyl	
Levulinic			Acetoin	

Source: Magoffin and Hoseney (1974).

sulfuric and hydrochloric acids in the doughs. These acids, together with carbonic acid, further lower the pH which, in turn, significantly influences gluten hydration and swelling, the reaction rate of enzymes in dough, oxidation-reduction reactions, and various chemical reactions (Magoffin and Hoseney 1974).

Reductases produced by *S. cerevisiae* also affect dough rheological properties by acting through intermediate substrates found in dough (Reed and Peppler 1973).

SPOILAGE OF BAKERY PRODUCTS

The microbial spoilage of bakery foods is caused mainly by molds and occasionally by bacteria (Dennington 1941-42; Brachfeld 1969; Barrett 1970; Pyler 1973). Contamination may be due to microflora naturally present in or on ingredients used to prepare bakery foods, or to extraneous sources such as air and packaging materials to which the products are exposed after processing. *Rhizopus stolonifer* (*R. nigricans*) produces a cottony white mycelium with black sporangia and is often referred to as the "black bread mold." *Neurospora sitophila* (perfect state, *Monilia sitophila*) is known as the "red bread mold" as a result of its prolific formation of pink to red conidia and mycelium. Many penicillia and aspergilli can grow at lower water activities (a_w) (see Chap. 2) and are associated with spoilage of a variety of bakery products. *Mucor* and *Geotrichum* spp., in addition to numerous other molds, are occasional causes of decay. Two yeasts, *Saccharomycopsis* sp. (formerly *Endomycopsis fibuligera*) and *Trichosporon variabile*, are responsible for "chalky bread," a condition resulting from white, chalk-like spots of growth developing in bread crumb. Problem fungi are generally high amylase producers, thriving on the starchy substrate provided in bakery foods.

Considering bakery products overall, bacterial spoilage occurs less often than mold spoilage. However, the implications of bacterial spoilage as a public health hazard are probably greater. *Staphylococcus aureus* growth and toxin production may occur in cream-filled bakery products not stored at refrigerated temperatures. Infections caused by *Salmonella* contamination of custards and cream pies and fillings have also been documented. Ingredients such as eggs, egg products, milk, milk products, soy flour, dried yeast, coconut, and nut meats, among others, have been found to be sources of *Salmonella* in bakery products (Silliker 1969). "Rope" is a bacterial infection caused by a mucoid variant of *Bacillus subtilis* (formerly called *B. mesentericus*) in bread and rolls. The bacterial spores survive baking temperatures and grow out, digesting protein and starch to make the crumb structure soft and sticky with a brown discoloration. The contaminated bread has an odor characteristic of overripe cantaloupe. *Serratia marcescens* is responsible for "bloody bread," a defect resulting from red pigment formation by the organism. Bacterial spoilage, if it occurs, generally precedes mold spoilage. Since bakery product mycology is the topic of interest here, bacterial contamination and spoilage will not be considered further in this chapter.

Sources of Fungal Contamination

Ingredients.—Generally, bread and other bakery products out of the oven are free of viable vegetative fungal cells and spores due to thermal inactivation during the baking process. However, glaze, icing, nut meat, sugar, or spice applications to products after baking may introduce viable molds and yeasts which could proliferate under suitable storage conditions and cause spoilage. Also, although heat inactivation of fungi is usually achieved during baking, fungal deterioration of ingredients prior to incorporation into product formulae may result in undesirable organoleptic properties of the finished product. Potential sources and causes of fungal contaminants and deterioration are outlined in the following sections.

Flour and Meal.—As pointed out in Chap. 7, the moisture content of grains and oilseeds coupled with the temperature at which they are stored are the two most important factors influencing the growth of fungal contaminants. If such conditions are suitable, the microflora indigenous to seeds may proliferate and produce spores or other structures which are generally more resistant to heat or chemicals than are vegetative cells. During milling, oxidizing agents may be used to bleach flours. This treatment, in addition to the physical processes of dry-cleaning and washing of seeds prior to milling, and followed by sifting, substantially reduces fungal populations of finished flours and meals. Such treatments, however, do not necessarily totally remove or inactivate mycotoxins which may have been produced by fungi on grains and oilseeds during storage. Several fungi known to produce mycotoxins will grow readily on grains and oilseeds (Fig. 8.9). Constant monitoring for the presence of mycotoxins in flours and meals by millers and bakery food manufacturers is therefore recommended in order to control the risk of contamination in marketed products.

Milled grains and oilseeds are extremely receptive to mold growth. Moisture uptake may occur as a result of exposure to adverse weather, storage, or transport conditions, enabling spores surviving milling to germinate and grow. Airborne fungal spores may also contaminate and grow on flours and meals after milling. In turn, airborne spores from these sources can contaminate bakery products as they leave the

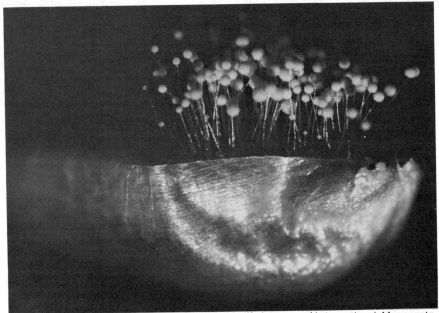

Courtesy of Dr. C. M. Christensen, University of Minnesota
FIG. 8.9. *ASPERGILLUS OCHRACEUS* CONIDIOPHORES GROWING FROM THE
GERM OF A WHEAT KERNEL THAT WAS CUT IN HALF AND PLACED ON AGAR.

oven, but before packaging. Thus, with the possible exception of oven
residence time, at no stage in the handling of grains and oilseeds, their
flours and meals, or products produced from them, is there complete
freedom from the risk of fungal contamination and growth.

Sugar.—The susceptibility of raw sugar to fungal attack depends on
the composition of molasses film on sugar crystals and, in particular,
upon its a_W (Eskin *et al.* 1971). The a_W of most raw sugars lies between
0.60 and 0.75, thus enabling xerotolerant molds and osmophilic yeasts
to grow.

Biodeterioration of raw sugars, molasses, and corn syrups is usually
caused by invert**ase**-producing yeasts and molds. Inversion of sucrose
to glucose and fructose results in changes in specific rotation of
substrate solutions as well as sweetness and functional properties.
Aspergillus, Penicillium, Monilia, Scopulariopsis, Alternaria, and
Cladosporium spp. of molds have been associated with deterioration
of raw cane sugars. Yeast species likely to cause problems include
Saccharomyces, Torulopsis, Candida, Schizosaccharomyces,
Hansenula, and *Pichia.* The consequence of sugar or sugar syrup
spoilage may be little more than higher yeast and mold counts on

TABLE 8.5

FUNGI ASSOCIATED WITH SPOILAGE OF FATS AND OILS IN FOODS

Type of Food	Spoilage	Spoilage Fungus
Cream	Methyl ketones, free fatty acids	Geotrichum candidum
Butter	Free fatty acids	Cladosporium suaveolins, G. candidum
Margarine	Rancidity, methyl ketones	Cladosporium butyri, Candida lipolytica
Lard	Free fatty acids	Paecilomyces aureocinnamoneum
Palm oil	Free fatty acids	Aspergilli, P. varioti
Coconut oil	Methyl ketones	Margarinomyces bubaki
Peanut oil	Free fatty acids	Aspergilli
Corn oil	Free fatty acids	Aspergillus tamarii
Rapeseed oil	Lipoxidation	A. niger
Olive oil	Rancidity	A. repens
Oats	Bitterness	A. restrictus
Wheat	Soapiness	Paecilomyces varioti
Barley	Free fatty acids	Monilia acremonium

Adapted from Eskin et al. (1971).

finished bakery products if, for example, the contaminated sugar is added to the baked product in the form of icing or powder. Use of extensively deteriorated sugars and syrups, however, can affect the organoleptic and textural characteristics of bakery foods.

Shortening, Oils, and Fats.—The lipolytic activities of fungi on the triglycerides of oils and fats used in bakery product formulations may cause rancidity, acidity, bitterness, soapiness, and other off-flavors. Such activities may occur in seeds or other plant parts from which oils are derived, during storage of plant and animal oils and fats, or in the finished bakery food. Table 8.5 lists some types of food spoilage associated with fungal degradation of fats in foods.

Nuts, Raisins, Spices, Etc.—As stated earlier, nearly all fungi are killed during baking processes given to most bakery foods. However, incorporation of nut meats, raisins, and other dried fruits into bakery products such as fruitcakes, which are subjected to relatively low baking temperatures, represents a potential problem with respect to mold spoilage. Fruitcakes are often prepared months in advance of consumption and stored at room temperature. A desirable a_W range for fruitcakes is from 0.70 to 0.85, since lower a_W tends to reduce palatability. These conditions are adequate for supporting the growth of fungi which may be initially present on or in nut meats and dried fruits, and which may survive baking. Nut meats and dried fruits, in addition to spices, also represent a potential source of contamination when added to food items after baking.

Although spices often contain a variety of fungal contaminants, some may actually inhibit the growth of fungi in bakery foods. Bullerman (1974) found that cinnamon, at concentrations ranging from 0.02 to 20%, inhibited *Aspergillus parasiticus* growth and aflatoxin production in various degrees in yeast extract broth. The inhibitory effect was greater on aflatoxin production than on mold growth, with cinnamon levels of 2% inhibiting aflatoxin production by 97 to 99%. These studies were conducted as a result of observations that rye and whole wheat breads would support *A. parasiticus* growth and aflatoxin production better than would raisin bread containing cinnamon.

Other Sources.—Contamination of bakery products can occur upon exposure to airborne fungal spores during the cooling period between baking and packaging. Contact with unsanitary machinery and packaging materials may also result in transfer of viable fungi to baked foods.

Control of Fungal Growth

Antimicrobial Agents.—In order to extend the shelf-life of bread and other bakery products, microbial spoilage, particularly mold growth, should be eliminated or reduced. There are basically two ways to achieve this purpose. One is, of course, to adopt a strong sanitation program and good quality control measures. The other involves the application of antimicrobial agents.

The use of vinegar as an antimicrobial additive can be traced far back into history. The early work of Kirbey *et al.* (1937) showed that acetic acid had a marked toxicity for *Aspergillus niger*, the active agent being the undissociated acid molecule. The effects of various acids at the same pH value differ considerably. Fatty acids, as a rule, are much more toxic to molds than are mineral acids and other organic acids such as lactic, citric, and tartaric.

Propionic Acid and Its Salt.—The discovery that certain lower fatty acids exert an antimicrobial effect on molds led Hoffman *et al.* (1939) to patent the use of propionic acid and its salts as mold inhibitors in bread. Propionates were selected because the higher homologs, though they had higher antimicrobial activity, gave baked goods undesirable odors and flavors. Propionic acid and its salts were further found by O'Leary and Kralovec (1941) to be effective against *Bacillus mesentericus* (now considered to be a *B. subtilis* variant), the rope-causing bacterium. Because of their potent antimicrobial action on both mold and bacterial growth, sodium and potassium propionates have been the most widely used antimicrobial additives in the baking industry. These salts are free-flowing, readily soluble, and much easier to use than propionic acid. They are more effective in the lower pH range. Added at recommended levels, propionates have no appreciable influence on yeast fermentation or on the flavor of yeast- or chemically-leavened products.

Sodium Diacetate.—The other salts of lower fatty acids used in the baking industry are acetates and diacetate. The application of sodium diacetate and its usage levels in cakes have been discussed by Glabe (1940-41) and Chichester and Tanner (1968), respectively. Sodium diacetate is essentially dry acetic acid and is similar in its action to the propionates.

Sorbates.—Sorbic acid, unlike acetic or propionic acids, is a conjugated unsaturated fatty acid ($CH_3CH = CHCH = CH\text{-}COOH$). In 1945 Gooding secured a patent covering the use of unsaturated fatty acids as fungistatic agents. He reported that sorbic acid was particularly effective, surpassing sodium benzoate in certain applications. Subsequent studies by other investigators confirmed the antifungal properties of sorbic acid and established its commercial applications for controlling mold and yeast growth in foods, including bakery products (Wyss 1948; Grubb 1957).

Sorbic acid is only slightly soluble in water, while its salts, particularly potassium sorbate, are very soluble. The sorbates are more effective than sodium propionate in inhibiting the growth of molds in bread (DeSa 1966). Although they are extremely effective against yeasts and molds, they have limited antibacterial activity at the levels generally used in foods. Because sorbates also inhibit yeast growth, they are not directly used in yeast-leavened products (Chichester and Tanner 1968). Several means have been proposed to minimize the adverse effects of sorbates on yeast fermentation. One is to spray sorbic acid or its potassium salt on the bakery foods rather than to incorporate it into the dough. The other is to coat sorbic acid or its salt with a high melting point fat such as palmitic acid to reduce its inhibitory effect on yeast before baking (Melnick et al. 1961; Neu 1973).

Sodium Benzoate and Parabens.—Sodium benzoate has long been used as an antimicrobial additive for foods. It becomes active when converted to the acid form. The pH range for optimal microbial inhibition by benzoic acid is 2.5 to 4.0, which is lower than that of propionic or sorbic acids. Therefore, its use in bakery foods is limited to fruit fillings, jellies, and jams. Under regulations of the United States Food and Drug Administration, sodium benzoate and benzoic acid are allowed to be used in bakery foods at levels up to 0.1%.

Parabens are alkyl esters of p-hydroxybenzoic acid. These compounds have essentially the same spectrum of antimicrobial activity as benzoic acid. They are, however, effective over a wider pH range. Generally, a mixture of parabens, sometimes with benzoate, is used to improve keeping qualities of cakes (particularly fruitcakes), pie crusts, icing, toppings, and fillings because their antimicrobial effects are additive (Chichester and Tanner 1968). Like sorbates, parabens exert an inhibitory effect on yeast and are therefore not used in yeast-leavened foods.

Antimicrobial agents, as discussed above, have some overlapping in their activity spectra. Propionates are active against molds and essentially inactive against yeasts at levels commonly used in bakery products. Although they are not so effective against bacterial growth, they do inhibit the growth of rope bacteria in bread. Sorbates are active against bacteria, yeasts, and molds, but they are used mainly against the latter two. The benzoates and parabens have the widest range of activity against spoilage bacteria, molds, and yeasts. Their activity within a low pH range limits their use in baked foods. Table 8.6 summarizes the application of these additives to bakery foods (Chichester and Tanner 1968).

Mode of Action and Levels of Use.—The exact mechanism of action of most of the above antimicrobial agents is not fully understood. These agents could be adsorbed onto the cell wall of the microorganism to prevent it from absorbing nutrients. They could also block one or more enzyme systems to inhibit some specific metabolic functions of the organism.

Several antimicrobial additives are legally permitted for use in the food industry in the United States and many other countries. However, levels vary from country to country. Propionates are not mentioned among permitted additives in France, Germany, and Portugal, although their use is legal in the United States; parabens are excluded in Belgium, France, The Netherlands, and Turkey, but are permitted in some other countries (Chichester and Tanner 1968).

In standard breads, rolls, and buns, propionates and sodium diacetate are permitted to be used at levels not more than 0.32% and 0.4% by weight of flour, respectively. Suggested levels of propionates and sorbates in bakery products as recommended by Brachfeld (1969) are listed in Tables 8.7 and 8.8.

Other Means.—In addition to the use of antimicrobial agents, a number of other methods can also be used to prevent or reduce mold spoilage in bakery products. Ultraviolet rays can be used to sterilize the atmosphere to which baked products are exposed after baking (Nagy 1948). Microwave treatment for the sterilization of wrapped bread has also been studied. Burg and Schweiz (1968) reported the

TABLE 8.6

APPLICATION OF PRINCIPAL ANTIMICROBIAL ADDITIVES TO
BAKERY FOODS

Bakery Food	Benzoic Acid and Sodium Benzoate	Methyl and Propyl Parabens	Sorbates	Propionates	Acetates and Diacetates
Yeast-leavened				+	+
Chemically leavened		+	+	+	+
Pie crust and pastries		+	+	+	+
Pie filling	+	+	+	+	

TABLE 8.7

SUGGESTED LEVELS OF CALCIUM OR SODIUM PROPIONATE IN
BAKERY FOODS

Bakery Food	Level of Propionate[1]
White breads, buns, and rolls	2.5-5.0 oz per 100 lb flour
Dark breads, whole or cracked wheat or rye breads, buns, and rolls	3.0-6.0 oz per 100 lb flour
Angel food cake	1.5-3.5 oz per 100 lb batter
Cheese cake	2.0-4.0 oz per 100 lb batter
Chocolate or Devil's food cake	5.0-7.0 oz per 100 lb batter
Fruitcake	2.0-6.0 oz per 100 lb batter
Pound cake, white or yellow layer cake	4.0-6.0 oz per 100 lb batter
Pie crust	2.0-5.0 oz per 100 lb dough
Pie fillings	2.0-5.0 oz per 100 lb filling

[1]1 oz = 28.35 gm;   100 lb = 45.4 kg.

usefulness of sterilizing bread by heating for 45 to 90 sec in a
microwave sterilizer with a holding capacity of 181 to 272 kg.

Aside from sterilization, control of environmental factors such as
humidity and temperature is important. Before packaging, bakery
products should be cooled to a temperature which will prevent water
condensation within the container. Schulthesis and Spicher (1974)
recently conducted a study to examine the influence of moisture on the
growth of mold present in the air of bakery and storage rooms.

TABLE 8.8

SUGGESTED LEVELS OF SORBIC ACID OR POTASSIUM SORBATE IN
BAKERY FOODS

Bakery Food	Percent in Batter	Oz per 100 lb Batter[1]
Angel food cake	0.03 - 0.06	0.5-1.0
Cheese cake	0.09 - 0.125	1.5-2.0
Chocolate cake	0.09 - 0.125	1.5-2.0
Devil's food cake	0.30	5.0
Fruitcake	0.05 - 0.10	0.8-1.5
Pound, yellow or white layer cakes	0.075- 0.10	1.2-1.5
Cake mixes	0.05 - 0.10	0.8-1.5
Fillings, fudges, icings, and toppings	0.05 - 0.10	0.8-1.5
Pie crust dough	0.05 - 0.10	0.8-1.5
Pie fillings	0.05 - 0.10	0.8-1.5
Doughnut mixes	0.03 - 0.08	0.5-1.25

[1] 1 oz = 28.35 g;   100 lb = 45.4 kg.

Cultures inoculated onto the crumb of toasting bread were stored at 75 to 100% relative humidity and 10° to 35°C. They found that, in general, the latent period becomes shorter and the number of microorganisms increases as a_w increases. The influence of moisture on mold growth becomes more marked as the temperature nears the optimum. It becomes more critical to control fungal spoilage of bakery foods during humid summer days and in tropical climates.

Contribution No. 78-106-A, Department of Grain Science and Industry, Kansas Agricultural Experiment Station, Manhattan, KS 66506.

REFERENCES

ANON. 1958. The newest of the continuous dough-making systems. Baker's Dig. 32, No. 6, 49-52.

ANON. 1974. Bakery trends 1974. Bakery Prod. Mark. 9, No. 6, 51-67.

BAKER, J. C. 1954. Continuous processing of bread. Proc. Am. Soc. Bakery Eng. 65-79.

BAKER, J. C., and MIZE, M. D. 1941. The origin of the gas cell in bread dough. Cereal Chem. 18, 19-34.

BARRETT, F. F. 1970. Extending the keeping quality of bakery products. Baker's Dig. 44, No. 4, 48-49, 67.

BARRETT, F. F. 1975. Enzyme uses in the milling and baking industries. In Enzymes in Food Processing. G. Reed (Editor). Academic Press, New York.

BRACHFELD, B. A. 1969. Antimicrobial food additives. Baker's Dig. 43, No. 5, 60-65.

BUCKHEIT, J. T. 1971. Yeast—its controlled handling in the bakery. Baker's Dig. 44, No. 1, 46-49, 60.

BULLERMAN, L. B. 1974. Inhibition of aflatoxin production by cinnamon. J. Food Sci. 39, 1163-1165.

BURG, F., and SCHWEIZ, M. D. B. 1968. Microwave sterilization of sliced bread. Brot Gebaec. 22, 58-60. (German)

CHICHESTER, D. F., and TANNER, F. W. 1968. Antimicrobial food additives. In Handbook of Food Additives. T. E. Furia (Editor). Chemical Rubber Co., Cleveland, Ohio.

COLLYER, D. M. 1966. Fermentation products in bread flavor and aroma. J. Sci. Food Agric. 17, 440-445.

COOPER, E. J., and REED, G. 1968. Yeast fermentation. Baker's Dig. 42, No. 6, 22-29, 63.

COTTON, R. H., and PONTE, J. G., JR. 1974. Baking industry. In Wheat Production and Utilization. G. Inglett (Editor). AVI Publishing Co., Westport, Conn.

DENNINGTON, A. R. 1941-42. The bakers' summer battle against mold. Baker's Dig. 16, 207-209.

DeSA, C. 1966. Sorbic acid, its use in yeast-raised bakery products. Baker's Dig. 40, No. 6, 50-52.

DWORSCHAK, E., MOLNAR, E. B., and SZILLI, M. 1973. Effect of fermentation promoting additives on the nutritive value of bakery products. Lebensm. Wiss. Technol. 6, 14-18.

ESKIN, N. A. M., HENDERSON, H. M., and TOWNSEND, R. J. 1971. Biochemistry of Foods. Academic Press, London.

EUVERARD, M. R. 1967. Liquid ferment systems for conventional dough processing. Baker's Dig. 41, No. 5, 124-129.

FANCE, W. J. 1966. The Student's Technology of Breadmaking and Flour Confectionery. Routledge and Kegan Paul, London.

GLABE, E. F. 1940-41. Advances in the control of bread diseases. Baker's Dig. 15, 221-223.

GOODING, C. M. 1945. U. S. Pat. 2,379,294. June 26.

GRUBB, T. C. 1957. Symposium on antimicrobial preservatives. Bacteriol. Rev. 21, 251-254.

HARRISON, J. S. 1971. Yeast in baking: Factors affecting changes in behavior. J. Appl. Bacteriol. 34, 173-179.

HOFFMAN, C., DALBY, G., and SCHWEITZER, T. R. 1939. U. S. Pat. 2,154,449. Apr. 18.

JACKEL, S. S. 1969. Fermentation flavors of white bread. Baker's Dig. 43, No. 5, 24-25, 28, 64.

JACOB, H. E. 1970. Six Thousand Years of Bread. Greenwood Press, Westport, Conn.

JENKINS, S. M. 1975. Bakery Technology, Book 1, Bread. Lester and Orpen, Toronto.

KILBORN, R. H., and TIPPLES, K. H. 1968. Sponge and dough type bread from mechanically developed doughs. Cereal Sci. Today. 13, No. 1, 25-28, 30.

KIRBEY, G. W., FREY, C. N., and ATKINS, L. 1937. Further studies on the growth of bread molds as influenced by acidity. Cereal Chem. 14, 865-878.

LABUZA, T. P., and JONES, K. A. 1973. Functionality in breadmaking of yeast protein dried at two temperatures. J. Food Sci. 38, 177-178.

MAGA, J. A. 1974. Bread flavor. CRC Crit. Rev. Food Technol. 5, 55-142.

MAGOFFIN, C. D., and HOSENEY, R. C. 1974. A review of fermentation. Baker's Dig. 48, No. 6, 22-23, 26-27.

MATZ, S. A. 1972. Bakery Technology and Engineering, 2nd Edition. AVI Publishing Co., Westport, Conn.

McNEIL, W. H. 1967. World History. Oxford Univ. Press, New York.

MELNICK, D., VAHLTEICH, H. W., and BOHN, R. T. 1961. U. S. Patent 2,997,394. Aug. 22.

NAGY, R. 1948. Control of fungi in bakery plants. Baker's Dig. 22, No. 3, 47-48, 51.

NEU, H. 1973. Mold prevention in bread by sorbic acid-palmitic acid anhydride. Dtsch. Lebensm. Rundsch. 69, 401-404.

NG, H. 1976. Growth requirements of San Francisco sour dough yeasts and bakers' yeast. Appl. Environ. Microbiol. 31, 395-398.

O'LEARY, D. K., and KRALOVEC, R. D. 1941. Development of B. mesentericus in bread and control with calcium acid phosphate or calcium propionate. Cereal Chem. 18, 730-741.

PEPPLER, H. J. 1960. Yeast. In Bakery Technology and Engineering, 2nd Edition. S.A. Matz (Editor). AVI Publishing Co., Westport, Conn.

PHAFF, H. J., MILLER, M. W., and MRAK, E. M. 1966. The Life of the Yeasts. Harvard Univ. Press, Cambridge, Mass.

POMERANZ, Y., and SHALLENBERGER, J. A. 1971. Bread Science and Technology. AVI Publishing Co., Westport, Conn.

POMPER, S. 1969. Biochemistry of yeast fermentation. Baker's Dig. 43, No. 2, 32-38.

PONTE, J. G., JR., GLASS, R. L., and GEDDES, W. F. 1960. Studies on the behavior of active dry yeast in breadmaking. Cereal Chem. 37, 263-279.

PYLER, E. J. 1969. Enzymes in baking—theory and practice. Baker's Dig. 43, No. 1, 46-52.

PYLER, E. J. 1973. Baking Science and Technology. Siebel Publishing Co., Chicago.

REED, G., and PEPPLER, H. J. 1973. Yeast Technology. AVI Publishing Co., Westport, Conn.

REISS, J. 1975. Mycotoxins in foodstuffs. V. The influence of temperature, acidity, and light on the formation of aflatoxins and patulin in bread. Eur. J. Appl. Microbiol. 1, 183-190.

ROBINSON, R. J., LORD, T. H., JOHNSON, J. A., and MILLER, B. S. 1958. The aerobic microbiological population of pre-ferments and the use of selected bacteria for flavor production. Cereal Chem. 35, 295-305.

SCHULDT, E. H., and SEELEY, R. D. 1966. Bulk yeast. Baker's Dig. 40, No. 2, 42-44.

SCHULTHESIS, J., and SPICHER, G. 1974. Influence of processing techniques on mold growth on bread and baked goods. Getreide Mehl Brot. 28, 288-296. (German)

SEILING, L. 1969. Equipment demands of changing production requirements. Baker's Dig. 43, No. 5, 54-59.

SHEPPARD, R., and NEWTON, E. 1957. The Story of Bread. Routledge and Kegan Paul, London.

SILLIKER, J. H. 1969. Some guidelines for the safe use of fillings, toppings, and icings. Baker's Dig. 43, No. 1, 51-54.

STORCH, J., and TEAGUE, W. D. 1952. Flour for Man's Bread. Univ. of Minnesota Press, Minneapolis.

TADAYON, R. A. 1976. Characteristics of yeasts isolated from bread doughs of bakeries in Shiraz, Iran. J. Milk Food Technol. 39, 539-542.

TANAKA, Y., KAWAGUCHI, M., and MIYATAKE, M. 1976. Studies on injury of yeast in frozen dough. 2. Effect of ethanol on frozen storage of Bakers' yeast. J. Food Sci. Technol. (Tokyo) 23, 419-424.

TANAKA, Y., and SATO, T. 1969. Fermentation of fructosides in wheat flour by bakers' yeast. J. Ferment. Technol. 47, 587-595.

TANG, R. T., ROBINSON, R. J., and HURLEY, W. C. 1972. Quantitative changes in various sugar concentrations during breadmaking. Baker's Dig. 46, No. 4, 48-55.

TRUM, G. W. 1964. The AMF continuous bread pilot plant. Cereal Sci. Today 9, 248-254.

WADE, P. 1972. Technology of biscuit manufacture: Investigation of the role of fermentation in the manufacture of cream crackers. J. Sci. Food Agri. 23, 1021-1034.

WHITE, J. 1954. Yeast Technology. John Wiley and Sons, New York.

WYSS, O. 1948. Microbial inhibition by food preservatives. Adv. Food Res. 1, 373-393.

YANEZ, E., WULF, H., BALLESTER, D., FERNADEZ, N., GATTAS, V., and MONCKEBERG, F. 1973. Nutritive value and baking properties of bread supplemented with Candida utilis. J. Sci. Food Agric. 24, 519-525.

9

Traditional Fermented Food Products

L. R. Beuchat

F ermentation was among the first methods used by man to preserve foods. Perhaps by accident, the addition of salt to food
was observed to prevent food from spoiling and causing illness.
One can speculate that early man also observed improved flavor,
aroma, and texture in aged oilseeds, grains and fleshy roots to which
salt and moisture had been added. Today we recognize these changes
to be due to fermentation by nontoxic fungi and bacteria. Traditional
fermented foods are especially popular in the Far East. Fermented
grains and starchy roots are an important part of diets of people in
parts of Africa and Latin America.

Several factors are responsible for the continued popularity of
traditional fermented foods. First, there are millions of people living in
tropical, subtropical, and temperate regions of the world who do not
enjoy the luxury of supermarkets or electricity. Many do not have
access to commercially processed foods, or refrigerators in which to
preserve their own. These people rely upon fermentation to provide
variety to diets consisting largely of grains and vegetables, and to
extend the keeping time of oilseeds, grains, and roots. Fungi used in
food fermentation processes modify original materials organoleptically, physically, and nutritionally. In parts of the world where
protein in the form of meat and dairy products is not available,
fermented foods are used as flavoring agents for otherwise bland

vegetable diets. Microbial hydrolytic enzymes digest plant constituents and presumably enhance the digestibility of raw materials. In a few instances, fermented products are used to add color, and thus appeal, to nonfermented foods.

Procedures for preparing some traditional fermented foods are outlined below. Processes were selected for inclusion based on their widespread use and, to some extent, on the availability of reliable information pertaining to the microorganisms involved and biochemical changes which occur. Detailed discussions of traditional fermented foods are given in extensive reviews (Yokotsuka 1960; Hesseltine 1965; Gray 1970; Kwon 1972; Iljas et al. 1973; Yong and Wood 1974; Beuchat 1976). A list of fermented foods from around the world is given in Table 9.1.

KOJI

The word "koji" is an abbreviation of the Japanese work "kabitachi," meaning "bloom of mold" (Yong and Wood 1974). Koji applies to molded masses of cereals, legumes, or flours of these seeds which serve as sources of enzymes and in some cases as inocula for larger quantities of nonfermented materials. Koji, then, refers not to the end product of a traditional fermentation process but rather to the first stage in a process usually involving a second fermentation. The use of koji in traditional food fermentations is analogous to the use of malt in alcoholic fermentation of grain in western nations.

Specific kojies are prepared for various types of fermentations. Desired levels of proteolytic, lipolytic, and amylolytic enzymes in kojies are attained through proper selection of molds and yeasts and through variations in substrate and incubation conditions. Tane koji is an inoculum, usually containing high numbers of fungal spores, used to start any type of koji. The term koji is sometimes used to designate the starter (tane koji) as well as the whole mixture of soybeans and wheat used in shoyu (soy sauce) manufacture. The reader unfamiliar with shoyu fermentation is often confused by this terminology. Other products similar to koji are known by other names (Hesseltine 1965). Ragi is made with rice flour and is used to manufacture arrak (arak, arack), tapé ketan (tapej), brem, and tapé ketella (peujeum) in Indonesia. Kyoku-shi, chiu-niang, and chou are Chinese terms for koji (Pederson 1971). Methods for preparing specific kojies are detailed below with descriptions of traditional fermented foods.

SOYBEANS

Shoyu

The brewing of shoyu or sho (soy sauce) goes back at least 1000

TABLE 9.1

SOME FERMENTED FOODS OF FUNGAL ORIGIN

Product	Geography	Substrate	Microorganism(s)	Nature of Product	Product Use
Ang-kak	China, Southeast Asia, Syria	Rice	Monascus purpureus	Dry red powder	Colorant
Chee-fan	China	Soybean whey curd	Mucor sp., Aspergillus glaucus	Solid	Eaten fresh, cheese-like
Chinese yeast	China	Soybeans	Mucoraceous molds and yeasts	Solid	Eaten fresh or canned, used as side dish with rice
Dawadawa	West Africa	African locust bean	—	Solid, sun-dried	Eaten fresh
Gari	West Africa	Cassava	Corynebacterium manihot, Geotrichum candidum	Wet paste	Eaten fresh as staple
Hamanatto	Japan	Whole soybeans	Aspergillus oryzae	Beans retain individual form, raisin-like	Flavoring agent for meat and fish, eaten as snack
Idli	Southern India	Rice and black gram	Lactic bacteria, Torulopsis candida and Trichosporon pullulans	Spongy, moist	Bread substitute
Injera	Ethiopia	Teff, or maize wheat, barley, sorghum	Candida guilliermondii	Bread-like, moist	Bread substitute
Kaanga-kopuwai	New Zealand	Maize	Bacteria and yeasts	Soft, slimy	Eaten as vegetable
Ketjap	Indonesia	Black soybeans	Aspergillus oryzae	Syrup	Seasoning agent

Name	Country	Substrate	Microorganisms	Texture	Use
Lao-chao	China	Rice	Rhizopus oryzae, R. chinensis, Chlamydomucor oryzae, Saccharomycopsis sp.	Soft, juicy	Eaten as such or combined with eggs, seafood
Meitauza	China, Taiwan	Soybean cake	Actinomucor elegans	Solid	Fried in oil or cooked with vegetables
Meju	Korea	Soybeans	Aspergillus oryzae, Rhizopus spp.	Paste	Seasoning agent
Minchin	China	Wheat gluten	Paecilomyces, Aspergillus, Cladosporium, Fusarium, Syncephalastrum Penicillium, Trichothecium spp.	Solid	Condiment
Miso	Japan, China	Rice and soybeans or rice and other cereals such as barley	Aspergillus oryzae, Lactobacillus bacteria, Saccharomyces rouxii	Paste	Breakfast food, soup base, seasoning
Ogi	Nigeria, West Africa	Maize	Lactic bacteria, Cephalosporium, Fusarium, Aspergillus, Penicillium spp., Saccharomyces cerevisiae, Candida mycoderma (C. valida or C. vini)	Paste	Staple, eaten with vegetables
Oncom	Indonesia	Peanut press cake	Neurospora sitophila, less often Rhizopus oligosporus	Solid	Roasted or fried in oil, used as meat substitute
Poi	Hawaii	Taro corms	Lactobacillus bacteria, Candida vini (Mycoderma vini), Geotrichum candidum	Semisolid	Side dish with fish, meat

TABLE 9.1 (Continued)

Product	Geography	Substrate	Microorganism(s)	Nature of Product	Product Use
Shoyu	Japan, China Philippines, other parts of Orient	Soybeans and wheat	Aspergillus oryzae or A. soyae, Lactobacillus bacteria, Saccharomyces rouxii (Zygosaccharomyces spp.)	Liquid	Seasoning for meat, fish, cereals, vegetables
Sufu	China, Taiwan	Soybean whey curd	Actinomucor elegans, Mucor hiemalis, M. silvaticus, M. subtilissimus	Solid	Soybean cheese
Tao-si	Philippines	Soybeans plus wheat flour	Aspergillus oryzae	Semisolid	Seasoning agent
Taotjo	East Indies	Soybeans plus roasted wheat meal or glutinous rice	A. oryzae	Semisolid	Condiment
Tapé	Indonesia and vicinity	Cassava or rice	Saccharomyces cerevisiae, Hansenula anomala, Rhizopus oryzae, Chlamydomucor oryzae, Mucor sp., Endomycopsis fibuligera	Soft solid	Eaten fresh as staple
Témpé	Indonesia and vicinity, Surinam	Soybeans	Rhizopus spp., principally R. oligosporus	Solid	Fried in oil, roasted, or used as meat substitute in soup

years in Japan. It probably started as a result of the introduction of Buddhism from China and the change to a vegetable diet in 552 A.D. (Hesseltine 1965). Several types of shoyu are available in the Far East. Fish are the raw materials of shoyu in Thailand and Vietnam, soybeans are the main ingredient in China and Korea, and equal amounts of soybeans and wheat (Triticum aestivum) are used in Japan (Yokotsuka 1972). Good quality chinese chau yau (shoyu) has high specific gravity and viscosity, high nitrogen content, a dark brown color, and may have cane sugar added. The opposite is true of Japanese-type shoyu, which has low viscosity and nitrogen content, and a slightly lighter reddish brown color. The alcohol content of genuine Japanese shoyu is higher than that of shoyu from other Far East countries. Ninety percent of Japanese shoyu produced is of the darker koikuchi type. About 10% of the Japanese production is of a lighter colored usukuchi type. Small amounts of tamari shoyu are produced. Koikuchi shoyu is made from soybeans and wheat, usukuchi from soybeans, wheat, and rice, and tamari from soybeans alone. More than 4000 Japanese manufacturers produce over 1 million kl of shoyu per year. This translates into a relatively constant annual per capita consumption in Japan of about 10 liters. Statistics are not available on soy sauce production in the People's Republic of China, Taiwan, and other nations of Southeast Asia.

The first step in making shoyu requires preparing the tane koji or seed koji. This is accomplished by inoculating steamed, polished rice or a mixture of moistened wheat bran and soybean flour with Aspergillus oryzae or A. soyae. The mixture is spread in small wooden boxes or trays and cultured in a warm room for 72 hr. The seed koji is cooled twice by hand mixing.

At the same time tane koji is being cultured, defatted soybeans and wheat are being readied for inoculation. Soybeans are soaked in running water for 10 to 15 hr at room temperature and then cooked under 10 to 12 psi (980 to 1177 g/cm^2) pressure for about 1 hr. This process increases soybean weight by about 210% and volume by 220%. Extended cooking times may cause decreases in total, amino, ammonia, and tannin-precipitable nitrogen, acidity, volatile acids, and glycerin (Yokotsuka 1960). Immediately after cooking, soybeans are cooled to below 30°C. Wheat is prepared by roasting and crushing. Degradation of lignin and glycosides to yield vanillin, ferulic acid, vanillic acid, and 4-ethylguaiacol during the roasting process is believed to contribute to shoyu aroma and flavor.

To prepare koikuchi-type shoyu, 0.1 to 0.2% of tane koji is added to a mixture of 55% treated soybeans and 45% roasted, crushed wheat. The resultant mixture is known as "culturing koji" or just koji by those familiar with shoyu manufacture. It contains about 45% moisture. Koji is spread in 2- to 5-cm layers on large porous bamboo trays. The cultures are then placed in humidity-controlled rooms maintained at 25° to 35°C for about 50 hr (Yokotsuka 1960, 1972). During this period, extensive production of yellowish-green conidia occurs. Mature koji is mixed with brine containing 17 to 22% salt to form the mash, or moromi. The volume of salt water is 110 to 120% of the weight of the raw material. The mash is allowed to ferment in wooden or concrete containers for 3 to 4 months, if warmed, or about 1 year at ambient temperatures. The mash is occasionally stirred with a rod or with compressed air. A rapid decrease in mash pH from 6.0 to 7.0 initially to about 4.5 occurs. Alcohol fermentation begins at this point (Yokotsuka 1960). When mash is satisfactorily fermented, the liquid is separated using a hydraulic press (Fig. 9.1) or less often by siphoning. Raw shoyu is pasteurized at 70° to 80°C, followed by clarification with alum or kaolin. In Japan, benzoic acid or butyl-p-hydroxybenzoate is legally added to refined shoyu, but the trend is toward aseptic bottling without the addition of chemical preservatives (Yokotsuka 1972; Yong and Wood 1974). A typical process for making shoyu is illustrated in Fig. 9.2.

Flavor of shoyu is attributed largely to activities of microorganisms which grow in the mash. Neutral and acid proteases and peptidases of *A. oryzae* or *A. soyae* hydrolyze soybean and wheat proteins to form peptides and free amino acids. These components, especially glutamic acid, contribute to taste. Amylase and lipase are also produced. Once brine is added to the koji, vegetative *Aspergillus* cells are killed. Breakdown products in koji serve as nutrients, initially for lactic acid bacteria and later for osmophilic yeasts in the mash. Salt tolerant, tetrad-forming cocci found in mash have been named *Sarcina hamaguchiae*, *Tetracoccus soyae*, *Pediococcus acidilactici* var. *soyae*, and *Pediococcus soyae* (Yong and Wood 1974). *Lactobacillus delbrueckii* is also active in the mash. After bacteria produce organic acids to reduce the pH to about 4.5, salt-tolerant osmophilic yeasts begin to grow rapidly. *Saccharomyces soja*, *Zygosaccharomyces japonicus*, *Z. soyae*, *Z. major*, *Z. salsus*, *Pichia farinosa*, *Torulopsis famata* (now *T. candida*), *Candida polymorpha* (now *C. diddensii*), and *Mycoderma*, *Torula*, and *Monilia* species have been reported to grow in mash (Onishi 1963; Yong and Wood 1974). *Z. major* and *Z. soyae* impart desirable flavor characteristics through their fermentation. Pellicle- or film-forming yeasts, such as *Z. salsus*, *Z. japonicus*, and *Pichia*, ring-forming *Torulopsis*, and certain bottom yeasts belonging to *Zygosaccharomyces* are harmful to the keeping quality of shoyu. *Z. major*, *Z. soyae*, *Z. salsus*, and *Z. japonicus* are now classified as one species, *Saccharomyces rouxii*. Most *Zygosaccharomyces* spp. are now classified as *S. rouxii* (Lodder 1970). One can only as-

Courtesy of Kikkoman Foods
FIG. 9.1. HYDRAULIC PRESSES USED FOR EXPELLING LIQUID FROM MASH IN THE MANUFACTURE OF SHOYU (SOY SAUCE)

sume that reference in the literature to the desirability of S. rouxii in shoyu mash fermentation is directed toward characteristics of Z. major and Z. soyae. These three yeasts are referred to as soy yeasts. Their biochemical activities in shoyu mash, although not thoroughly understood, are greatly affected by salt concentration, pH, and temperature (Onishi 1963). S. rouxii ferments glucose and maltose but not galactose, sucrose, and lactose. Furfuryl alcohol, a flavor-producing component in shoyu, is formed by S. rouxii by reduction of furfural present in koji. Torulopsis spp. produce alkyl phenols, providing characteristic flavor to mash, but S. rouxii does not. Peptides, amino acids and derivatives (especially sodium glutamate

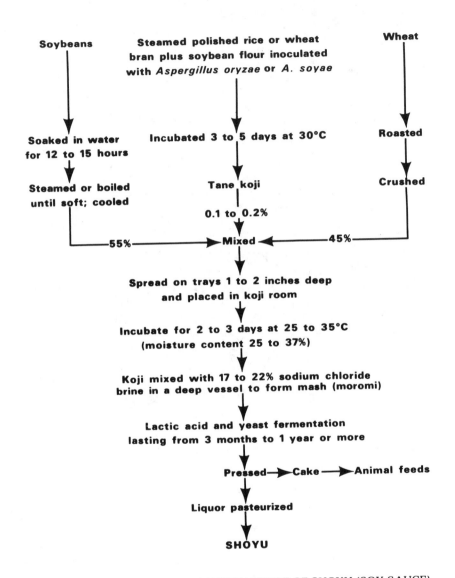

FIG. 9.2. FLOW SHEET FOR THE MANUFACTURE OF SHOYU (SOY SAUCE)

TABLE 9.2

PROXIMATE COMPOSITION OF SOME FERMENTED SOYBEAN PRODUCTS

	Product		
Component (%)	Shoyu	Miso[1]	Témpé
Moisture	62-70	41.2-52.3	55-65
Protein	0.2-0.6	8.3-23.5	6-11
Oil	0-0.2	1.6-10.5	9-12
Sugar	2.0-8.2	3.2-20.4	0.1-0.6
Starch and dextrin	1.0-1.2	3.2-10.0	0-1.0
Fiber	0	1.6- 4.0	3.1-5.8
Mineral matter[2]	1.8-2.0	0.9- 2.9	0.2-0.4
Sodium chloride	16-20	4.0-12.8	0-0.2
Amino nitrogen	0.5-0.7	0.2- 0.8	1.1-2.0
Total acids	0.5-1.3	0.6- 2.5	0.1-0.5
pH	4.5-4.7	5.1- 5.3	7.2-7.5

[1]Adapted from Shibasaki and Hesseltine (1962).
[2]Excludes sodium chloride.

formed from glutamic acid), ethyl esters of fatty acids, polyalcohols, and aldehydes also impart desirable flavor and aroma to shoyu. The proximate composition of Japanese shoyu is shown in Table 9.2.

The color of shoyu is due to products of amino-carbonyl reactions. Color formation in mash proceeds nonoxidatively. Color change after removing the seal from bottled shoyu is caused by oxidative reactions.

The production of aflatoxins by aspergilli used in shoyu preparation has been totally denied by many researchers. Some aspergilli widely used in Japanese food industries have been demonstrated to be capable of producing other kinds of so-called mycotoxins such as aspergillic, kojic, β-nitro-propionic, oxalic and formic acids (Yokotsuka 1972).

In order to lower production costs and obtain a product with more uniform quality, a process for making "chemical" shoyu has been established. Soybean meal and wheat are hydrolyzed by refluxing with boiling hydrochloric acid for 12 to 16 hr. When a maximum concentration of amino acid nitrogen is reached, the preparation is adjusted to pH 4 with sodium hydroxide. There is no market for the inferior flavored chemical shoyu in Asia, although it finds a market in Europe and North America. Still another type of shoyu is called shinshiki. This shoyu is made by first partially hydrolyzing raw materials with acid and then fermenting the hydrolysates with bacteria and yeasts.

Miso

Miso is a semisolid fermented food made in Japan from soybeans alone or from a mixture of soybeans and rice or barley. These three types of miso are further classified on the basis of taste into sweet, medium salty, and salty groups. Groups are divided by color into white, light yellow, red, and brown (Shibasaki and Hesseltine 1962; Ebine 1971). White miso, preferred in western Japan, is very sweet and has a low salt content. Edo miso is light reddish brown, has a sweet taste and low salt concentration, and is preferred in the Tokyo area. A reddish brown product with high salt content, sendai miso, is popular in northern Japan. Shinshu miso, produced in larger quantities than any other type of miso, is popular in Tokyo and central Japan. Mame miso requires the longest fermentation time (2 years) and is prepared almost entirely from soybeans. This deep reddish brown, distinctively flavored miso is preferred in Nagoya and parts of central Japan. The prime function of miso as a food is its flavoring qualities. Miso is often consumed at breakfast along with rice. It is used as an ingredient in soup which might also contain vegetables, tofu, and seaweed (Shibasaki and Hesseltine 1962). Two types of miso marketed in the United States are shown in Fig. 9.3.

Miso is prepared by a two-step fermentation process. The first step involves producing a koji from polished rice. In the second stage, the koji is combined with steamed soybeans or a mixture of soybeans and rice or barley. Salt and inoculum in the form of miso from a previous batch are then added to complete the fermentation mix.

Polished rice is used most frequently in preparing miso koji; however, barley and soybeans are also used in some districts in Japan. Brown rice is not suitable because the surface of the grain is hard and contains wax, thus inhibiting the growth of inoculum (Shibasaki and Hesseltine 1962). After soaking the rice in water overnight or until the moisture content increases to about 35%, excess water is removed and the swollen rice is steamed for about 40 min in a closed cooker or 1 hr in an open kettle. Rice is then spread in trays, cooled to 35°C, inoculated with about 0.1% of tane koji, and mixed. The preparation of tane koji for miso is similar to that described for shoyu, except A. oryzae strains should possess high amylolytic, as well as proteolytic activity, since hydrolysis of large quantities of rice starch is necessary during the second stage of fermentation.

Temperature, moisture, and aeration are important factors in the production of miso koji. Satisfactory mold growth and production of large amounts of desirable enzymes are attained at 28° to 35°C. Overheating results in growth of undesirable bacteria which may have contaminated the koji during incubation. Rice should be moist enough for A. oryzae to grow but not so moist that bacteria will grow.

Courtesy of Dr. C. W. Hesseltine, USDA
FIG. 9.3. TWO TYPES OF MISO MARKETED IN THE UNITED STATES

Humidity should be high enough to prevent the rice from drying, and ventilation should supply adequate oxygen and dissipate carbon dioxide from the atmosphere. During the 40- to 50-hr fermentation period, koji should be thoroughly stirred at least twice. Rice is completely covered with white mycelium at the end of the incubation period. It is important to terminate incubation before yellowish green conidia are produced. Growth beyond this stage may result in undesirable flavors. At the end of incubation, salt is added to the cooled koji at a level calculated to be necessary to make a batch of miso.

Concurrently with the preparation of rice koji, whole soybeans are washed, soaked in water overnight, drained, and steamed (Hesseltine 1965). Soak-water should be changed at least once, since rapid fermentation by spore-forming bacteria may occur. Salted koji and water containing miso inoculum are added to the cooled cooked soybeans and the mixture is blended, mashed slightly, and placed in large wooden or concrete tanks for the second stage of fermentation. Fermentation is allowed to progress at 28°C for about seven days at which time the temperature is raised to 35°C. Incubation may continue for several months.

The ferment soon becomes anaerobic, inhibiting the growth of A.

oryzae. At the same time, a salt content of 4 to 13% limits the growth of potentially pathogenic bacteria. Early in the second stage of fermentation enzymes from the koji act on soybean and rice constituents. Starch is digested by amylase and maltase to dextrin, maltose, and glucose. Protease acts on soybean protein to yield polypeptides, peptides, and free amino acids which enhance the palatability of miso. Glycinin, the main soybean protein, contains almost 20% glutamic acid. Upon liberation of this amino acid, sodium glutamate, a seasoning agent, is formed. Koji lipases free soybean lipid fatty acids, which form fatty acid ethyls and contribute to miso aroma late in the fermentation period.

Many bacteria and yeasts are responsible for flavor, texture, aroma, and color development in miso. *Pediococcus halophilus, Pediococcus pentosaceus, Streptococcus faecalis,* and other lactic bacteria grow in koji and eventually produce lactic acid early during the second fermentation stage. *Saccharomyces rouxii, Debaryomyces, Hansenula, Pichia,* and *Torulopsis* yeast species are introduced via the addition of inoculum from a previous batch of miso. These osmophilic yeasts produce ethyl, butyl, and amyl alcohols which contribute to the pleasant aroma of miso. Esters produced by the reaction of these alcohols with organic acids also enhance aroma development.

When the fermentation is completed, miso is allowed to age for about two weeks at room temperature before it is ready for consumption. It may be ground into a paste having consistency and appearance similar to peanut butter (Hesseltine 1965). A typical process for making miso is outlined in Fig. 9.4.

The proximate composition of miso is listed in Table 9.2. The wide range in levels of some components results from various ratios of soybeans and rice used in miso preparation and from the extent of substrate digestion.

Natto

Natto is a popular soybean food in Japanese diets. A similar product is called tu su by the Chinese and tao-si by the Filipinos (Hesseltine and Wang 1967). It is used as snacks and for adding flavor to cooked foods such as boiled rice, fish, beef, and lobster.

Traditionally, three kinds of natto are produced in Japan and each has its own method of preparation (Kiuchi et al. 1976). All products have as part of their formulation whole soybeans, and all have different appearance and flavor, depending upon the microorganism used. Itohiki-natto is produced using *Bacillus natto,* a strain of *Bacillus subtilis.* In this type of natto the soybean surface is covered with a viscous polymer of glutamic acid. Yukiwari-natto is made by mixing itohiki-natto with rice koji and salt, and aging at 25° to 30°C

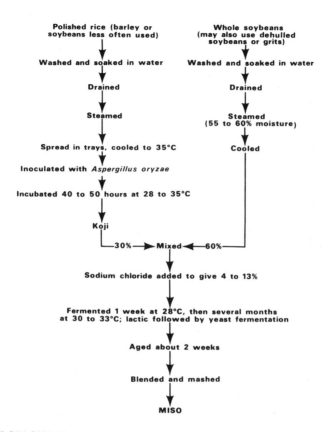

FIG. 9.4. FLOW SHEET FOR PREPARING MISO

for about two weeks. The people in the Hamanatsu vicinity of Japan ferment whole soybeans using *A. oryzae* from soybean koji to produce hama-natto (hamanatto) (Hesseltine 1965). To prepare hama-natto, soybeans are soaked in water for 4 hr and then steamed without pressure for 10 hr. Cooled beans are thoroughly inoculated with koji prepared from roasted wheat or barley and placed in a tray. After about 20 hr, beans become sticky and are covered with green mycelium. The fermented beans are dried in the sun to reduce the moisture to 12%. Beans are then combined with strips of ginger in wooden baskets, covered with salt water, and allowed to age under pressure for 6 to 12 months. The reddish colored product turns black when dried. Hamanatto contains about 38% moisture, 25% protein, 25% carbohydrates, and 11 to 13% salt. The finished product is shown in Fig. 9.5.

The flavor of natto is similar to miso and shoyu, but sweeter.

Courtesy of Dr. C.W. Hesseltine, USDA
FIG. 9.5. NATTO PREPARED FROM WHOLE SOYBEANS

Proteolytic, amylolytic, and lipolytic activities of microorganisms result in extensive breakdown of soybean constituents. In addition to *B. natto* and *A. oryzae*, other microorganisms such as *Micrococcus*, *Streptococcus*, and *Pediococcus* may also be responsible for unique flavor development. The harsh taste associated with hama-natto is probably due to the high level (12%) of free fatty acids (Kuichi *et al.* 1976).

Sufu

Sufu is a fermented soybean curd product made in China. Other synonyms for sufu are tosufu, fu-su, fu-ru, toe-fu-ru, tou-fu-ru, teou-fu-ru, fu-ju, fu-yu, and foo-yue (Wang and Hesseltine 1970). In the western world sufu is known as Chinese cheese or bean cake.

Preparation of sufu involves three steps. Soybean milk and curd are prepared first. The curd is then inoculated with the appropriate mold. Lastly, the freshly fermented curd is brined and aged.

In sufu preparation the soybeans are washed, soaked, and ground in a mill. Milk is pressed from the mash and heated to boiling. Heating serves to inactivate enzymes and trypsin inhibitors and lessens some of the beany flavor. Calcium sulfate or magnesium sulfate is used to curdle the soybean milk, and excess whey is removed by pressing. The resulting soft curd is called tofu. This bland-tasting product may be eaten directly or cooked with other foods such as meat, fish, or vegetables. Tofu contains about 83% water, 10% protein, and 4% lipid (Wang and Hesseltine 1970).

To prepare sufu from tofu, the curd is first cut into cubes approximately 2 cm square, dipped in solution containing 2 to 6% salt and 0.8 to 2.5% citric acid, and surface pasteurized in a hot-air oven (Hesseltine 1965; Wang and Hesseltine 1970). This process helps to control the growth of bacteria. Cubes are placed separately on a perforated tray and inoculated by coating their surfaces with a pure culture of a mucoraceous fungus. *Actinomucor elegans* is preferred; however, *Mucor hiemalis, M. silvaticus* and *M. subtilissimus* also yield satisfactory sufu. After 3 to 7 days of incubation at 20°C, cubes are covered with white or yellowish-white mycelium. Freshly molded cubes are known as pehtzes. The mucoraceous fungi used for sufu preparation exhibit strong protease and lipase activity. Soybean proteins are digested to peptides and free amino acids. Carbon in lipids may be used as an energy source.

The final step in preparing sufu requires brining the pehtzes in a solution containing 5 to 12% salt and 10% ethyl alcohol, often added as rice wine or distilled liquor. In addition to preventing microbial growth, ethanol reacts with free fatty acids to form pleasantly aromatic esters. Brining is complete after 40 to 60 days or longer. Sufu is then bottled with the brine and marketed. Cubes of sufu after removal from brine are shown in Fig. 9.6.

The color or flavor of sufu may be modified by adding ingredients to the brine. Hon-fang (hon-fan) or red sufu is made by adding red rice (ang-kak). Tsui-fang (tsue-fan) literally means "drunken cheese" and can be produced by adding fermented rice mash or a large amount of wine to the brine (Smith 1961; Wang and Hesseltine 1970).

A type of Chinese cheese similar to sufu is chee-fan (Smith 1961). Soybean curd cubes are inoculated with *Mucor*, salted, and incubated for about one week. *Aspergillus glaucus* may also take part in the fermentation. The cubes are then submerged for about 1 year in yellow wine.

Meitauza

A fermentation product of the solid waste material from the manufacture of tofu is known as meitauza (Hesseltine 1965). Residual solids from soybeans from which milk has been extracted are pressed into cakes. During 10 to 15 days of incubation at a cool temperature, cakes become covered with white mycelium of *Actinomucor elegans*. After fermentation, the cake is sun dried for a few hours. It is fried in oil or cooked with vegetables before eating.

Témpé

A popular fermented soybean product in Indonesia, New Guinea, and Surinam is témpé (témpéh). Kedelee or kedele, meaning soybean, is used to differentiate témpé made using soybeans from témpé

Courtesy of Dr. C. W. Hesseltine. USDA

FIG. 9.6. SUFU (CHINESE CHEESE) REMOVED FROM BRINE. Cubes are approximately 2 cm².

bongkrek, a product prepared from coconut press cake (copra). Témpé kedelee is preferred to the cheaper témpé bongkrek.

Several variations exist for preparing authentic témpé (Iljas et al. 1973). One method consists of parboiling soybeans and soaking them in water for 2 to 3 days. The beans are drained, pressed, heated slightly, and spread into wooden frames. The mash is inoculated with témpé fungus by addition of a portion of a previous batch, wrapped in banana or other suitable leaves, and left to ferment at 31° to 40°C for one day. Mixing may be necessary for the first day in order to satisfactorily distribute mycelium. A second method is more elaborate. Soybeans are washed and boiled, then transferred to cold water to soak for one day. The skins are removed and the cotyledons are boiled again and steamed. Concurrently, the inoculum is prepared by wrapping older témpé in teak or hibiscus leaves and allowing the leaves to dry for two days. To inoculate, the leaves are crushed and sprinkled over the cooled soft soybeans. The mixture is wrapped in banana leaves and fermented a day or two. Still another method involves removing hulls from soaked soybeans by treading on the beans at the edge of a stream or river. Freed seed coats are swept away in the current. Hulled beans are then boiled before fermenting in banana leaves. Témpé may also be prepared from soybean grits. Whatever the method, fermentation is judged complete when soybeans are thoroughly bound together by white mycelium (Fig. 9.7).

In the production of témpé, a number of Rhizopus spp. and strains are involved. Hesseltine (1965) listed forty strains belonging to six species which are acceptable for témpé production. The six species are Rhizopus oligosporus, R. stolonifer, R. arrhizus, R. oryzae, R. for-

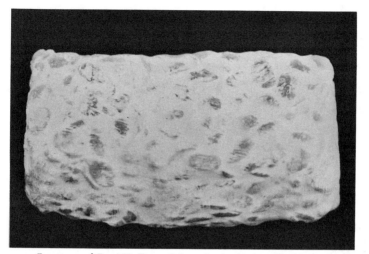

Courtesy of Dr. KO Swan Djien, Agric. Univ., The Netherlands
FIG. 9.7. TÉMPÉ CAKE. The mycelium of R. oligosporus binds the soybeans together into a white, compact cake.

mosaensis, and R. achlamydosporus. R. oligosporus is most often used in Indonesia. It is characterized by having no striations on irregularly-shaped sporangiospores (Hesseltine 1965). The sporangiophores are short, unbranched, and arise opposite rhizoids which are reduced in length.

R. oligosporus produces a number of proteases which digest soybean proteins during fermentation. Water-soluble crude protein increases over tenfold as a result of fermentation (Van Buren et al. 1972), indicating accumulation of peptides and free amino acids. R. oligosporus utilizes xylose, glucose, galactose, trehalose, cellobiose, and soluble starch, but not stachyose, raffinose, or sucrose. Hemicellulose decreases during fermentation. The fungus exhibits strong 1,3-lipase activity. Over 30% of the triglycerides are hydrolyzed during a 3-day fermentation period. Even with these strong hydrolytic activities, the amino acid and fatty acid composition remains relatively constant throughout fermentation. Fiber content may increase due to the development of mycelium.

Témpé has strong antioxidant acitivity. Di- and trihydroxyisoflavones produced during fermentation presumably preserve vitamin E in soybeans. Thiamine is decreased in soybeans as a result of heating and utilization by R. oligosporus. Riboflavin, niacin, and vitamins B-6 and B-12 are increased.

Témpé may be more digestible than cooked soybeans due to

decreased hemicellulose content and extensive solubilization of proteins. Protein efficiency ratios (PER) show little or no change during fermentation.

Témpé is highly perishable and, consequently, must be consumed shortly after an acceptable degree of fermentation is reached. Ammonia is produced as fermentation continues at ambient temperature, giving témpé an unpleasant flavor and odor. Tempe is sliced, fried in oil, and eaten hot. It is also cut into cubes and used as a substitute for meat in soups, or it may be roasted. The proximate composition of témpé is listed in Table 9.2.

PEANUTS

Oncom

Fermented peanut press cake is prepared in Indonesia where it is called oncom (ontjom or lontjom). After oil has been extracted from peanuts, the press cake, termed as boongkil, is broken into small pieces and soaked in water for about a day (Hesseltine and Wang 1967). Technical grade press cake contains less than 1% oil while village products may contain substantial amounts. In either case, oil which rises to the soak-water surface is removed and the press cake is steamed and pressed into molds about 3 × 10 × 20 cm in size (Hesseltine 1965). The molds are then placed in a bamboo frame covered with banana leaves and inoculated with either Neurospora sitophila or Rhizopus oligosporus. N. sitophila is the mold of choice. It is a common fungus in woody material under rain forest conditions shortly after sterilization or pasteurization has occurred, such as in the burning of forest land (Stanton 1971). Both fungi are usually available from previous batches of oncom. After 1 to 2 days of standing in a shady location, the fungus invades the peanut mass. N. sitophila results in an orange to red product owing to prolific formation of colored conidia. R. oligosporus produces a white oncom (van Veen et al. 1968). Constant aeration is important in the preparation of good oncom, as are temperature, moisture, and degree of granulation of the press cake (Ochse 1931). An added starch source such as tapioca, potato, or potato peels appears to enhance the fermentation process. The finished product averages 70% moisture, 3 to 9% oil, 20 to 30% crude protein (nitrogen × 6.25), about 4% carbohydrate, 1% ash, and 2% fiber (van Veen et al. 1968). The finished product is shown in Fig. 9.8.

Oncom has a fruity or somewhat alcoholic flavor. When fried, it takes on a mincemeat flavor. Oncom may also be roasted, covered with boiling water, and seasoned with salt or sugar before eating. It may also be roasted, cut into pieces, and covered with ginger sauce (Hesseltine 1965).

Courtesy of Dr. KO Swan Djien, Agric. Univ., The Netherlands
FIG. 9.8. ONCOM PREPARED FROM PEANUTS. Left—Vendor peddling oncom and other foods. To prevent rising of temperature, the red-orange oncom cakes are transported in the open air. Right—Close-up showing oncom covered with powdery *Neurospora* conidia.

Enzyme production by *N. sitophila* and *R. oligosporus* results in substantial changes in the peanut substrate. Proteinase and peptidase activities of *N. sitophila* are greatest on the albumin fraction. The extent to which proteins are solubilized varies, depending upon oxygen supply and temperature during fermentation. No significant changes in amino acid profiles have been noted as a result of peanut fermentation.

N. sitophila and *R. oligosporus* are active lipase producers. The distribution of free fatty acids in fermented and nonfermented peanuts has been reported (Beuchat and Worthington 1974). Disproportionately higher levels of saturated fatty acids are found in the free fatty acid fraction, indicating the action of 1,3-lipases, since saturated acids are located primarily in the 1,3 position of peanut triglycerides.

The carbohydrate fraction of peanut press cake is comprised largely of cellulose and simple oligosaccharides. Hemicellulosic-type materials may be partially solubilized by mold enzymes, resulting in a softening in substrate texture. This, in turn, may enhance the digestibility of peanut press cake. *N. sitophila* has strong α-galactosidase activity and essentially eliminates sucrose, raffinose, and stachyose during the one-day fermentation period (Worthington and Beuchat 1974). *R. oligosporus*, on the other hand, may utilize small amounts of stachyose but cannot utilize sucrose and raffinose. Although the stachyose and raffinose content of peanuts is considerably lower than that of most other legumes, often present in only trace quantities, it is possible that the absence of these oligosaccharides in oncom prepared with N. *sitophila* could enhance press cake digestibility (Beuchat 1976). Stachyose and raffinose are often associated with flatulence-causing factors in legume seeds.

The protein efficiency ratios (PER) of fermented peanuts are not increased over properly heat-treated raw ingredients (van Veen *et al.* 1968; van Veen and Steinkraus 1970; Quinn *et al.* 1975). Riboflavin and niacin levels are increased in peanuts as a result of fermentation with N. *sitophila*, while thiamine remains unchanged or decreases slightly. Pantothenate concentration appears to be unchanged during the fermentation process.

RICE

Lao-chao

Lao-chao is a Chinese fermented rice product. Its preparation and characteristics are described by Wang and Hesseltine (1970). Glutinous rice is first steamed, cooled, and then mixed with a commercial starter known as chiu-yüeh or peh-yüeh. After 2 to 3 days at room temperature, the rice becomes soft, juicy, sweet, and slightly alcoholic. It may be consumed as such or combined with eggs or seafoods. Unlike many other fermented foods which are salty in taste, lao-chao is a sweet dish having a distinct tartness and fruity aroma. Mucoraceous fungi, including *Rhizopus oryzae*, *R. chinensis* and *Chlamydomucor oryzae*, and a yeast, *Endomycopsis*, have been isolated from a commercial starter obtained from Taiwan (Wang and Hesseltine 1970). *Endomycopsis* (now *Saccharomycopsis*) is one of the few yeasts capable of producing amylases to utilize rice starch. Because it is highly oxidative and the fermentation condition is semianaerobic, the yeast may not efficiently utilize starch. The formation of fragrant esters undoubtedly results from the mixture of acids and alcohols produced by *Rhizopus* under anaerobic conditions.

Ang-kak

Red rice, or ang-kak (ang-khak, ankak, anka, ang-quac, beni-koji, aga-koji), is used for coloring foods such as fish and Chinese cheese, and for manufacturing red wine in the Orient (Palo *et al.* 1961; Hesseltine 1965). Certain strains of *Monascus purpureus* are selected to make ang-kak. Strains are used which produce a dark red pigment, monascorubrin, throughout the rice grains at a low enough moisture level to prevent individual grains from sticking together. Varieties of glutinous rice are unsatisfactory because grains tend to stick together when adjusted to the desired moisture content. To prepare ang-kak, rice is steamed, cooled, and inoculated with M. purpureus from a previous batch. Growth of the organism progresses at 25° to 30°C over about a three-week period. During this time, rice grains change from white, through pink and red, to deep purplish red. The ferment is then dried and may be pulverized before using as a food colorant.

Idli

Idli is a fermented food of southern India. It is prepared principally from rice and black gram (Phaseolus mungo) flours (Batra and Millner 1974). Idli is prepared by incubating moist dough made from various proportions of ground rice and decorticated black gram, known as dahl. Cashew nuts, ghee, salt, pepper, ginger, and cumin may be added as seasoning (Steinkraus et al. 1967). Pediococci, streptococci, and Leuconostoc spp. of bacteria, together with Torulopsis candida and Trichosporon pullulans yeasts, are natural microflora of rice and dahl which are largely responsible for idli fermentation. After about 24 hr of incubation, the dough is spongy with a honeycomb structure as a result of entrapment of gases produced by yeasts and bacteria during fermentation. The product is cooked by steaming and served hot. Idli is characteristically sour due to the lactic acid produced by micro-organisms.

MAIZE

Ogi

Maize is principally eaten in black Africa in the form of a sour meal (Akinrele et al. 1969; Akinrele 1970). Ogi, as it is called by the Yorubas in the western state of Nigeria, is the first native food given to babies at weaning. To prepare ogi, kernels are steeped in water for 1 to 3 days. When steeping is continued for three days, old liquor is replaced daily with fresh water. Following steeping, kernels are wet-milled and sieved with water through a screen to remove fiber, hulls, and a portion of the germ. The filtrate is then allowed to settle and ferment. The sediment (ogi) is usually marketed as a wet cake wrapped in leaves. It may also be diluted with water to 8 or 10% solids and boiled into a pap (agidi) or cooked to yield a stiff gel (eko) before eating.

Molds naturally present on maize during early steeping were identified by Akinrele (1970) as Cephalosporium, Fusarium, Aspergillus, and Penicillium spp. Bacteria prevalent in the fermenting mash are Corynebacterium sp., Aerobacter cloacae, and Lactobacillus plantarum. The beginning of the souring period is marked by rapid proliferation of Saccharomyces cerevisiae. Candida valida, a film former, is predominant near the end of the souring period. Both yeasts are presumably responsible for substantial flavor development in ogi.

The Bantu (South African) equivalent of ogi contains wheat bran and is known as maheivu. It has a thinner consistency and is consumed as a beverage. Sour maize dough in Ghana is called kenkey. Its preparation is similar to ogi. After three days of fermentation, pastes are wrapped in banana leaves before boiling or steaming.

Kaanga-kopuwai

Literally translated by the Maoris of New Zealand, kaanga-kopuwai means "maize soaked in water" (Yen 1959). It is also termed kaang-pirau (rotten corn) and kaanga-wai (water corn). Mature whole cobs of corn are placed unhusked in a jute sack and submerged in water. Fermentation time required for mature hard-grained cobs may be up to three months. The corn is ready for use when soft. Kernels are often slimy to the touch and distinctly aromatic. Little is known of the microflora responsible for fermenting kaanga-kopuwai. Yeasts and bacteria are undoubtedly involved.

Injera

Although injera is usually made from teff (Eragrostis tef), it may also be prepared from maize, wheat, barley, sorghum, or a mixture of these cereals (Stewart and Getachew 1962). Injera is an Ethiopian bread-like product similar in texture to pancakes in the United States. To prepare injera, flour is mixed with water in a container called a bohaka. This container is not thoroughly washed between batches, since part of the fermented paste from the previous batch serves as an inoculum for the new flour. Thin yellowish fluid saved from a previous fermentation is added to the watery paste and the mixture is allowed to ferment from 30 to 72 hr before baking. Stewart and Getachew (1962) described three types of injera. A portion of the fermented paste may be mixed with three parts of water and boiled, resulting in an absit. The absit is then mixed with another portion of fermented dough. The finished product has a thin, clean, nonpowdery appearance and a sour taste. Aflegna is a sweet-tasting, thick injera made from relatively unfermented paste 12 to 24 hr old. Komatata injera is made from over-fermented paste and is very sour.

Satisfactory flavor and texture characteristics of injera depend upon mixed-culture fermentation by fungi and bacteria. Indications are that a yeast, Candida guilliermondii, is the primary organism in this process. Its strong amylase activity, coupled with its ability to degrade simple sugars to alcohols and acids, contributes to flavor and texture development.

CASSAVA

Tapé

Tapé is an Indonesian delicacy with a sweet-acid taste and mild alcoholic flavor (Djien 1972). It is prepared by fermenting cassava

(Manihot utilissima) root or glutinous rice (Oryza sativa glutinosa). Cassava is also known as manioc, mandioca, aipum, yuca, cassada, or tapioca. Fermented cassava is named tapé ketella (Indonesian), tapé telo (Javanese), or peujeum (Sudanese), and fermented glutinous rice is tapé ketan.

The first step in the manufacture of tapé from either cassava or rice is to prepare the ragi (starter). In Indonesia, ragi is made with rice flour and is sold as white, somewhat flattened balls about 3 cm in diameter (Fig. 9.9) (Hesseltine 1965). Yeasts thought to be present in tapé ragi include Saccharomyces cerevisiae and Hansenula anomala. The mucoraceous molds Rhizopus oryzae and Chlamydomucor oryzae contribute substantial quantities of amylase which eventually saccharify cassava or rice starch during the second stage of fermentation.

To prepare tapé from cassava, roots are cut into pieces, smeared with powdered ragi, and either wrapped in banana leaves or placed unwrapped in a tray for 5 to 7 days (Hesseltine 1965). Glutinous rice is also combined with ragi. However, fermentation time is only 2 to 3 days. In either case, ragi contains enzymes and viable yeast and mold cells required to initiate and carry out the fermentation process. Both types of tapé are soft in texture. Sufficient acid is produced to reduce the pH to 4.

Several molds have been isolated from tapé (Djien 1972). Chlamydomucor oryzae converts starches to sugars while Mucor and Rhizopus spp. possibly play secondary roles in the fermentation process. A yeast, Endomycopsis fibuligera (Saccharomycopsis sp.) converts sugars to alcohols and flavor components. Bacteria are also thought to be involved in the fermentation of tapé.

Gari

Fermented cassava is one of the staple foods of people of the rain forest belt of West Africa. Collard and Levi (1959) described the traditional preparation of gari from cassava root. Both the corky outer peal and the thick cortex are removed and the body of the root is grated by hand on homemade raspers. Most of the juice is removed from the grated root and the pulp is packed in bags and allowed to ferment for 3 to 4 days.

Fermentation of cassava proceeds in two stages (Akinrele 1964; Collard and Levi 1959), during which the mash is gradually sterilized against adventitious microbial growth. During the first stage, starch is hydrolyzed and degraded to lactic, formic, and gallic acids by Corynebacterium manihot, a bacterium. When the pH of the ferment has dropped to about 4, a fungus, Geotrichum candidum, proliferates and produces a variety of aldehydes and esters. Gaseous hydrocyanic acid is liberated during fermentation at low pH through spontaneous

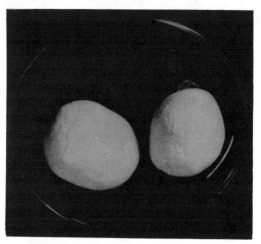

Courtesy of Dr. C. W. Hesseltine, U.S.D.A.

FIG. 9.9. RAGI, AN INDONESIAN FERMENTED RICE PRODUCT USED AS A STARTER (KOJI) IN THE PREPARATION OF TAPÉ. Rice balls are approximately 3 cm in diameter.

hydrolysis of linamarin, a poisonous cyanogenic glucoside found in fresh cassava. It is also believed that some of the formic acid breaks down by a hydrogenase system to form carbon dioxide and probably hydrogen, rendering the medium anaerobic (Akinrele 1964).

The use of fermented cassava as a staple in parts of Africa, Asia, and South America has given rise to widespread malnutrition because of its low protein content. In an attempt to improve the nutritional quality of fermented cassava, researchers at the Tropical Products Institute in England have explored the feasibility of using fungi to convert nitrogen from added ammonium salts to protein nitrogen (Stanton and Wallbridge 1969). Rhizopus spp. have been demonstrated to increase the protein content of cassava to 3.25%, which represents a six- to seven-fold increase over unfermented material.

TARO

Poi

Taro (Colocasia esculenta) is an important food crop in many tropical countries. In Hawaii it is the principal source of a fermented food called poi (Allen and Allen 1933). The manufacture of poi consists of two major processes. Taro corms are first cooked or steamed, peeled, and ground or pounded to a fine consistency. This is then mixed with water at which point it is known as fresh poi. Three kinds of poi may result, depending upon the amount of water added. Based on consistency, poi is designated as"one finger," "two finger," or "three finger," signifying the number of fingers required for a satisfactory helping. The second phase of poi preparation involves fermentation at room temperature for 1 to 3 days or longer. As fermentation progresses, texture changes from a sticky mass to one having a more watery and fluffy consistency.

Poi is commercially marketed in glass jars and plastic bags (Fig. 9.10). The same general procedures followed in preparing poi in the home are carried out by commercial production facilities. Corms are first pressure cooked, washed, and trimmed. Material is then passed through a colloid mill and put through a fine screen to remove some of the fiber before packaging. Fermentation occurs in the package as evidenced by increased acidity and a change of color from purplish grey to pinkish rose. The finished product is about 34% solids. It is combined with water to yield the desired consistency before serving.

Bacteria are predominant during early fermentation. *Lactobacillus delbrueckii, L pastorianus, L. pentaceticus, Streptococcus lactis,* and *S. kefir* produce large quantities of lactic acid and moderate quantities of acetic, propionic, succinic, and formic acids (Allen and Allen 1933). The pH is lowered to about 5 over a six-day fermentation period. *Candida (Mycoderma) vini* and *Geotrichum candidum* are exceedingly prevalent in the latter stages of fermentation. These fungi are thought to be responsible for imparting the pleasant fruity odors and flavors characteristic of older poi.

CACAO BEANS

Cocoa, chocolate, and chocolate liquor are products derived from cacao fruits *(Theobroma cacao)*. The tree-ripened fruits, called pods, are picked and broken or cut open, after which the seeds, called beans, are removed. The beans are placed in sweatboxes at a depth of 20 to 90 cm or in pits and covered with banana or plantain leaves. Sacking and boards are also sometimes used to cover the beans during a 2- to 12-day fermentation period. The duration of the process varies in different countries and with different bean cultivars. At 1- to 2-day intervals throughout the fermentation beans are thoroughly mixed and then transferred to another box and covered again. As a result of yeast and bacterial growth the temperature of the bean mass rises to 45° to 50°C during the first week and then declines (Roelofsen 1958). The thick, white endocarp (pulp) surrounding the seed is degraded during fermentation and the acid liquor is drained from the mass. After fermentation the moisture content of the beans must be reduced from about 60% to less than 7.5% before bagging.

Characteristic chocolate flavor is attributed to microbiological

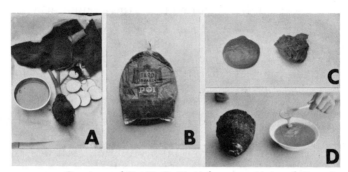

Courtesy of Dr. T. O. M. Nakayama, Univ. of Georgia

FIG. 9.10. TYPES AND CONSISTENCY OF POI. (A) Nonfermented fresh poi prepared from taro corm; (B) packaged fermented poi; (C) poi having thinner consistency (left) after blending with water; and (D) an illustration of size of taro corm and consistency of poi.

activities during fermentation as well as to unrelated chemical reactions which occur during the curing process. However, the flavor components of cocoa and chocolate, and the microorganisms responsible for their development are not fully understood. The ecological sequence of microorganisms during pulp fermentation consists of yeasts, followed by lactic acid bacteria, then acetic acid bacteria, and finally spore-forming bacilli (Forsyth and Quesnel 1963; Ostovar and Keeney 1973). The extent of activity by each of these groups is greatly affected by the degree of aeration. Yeasts commonly found in fermenting cacao bean pulp are *S. cerevisiae* and other *Saccharomyces* spp., *Candida krusei*, *Pichia farinosa*, *P. fermentans*, *Kloeckera apiculata*, *Metschnikowia pulcherrima*, *Hansenula anomala*, and *Schizosaccharomyces pombe*. Aerophilic yeast prefer the outer layers of the fermenting beans and semianaerobic ones dominate in the center.

Cacao beans are susceptible to spoilage during and after fermentation. *Aspergillus*, *Mucor*, *Penicillium*, and *Rhizopus* spp. develop on the surface of fermenting heaps which have been turned infrequently or not at all. (Roelofsen 1958). *Aspergillus*, *Mucor*, and *Penicillium* spp. commonly grow or cured cacao beans which have been mishandled or improperly dried.

REFERENCES

AKINRELE, I. A. 1964. Fermentation of cassava. J. Sci. Food Agric. 15, 589-594.

AKINRELE, I. A. 1970. Fermentation studies on maize during the preparation of a traditional African starch-cake food. J. Sci. Food Agric. 21, 619-625.

AKINRELE, I. A., MAKANJU, A., and EDWARDS, C. C. 1969. Effect of soya flour on the latic fermentation of milled corn. Appl. Microbiol. 17, 186-187.

ALLEN, O. N., and ALLEN, E. K. 1933. The manufacture of poi from taro in Hawaii: With special emphasis upon its fermentation. Hawaii Agric. Exp. Stn. Bull. 70.

BAI, R. G., PRABHA, T. N., RAMACHANDRA RAO, T. N., SREEDHARA, V. P., and SREEDHARA, N. 1975. Studies on tempeh fermented soybean food. I. Processing and nutritional evaluation of tempeh from a mixture of soybean and groundnut. J. Food Sci. Technol. 12, 135-138.

BANIGO, E. O. I., DE MAN, J. M., and DUITSCHAEVER, C. L. 1974. Utilization of high-lysine corn for the manufacture of ogi using a new, improved processing system. Cereal Chem. 51, 559-572.

BATRA, L. R., and MILLNER, P. D. 1974. Some Asian fermented foods and beverages, and associated fungi. Mycologia 66, 942-950.

BEUCHAT, L. R. 1976. Fungal fermentation of peanut press cake. Econ. Bot. 30, 227-234.

BEUCHAT, L. R., and BASHA, S. M. M. 1976. Protease production by the ontjom fungus, Neurospora sitophila. Eur. J. Appl. Microbiol. 2, 195-203.

BEUCHAT, L. R., and WORTHINGTON, R. E. 1974. Changes in the lipid content of fermented peanuts. J. Agric. Food Chem. 22, 509-512.

BEUCHAT, L. R., YOUNG, C. T., and CHEERY, J. P. 1975. Electrophoretic patterns and free amino acid composition of peanut meal fermented with fungi. Can. Inst. Food Sci. Technol. J. 8, 40-45.

CARENBERG, C. O. 1968. Traditional fermentation processes used in the production of protein-rich foods. Sven. Kem. Tidskr. 80, 252-258.

CHAH, C. C., CARLSON, C. W., SEMENIUK, G., PALMER, I. S., and HESSELTINE, C. W. 1975. Growth-promoting effects of fermented soybeans for broilers. Poult. Sci. 54, 600-609.

COLLARD, P., and LEVI, S. 1959. A two-stage fermentation of cassava. Nature (London) 183, 620-621.

DJIEN, K. S. 1972. Tapé fermentation. Appl. Microbiol. 23, 976-978.

DJIEN, K. S., and HESSELTINE, C. W. 1961. Indonesian fermented foods. Soybean Dig. 22, No. 1, 14-15.

DWIDJOSEPUTRO, D., and WOLF, F. T. 1970. Microbiological studies of Indonesian fermented foodstuffs. Mycopath. Mycol. Appl. 41, 211-222.

EBINE, H. 1971. Miso. In Conversion and Manufacture of Foodstuffs by Microorganisms. T. Kawabata, M. Fujimaki, and H. Mitsuda (Editors). Saikon Publishing Co., Tokyo.

FORSYTH, W. G. C., and QUESNEL, V. C. 1963. The mechanism of cacao curing. Adv. Enzymol. 25, 457-592.

GRAY, W. D. 1970. The use of fungi as food and in food processing. CRC Crit. Rev. Food Technol. 1, No. 2, 225-329.

HACKLER, L. R., STEINKRAUS, K. H., VAN BUREN, J. P., and HAND, D. B. 1964. Studies on the utilization of tempeh protein by weanling rats. J. Nutr. 82, 452-456.

HARRIS, R. V. 1970. Effect of Rhizopus fermentation on the lipid composition of cassava flour. J. Sci. Food Agric. 21, 626-627.

HESSELTINE, C. W. 1965. A millennium of fungi, food and fermentation. Mycologia 57, 149-197.

HESSELTINE, C. W. 1972. Solid state fermentations. Biotechnol. Bioeng. 14, 517-532.

HESSELTINE, C. W., and WANG, H. L. 1967. Traditional fermented foods. Biotechnol. Bioeng. 9, 275-288.

ILJAS, N., PENG, A. C., and GOULD, W. A. 1973. Tempeh. An Indonesian fermented soybean food. Ohio Agr. Res. Dev. Cent. Hortic. Ser. No. 394.

ITO, K. 1970. Tasting effect of miso as a seasoning. 5. Experiments on the protein of rice-koji fermented miso. Jpn. Soc. Food Nutr. J. 23, 56-63. (Japanese)

JURUS, A. M., and SUNDBERG, W. J. 1976. Penetration of Rhizopus oligosporus into soybeans in tempeh. Appl. Environ. Microbiol. 32, 284-287.

KINOSITA, R., ISHIKO, T., SUGIYAMA, S., SETO, T., IGARASI, S., and GOETZ, I. E 1968. Mycotoxins in fermented foods. Cancer Res. *28*, 2296-2311.

KIUCHI, K., OHTA, T., ITOH, H., TAKABAYASHI, T., and EBINE, H. 1976. Studies on lipids of natto. J. Agric. Food Chem. *24*, 404-407.

KWON, T. W. 1972. Fermented foods in Korea. Annotated bibliography (1917-1971). Korean Inst. Sci. Technol., Seoul.

LODDER, J. 1970. The Yeasts: A Taxonomic Study, North-Holland Publishing Co., Amsterdam.

LOPEZ, A., and QUESNEL, V. C. 1973. Volatile fatty acid production in cacao fermentation and the effect on chocolate flavour. J. Sci. Food Agric. *24*, 319-326.

MARTINELLI, A., and HESSELTINE, C. W. 1964. Tempeh fermentation: package and tray fermentations. Food Technol. *18*, 761-765.

MURATA, K., IKEHATA, H., and MIYAMOTO, T. 1967. Studies on the nutritional value of tempeh. J. Food Sci. *32*, 580-586.

OCHSE, J. J. 1931. Vegetables of the Dutch East Indies. Archipl. Drukkerij, Bruitenzorg, Java.

ONISHI, H. 1963. Osmophilic yeasts. Adv. Food Res. *12*, 53-94.

OSTOVAR, K., and KEENEY, P. G. 1973. Isolation and characterization of microorganisms involved in fermentation of Trinidad's cacao beans. J. Food Sci. *38*, 611-617.

PALO, M. A., VIDAL-ADEVA, L., and MECEDA, L. 1961. A study on ang-kak and its production. Philipp. J. Sci. *89*, 1-22.

PEDERSON, C. S. 1971. Nutritious fermented foods of the Orient. In Microbiology of Food Fermentations. C. S. Pederson (Editor). AVI Publishing Co., Westport, Conn.

QUINN, M. R., BEUCHAT, L. R., MILLER, J., YOUNG, C. T., and WORTHINGTON, R. E. 1975. Fungal fermentation of peanut flour: Effects on chemical composition and nutritive value. J. Food Sci. *40*, 470-474.

RAJALAKSHMI, R., and VANAJA, K. 1967. Chemical and biological evaluation of the effects of fermentation on the nutritive value of foods prepared from rice and grams. Br. J. Nutr. *21*, 467-473.

REINHOLD, J. G. 1975. Phytate destruction by yeast fermentation in whole wheat meals. J. Am. Diet. Assoc. *66*, 38-41.

ROELOFSEN, P. A. 1958. Fermentation, drying, and storage of cacao beans. Adv. Food Res. *8*, 225-296.

RUSMIN, S., and KO, S. D. 1974. Rice-grown *Rhizopus oligosporus* inoculum for tempeh fermentation. Appl. Microbiol. *28*, 347-350.

SHIBASAKI, K., and HESSELTINE, C. W. 1962. Miso fermentation. Econ. Bot. *16*, 180-195.

SMITH, A. K. 1961. Oriental methods of using soybeans as food. U. S. Dep. Agric., Agric. Res. Serv. *71-72*.

SMITH, A. K., RACKIS, J. J., HESSELTINE, C. W., SMITH, M., ROBBINS, D. J., and BOOTH, A. N. 1964. Tempeh: Nutritive value in relation to processing. Cereal Chem. *41*, 173-181.

STANTON, W. R. 1971. Microbially produced foods in the tropics. In Conversion and Manufacture of Foodstuffs by Microorganisms. T. Kawabata, M. Fujimaki, and H. Mitsuda (Editors). Saikon Publishing Co., Tokyo.

STANTON, W. R., and WALLBRIDGE, A. 1969. Fermented food processes. Process Biochem. *4*, No. 4, 45-51.

STEINKRAUS, K. H., HWA, Y. B., VAN BUREN, J. P., PROVVIDENTI, M. I., and HAND, D. B. 1960. Studies on tempeh—an Indonesian fermented soybean food. Food Res. *25*, 777-788.

STEINKRAUS, K. H., LEE, C. Y., and BUCK, P. A. 1965. Soybean fermentation by the ontjom mold *Neurospora*. Food Technol. *19*, 1301-1302.

STEINKRAUS, K. H., VAN VEEN, A. G., and THIEBEAU, D. B. 1967. Studies on idli—An Indian fermented black gram-rice food. Food Technol. 21, 916-919.

STEWART, R. B., and GETACHEW, A. 1962. Investigations of the nature of injera. Econ. Bot. 16, 127-130.

SUNDHAGUL, M., DAENGSUBHA, W., and SUYANANDANA, P. 1975. Thailand's traditional fermented food products: a brief description. Thai J. Agric. Sci. 18, 205-219.

SUNDHAGUL, M., SMANMATHUROJ, P., and BHODACHAROEN, W. 1972. Thua-nao, a fermented soybean food of northern Thailand. Part 1. Traditional processing method. Thai J. Agric. Sci. 5, 43-46.

ULLOA, S. C., and ULLOA, M. 1973. Fermented food preparations of corn consumed in Mexico and other Latin American countries. Rev. Soc. Mex. Hist. Nat. 34, 423-457. (Spanish)

VAN BUREN, J. P., HACKLER, L. R., and STEINKRAUS, K. H. 1972. Solubilization of soybean tempeh constituents during fermentation. Cereal Chem. 49, 208-211.

VAN VEEN, A. G., GRAHAM, D. C. W., and STEINKRAUS, K. H. 1968. Fermented peanut press cake. Cereal Sci. Today 13, 96-98.

VAN VEEN, A. G., and STEINKRAUS, K. H. 1970. Nutritive value and wholesomeness of fermented foods. J. Agric. Food Chem. 18, 576-578.

WAGENKNECHT, A. C., MATTICK., L. R., LEWIN, L. M., HAND, D. B., and STEIN-KRAUS, K. H. 1961. Changes in soybean lipids during tempeh fermentation. J. Food Sci. 26, 373-376.

WANG, H. L. 1967. Release of proteinase from mycelium of Mucor hiemalis. J. Bacteriol. 93, 1794-1799.

WANG, H. L., and HESSELTINE, C. W. 1970. Sufu and lao-chao. J. Agric. Food Chem. 18, 572-575.

WANG, H. L., SWAIN, E. W., and HESSELTINE, C. W. 1975. Mass production of Rhizopus oligosporus spores and their application in tempeh fermentation. J. Food Sci. 40, 168-170.

WORTHINGTON, R. E., and BEUCHAT, L. R. 1974. α-Galactosidase activity of fungi on intestinal gas-forming peanut oligosaccharides. J. Agric. Food Chem. 22, 1063-1066.

YEN, D. E. 1959. The use of maize by the New Zealand Maoris. Econ. Bot. 13, 319-327.

YOKOTSUKA, T. 1960. Aroma and flavor of Japanese soy sauce. Adv. Food Res. 10, 75-134.

YOKOTSUKA, T. 1972. Some recent technological problems related to the quality of Japanese shoyu. In Fermentation Technology Today. G. Terui (Editor). Soc. Ferm. Technol. Jpn., Osaka.

YONG, F. M., and WOOD, B. J. B. 1974. Microbiology and biochemistry of soy sauce fermentation. Adv. Appl. Microbiol. 17, 157-194.

10

Alcoholic Beverages

T. W. Humphreys and G. G. Stewart

T he use of yeasts immediately comes to mind when alcoholic beverages are considered because in large part the production of such beverages involves the use of these microorganisms. However, yeasts are not used exclusively, and in some instances other fungi and also bacteria have their role to play. Nevertheless, the yeasts are both quantitatively and economically the most important group of commercially exploited microorganisms.

The total amount of yeast produced annually, including that formed during brewing, distilling, baking, and wine making, is of the order of a million metric tons. Ethanol made for all purposes by fermentation processes involving yeast totals at least 1.5 million metric tons. The benefit of such activities to national exchequers through taxation is counted in thousands of millions of dollars. The excise regulations governing the taxation, marketing, and consumption of alcoholic beverages vary greatly from country to country and indeed from state to state and province to province; consequently no discussion of this diverse subject will be undertaken here.

The industrial uses of yeasts can be classified into three groups according to the relationship of the product to the biochemistry of the organism (Harrison and Rose 1970):

(1) Cell constituents
      Dry whole cells—fodder, food
      Macromolecular constituents—lipids, proteins, enzymes, nucleic acids

Extraction compounds—coenzymes, vitamins
Breakdown products—amino acids, purines, pyrimidines
(2) Excretion products
Beer, wine, cider, spirits, glycerol, carbon dioxide
(3) Enzyme-substrate interaction
Whey utilization—*Kluyveromyces (Saccharomyces) fragilis*
Maltotriose production—*Saccharomyces uvarum*

Most commercial applications fall into the second group comprising excretion compounds such as ethanol, glycerol, carbon dioxide, and the numerous flavoring congeners that contribute to the the flavor of alcoholic beverages and leavened products. This chapter will be restricted to a discussion of alcoholic beverages and the role that yeasts play in their manufacture together with a consideration of the use of brewery yeasts for food and fodder purposes. As a result of space limitations a detailed discussion of all the applications of yeast for producing alcoholic beverages is not possible. Consequently, the use of yeast in brewing will receive in-depth consideration and the other applications will be reviewed briefly.

HISTORICAL SURVEY OF ALCOHOLIC BEVERAGES

Beer

The application of yeasts for preparing alcoholic beverages goes back far beyond the dawn of recorded history. It is well known that beer was produced by the Sumerians before 7000 B.C. and wine by the Assyrians in 3500 B.C. Greek tradition has it that the god Dionysius fled from Mesopotamia in disgust because its people were so addicted to beer. Certainly, it appears that 40% of the Sumerian grain yield was used for beer production. An ordinary temple workman received a ration of about one liter a day, and senior dignitaries five times as much, some of which they probably used as currency (Tannanhill 1973).

Beer preparation evolved from a particular method of bread making. The neolithic housewife had learned how to make raw grain digestible by leaving it to sprout. She went on to discover that bread made from sprouted dried grain kept better than bread made from conventional flour. A special dough was made from sprouted dried grains and partially baked; the "loaves" were then broken up, soaked in water, and the mixture allowed to ferment for about a day. The liquor was then strained off and the beer was ready to drink. Consequently, beer making was specifically related not to raw sprouted grain but to baked bread.

Bread beer was also produced in the Egypt of the Pharaohs, where it was regarded as a holy gift from Osiris, the god of the dead. It was considered to be an enviable sign of wealth if one could purchase enough beer to become intoxicated. Even children in Egypt at the time were given beer to take with them to school; beer was considered to possess a therapeutic value in preventing stomach and kidney diseases. In the Israel of the Old Testament and Caesarean Rome beer was also drunk, although in Rome, unlike wine, it was forbidden to women. Throughout the whole of this period, brewing remained a hit-or-miss operation. The presence of the fermentation-causing microorganisms was largely fortuitous, although brewers were observant enough to realize that their old beer jars (full of cracks and crevices—a splendid home for yeasts and bacteria) produced a better beer than new ones.

The Celtic people living in central Europe before the invasion of the Germanic tribes made a beer that probably tasted somewhat better than bread beer, and the Germanic people probably learned brewing from them; in turn, the Huns learned the process from the Germanic people. It is said that Attila, the famous leader of the Huns, was convinced that his enormous personal consumption of beer was largely responsible for his victories. Beer was an extremely popular beverage at this time (5th century A.D.), but its taste must have been very dull because it lacked what is now considered to be the essential flavor of beer, the hops. Hops had been known as a drug since before the 8th century A.D. and were believed to be a good remedy against a number of diseases, including angina pectoris, sour stomach and kidney stones. It was during the Middle Ages that hops were used as an addition to beer, as were many other flavors and spices in different "medicinal beers." Hops have now won such a complete victory over other spices that a drink without hops would not be called beer, although nowadays the enormous rise in their cost together with a consumer preference for lighter beers has resulted in a considerable reduction in the utilization of hops in beer.

Early Americans, like their contemporaries in Europe, had a strong aversion to water. This was not altogether surprising since much fresh water within range of human habitation was, if not actually poisonous, very nearly undrinkable. One of the 17th century colonists' earliest concerns had been to organize a supply of fermented drinks. Their initial experiments in growing barley and hops in New England had proved disappointing. However, it was soon discovered that very potable brews could be made from pumpkins, maple sugar, and persimmons. The flavor was not the same as real beer, but the effect was. Apple orchards next sprang up in New England and Pennsylvania and the colonists' intake of fermented (hard) cider rapidly reached gargantuan proportions. Once Pennsylvania was settled, however, it proved to be good barley and hop country and, with the immigration of German and Dutch brewmasters, beer brewing soon became a well-developed industry in North America.

A Dutchman, Antonie van Leeuwenhoek, is credited with being the first man to have observed yeasts microscopically. In 1680 he sent descriptions and drawings of yeast cells to the Royal Society in London. Even though van Leeuwenhoek seems to have seen yeasts in the sediment of wines, the role of yeasts in the manufacture of alcoholic beverages was not recognized until long after this discovery. Leeuwenhoek was, unfortunately, not a scientist—he was a draper by profession—and it was left to the botanists of the day to lay the firm foundations of yeast microbiology.

The scientific theories of the early 19th century recognized the presence of yeasts in alcoholic fermentation, but the yeast was considered a complex and lifeless chemical catalyst. Cogniard-Latour in 1837 demonstrated that beer contains spherical bodies that are able to multiply and which belong to the vegetable kingdom. In this, he received support from Schwann, who termed yeast "Zuckapily" or "sugar fungus" from which the name "Saccharomyces" originates. This cellular or vitalistic theory of fermentation was, however, vehemently and occasionally mockingly attacked by a trio of chemists—Liebig, Wohler, and Berzelius. The euphoria which stemmed from the success of the newly recognized branch of organic chemistry in explaining hitherto complex and mysterious organic processes convinced this group that chemical reactions, rather than the activities of living cells, could perfectly well explain the alcoholic fermentation of sugars.

It fell to Louis Pasteur to prove that fermentation is due to living cells. Pasteur in his publications Etudes sur le Vin (Pasteur 1866) and Etudes sur la Bière (Pasteur 1876) began a revolution in the fermentation industries. The essence of this revolution was the application of scientific principles to the fermentation industries. Pasteur, although trained as a chemist, became a microbiologist, applied the techniques of chemical engineering, and, in the evaluation of his results by groups of tasters, he approached modern techniques used for the sensory evaluation of foods.

From the foregoing it can be seen that the association of yeasts with the science of biochemistry is an old and well-established one, and in fact biochemistry was born out of yeast technology. The Buchner brothers, working in Germany in 1897, were interested in preparing extracts of yeast for medicinal purposes and to this end they ground

brewers' yeast with sand and then squeezed out the juice. In an effort
to preserve their extract they added large amounts of sucrose, since it
was known that solutions containing high concentrations of sugar are
less prone to microbial infection. To their surprise, the sugar was
rapidly fermented by the yeast juice. This chance encounter
precipitated a whole series of studies directed toward elucidating the
nature of the steps in the fermentation of glucose to ethanol and
carbon dioxide. This work spanned a number of years and is
associated with the names of many outstanding biochemists.

Wine

There are many picturesque tales concerning the origins of wine,
but what almost certainly happened was that a container of grapes
was neglected, they fermented, and some inquisitive person tasted
the fermented juice and found that it was a very palatable beverage.
The wild vine flourished in the Caucasus, and it was probably in this
region that grapes were first brought under cultivation. By 3000 B.C.
viticulture had reached both Mesopotamia and Egypt. In Egypt, wine
was first used almost entirely for temple rituals and it was not until
Greek influence began to be felt in the first millennium B.C. that
private vineyards became common and wine found its place as a
popular drink. However, Egyptian temple vintners had become
experts long before then, and it is possible that the Greeks simply re-
exported to secular Egypt the knowledge that they had earlier
imported from priestly Egypt.

In the Mediterranean during the Greek Golden Age, many countries
produced their own *ordinaires*, but the rich insisted on importing the
scarce and expensive vintages of Lesbos and Chios. The great wines
appear to have been sweet and, since both the Greeks and Romans
followed the Egyptian custom of drinking their wines well diluted
with water, the finer vintages were often kept until they were thick
and sticky as honey.

The wine was fermented in vats smeared inside and out with resin,
which gave the wine a characteristic flavor, and then filtered into goat
or pigskins if it was intended for local consumption, or into clay
vessels for export. Obviously fermentation was not a scientifically
controlled process and the wines of the ancient world did not keep
well unless preservatives of one form or another were added. Each
region had its own formula. In one region a brew of herbs and spices
which had been mixed with condensed seawater and matured for
some years was used, while a later Roman recipe favored the addition
of liquid resin mixed with wine ash to the grape juice before
fermentation. Filled wine jars were often kept to mature in a loft

where wood was seasoned and meat smoked. Although reasonable smoking was thought to improve a wine, all Romans with pretensions to good taste were united in vilifying those French vintners who oversmoked their wines in order to make them appear older than they were.

Greek wines went out of international favor after the appearance of the first great Italian vintages, the Opimian. In the centuries that followed, many other Italian wines, including Falernian, became household names and the competition proved to be too great for Greece. Italian vineyards were able to produce over 7500 liters of wine per acre (about 18,500 l/hectare) which was far more than that of Greece where vineyards were unproductive because of old-fashioned methods. Further, as the power of Rome expanded, the taste for Italian wine (even the vine itself) was carried to many new lands. For instance, grape culture had spread far down the Rhine by 200 A.D. and to France about the same period.

During the centuries after the fall of Rome no one country immediately took over the Italian role. As time progressed, however, France became the wine-producing country *par excellence,* and the names of Beaune, St. Emilion, Chablis, and Epennay can be found in manuscripts that date from the 13th century. The Hansa towns, Flanders, and England were among the steadiest markets for the wines of Gascony (Bordeaux), whose pale rosés, known as claret, were drunk only weeks old. By the end of the Middle Ages, Spanish and Portuguese wines had come into fashion, as had the red vernaccia of northern Italy, Malmsay from Crete, Melagne, Madeira, and the Rhenish wines from Germany.

Wherever Europeans have settled they have attempted viticulture. Failing this, they have imported European wines for their needs. The growth of viticulture outside Europe has been an extremely complex process; consequently, it will only be considered in a very superficial manner and no attempt has been made to discuss the chronological order of such developments.

South America.—Viticulture was taken to South America by the Spanish and Portuguese colonists during the early 16th century and today there are vineyards in Peru, Chile, Argentina, Bolivia, Uruguay, Paraguay, and Brazil.

Australia.—Vines were first brought to Australia in 1788. Several small vineyards were planted and later considerably expanded. Wine was shipped to England in 1822 where it received a silver medal for its quality.

Africa.—In Africa, the wine industry has developed in the north, i.e., Morocco, Algeria, Tunisia, and Egypt, and in the far south in the Union of South Africa. The Moroccan wine industry dates back to the Roman period. However, the modern period dates from the French occupation of 1912. The Algerian wine industry was developed by the French as an adjunct to their own low-alcohol products. Wine was introduced into South Africa in 1672 and a sweet muscatel, Constartia, achieved considerable fame in Europe in the 18th century. Today total wine production ranges up to nearly 400 million liters and about 240,000 acres (about 100,000 hectares) are planted with vines. Much of the wine is intended for export to England and elsewhere.

North America.—The vineyards of North America may be classified into three regions.

Eastern U. S.—During the Colonial period numerous attempts were made to introduce the European grape to the colonies along the Atlantic seaboard. Winter killing, high summer humidity, phylloxera, and virus diseases defeated all of these efforts. It was not until the end of the 18th century that serious attention was turned to domesticating varieties of a number of local species.

California. —The original vineyards were planted at the Missions, primarily with the Mission grape from Baja California, in order to produce wines for religious and domestic use. Table wines from the Mission grape are flat and spoil easily and this fact was particularly responsible for the generally low quality of California table wines prior to 1880.

The next step in the development of the wine industry of this region was the importation of new grape varieties. Wine makers of diversified experience from such European countries as France, Germany, Italy, Britain, and Spain were active in California. The climatic conditions of California however, were quite different from those of Europe; for example, the time of harvest in Europe was 2 to 6 weeks too late for California conditions, and the temperatures of fermentation and storage were also higher in California, leading to problems of bacterial spoilage. Indeed Amerine *et al.* (1972) remarked, "the wonder is that so many sound wines were produced in the pre-1900 period."

The development of the California wine industry has been impeded by a continuing pattern of years of high production and low prices followed by years of lower production but higher prices. This has had a very restraining influence on the industry. Another historical factor that has had a profound effect on the development of the California wine industry has been the development of the vineyard-winery company with emphasis on quality. Although prohibition eliminated most of them, some survived and their historical stature is of considerable value to the industry as a whole.

The impact of industrialization was gradually felt by the California wine industry in the post-prohibition period. These influences are too recent for complete evaluation but have led to much centralization, a trend which will continue. Among the advances of the first 25 years in the post-prohibition era, Gallo (1958) has noted a development of slightly sweet, red table wines, creation of the rosé market, cold fermentation, early bottling of white wines, sterile filtration, flor sherry, ion-exchange, better maturity standards, and production of wine vinegar.

Canada.—Wine produced from domestic grapes is largely restricted to the Niagara Peninsula, where winter temperatures are moderated by Lake Ontario and Lake Erie, and to the Okanozan region of British Columbia. A number of grape varieties which are suited to cultivation have been developed. These include a number of varieties of *Vitis labrusca* such as Concord, Niagara, and Delaware. They all resemble each other with a more or less strong and distinctive aroma believed to be partially due to methyl arthanilate. The odor is commonly said to be "foxy." Attempts are currently being made to introduce weather-resistant hybrids of *V. vinifera* with some modicum of success, although only time will tell. In Ontario, development of the industry is seriously impeded by government regulations restricting the import of grapes from outside the province.

Spirits

The manufacture of alcohol via the distillation of fermented carbohydrate-containing materials must be one of the earliest examples of the preparation by indirect means of a chemical compound in almost the pure state. Although methods for such processes have been documented since the time of the ancient Egyptians (Simmonds 1919), distillation methods only seem to have been put to extensive use since the 1st century A.D. Nevertheless, sautchoo was produced in China long before the Christian era and arrack was made in India as early as 800 B.C. Later, distillation of fermented grain liquor was practiced in Ireland and Scotland. The distillation of wine to produce brandy appears to have been carried out in the 9th century and became a sizeable industry in France by the 14th century.

Fundamentally, the distillation process consists of converting a liquid into vapor through the application of heat and then condensing the vapor. It can be used either to remove liquid from a partially moist substance, leaving the latter dehydrated, or to extract a pure liquid essence from particular types of solids. Alchemists discovered in the Middle Ages that distilling could do more than merely separate liquids from solids. With careful temperature control and rapid cooling of the vapor it was possible to separate one liquid from another. The alcohol produced from distilled wine, and later from distilled ales, was first known as "aqua vitae" (water of life) or in the Gaelic as "uisge beatha" which was corrupted via the abbreviated uisge to whiskey.

By the 16th century spirits were being distilled in northern Europe mainly from fermented grains, and local variants such as akvavit, schnapps, and gin (the name gin being derived from genever, or juniper, with which the drink was, and still is, flavored) began to appear in continental Europe and whiskey appeared in Ireland and Scotland.

Scottish and Irish settlers introduced distilling to North America by the 18th century and the product, although lacking many characteristics of the "real" product, was preferable to many palates when compared to the early rye whiskey or the corn versions developed in Bourbon County, Kentucky. The most popular drink in 18th century America, however, was rum. It was estimated that immediately prior to the War of Independence, colonists were consuming over 14 liters per capita per annum, women and children included. Indeed, it has been argued that it was not the British tax on America's tea that precipitated the altercation, but the Molasses Act of 1733, which imposed a heavy tax on sugar and molasses coming from everywhere except the British sugar islands in the Caribbean.

The rum trade was founded on molasses, the residue after the juice of the sugar cane is boiled to produce sugar crystals. For decades the shipowners of New England found it profitable to sail with a cargo of rum to the slave coast of Africa and exchange it for slaves, whom they transported back to the West Indies for sale to plantation owners. In the West Indies, the merchants replaced the slave cargo with molasses which they took home to be fermented and distilled into rum.

BEER BREWING

By classicial definition, beer is a fermented alcoholic beverage made from malted cereals, water, hops, and yeast. Only one country, West Germany, still abides by this strict definition. In that country, the Bavarian Purity Law that was establishd in the 16th century is still enforced and only the previously mentioned four ingredients are allowed if the product is to be called beer and sold as such in West Germany. All other countries nowadays permit considerable digression from this. For example, the use of unmalted cereals such as corn, rice, and wheat to supplement the more expensive malt is now common practice in North America and most European countries. The addition of enzymes at various points of the process is also permitted and certain countries allow the use of antifoams during fermentation. Various types of "beer" are designated according to color and taste. Lager beer is rather high in alcohol and low in hops. Bock beer is heavy, dark in color, and high in alcohol. Weiss beer is made mainly from wheat, is tart, has a marked malt and hop flavor, contains natural fermentation gas, and is often turbid. Ale is pale in color, high in alcohol and generally contains more hops than does lager. Porter is a dark ale brewed from highly roasted malt, while stout is a strong porter having a pronounced sweet malt flavor.

Malting

The manufacture of beer is a biological process whereby agricultural products, such as barley and hops, are converted into beer by control of the biochemical reactions in malting, mashing, and fermentation. It has been suggested that malting and mashing may not always be a part of the manufacture of beer (Lewis 1968), whereas fermentation has no prospect of being replaced at the present time or in the foreseeable future. The process whereby barley is converted into malt is a complex biochemical process. Barley grains are steeped in water at 10° to 15°C and then germinated at 15° to 20°C for 4 to 7 days. Sprouts or germs are removed, leaving the malt, a reservoir of such enzymes as α- and β-amylases and proteases together with their respective substrates. Malt is a remarkably efficient, readily storable, prepackaged convenience form of the enzymes and substrates necessary for producing the mash and eventually the beer fermentation medium, wort. In the past 15 to 20 years, much research has been carried out on malt replacement systems. Toward this end, the ability of proteases and amylases from microbial sources such as *Bacillus subtilis* and *Aspergillus* spp. to hydrolyze the carbohydrates and proteins from such cereals as unmalted barley and by this means produce a wort has been intensively investigated (Wieg 1970). Although such studies have generally failed commercially, there are enzyme mixes commercially available which find a market where the price of malt is high. It is doubtful that malt will be completely replaced as an enzyme source in the foreseeable future; however, cereals (e.g., corn, wheat, starch, and rice) are used as malt adjuncts, thus utilizing the surplus enzyme potential of malt.

Mashing

In the first stage of the actual brewing process (Fig. 10.1), the malt is ground in a suitable mill composed of from 4 to 6 rollers set in such a way that the husk of the grain is cracked and broken open, thus exposing the starchy endosperm which becomes a course flour. The ground malt is mixed with hot water and mashed for a fixed period of time in a vessel known as a mash tun. The process of mashing need not be considered in detail here; suffice it to say that it is an enzymatic process whereby most of the nonsoluble unfermentable carbohydrates and proteins are hydrolyzed to soluble fermentable materials. An initial temperature of 38° to 50°C is maintained to enhance protease activity followed by a gradual rise in temperature to 65° to 70°C to favor amylases. The temperature is finally increased to 75°C or above to inactivate the enzymes. The reader is referred to the descriptions of Hough et al. (1971) for detailed consideration of the mashing process.

Extraction and Boiling of the Wort

After mashing, the aqueous extract or wort is filtered from the malt grains using a lauter tun and complete extraction of soluble material is achieved by percolating the spent grains with hot water sprayed onto the bed by a revolving arm. The wort is collected in a wort boiling vessel (kettle) and boiled with hops for 60 to 90 min. During the boiling, the enzymes are inactivated, proteins are coagulated and precipitated from the wort as a "break," and the wort is sterilized and concentrated to a degree. There is an initial precipitation of protein during boiling (the hot break) and a second precipitation as the wort cools (the cold break). During boiling complex changes occur in the compounds extracted from the hops. Humulone (α-bitter acid) and lupulone (β-bitter acid) may be converted to soft resins by oxidation and polymerization. These acids possess antiseptic properties and, together with essential oils in hops, impart characteristic flavors to beer. Tannin of the hops is converted to phlobaphene which complexes with protein during boiling to form a precipitate and settle out. The hopped wort is separated from the trub (spent hops and coagulated protein) by filtration or centrifugation and is then cooled prior to fermentation.

Fermentation

The two main types of beer are lager and ale. These are fermented

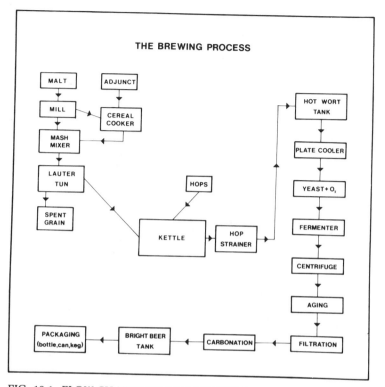

FIG. 10.1. FLOW CHART FOR THE BREWING PROCESS

with strains of *Saccharomyces carlsbergensis* (*uvarum*) and *Saccharomyces cerevisiae*, respectively. *S. carlsbergensis* and *S. logos* are now classified by taxonomists as one species, *uvarum* (Lodder 1970). Many brewers still recognize *S. carlsbergensis* as having characteristics sufficiently distinct to set it apart from *uvarum*. Hence, *S. carlsbergensis* will be used here. Traditionally, lager is produced by bottom-fermenting yeasts. This means that at the end of fermentation bottom yeasts flocculate and collect on the bottom of the fermentor. Top-fermenting yeasts used in the production of ale tend to be somewhat less flocculent and loose clumps of cells are carried to the fermenting wort surface adsorbed to carbon dioxide bubbles. Consequently, top yeasts are collected for reuse from the surface layer of the fermenting wort (a process called skimming), whereas bottom yeasts are collected (or cropped) from the fermentor bottom. The starting temperature for bottom-fermentation is 6° to 15°C and is complete in 7 to 12 days; top-fermentation is initiated at 18° to 22°C and is complete in 5 to 7 days.

Originally, the secrets of using bottom-cropping yeasts were held by Bavarian brewers, notably in Munich, and until the middle of the 19th century the rest of the world used top-cropping yeasts. The yeast and fermentation techniques were smuggled to Czechoslovakia by a Bavarian monk in 1842 and so helped establish Pilsen as a premier brewing center. Only 3 years later a Danish brewer, Jacobsen, took bottom fermentation techniques from Munich to Copenhagen and improved Danish beer so that Copenhagen also became a world-renowned center of brewing. About the same time, bottom-cropping yeasts were introduced into Pennsylvania and spread throughout the United States, largely due to the immigration of German and Danish brewmasters. After the spread of bottom fermentation, the traditional top-fermentation techniques were largely discarded except in the British Isles. However, a proportion of the beer produced in Australia, Belgium, Canada, the United States, and West Germany is of this type; indeed, in the last few years this proportion has increased. Some brewers in the United States producing both lager and ale use only one yeast, a strain of S. carlsbergensis, for both beer types, thereby eliminating the problems of keeping two yeast cultures separate; ales produced in this manner are designated "bastard ales." A lager yeast used to ferment ale is usually not reused because of enhanced autolysis due primarily to the higher fermentation temperature of ale as compared to lager beer.

Characteristics of Yeast Strains.—It was in 1880 that Emil Christian Hansen, working in Copenhagen, devised methods of isolating single cells of brewery yeasts by repeated dilution of a yeast suspension. He was therefore in a position to separate the component strains of a yeast mixture and study them in isolation. The technique provided an opportunity to free the yeasts from attendant bacteria and wild yeasts and also furnished the brewer with a pure culture. As a result of Hansen's studies, the practice of using a pure strain in lager production was soon adopted, but in ale-producing regions this radical innovation encountered severe opposition, the method being regarded merely as a means of reducing infection by wild yeasts and bacteria. Over the years, however, the use of pure yeast strains has increased in ale-producing areas, and it is interesting to note that it was reported by Hough (1959) that of 39 cultures in commercial use in Britain, 12 contained only a single strain, 16 had two major strains, and the rest had three or more components. If a similar survey were conducted today, it is probable that the percentage of pure strains being used for ale production would be considerably higher.

The use of pure cultures in brewing naturally involves considerable attention to the management of the yeasts employed. Pure culture brewing, although it removes some of the anxieties about the changing characteristics of the yeast culture (i.e., changes in the proportion of the yeast strains one to another, thus leading to possible changes in fermentation characteristics) and the invasion of gross infection, it does not eliminate all the brewer's anxieties. Even with the yeast strain derived from a single cell or from an ascus, mutation is possible and should always be guarded against.

The taxonomic characteristic used to distinguish the two species of yeast (Lodder 1970) used by brewers is that S. cerevisiae can ferment only one-third of the raffinose molecule, whereas S. carlsbergensis (uvarum) can ferment raffinose completely (Fig. 10.2). This distinction coincides in many cases with the ability of S. cerevisiae to form a yeast "head" toward the end of fermentation, whereas strains of S. carlsbergensis fall to the bottom of the fermentor. This coincidence is by no means universal, and it is common to find strains of S. cerevisiae that behave as bottom yeasts, just as it is possible, in some rare circumstances, to find strains of S. carlsbergensis that form a head during fermentation. The formation of the yeast head depends on the fermentation temperature and an adequate depth of liquid in the vessel, as well as on the yeast strain itself.

Another major difference between the two species is one that is germane in the brewing context, i.e., the optimum temperature for growth of the two species is different. S. carlsbergensis has an optimum of approximately 20° to 24°C whereas for S. cerevisiae it is

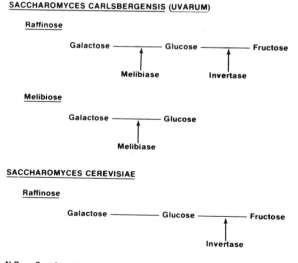

FIG. 10.2. TAXONOMIC DIFFERENCES BETWEEN *SACCHAROMYCES CARLS-BERGENSIS (UVARUM)* AND *SACCHAROMYCES CEREVISIAE*

approximately 28° to 32°C. As a consequence of this, lagers are fermented at a much lower temperature (6° to 15°C) than ales (18° to 22°C). These differing incubation temperatures result in an altered overall metabolism and different excretion products with the end effect being beers with contrasting flavors.

It is at the strain level that interest in brewing yeasts centers. At last count there were at least 1000 separate strains of S. cerevisiae— these strains may be brewing, baking, wine, distilling, or laboratory cultures. There is a problem classifying such strains in the brewing context; the minor differences between strains that the taxonomist dismisses are vitally important to brewers. In order to distinguish between strains of brewing yeasts, many different systems have been devised (Stevens 1966; Campbell 1971; Gilliland 1971; Stewart 1972; Thorne 1975). One of the simplest and yet most suitable methods for this purpose is the Giant Colony Morphology Method (Richards 1967). This method involves inoculating the yeast culture onto a wort/gelatin medium and examining the colonial morphology of the culture after incubating at 20°C for 3 to 4 weeks in a humid atmosphere. Experiments in this laboratory have shown that all strains of S. cerevisiae studied have characteristic colony morphologies (Fig. 10.3); however, strains of S. carlsbergensis have uniform colonial morphologies (Fig. 10.4). It should be emphasized that this is a method of distinguishing between and among yeast strains and not a classification method. Indeed it has been reported (Walkey and Kirsop 1969) that there is no relationship between the fermentation characteristics of a yeast strain and its morphology. However this may be, the colonial morphology must be determined by one or more biochemical parameters which in turn will be controlled by certain genetic factors. It therefore follows that strains having similar morphologies must have certain metabolic characteristics in common.

In order to achieve beer of high quality, the yeast culture must be effective in removing the desired nutrients from the growth medium, it must impart the required flavor to the beer, and finally the microorganisms themselves must be effectively removed from the fermented wort after they have fulfilled their metabolic role (Stewart 1975A). This definition is very wide-ranging and grossly unspecific. The selection of a yeast strain for brewing purposes is very subjective and will depend to a considerable degree on the type of brewing envisaged—ale or lager, top- or bottom-cropping, batch or continuous; if continuous is preferred to batch, what type of continuous? Another important question is whether the flocculation properties of the strain are sufficient for the strain to remove itself from suspension, or if centrifugation is needed. The composition of

FIG. 10.3. GIANT COLONIES OF *SACCHAROMYCES CEREVISIAE* STRAINS

the wort is also a vital feature. Wort composition and the properties of the yeast strain used to ferment that wort are the prime determinants of beer quality. It is difficult to disassociate these two factors because any variation in wort composition will affect yeast quality which in turn will influence the flavor attributes of the final beer.

The behavior, performance, or quality of a yeast strain is influenced by two sets of determining factors—these are the nature-nurture effects (Stewart *et al.* 1975). The nurture influences are all the environmental factors, collectively (i.e., the phenotype), to which the yeast is subjected from inoculation onwards, as distinguished from its nature or heredity. The nature influence is the genetic makeup (i.e., the genotype) of a particular yeast strain.

The most important environmental influence upon yeast behavior is the composition of the wort; however, it is by no means the only one. Other environmental features that will influence yeast behavior are the fermentation temperature, the wort oxygen level at inoculation,

FIG. 10.4. GIANT COLONIES OF *SACCHAROMYCES CARLSBERGENSIS (UVAR-UM)* STRAINS

the type of fermentation (i.e., batch or continuous), the dimension of the fermentor (e.g., vertical or horizontal), the yeast inoculation (or pitching) rate, the condition of the yeast culture at pitching, and finally the level of microbial contamination (i.e., bacteria or wild yeasts), in the yeast culture. All these factors will influence the behavior of the yeast and such behavior will have a profound effect upon beer flavor and quality.

Brewers' wort is a very complex growth and fermentation medium consisting of simple sugars, dextrins, amino acids, peptides, vitamins, ions, nucleic acids, and other constituents too numerous to mention. If, perchance, a yeast strain does not have the ability to take up and metabolize any of the wort nutrients, serious defects will occur during fermentation which in turn will affect the flavor of the product. In the batch fermentation of brewers' wort, individual sugars are removed in a distinct order (Phillips 1955). The precise sequence may vary from one strain of yeast to another but for most strains of S. cerevisiae and S. carlsbergensis the general order can be stated to be sucrose, glucose, fructose, maltose, and maltotriose (Fig. 10.5). For most situations in brewing the attenuation limit of a fermentation is only achieved when all of these five sugars have been taken up by the yeast. It can happen, however, that a strain will lose its ability to take up one or more of these sugars (Gilliland 1969), and invariably it is a loss of ability to ferment either maltotriose or maltose (Fig. 10.6).

The major fermentable sugar in wort is maltose. Consequently, a detailed knowledge of factors affecting maltose uptake into the yeast cell is of vital importance when considering yeast quality and performance. The uptake of maltose is an active transport process (Halvorson and Spregelman 1952) and two separate metabolic steps are involved prior to the sugar's entry into the common carbohydrate metabolic routes of the cell: (1) maltose permease acts to transport the maltose across the cell membrane into the cytoplasm, and (2) maltase (α-glucosidase) hydrolyzes the maltose into two glucose units which are then phosphorylated (Fig. 10.7).

Although knowledge of the rate of sugar uptake by various strains of brewers' yeast is of considerable value, of equal importance is an account detailing the fate of sugars once they are inside the cell (Stewart 1973). The type and concentration of excretion products that occur in the fermenting wort will depend on the overall metabolic balance of the yeast culture (Fig. 10.8). There are many factors that can alter this balance and hence the flavor of the product: these include the yeast strain, incubation temperature, adjunct level, wort pH and buffer capacity, and the wort gravity (Pfisterer and Stewart 1975).

Just as important in a discussion of brewers' yeast strains and their carbohydrate metabolism in wort is the manner in which they utilize nitrogenous compounds and, in particular, amino acids. Nitrogenous compounds are vital factors in the brewing process due to the fact that they play key roles in foam stability, yeast nutrition, biological stability, and flavor. Wort contains 19 amino acids and, under brewery fermentation conditions, brewers' yeast takes them up in an orderly manner, different amino acids being removed at various points in the fermentation cycle. Consequently, amino acids can be grouped according to the order of their removal from wort (Jones and Pierce 1964) (Table 10.1). Under anaerobic conditions (those encountered in brewing situations) proline, the most plentiful amino

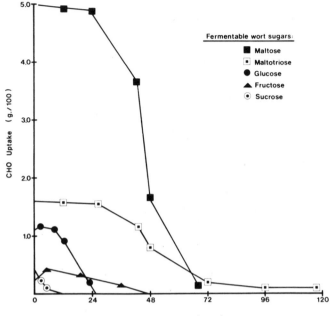

FIG. 10.5. UPTAKE OF WORT SUGARS BY A TYPICAL BREWERS' YEAST STRAIN

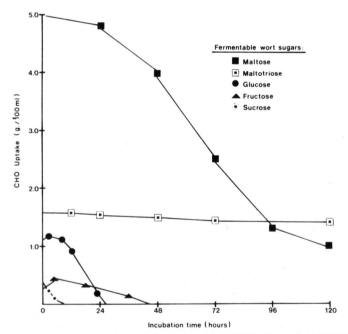

FIG. 10.6. UPTAKE OF WORT SUGARS BY A MALTOTRIOSE-NEGATIVE YEAST STRAIN

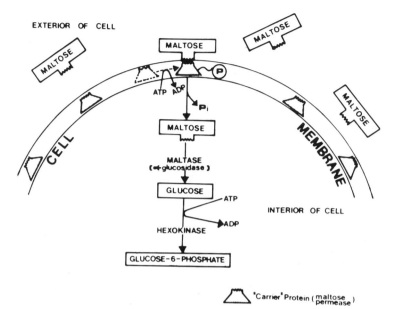

FIG. 10.7. SUGGESTED TRANSPORT MECHANISM OF MALTOSE BY BREWERS' YEAST

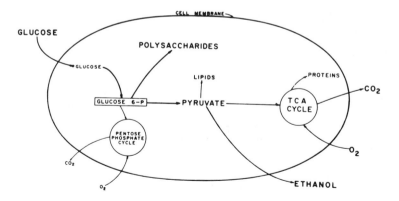

FIG. 10.8. SCHEMATIC OUTLINE OF GLUCOSE METABOLISM IN THE YEAST CELL

TABLE 10.1

ORDER OF ABSORPTION OF AMINO ACIDS FROM WORT
BY BREWERS' YEAST

| A | B | C | D |
| | | | (Only Slowly |
(Immediately Absorbed)	(Absorbed Gradually During Fermentation)	(Absorbed After a Lag)	Absorbed After 60 hr)
Arginine	Histidine	α-Alanine	Proline
Asparagine	Isoleucine	Ammonia	
Aspartic acid	Leucine	Glycine	
Glutamic acid	Methionine	Phenylalanine	
Lysine	Valine	Tryptophan	
Serine		Tyrosine	
Threonine			

Source: Jones and Pierce (1964).

acid present in wort, is scarcely assimilated. While 95% of the other amino acids have disappeared by the end of the fermentation, there is a considerable amount of proline in the finished product (ca. 200 to 300 µg per ml). Under aerobic laboratory conditions, however, proline is assimilated after exhaustion of the other amino acids.

The inability of *Saccharomyces* to assimilate proline under brewery conditions is the result of several phenomena (Bourgeois and Thouvenot 1972) (Fig. 10.9). As long as other amino acids or ammonium ions are present in the medium, the activity of proline permease (the enzyme that catalyzes the transport of proline across the cell membrane) is repressed; as a consequence of this permease repression, proline adsorption is slight. The first catabolic reaction of proline, once it is inside the cell, involves proline oxidase which requires the participation of cytochrome c and molecular oxygen. By the time all the other amino acids have been assimilated, however, thus removing the repression of the proline permease system, conditions are strongly anaerobic and, as a consequence, the activity of proline oxidase is inhibited and proline uptake does not occur anyway.

As stated previously, the flocculation properties of a yeast strain are vital to its selection for brewing purposes. The purpose of yeast flocculation is to agglomerate otherwise discrete microorganisms into flocs so that under quiescent conditions they will be of sufficient mass to settle out of suspension. Under other circumstances appropriate strains will be carried to the surface of the fermenting medium adsorbed to carbon dioxide bubbles. Many definitions of yeast flocculation occur in the literature (Rainbow 1970; Stewart

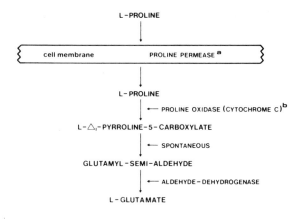

L-PROLINE

cell membrane   PROLINE PERMEASE [a]

L-PROLINE

←— PROLINE OXIDASE (CYTOCHROME C)[b]

L-△₁-PYRROLINE-5-CARBOXYLATE

←— SPONTANEOUS

GLUTAMYL-SEMI-ALDEHYDE

←— ALDEHYDE-DEHYDROGENASE

L-GLUTAMATE

[a] Enzyme activity repressed by other amino acids and ammonium ions.

[b] Presence of molecular oxygen is required for the activity of this enzyme.

FIG. 10.9. MECHANISM OF PROLINE UPTAKE BY BREWERS' YEAST

1975A), but the following would appear to define the phenomenon most succinctly: flocculation is the phenomenon wherein yeast strains adhere in clumps and sediment rapidly from the medium in which they are suspended.

Yeast flocculation is influenced by both the genetic makeup of the strain and also by a number of environmental factors. It is beyond the scope of this text to discuss in any detail this complex phenomenon and for further information the reader is referred to the publications of Stewart (1972; 1974; 1975 A,B), Stewart et al. (1973A,B; 1975) and Day et al.(1975).

Although published data relating to a particular yeast strain together with laboratory and pilot plant studies can be of assistance in the selection and characterization of a prospective production yeast strain, these cannot replace practical experience gained with production trials using the actual conditions and facilities in which the yeast will be required to ferment. Such conditions could be batch or continuous. If batch, there are a number of fermentor types available (horizontal or vertical, open or closed) and if continuous, the choice of type is equally as great and confusing. For detailed information on this topic the reader is referred to reviews by Rainbow (1970), Button (1971), Hough and Button (1972),and Stewart (1974).

The yeast is the only microorganism required for the production of good quality beer. Accordingly, one might conclude that any other microorganism which becomes involved in the process, at any stage, is a contaminant (Kaye 1956). Generally speaking, this is true. However, it should be pointed out that some brewers firmly believe that certain bacteria living in association with their pitching yeast are responsible for the unique flavor of their product. In such cases a deliberate effort is made to perpetuate the cultural association in the appropriate balance required to maintain continuity of product flavor (Hough et al. 1971). Obviously this may be more difficult to achieve than with a pure yeast culture, especially if the specific identity of the bacteria involved is unknown.

Spoilage and Sanitation.—Most modern brewers give as high priority to pure yeast culture maintenance as they do to environmental sanitation. Relaxation of effective procedures may quickly give rise to considerable problems because contaminating microorganisms, although kept under control, are never completely eliminated from the environment. With respect to the environment, breweries are really no different from any other manufacturing establishment producing edible products. They do differ from most, however, in that commonly recognized pathogenic microorganisms are rarely of concern (Kaye 1956; Hough et al. 1971), mainly because of the nature of the raw materials, the manner of processing, and the limiting environmental characteristics of the final product (i.e., low pH, high alcohol concentration, and carbon dioxide tension). An exception to this is the rather unlikely possibility that significant levels of toxic metabolic products from certain fungi may pass from infected raw materials into finished products. However remote, such a possibility has been demonstrated experimentally (Krogh et al. 1974; Chu et al. 1975; Nip et al. 1975) and indicates the importance of rigid quality control of raw materials since currently there is no practical way of detoxifying a contaminated finished product.

Historically, prophylaxis has generally superseded detailed taxonomic interest in brewery bacteria (Ault 1965), probably because breweries are traditionally production oriented. Despite this, it is clear that in addition to restricting the involvement of pathogens, brewing environments impose considerable limitations on the spectrum of significant spoilage organisms. Bacterial contamination is relatively easy to distinguish by observation of morphological differences microscopically or, if the level of infection is low, by appropriate culturing. Contamination of pitching yeasts by wild yeasts is less readily distinguishable and usually requires more discriminating procedures (Kaye 1956).

Although spore-forming bacteria, including *Clostridium* spp., may be involved in the spoilage of brewery by-products such as spent grains, all bacteria of significance in the liquid phase of the process are non-sporeformers. These may be involved in a wide variety of problems in wort, including acidification, acetification, pH elevation, incomplete fermentation, ropiness, and slow run-off time. Such infection may also be directly or indirectly responsible for various off-odors and biological hazes in finished beer (Kaye 1956; Hough et al. 1971).

To facilitate identification of contaminating bacteria, they may be divided into two broad groups, depending upon their reaction to Gram's stain, and further into subgroups based on their morphology, motility, catalase reaction, and oxygen requirements (Fig. 10.10). The Gram-positive group includes certain species of the genera Lactobacillus, Pediococcus, and Micrococcus. Genera belonging to the Gram-negative group include Acetobacter, Zymomonas, Aceto-monas, and Flavobacterium, as well as several genera of the family Enterobacteriaceae, including Enterobacter, Klebsiella, Citrobacter, Hafnia, and Serratia. Of the various genera (Buchanan and Gibbons 1975), Lactobacillus is usually regarded as the most troublesome (Ault 1965) because its species represent a potential spoilage hazard at virtually all stages of production, including finished beer. The remaining genera are less versatile under brewery conditions and consequently their spoilage potential is, to varying degrees, more limiting. Zymomonas anaerobia, for example, is known to be particularly troublesome in British breweries which have undergone recent construction (Richards 1970). This organism is a common soil contaminant in that region; presumably it becomes airborne under these conditions and subsequently finds its way into the process. The stages during which the various genera may flourish in the brewing cycle have been reviewed in detail by Ault (1965). More recent literature (Hough et al. 1971; Priest et al. 1974) supports the view that the enterobacteria may have a more profound effect on the fermentation, flavor, and aroma of beer than indicated earlier by Ault (1965).

A variety of approaches is available for detecting and differentiating the various brew contaminants, some of which have been adopted as standard practice in the industry. These include selective and differential culture media (Scherrer et al. 1969; Brenner et al. 1970; Brenner and Hsu 1971; Van Engel 1971; Middlekauff and Sondag 1972; Kovecses et al. 1973; Solberg and Clausen 1973; Lin 1974A,B; Lee et al. 1975), either alone or in combination with centrifugation or millipore filtration, depending upon the expected cell density (Glenister 1970; Kovecses and Van Gheluwe 1972; Lin 1975), as well as various serological techniques (Dolezil and Kirsop 1975) and impedance measurement (Harrison et al. 1974).

The most effective means of preventing spoilage is to control

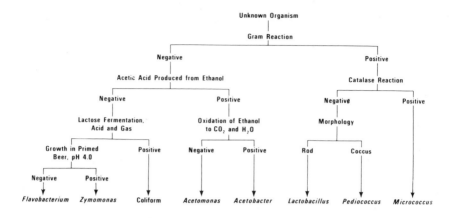

FIG. 10.10. IDENTIFICATION SCHEME FOR BACTERIAL CONTAMINANTS FOUND IN BREWERIES

infection. This can be accomplished most effectively by developing and maintaining a comprehensive cleaning and disinfecting program appropriate for each manufacturing establishment. Often this is conducted in cooperation with a reliable cleaning and disinfectant supply house. Contrary to popular opinion, looking clean is not necessarily being clean. No part of any equipment in direct contact with the product should be inaccessible. Except when using detergent-type sanitizers, cleaning should precede sanitizing and the agents should not leave unfavorable residues in the product. In order to reduce labor costs and to improve efficiency, the trend in the industry is towards in-place cleaning equipment, centralized cleaning systems, and automation (Reap 1965; Borker 1966; Cooper and Lloyd 1971; Stanton 1971).

The bacterial count of pitching yeast may be reduced by treatment with acidified ammonium persulfate (Bruch et al. 1964) as well as with dilute acids including phosphoric, sulfuric, and tartaric. Although acid treatment is very effective in reducing bacterial infection, it also has a deleterious effect upon the yeast culture and sluggish fermentations can be expected in the first few cycles after washing. Sulfur dioxide has been used in the past for the control of wort bacteria and indeed is still used in some breweries (Ault 1965). The most common procedure for microbiological control in packaged beer continues to be heat pasteurization. Alternative procedures to heat pasteurization have been investigated for a variety of reasons including the "adverse" effect of heat on beer flavor, more efficient use of space, and the elimination of heat and humidity from working areas—this latter point finding considerable favor with the labor force. Such alternative procedures are commonly referred to as "cold pasteurization" and include the use of chemical agents such as diethyl pyrocarbonate (DEPC) (Cuzner et al. 1971; Larrouquere 1965), WS-7 (Strandskov and Bockelmann 1965; Itriago-Gimon and Lewis 1970; Kozulis et al. 1971), and propylgallate (Loncin et al. 1970), alone or in combination, as well as millipore filtration followed either by aseptic packaging or in conjunction with any of the preceding preservatives. Official permission for the use of chemical preservatives varies among different countries and may change as new information on safety accumulates, e.g., DEPC was barred in Canada in 1973.

Aging

Following centrifugation of the ferment, the "green" beer is stored at 0° to 2°C for one to several weeks to permit unstable proteins, yeast, resins, and other soluble materials to settle out. Esters are formed as the beer matures, and harshness disappears to yield a mellow, smooth flavor. The beverage is usually chillproofed at this point to prevent haze development at lower temperatures during future handling and storage.

Finishing

After filtration, usually through diatomaceous earth, the stored beer is carbonated to a final carbon dioxide level of 0.45 to 0.52%. It is cooled, clarified, and packaged in cans, bottles, barrels, or kegs. Canned and bottled beers are pasteurized at about 60°C.

WINE

"Wine," without any qualification, means the fermented juice of the grape. This fruit is grown in most countries of the world between the 30th and 50th parallels north and south of the equator. Most commonly wine is obtained from *Vitis vinifera* or from hybrids between its many strains. In North America the choice is much wider; strains of *V. labrusca*, *V. rotundifolia*, and *V. rupestris* are also used, as well as interspecific hybrids between these and *V. vinifera*. The flavor of wine is conditioned not only by the strain of grape but also by the method of cultivation, harvesting, processing, fermentation, and storage to which it is subjected.

In 1974, 1.28 billion liters of wine were consumed in the United

States, 80% of this being domestically produced. This consumption is expected to continue to increase over the next few years and indeed global consumption of inexpensive table wines is increasing rapidly. In consequence, a very large area of new vineyards has been planted and production methods have been greatly improved. The vineyards now being planted must be amenable to the use of mechanical harvesting machines. The selection of site is based on the concept of allowing the grapes to achieve maximum yield consistent with quality; there is, however, no incontrovertible scientific evidence of a relationship between the two parameters. The difficulties arise partly in overcoming the problem of supply of experimental material and partly on agreement over the need to assess "quality" subjectively. This latter item is one area where the skilled taster remains unchallenged, the function being to define the quality of the wine and ensure that the blend conforms to the brand style.

Wine Making

Briefly, to make wine, grapes are crushed and the juice (must) is allowed to ferment in vats for the length of time needed for the particular type of wine required. After fermentation the wine is drawn off into casks or tuns to age until suitable for drinking. During this period chemical changes occur which determine the flavor and aroma of the wine. An important part of wine making is the refining of the wine to remove tannins, pectins, proteins, and phenolics, all of which affect the quality of the finished product.

The most common classes of wine are dry or sweet, depending on taste and the percentage of sugar remaining or added after fermentation. The wines may be still, such as claret, Rhine, or sauterne; sparkling, which refers to those containing natural carbon dioxide such as champagne and sparkling Burgundy; or fortified, such as port, sherry, and Madeira. This last group are wines to which spirits (usually brandy) have been added to increase the alcohol content.

Table wines, with few exceptions, can contain between 8.5 and 14% (vol/vol) ethanol. White wine is made by pressing white- or black-skinned grapes as soon as they are crushed. Red wines need the pigment in the skin cells of colored grapes, which can be solubilized by alcohol (fermentation on the skins), heat (thermo-vinification), or carbon dioxide (macération carbonique). Rosé wines can be made from red-fleshed (Leinturier) grapes or by allowing shorter time intervals for the color extraction procedure to take place.

In the early days of mechanization of the wine-making process, emphasis was mainly on the speed and labor-saving qualities of the equipment. While these two properties are still important, more emphasis is now being given to the prevention of any unnecessary reduction in product quality. This has been made possible by a greater knowledge of grape physiology and yeast biochemistry. For instance, crushers for white grapes just break the skins without violent rupture of the seeds and stalks, and the presses liberate juice without prolonged holding periods. Failure to observe these precautions in the past caused excessive quantities of phenolics and oxidizing enzymes to be liberated into the juice. For more detailed information on wine making, the reader is referred to texts and reviews by Amerine *et al.* (1972) and Beech and Timberlake (1975).

Wine Yeasts.—Wherever one finds sugar in nature, yeasts are also found during the growing period, especially beneath plants whose sugary fruits have fallen to the ground. Indeed, it was Hansen who, as well as making notable observations on brewery yeasts, showed that wine yeasts were invariably present in the soil of vineyards. Hansen's use of pure culture techniques in brewing were introduced into the wine industry between 1880 and 1890. However, Bersch used a kind of yeast culture as early as 1878. Grapes were crushed several days before the main harvest. The juice was allowed to ferment naturally in a warm room and this was used as a starter culture. The first distribution of pure yeast cultures for wine making was apparently made from Geisenheim-am-Rhein in 1890. In France, pure yeast cultures were used in Bordeaux as early as 1894. Jacquemin (1893) believed that the bouquet of wine was solely due to yeasts, and in particular those yeasts isolated from premium quality vineyards. To develop special bouquet he used grape leaves and his pure yeast cultures. His cultures were widely distributed and at least two of his wine yeasts (Champagne and Burgundy) are still maintained in the Department of Viticulture and Enology, University of California, Davis.

Since pure yeast cultivation was started, an enormous number of strains have been accumulated. Originally strains were isolated from many different localities, but it was found later that the locality of origin had no bearing on the characteristic properties of the different strains. The term "pure yeast culture," as used in wine making, is not precise, since grape must is not sterile before addition of yeast starter. It is assumed, however, that the use of sulfur dioxide and large amounts of yeast starter overwhelms the natural flora and results in a fermentation by essentially one kind of yeast. Nevertheless, wild sulfite-resistant yeasts, especially *Saccharomyces* spp., can still survive. Pure yeast cultures are of universal value to the wine industry and, combined with the inhibiting effect of sulfur dioxide, have made wine spoilage rare. The use of sulfur dioxide and pure yeast cultures is now common in many parts of the world, and pure yeast cultures are widely available. No fewer than eight state agencies distribute them in West Germany. Cultures are available worldwide from l'Institut Pasteur, the America Type Culture Collection, and several firms in California.

Under natural conditions, musts often appear to be fermented by a succession of organisms and, as a consequence, one might expect that the use of a proper mixture of yeasts would be beneficial. Indeed, as in the case of certain beer manufacturers, some wine producers, particularly in Europe, still do not accept the monoculture concept and believe that the distinctive flavor and bouquet of their product reflect the interactions of a mixed population. There can be little doubt, as explained earlier, that the dynamics of such a mixed population are more difficult to control and the results less predictable than in the case of a single culture. A detailed study of the microflora occurring naturally in must has been discussed by Kunkee and Amerine (1970).

Wine yeasts are those yeasts which function well at the relatively high acidity of grape juice, are resistant enough to allow the formation of greater than 10% (vol/vol) ethanol, and can adapt to the low concentrations of sulfur dioxide added as an antiseptic. It was early established that the primary agent of alcoholic fermentation in wine was *S. cerevisiae* Hansen var. *ellipsoideus*; however, yeast taxonomists prefer to classify this yeast supply by its specific name of *S. cerevisiae* (Lodder 1970). Other species of *Saccharomyces* which have often been classified as wine yeasts include *S. willianus* (now *S. cerevisiae*; Lodder 1970), *S. fermentati*, and *S. bayanus*.

Studies on yeast strains have unfortunately very often been conducted under variable and nonspecified conditions of must composition. Furthermore, many studies have been conducted under laboratory conditions where surface-to-volume ratios and conditions of aeration are quite different from those used in wineries. However, a number of characteristics can be used as a basis for study: (1) high ethanol-producing ability [18 to 20% (vol/vol)]; (2) cold resistance (fermenting at 4°C); (3) resistance to sulfite; (4) resistance to ethanol (i.e., ability to start a new fermentation at 8 to 12% ethanol); (5) resistance to tannin (for red wine production); (6) ability to produce sparking wine (i.e., ferment wine under pressure at low temperature to dryness so that it settles rapidly with the formation of no permanent deposit on the glass); (7) film formation; (8) resistance to high concentrations of sugar (i.e., ability to start fermentation at greater than 30% wt/vol); (9) heat resistance (i.e., ability to ferment at 30° to 32°C); and (10) production of low volatile acidity (mostly accounted for as acetic acid). In addition, there are numerous reports of strain variability in the formation of acids, esters, high alcohols, hydrogen sulfide, glycerol and 2,3-butylene glycol.

Fermentation.—Basically, wine, like beer, only requires a pure culture of yeast as a microbiological tool for its production. The use of appropriate acid-tolerant bacteria to reduce the acidity of certain wines via the so-called malolactic fermentation, in order to make them more mellow, may be regarded as an exception to this broad statement. This process simply involves the conversion of malic acid to lactic acid. Although most commonly employed for reducing the acidity of red table wines produced from grapes grown in cool climates, malolactic conversion may also be applied to white wine and ciders for organoleptic improvement. A variety of organisms including both lactobacilli and lactic cocci have been implicated in this conversion (Amerine et al. 1972). While some wine producers rely on natural bacterial contaminants in the wine to effect the conversion, others maintain special cultures for this purpose or at least to augment those already present. Sensitivity to sulfur dioxide must, of course, be taken into consideration regardless of the procedures followed.

Whereas in brewing an extended period of boiling is commonly used to "sterilize" wort prior to pitching with selected yeast, such severe treatment is precluded in wine making because of its detrimental effect on the must. Instead, the commonly accepted practice is to add sulfur dioxide in an approved form to the crushed grapes at the rate of 50 to 200 ppm a couple of hours or so prior to pitching in order to destroy or inhibit undesirable bacteria and yeasts. In general, the poorer the initial quality of the grapes and the higher the temperature of the must, the more sulfur dioxide required. Less than 150 ppm is usually needed and it should be emphasized that since the pitching yeast is also somewhat sensitive to sulfur dioxide, the higher the initial level, the longer inoculation may have to be delayed until the residual sulfur dioxide diminishes to a tolerable level. If the delay is too long, sulfur dioxide-resistant contaminants may develop. It should be emphasized that sulfur dioxide, at any stage of wine production, is neither an adjunct to, nor a substitute for, the good manufacturing practices expected in any food manufacturing establishment. In this respect, the approach to the production of high-quality wine should be little different than described earlier for beer.

The finest wines are produced at fermentation temperatures below 30°C. A range of 20° to 32°C is satisfactory for red wines and 10° to 21°C for white wines during the primary (active) fermentation period. Usually red wines ferment for 3 to 6 days and white wines for 1 to 2 weeks at these temperatures before the fermented juice is drawn off from the residues (pomace) and stored in a second tank for 7 to 11 days at 20° to 30°C.

Racking, Aging and Packing.—"Racking" is a term used to describe the drawing off of the wine from the lees or sediment. Racked wine is then stored and aged to allow for clearing and flavor and color development. Alcohols and acids give rise to esters, which are important to aroma and bouquet.

Following clarification and/or filtering, wines are packaged in oak barrels for bulk sale or in bottles or cans. Small- and medium-sized containers are pasteurized for about 30 min at 60°C.

Spoilage.—Because of the lower pH (less than 4) encountered at virtually all stages of wine production and the generally higher ethanol content of the finished product, more limiting conditions are imposed on potential spoilage organisms during wine making than during brewing. Consequently, although the types of spoilage encountered are similar, the range of significant organisms involved is narrower and is encompassed essentially by the *Lactobacillus*, *Leuconostoc* and *Acetobacter* genera. *Pediococcus* should probably be included also, although it is of minor significance. Although there is always risk of generalizing, certainly *Lactobacillus* has the greatest spoilage potential in most wines because of the versatility of this genus with respect to pH and oxygen requirements and its greater ethanol tolerance than *Leuconostoc*. *Acetobacter* is of minor concern in wine in North America, although acetification may occur in the floating mat of skins (cap) peculiar to the production of red wine. An obvious solution to the problem is to keep the cap wet or submerged to reduce the rate of oxygen transfer. Previously, the acetification process has invariably been linked with the inclusion of air, thus permitting *Acetobacter* to proliferate. It has been shown, however, that many different organisms may be involved, including both wild and fermenting yeasts (Amerine *et al.* 1972). *Acetobacter* is the cause of considerable economic loss due to spoilage in the cider industry in the United Kingdom (Beech 1972).

The spoilage hazard presented by any organism will, as already implied, depend on the expertise employed in manufacturing. An exception to this is the preprocess spoilage of raw fruit by molds. In such instances, there is the attendant risk that toxic metabolic by-products (mycotoxins) may persist through the process and appear in significant amounts in the finished product. Although there is some indication that wine is free of mycotoxins (Lemperle *et al.* 1975), the possibility that mycotoxins may be produced in sufficient amounts in raw fruit to be detectable in the juice has been clearly demonstrated through investigations on the patulin content of fresh apple juice (Wilson and Nuovo 1973). Although patulin apparently disappears during fermentation (Harwig et al. 1973), the exact mechanism by which this occurs is not understood and deserves further consideration. It is thought that -SH groups contained in sulfur amino acids might be implicated.

The propensity toward the biological spoilage of finished wine is determined by a number of interrelated factors. In general, the lower the pH and available sugar content together with the higher percentage ethanol and carbon dioxide pressure (where applicable), the less susceptible the wine is to spoilage. Unfortunately, none of these factors alone or in combination is usually sufficient to afford complete protection. Low-alcohol wines containing a high residual fermentable sugar are notoriously unstable. Even in high-alcohol wines, such as sherry, spoilage by alcohol-resistant bacteria is not uncommon in certain areas of the world (Amerine et al. 1972). Consequently, additional means of protection are generally used, such as heat pasteurization either before or after bottling, or filter sterilization. Sulfur dioxide is commonly added for insurance purposes, particularly to low-alcohol wines containing residual fermentable sugar where it also functions as an antioxidant. Sorbic acid or potassium sorbate is allowed in some countries such as Canada (200 ppm as acid) but its use may lead to the development of an off-odor during storage. Diethyl pyrocarbonate (DEPC) has also been investigated as a pasteurizing agent to be added just prior to bottling, but its safety has been challenged and its regulatory status is under question. A major deficiency of a product such as DEPC is that it dissipates rapidly to yield ethanol and carbon dioxide, thus leaving no effective residue to control organisms which may survive the initial treatment. For additional information on spoilage, filtration, and pasteurization of wine see Chap. 3.

REGIONAL NONDISTILLED ALCOHOLIC BEVERAGES

Together with the major nondistilled alcoholic beverages, i.e., beer and wine, there are many minor fermented beverages that are regionalized in origin and consumption. Nevertheless, they are of interest to the student of zymology and deserve, albeit superficially, mention in a text of this nature.

Saké

Saké brewing has been a part of Japanese life from earliest times. It is generally believed that the technique originated in China, but comparison of the production processes for saké and Chinese alcoholic beverages shows marked differences, especially in respect to the microorganism concerned. According to earliest records, saké was originally brewed from rice that had been chewed to achieve saccharification, followed by natural fermentation. Saké brewed in this way was used as a sacred wine in the worship of the Shinto gods. This association with religion, Shintoism and Buddhism, has caused a deep intertwining of saké with the traditions and social customs of Japan. Thus today, saké is served at ceremonies and celebrations of all kinds. There is, however, a trend away from saké toward beer consumption in Japan. Although the consumption of both alcoholic beverages is increasing in response to a growing population, consumption of beer per capita is far exceeding that of saké. Nevertheless, the techniques of the saké brewing industry have developed in response to experience and, in modern times, as a result of research. Saké brewing is interesting in the context of this text because, as well as yeasts, other types of fungi are used in the process.

Production.—Saké is manufactured in large and small factories throughout Japan. In the smaller factories, because brewing requires low ambient temperatures, saké is brewed only in the winter. In the larger factories, refrigeration now permits saké manufacture all year round, more closely in line with the consumption of the beverage. The saké industry is very fragmented, their being over 3500 separate companies manufacturing the beverage in Japan compared with only four major beer brewing companies. Saké is a rice wine, clear and pale yellow in color and brewed to an alcohol content that may vary between 14 and 20% (vol/vol). Its aroma is characteristic and owes much to the koji (saccharifying agent) used in its preparation (see Chap. 9). On the palate the beverage gives ample evidence of its alcohol content with no astringency, little acidity, and slight sweetness. In Japan it is often served warm, especially in winter.

There are numerous factors associated with the manufacture of saké which, while individually not unusual, in combination render the commercial manufacture of this beverage a unique process (Ohwaki and Lewis 1970). These factors are: (1) the use of mold enzymes to achieve saccharification and proteolysis; (2) a greater or lesser dependence on the development of natural fermentation, and hence the absence of aseptic techniques; (3) saccharification and fermentation proceed simultaneously; (4) the fermentation is performed in the presence of a high proportion of solid matter; and (5) the low temperature of fermentation.

The saké brewer has a wide glossary of terms to describe the various procedures in the manufacture of saké. In order to discuss the process adequately, it is necessary to make use of a few of them. The starting material for the brewing of saké is rice, which is polished, washed, and then steeped for several hours in water. Water enters the kernel readily, raising the moisture content to about 30% in a few hours. Modification of the structure of the starch granules of the kernels is then achieved by a short steaming procedure.

The koji is prepared by mixing spores of selected strains of *Aspergillus oryzae* with cooled moist rice. After about two days, the rice shows mold on the surface and in the interior channels and pores of the kernels. The rice kernels retain their structure since the amylase enzymes necessary for conversion of the starch to fermentable sugar are contained within the mycelium of the mold.

When the koji is ready, brewing water, yeast of a suitable strain, and steamed rice are added to it to prepare the seed yeast (shobu or moto). The classicial motos, called ki-moto or yamahai-moto, are acidified by a spontaneous lactic fermentation, but in the more modern procedure (sokujo-moto) acidification is accomplished by addition of commercial lactic acid. This acidification helps guard against infection of the brew by undesirable bacteria or yeasts.

Saké yeast (in this context an ecological group of yeasts responsible for saké brewing) has a high resistance to unfavorable and violently changing conditions (Kodama 1970). High viscosity and high concentrations of sugar, acid, and alcohol in the mash are all undesirable for growth of sake yeast itself, but the conditions prevent the mash from being invaded by contaminating microorganisms. Furthermore, by overcoming many of the above-mentioned difficulties, saké yeast gives the mash a good flavor and taste as well as a high alcohol concentration. On the basis of these characteristics, authentic strains of yeasts suitable for saké brewing have been selected which are invariably species of *S. cerevisiae* (Lodder and Kreger-van Rij 1952).

After the moto stage is completely developed, the moto is transferred to the moromi tank and koji, steamed rice, and water are added in three steps (sow, nake, tome) to make the main mash moromi. The moromi is a combination mashing procedure and fermentation; the starch of the rice is hydrolyzed by the amylases eluted from the koji, and the sugars so produced are fermented by the yeasts. This mash/fermentation is started at 10°C in open tanks. The combination of progressive hydrolysis of starch and slow fermentation at low temperature contributes to the high alcohol content of the saké. The moromi procedure usually takes 3 weeks for completion.

The moromi is dense and mashy because of the suspended matter such as koji and steamed rice, some of which remains undigested. This solid matrix serves to maintain the yeast population in suspension throughout the fermentation period. At the completion of fermentation the mash is filtered to remove solid matter as a compressed cake. The pale yellow product, fresh saké, is then further settled, filtered, pasteurized, stored, blended and bottled.

Saké brewing, like beer brewing, has long been carried out largely on the basis of traditional techniques which have been handed down and improved from generation to generation. Recent knowledge of the physiology, biochemistry, and ecology of the microorganisms involved enables confirmation of the fundamental soundness of the conventional methods of brewing. This scientific approach will also make possible further improvements in the technology of saké brewing by eliminating certain unnecessary procedures. New equipment for continuous steaming and filtration of mash and automatic koji preparation and cultivation of yeast have come into use. In addition, pneumatic systems and belt conveyors for transporting rice and rice-koji have replaced manual labor. Further, up-to-date research on the application of commercial enzyme preparations (obtained from various fungi and bacteria) and of submerged culture of koji fungi substituted for classical rice-koji will bring about quite dramatic changes in the saké brewing industry in the near future. Nevertheless, the brewing of saké by traditional methods to produce a refined product with its own individual characteristics will undoubtedly continue in a few breweries using elaborate manual techniques.

Murcha

Murcha is a mildly alcoholic rice beer prepared in India. It is also known as bakhar in the Cuttack district, pachwai in north India, ranu in Madhya Pradesh, and u-t-iat in Khasi Hills (Batra and Millner 1974). Small rice starch cakes which may also contain berries, roots, and leaves of wild native plants are sold in food markets. *Mucor rouxianus* saccharifies the starch in addition to fermenting several sugars in a water suspension of rice cake. A fermenting yeast, *Hansenula anomala* var. *schneggii*, is also involved in murcha brewing along with other *Mucor* spp.

Kvass

Malted barley and rye, together with rye flour, usually form the base for the alcoholic Ukrainian drink kvass (kwass). Peppermint may be added for flavoring. Dry rye bread, sugar, raisins, and orange flavoring are fermented to prepare another type of kvass. Teekvass is a tea beverage produced in Russia, supposedly by a symbiotic fermentation of a tea sugar solution by *Schizosaccharomyces pombe* and *Acetobacter xylinum* (Pederson 1971).

Busa

A fermented drink prepared from millet and sugar by the Tartars of Krim is called busa (Hesseltine 1965). *Lactobacillus delbrueckii* and *Saccharomyces* spp. are responsible for the fermentation. The busa of Turkestan is prepared from rice.

Thumba

Boiled millet is fermented for about 10 days after inoculating with small cakes mixed with roots of a jungle plant to produce thumba in Africa. *Endomycopsis fibuligera* is thought to be involved in the fermentation.

Kaffir Beer

Kaffir beer (not to be confused with kefir, a fermented milk beverage of the Balkan countries) is made by fermenting kaffir corn, a type of sorghum grown in parts of Africa (Novellie 1968). The process uses malted grain, but its role is minor to the action of saccharification by *Aspergillus flavus* and *Mucor rouxii* (Hesseltine 1965). Fermentation by yeasts and lactic acid bacteria results in increased levels of riboflavin and nicotinic acid. Consumption of kaffir beer by the Bantu tribes of South Africa, a people subsisting largely on maize diets, is thought to partially alleviate the disease pellagra.

Cider

Cider in England is the fermented juice of the apple; it is also spelled "cyder" in the countries of Devon and Norfolk. A similar drink, perry, is made from special pear varieties grown mainly in the English counties of Hereford and Gloucester and in Normandy and Switzerland. Perry made from dessert pears is not very attractive and is usually distilled to make a fruit brandy. Due to the economic importance of cider making in certain regions of England, a research group has examined many of the scientific aspects associated with the production of this beverage and a number of comprehensive reviews on this subject have been published (Beech and Davenport 1970; Beech 1972).

The process of cider making has been known for centuries. Basically it consists of grinding the fruit to pulp and expressing the juice under pressure. The freshly pressed juice may be allowed to ferment naturally with the organisms it has acquired from the fruit or pressing equipment, but in the United Kingdom and Switzerland sulfur dioxide is added in amounts up to 150 ppm to suppress all but the fermenting yeasts, i.e., *Saccharomyces* spp. With higher standards of factory hygiene, naturally occurring strains of fermenting yeasts can be rare in the juice, and a suitable culture yeast is added once the action of the sulfite is complete. The process of fermentation can take from 2 weeks to several months, according to the nutrient status of the juice and the fermentation conditions. In France the rate of fermentation is deliberately reduced by chilling to assist in the production of ciders that are sweet by virtue of some juice remaining unfermented. Cider is normally clarified and blended after fermentation and stored in airtight vessels. If the bacterial conversion of malic acid to lactic acid and carbon dioxide has not taken place during the yeast fermentation, it normally does so during storage, thereby imparting a more mature flavor to the cider. Finally, the beverage is sold in glass, plastic, or wooden containers after any necessary degree of sweetening and sterilizing treatment that may be required.

As in beer brewing, it behooves the cidermaker to be discriminating to ensure that the best yeast is chosen since it can affect many of the characteristics of the fermentation even if its influence on the flavor of the final product is more subtle. Beech (1969) has set out the following criteria for choosing a yeast: (1) freedom from infection by bacteria and other yeasts; (2) stability against formation of respiratory-deficient mutants and changes in flocculation pattern; (3) inability to produce excessive amounts of unwanted metabolic products such as hydrogen sulfide, diacetyl, and sulfite-binding compounds; (4) a degree of flocculence suitable for the fermentation system to be adopted; (5) no reliance on an external source of vitamins; and (6) the production of a suitable pattern of higher alcohols, esters and aldehydes.

Ginger Beer

An acidic, ginger-flavored, mildly alcoholic beverage called ginger beer is produced in England and is similar to a product called tibi prepared in Switzerland. The raw materials are water, sugar, and pieces of the ginger-beer plant root. Lactic acid bacteria and *Saccharomyces* are responsible for fermentation.

Kanji

A beer-like beverage common in households and marketplaces in India is called kanji (Batra and Millner 1974). In north India it is prepared with purple or occasionally orange cultivars of carrots plus beets and spices. The mixture is inoculated with a portion of a previous batch of kanji. In south India fermented rice water (torami) is the inoculum for the vegetable mixture. *Hansenula anomala*, *Candida guilliermondii*, *Candida tropicalis*, and *Geotrichum candidum* are reported to be involved in the fermentation process.

Toddy

Toddy is the European designation for the sweet fermented juice of tropical palms. In India there are three types of toddy (Batra and Millner 1974): (1) Sendi, from the palm; (2) tari, from the palmyra and date palms; and (3) nareli, from the coconut palm. *Geotrichum*, *Saccharomyces* and *Schizosaccharomyces* spp. of yeast are responsible for fermentation.

Pulque

The juice of the agave (century plant) is fermented in Mexico to produce a light-colored, thick, slightly alcoholic beverage called pulque. It may be sweetened with honey or fruit juice, or distilled to yield tequila. Species of the lactic acid bacteria, *Lactobacillus* and *Leuconostoc*, and the yeast *Saccharomyces* are involved in pulque fermentation.

DISTILLED SPIRITS

The spirit distillation industry comprises a heterogeneous assortment of manufacturing processes linked by yeasts as a common function (Harrison and Graham 1970). Distillery spirit is available in many forms, varying from pure ethanol to complex potable spirits. Nevertheless, they are all based on the same biochemical and physical principles and similar manufacturing stages. The economic importance of ethanol is considerable, the annual output of potable spirit being equivalent to about 20 million hectoliters of pure ethanol, and several million metric tons of industrial alcohol, of which about 1.2 million metric tons were made by fermentation processes in 1963. Yeasts therefore continue to play a significant part in ethanol production, despite competition from synthetic processes. In many countries, for instance, France, Scotland, and the West Indies, spiritous liquors represent a high proportion of the export business.

Whiskey

Fermentation.—All industrial processes based on the conversion of carbohydrates to ethanol and carbon dioxide require an aqueous medium of assimilable sugars, or complexes which can be broken down during the process to simple sugars, as, for example, by the action of enzymes on starches. Consequently, the first stage in the preparation of spirits such as whiskey from starchy materials (rye, corn, wheat, barley, etc.) is the conversion of starch into a dextrin-fermentable sugar medium (e.g., glucose, fructose, maltose, etc.). This is performed at relatively low temperatures so that the diastatic action will be maintained at a high level. The temperature, nevertheless, must be high enough to restrict the growth of a great variety of undesirable microorganisms. Also, in order to avoid undesirable fermentation, it is necessary to allow the wort to sour naturally or to add acid. The mashes are usually soured by inoculating with a culture of homofermentative lactic acid bacteria. A thermoduric strain of *Lactobacillus delbrueckii* is generally used. This strain will grow at 55°C and produce lactic acid as a major end product, thus lowering the pH of the medium from approximately 5.6 to 3.8 in 6 to 10 hr. Further, small amounts of volatile products are also produced which are considered to improve the flavor of the distilled spirit. When diastatic action and lactic acid production have reached a suitable level, the mash is heated rapidly to approximately 80°C, a temperature at which microorganisms will be rendered nonviable in the acid mash. It is then cooled to 25°C for fermentation and inoculated with a strain of *S. cerevisiae* in logarithmic growth phase.

Due to the complexities of the raw materials, fermentation, distillation, maturation, and subjective judgment of the properties of the final product, the choice of yeast cultures for the preparation of potable spirits is nearly as difficult as for brewing purposes. On the other hand, for industrial spirits, where only high ethanol content and possibly freedom from particular chemical contaminants is required, the problem can be approached in a straightforward scientific manner. In the latter case, simulated conditions of time, temperature, pH, etc., can be applied in the laboratory. By use of the raw materials commercially available and by careful analysis of the fermented liquor, the yield and quality of the spirit can be ascertained relatively quickly. Laboratory studies on the alcohol (which can vary from 5 to 18%) and sugar tolerance and the growth rate of a large number of yeast strains can be conducted with ease. Yeasts can be screened and selected in this way and, if required, new types can be produced by adaptation, breeding, or mutation.

Distillation.—After the completion of fermentation, the liquid is distilled to recover the alcohol and other volatiles. These volatiles are diverse (Harrison and Graham 1970). The quality of the distilled product will be determined by the nature of the fermentation, the ingredients used in the mash, and the aging process employed.

Aging.—The aging or maturing period is a most important stage in the production of whiskey and many other distilled liquors. The spirit for Scotch whiskey is matured for at least 3 years (frequently much longer) in oak containers. During this period the fiery nature of the new spirit is mellowed as a result of chemical reactions and extraction of substances from the wood. Both rye- and bourbon-fermented mashes are distilled in continuous stills and the spirit is matured in white oak or charred oak barrels, the type of barrel determining the type and quality of the product.

Other Distilled Spirits

As previously mentioned, there are a great number of alcoholic spirits produced from the various fermented beverages by distillation. Brandy is one of the oldest of the alcoholic spirits. It is ordinarily prepared from grape wine in many parts of the world, including Spain, Portugal, Greece, South Africa, the United States, Australia, Cyprus, Egypt, and Algeria, but especially in France, where the finest types come from the district of Cognac, north of Bordeaux, or from Armagnac. Other fermented fruit juices are also distilled, e.g., plum brandy or Slivovitz, apple brandy, cherry brandy, or kirsch. Rums are alcoholic spirits from fermented sugar substances such as molasses. They are traditionally manufactured in the cane-growing countries, particularly in the West Indies.

An East Indian alcoholic beverage made by the fermentation of rice inoculated with ragi (see Chap. 9) is arrak (arack, rak). Molasses, dates, or palm juices are sometimes added to the rice. Arrak is matured in teak casks. Okelehao, a type of arrak, is prepared in Hawaii by fermentation of rice, taro, and molasses mash, or by fermenting rice, molasses, and coconut milk (Pederson 1971).

BREWERS' YEAST

Food yeast production developed as an extension of processes yielding yeast crops intended for use in baking, brewing, or distilling (Bressani and Elias 1968). The usefulness of leftover yeast from these processes went unrecognized until the late 19th century when spent brewers' yeast was added to livestock feeds. A comprehensive review of early development in this field has been compiled by Braude (1942) who confirms that dried or wet brewers' yeast can take the place of other feed concentrates in maintaining and even improving animal growth, raising the milk supply of dairy cows, and increasing the egg-laying powers of poultry. Such improvements were so significant that during World War I Germany replaced over 60% of its imported protein concentrates with dried yeast. With the increasing demand for yeast as a fodder, German workers developed an interest in the possibility of supplementing available supplies coming from breweries. It has been reported that these supplies amounted to approximately 15,000 metric tons per annum in Germany in 1915 (Mateles and Tannenbaum 1968), and reliance was placed on Pasteur's observations that yeasts will grow in media in which the nitrogen source is of an inorganic nature.

The product obtained in this way was found to be equivalent to brewers' yeast nutritionally, with the added advantage that the taste of hops was absent. Economically this "mineral" yeast could not compete with the waste product from breweries, but this was of minor consequence during the war. With the advent of more settled economic conditions in the postwar era and with the consequent reduction in the cost of alternative protein concentrates, fodder yeast, including that from breweries, lost favor with agriculturists because of its higher price. A second wave of interest in yeast as a nutritional concentrate followed the realization in the 1930s that yeast constitutes an invaluable source of most, if not all, of the water-soluble vitamins of the B group.

The name "food yeast" dates from 1941, when a committee of the Royal Society of London investigated the possibility of large-scale production of yeast. The committee felt it desirable to adopt a name which would differentiate such yeast from other types—brewers', bakers', wine, fodder, etc. Today, large quantities of yeast slurries, by-products of breweries and distilleries, are sold as high-protein fodder in either the dry or wet form. In many countries fermentation industries produce large quantities of other types of yeasts, cultivated specifically to serve as protein sources in order to improve the quality of feed for farm animals. As long as sufficient quantities of

TABLE 10.2

ESSENTIAL AMINO ACID CONTENT OF BREWERS'
YEAST AND OTHER PROTEINS

Amino Acid	Protein Source[1]				
	Saccharomyces cerevisiae	Casein	Cotton-seed	Soybean	Egg
Tryptophan	96	84	74	86	103
Threonine	318	269	221	246	311
Isoleucine	324	412	236	336	415
Leucine	436	632	369	482	550
Lysine	446	504	268	395	400
Total sulfur amino acids	187	218	188	195	342
Phenylalanine	257	339	327	309	361
Valine	368	465	308	328	464
Arginine	304	256	702	452	410
Histidine	169	190	166	149	150

[1] Amino acid content as mg/g nitrogen.

legumes are produced which, after extraction, yield products that nutritionally are similar in quality to dried brewers' yeast (Table 10.2), this product must compete in price with soybean flour, cottonseed meal, and fish meal which have a world market value of approximately 26 cents per kg of protein. For this reason, the production of fodder yeast from molasses is usually too expensive, except possibly in areas close to cane sugar factories where blackstrap molasses is often an unwanted by-product.

In addition to economic considerations, the use of yeast protein in any quantity in the human diet has one serious drawback. It has long been recognized (Funk et al. 1916; Wintz 1916) that, owing to the high content of nucleic acid, a diet of yeast can increase urinary uric acid to serious levels. Man and higher apes lack the enzyme uricase, which catalyzes the oxidation of uric acid to the more soluble allantoin, so that individuals with a genetic tendency to primary overproduction of uric acid may have crystals of this substance deposited in joints (gout) and soft tissues (trophii) and the formation of stones in the urinary tract. It follows, therefore, that if yeast is to be used as a primary protein source for human consumption, the nucleic acid content will have to be reduced to a safe level. Edozian et al. (1970) fed Candida utilis to humans and found that the maximum intake must be in the range of 2 g of nucleic acid per day. A number of methods for reducing the nucleic acid content of yeasts have been published

(Mateles and Tannenbaum 1968; Maul *et al.* 1970; Canepa *et al.* 1972). Most of the methods involve heat treatment, with an initial heat shock phase at 90° to 100°C followed by incubation at 50° to 60°C for approximately 50 min. Some of the methods use exogenous ribonuclease along with the heat treatment. For further discussion, see Chap. 12.

REFERENCES

AMERINE, M. A., BERG, H. W., and CRUESS, W. V. 1972. The Technology of Wine Making, 3rd Edition. AVI Publishing Co., Westport, Conn.

AULT, R. G. 1965. Spoilage bacteria in brewing—a review. J. Inst. Brew., London 71, 376-391.

BATRA, L. R., and MILLNER, P. D. 1974. Some Asian fermented foods and beverages, and associated fungi. Mycologia 66, 942-950.

BEECH, F. W. 1969. The inter-relations between yeast strain and fermentation conditions in the production of cider from apple juice. Antonie van Leeuwenhoek J. Microbiol. Serol. 35, F11.

BEECH, F. W. 1972. English cydermaking. Prog. Ind. Microbiol. 11, 133-213.

BEECH, F. W., and DAVENPORT, R. R. 1970. The role of yeasts in cider making. In The Yeast, Vol. 3. A. H. Rose and J. S. Harrison (Editors). Academic Press, New York.

BEECH, F. W., and TIMBERLAKE, C. F. 1975. Production and quality evaluation of table wines. J. Inst. Brew., London 81, 454-465.

BORKER, E. 1966. Total quality control application to food processing. Tech. Q. Master Brew. Assoc. Am. 3, 76-80.

BOURGEOIS, C. M., and THOUVENOT, D. R. 1972. Attempt to obtain races of yeast which assimilate proline anaerobically. J. Inst. Brew., London 78, 270-278.

BRAUDE, R. 1942. Dried yeast as fodder for livestock. J. Inst. Brew., London 48, 206-212.

BRENNER, M. W., and HSU, W. P. 1971. A study of wild yeast and lactic acid bacteria in the brewery. Tech. Q. Master Brew. Assoc. Am. 8, 45-51.

BRENNER, M. W., KARPISCAK, M., STERN, H., and HSU, W. P. 1970. A differential medium for detection of wild yeast in the brewery. Am. Soc. Brew. Chem. Proc. 28, 79-88.

BRESSANI, R., and ELIAS, L. G. 1968. Safe and nutritious food supply. Adv. Food Res. 16, 1-103.

BRUCH, C. W., HOFFMAN, A., GOSINE, R. M., and BREUNE, M. W. 1964. Acidified ammonium persulfate has been used with success for washing yeast. J. Inst. Brew., London 70, 242.

BUCHANAN, P. E., and GIBBONS, N. E. 1975. Bergey's Manual of Determinative Bacteriology, 8th Edition. Williams and Wilkins Co., Baltimore.

BUTTON, A. H. 1971. Changes in fermentation techniques. Brew. Dig. 100, 111-116.

CAMPBELL, J. 1971. Comparison of serological and physiological classification of the genus Saccharomyces. J. Gen. Microbiol. 63, 189-198.

CANEPA, A., PIEBER, M., ROMERO, C., and LOHE, J. C. 1972. A method for large reduction of nucleic acid content of yeast. Biotechnol. Bioeng. 14, 173-199.

CHU, F. S., CHANG, C. C., ASHOOR, S. H., and PRENTICE, N. 1975. Stability of aflatoxin B, and ochratoxin A in brewing. Appl. Microbiol. 29, 313-316.

COOPER, T. F. S., and LLOYD, A. K. 1971. Automation in the brewery. Tech. Q. Master Brew. Assoc. Am. 8, 214-220.

CUZNER, J., BAYNE, P. D., and REHBERGER, P. 1971. Analytical methods for the application of diethyl dicarbonate to beer. Am. Soc. Brew. Chem. Proc. *30*, 116-127.

DAY, A. W., POON, N. H., and STEWART, G. G. 1975. Fungal fimbriae. III. The effect on flocculation in *Saccharomyces*. Can. J. Microbiol. *21*, 552-558.

DOLEZIL, L., and KIRSOP, B. H. 1975. An immunological study of some lactobacilli which cause beer spoilage. J. Inst. Brew., London, *81*, 281-286.

EDOZIAN, J. C., UDO, U. U., YOUNG, V. R., and SCHRIMSHAW, N. S. 1970. Effects of high levels of yeast feeding on uric acid metabolism of young men. Nature (London) *228*, 180.

ETO, M., and NAKAGAWA, A. 1975. Identification of a growth factor in tomato juice for a newly isolated strain of *Pediococcus cerevisiae*. J. Inst. Brew. London *81*, 232-236.

FLESCH, P., and HOLBACH, B. 1965. On the decomposition of L-malic acid. Arch. Mikrobiol. *51*, 401-413. (German)

FUNK, C., LYLE, W. G., and McCASKEY, D. 1916. Nutritive value of yeast, polished rice and white bread. J. Biol. Chem. *27*, 173-186.

GALLO, E. 1958. Outlook for a mature industry. Wines Vines *39*, No. 3, 27-28, 30.

GILLILAND, R. B. 1969. Yeast strain and attenuation limit. Eur. Brew. Convention Proc. 303-315.

GILLILAND, R. B. 1971. Classification and selection of yeast strains. *In* Modern Brewing Technology. W.P.K. Findlay (Editor). MacMillan Press, London.

GLENISTER, P. R. 1970. Some useful staining techniques for the study of yeast, beer and beer sediments. Am. Soc. Brew. Chem. Proc. *29*, 163-167.

GLENISTER, P.R. 1972. Some applications of fluorescence microscopy in the examination of beer sediments. Am. Soc. Brew. Chem. Proc. *31*, 33-35.

HALVORSON, H. O., and SPREGELMAN, S. 1952. Inhibition of enzyme formation by amino-acid analogs. J. Bacteriol. *64*, 207-221.

HANSEN, A. M. 1967. Microbiological Technology. Van Nostrand-Reinhold, Princeton, N. J.

HARRISON, J., WEBB, T. J. B., and MARTIN, P. A. 1974. The rapid detection of infection. Am. Soc. Brew. Chem. Proc. *33*, 76-79.

HARRISON, J. S. 1968. Biochemicals from yeast. Process Biochem. *3*, No. 6, 59-62.

HARRISON, J. S., and GRAHAM, J. C. J. 1970. Yeasts in distillery practice. *In* The Yeasts, Vol. 3. A. H. Rose and J. S. Harrison (Editors). Academic Press, New York.

HARRISON, J. S., and ROSE, A. H. 1970. Yeasts in distillery practice. *In* The Yeasts, Vol. 3. A .H. Rose and J. S. Harrison (Editors). Academic Press, New York.

HARWIG, J., SCOTT, P. M., KENNEDY, B. P. C., and CHEN, Y-K. 1973. Disappearance of patulin from apple juice fermented by *Saccharomyces* spp. Can. Inst. Food Sci. Technol. J. *6*, 45-49.

HESSELTINE, C. W. 1965. A millennium of fungi, food, and fermentation. Mycologia *57*, 149-197.

HOUGH, J. S. 1959. Flocculation characteristics of strains present in some typical British pitching yeasts. J. Inst. Brew., London *65*, 479-482.

HOUGH, J. S., BRIGGS, D. E., and STEVENS, R. 1971. Malting and Brewing Science. Chapman and Hall, London.

HOUGH, J. S., and BUTTON, A. H. 1972. Continuous brewing. Prog. Ind. Microbiol. *11*, 89-133.

HSU, W. P., TAPAROWSKY, J. A., and BREENER, M. W. 1975. Two new media for culturing brewery organisms. Brew. Dig. *50*, No. 7, 52-54, 56, 57.

ITRIAGO-GIMON, M., and LEWIS, M. J. 1970. Some effects of medium constituents on the growth of *Saccharomyces carlsbergensis* in the presence of n-heptyl p-hydroxy benzoate. Am. Soc. Brew. Chem. Proc. *29*, 168-175.

JACQUEMIN, G. 1893. Studies on the improvement of yeasts used in the production of alcoholic beverages. Imprimerie Nanceienne, Nancy. (French)

JERCHEL, D., FLESCH, P., and BAUER, E. 1956. Investigations on the decomposition of L-malic acid by *Bacterium gracile*. Liebigs Ann. Chem. *601*, 40-60 (German)

JONES, M., and PIERCE, J. S. 1964. Absorption of amino acids from wort by yeasts. J. Inst. Brew., London *70*, 307-315.

KAYE, N. 1956. Brewery Theory and Practice. Adland and Son, London.

KODAMA, K. 1970. Saké yeast. *In* The Yeasts, Vol. 3. A. H. Rose and J. S. Harrison (Editors). Academic Press, New York.

KOVECSES, F., MORRISON, N. M., and SLATER, G. 1973. Monitoring changes in yeast population during fermentation. Am. Soc. Brew. Chem. Proc. *32*, 40-42.

KOVECSES, F., and VAN GHELUWE, G. 1972. Membrane filtration: an efficient means of brewing quality control. Am. Soc. Brew. Chem. Proc. *31*, 38-41.

KOZULIS, J. A., BAYNE, P. D., and CUZNER, J. 1971. New technique for the cold sterilization of beer. Am. Soc. Brew. Chem. Proc. *30*, 105-115.

KROGH, P., HALD, B., GJERTSON, P., and MYKEN, F. 1974. Fate of ochratoxin A and citrinin during malting and brewing. Appl. Microbiol. *28*, 31-34.

KUNKEE, R. E., and AMERINE, M. A. 1970. Yeasts in wine making. *In* The Yeasts, Vol. 3. A. H. Rose and J. S. Harrison (Editors). Academic Press, New York.

LAINE, B. 1970. Hydrocarbon utilization by yeasts. Proc. 10th Int. Congr. Microbiol., 82-84. Mexico City.

LARROUQUERE, J. 1965. Reaction of diethyl pyrocarbonate with thiols. Extract Bull. Soc. Chim. Fr. (French)

LEE, S. Y., JANGAARD, N. O., COORS, J. H., FUCHS, C. M., and BRENNER, M. W. 1975. Lee's multi-differential agar (LMDA); a culture medium for enumeration and identification of brewery bacteria. Am. Soc. Brew. Chem. Proc. *33*, 18-35.

LEMPERLE, E., KENNER, E., and HEIZMANN, R. 1975. Investigations on the aflatoxin content of wine. Wein-Wiss. *30*, 82-86. (German)

LEWIS, M. J. 1968. Yeast and fermentation. Tech. Q. Master Brew. Assoc. Am. 5, 103-113.

LIN, Y. 1974A. Detection of wild yeasts in the brewery. III. A new differential medium. Am. Soc. Brew. Chem. Proc. *32*, 69-75.

LIN, Y. 1974B. Detection of wild yeasts in the brewery. Brew. Dig. *49*, No. 10, 38-40, 42, 44.

LIN, Y. 1975. Detection of wild yeasts in the brewery. IV. Use and effect of membrane filters. Am. Soc. Brew. Chem. Proc. *33*, 124-132.

LODDER, J. 1970. The Yeasts, 2nd Edition. North-Holland Publishing Co., Amsterdam.

LODDER, J., and KREGER-VAN RIJ, N. J. W. 1952. The Yeasts, 1st Edition. North-Holland Publishing Co., Amsterdam.

LONCIN, M., KOZULIS, J. A., and BAYNE, P. D. 1970. n-Octyl gallate, a new beer microbiological inhibitor. Am. Soc. Brew. Chem. Proc. *29*, 88-101.

MATELES, R. I., and TANNENBAUM, S. R. 1968. Single-Cell Protein. MIT Press, Cambridge, Mass.

MAUL, S. B., SINSKY, A. J., and TANNENBAUM, S. R. 1970. New process for reducing nucleic acid content of yeast. Nature (London) *228*, 181.

MERCIER, P. 1969. Trappist beer-production in the monastery. Wallerstein Lab. Commun. XXXII, 15-24.

MIDDLEKAUFF, J. E., and SONDAG, R. 1972. Studies on quality and quantity of end products during *Pediococcus* growth and metabolism. Am. Soc. Brew. Chem. Proc. *31*, 17-19.

NIP, W-K., CHANG, F. C., CHU, F. S., and PRENTICE, N. 1975. Fate of ochratoxin A in brewing. Appl. Microbiol. *30*, 1048-1049.

NOVELLIE, L. 1968. Kaffir beer brewing—ancient art and modern industry. Wallerstein Lab. Commun. XXXI. 17-32.

OHWAKI, K., and LEWIS, M. J. 1970. Saké brewing. Wallerstein Lab. Commun. XXXIII, 105-116.

PASTEUR, L. 1866. Etudes sur le Vin. Imprimeuss Imperials, Paris. (French)

PASTEUR, L. 1876. Etudes sur la Bière. Gauthier-Villars, Paris. (French)

PEDERSON, C. S. 1971. Microbiology of Food Fermentations. AVI Publishing Co., Westport, Conn.

PEPPLER, A. H. 1967. Microbiological Technology. Van Nostrand-Reinhold, Princeton, N. J.

PEPPLER, A. H. 1970. Food yeasts. In The Yeasts, Vol. 3. A. H. Rose and J. S. Harrison (Editors). Academic Press, New York.

PFISTERER, E., and STEWART, G. G. 1975. Some aspects on the fermentation of high gravity worts. EBC Proc., 255-267.

PHILLIPS, A. W. 1955. Utilization by yeasts of the carbohydrates of wort. J. Inst. Brew. London 61, 122-127.

PRIEST, F. G., COWBOURNE, M. A., and HOUGH, J. S. 1974. Wort enterobacteria—a review. J. Inst. Brew., London 80, 342-356.

RAINBOW, C. 1970. Brewer's yeasts. In The Yeasts, Vol. 3. A. H. Rose and J. S. Harrison (Editors). Academic Press, New York.

REAP, T. A. 1965. Some aspects of sanitation and housekeeping. Tech. Master Brew. Assoc. Am. 2, 58-60.

RICHARDS, M. 1967. The use of giant-colony morphology for the differentiation of brewing yeasts. J. Inst. Brew., London 73, 162-169.

RICHARDS, M. 1970. Brewery spoilage organisms. J. Inst. Brew., London 77, 95-96.

SAHA, R. B., SONDAG, R. V., and MIDDLEKAUFF, J. E. 1974. An improved medium for the selective culturing of lactic acid bacteria. Am. Soc. Brew. Chem. Proc. 32, 9-10.

SCHERRER, A., SOMMER, A., and PFENNINGER, A. 1969. Improved methods for detecting wild yeasts. Brauwissenschaft. 22, 191-195.

SIMMONDS, C. 1919. Alcohol, Its Production, Properties, Chemistry and Industrial Applications. MacMillan, London.

SOLBERG, O., and CLAUSEN, O. G. 1973. Classification of certain pediococci isolated from brewery products. J. Inst. Brew., London 79, 227-230.

STANTON, J. H. 1971. Sanitation techniques for the brewhouse, cellar and bottleshop. Tech. Q. Master Brew. Assoc. Am. 8, 148-151.

STEVENS, T. J. 1966. A method of selecting pure yeast strains for ale fermentations. J. Inst. Brew., London 72, 369-373.

STEWART, G. G. 1972. Recent developments in the characteristics of brewery yeast strains. Tech. Q. Master Brew. Assoc. Am. 9, 183-191.

STEWART, G. G. 1973. Some observations on maltose, fructose and glucose metabolism in Saccharomyces cerevisiae. Am. Soc. Brew. Chem. Proc. 31, 86-93.

STEWART, G. G. 1974. Some thoughts on the microbiological aspects of brewing and other industries utilizing yeast. Adv. Appl. Microbiol. 17, 233-264.

STEWART, G. G. 1975A. Yeast flocculation—laboratory findings and practical implications. Brew. Dig. 50, No. 3, 42-56.

STEWART, G. G. 1975B. Studies on the binding of alcian blue by flocculent and non-flocculent brewer's yeast strains. 4th Int. Symp. on Yeast and Other Protoplasts. Nottingham, England.

STEWART, G. G., and GARRISON, I. F. 1972. Some observations on co-flocculation in Saccharomyces cerevisiae. Am. Soc. Brew. Chem. Proc. 30, 118-129.

STEWART, G. G., RUSSELL, I., and GARRISON, I. F. 1973A. Further studies on flocculation and co-flocculation in Saccharomyces cerevisiae. Am. Soc. Brew. Chem. Proc. 31, 100-106.

STEWART, G. G., RUSSELL, I., and GARRISON, I. F. 1973B. The influence of peptides on the development of flocculation in certain brewer's strains of *Saccharomyces cerevisiae.* Fed. Eur. Brew. Soc. Special Meeting on Industrial Aspects of Biochemistry. Abstr. No. 102.

STEWART, G. G., RUSSELL, I., and GARRISON, I. F. 1974. Factors influencing the flocculation of brewer's yeast strains. 4th Int. Yeast Symp. Abstr. No. B32.

STEWART, G. G., RUSSELL, I., and GORING, T. 1975. Nature-nurture anomalies. Further studies in yeast flocculation. Am. Soc. Brew. Chem. Proc. *33,* 137-147.

STRANDSKOV, F. B., and BOCKELMANN, J. B. 1965. U.S. Pat. 3.175,912. Mar. 30.

TANNANHILL, R. 1973. Food in History. Stein and Day, New York.

THORNE, R. S. W. 1975. Brewing yeasts considered taxonomically. Process Biochem. *10,* No. 9, 17-28.

VAN ENGEL, E. L. 1971. *Pediococcus cerevisiae* in fermentation. Am. Soc. Brew. Chem. Proc. *30,* 89-95.

WALKEY, R. J., and KIRSOP, B. H. 1969. Performance of strains of *Saccharomyces cerevisiae* in batch fermentation. J. Inst. Brew., London *75,* 393-398.

WIEG, A. J. 1970. Technology of barley brewing. Process Biochem. *5,* No. 8, 46-48.

WIEG, A. J. 1973. Brewing adjuncts and industrial enzymes. Tech. Q. Master Brew. Assoc. Am. *10,* 79-86.

WILKES, H. G. 1969. Tibetan beer and other beverages of the Eastern Himalayas. Wallerstein Lab. Commun. XXXII, 93-100.

WILSON, D. M., and NUOVO, G. J. 1973. Patulin production in apples decayed by *Penicillium expansum.* Appl. Microbiol. *26,* 124-125.

WINTZ, H. 1916. Antitoxin content in the serum of tetanus patients. Muench. Med. Wochenschr. *62,* 1564-1566.

11

Edible Mushrooms

W. A. Hayes

Most of the edible species of mushrooms are members of the genus *Agaricus*, the type genus of the family Agaricaceae in the class Basidiomycetes (Basidiomycotina). Growing on an industrial scale is confined almost entirely to one species, *Agaricus bisporus*, the common edible mushroom which exists in the wild throughout the Northern Hemisphere, but outside the tropics and the arctic. *Volvariella volvacea*, known as the padi-straw mushroom or Chinese straw mushroom, has been cultivated by the Chinese for many centuries but is now cultivated on a small scale throughout the Orient. It is a tropical species and, with new and improved methods of cultivation which are based on some procedures used in *A. bisporus* culture, a considerable improvement over traditional methods of culture is possible. Its production is likely to increase over the next decade.

A popular natural edible species in East Asia is the shiitake mushroom. This mushroom, *Lentinus edodes*, is cultivated in forests on tree logs, especially in Japan, where it is highly regarded as a food which promotes good health.

Some other mushroom species are cultivated, but less extensively at present. The oyster mushroom, *Pleurotus oestreatus*, is grown in Italy and India, and the winter mushroom, *Flammulina velutipes*, is

TABLE 11.1
WORLD PRODUCTION OF EDIBLE MUSHROOMS IN 1975

Organism	Common Name	Quantity (Metric Tons)
Agaricus bisporus	Common cultivated mushroom	670,000
Lentinus edodes	Shiitake	130,000
Volvariella volvacea	Straw mushroom	42,000
Flammulina velutipes	Winter mushroom	38,000
Pholiota nameke	Nameko	15,000
Pleurotus spp.	Oyster mushrooms	12,000
Auricularia spp.	Jew's ear	5,700
Tuber spp.	Truffles	200

Source: Delcaire (1976).

cultivated in Japan and Eastern Europe. *Auricularia auricula-judae* and *Pholiota nameko* are grown on wood and sawdust in Japan and Taiwan, while the prized *Tuber melanosporum* (the black Perigold truffle) and other *Tuber* spp., collectively known as truffles, are cultivated in their natural habitats in France. Truffles belong to the class Ascomycetes (Ascomycotina). *Tremella fuciformis,* the white jelly fungus, has been artificially cultivated in the Far East.

The production figures for some of the above species for 1975 are given in Table 11.1 These figures, however, do not include edible species which are collected from their natural habitats and consumed for food. Although seasonal, mushroom collecting is a popular recreational activity in many European countries, India, North America and Africa. Artificial methods for mass culture of most of these species have not been established; however, some are known to form specific mycorrhizal associations with the roots of some tree species. The highly prized matsutake or pine mushroom (*Tricholoma matsutake*), for example, forms an association with pine roots. Our knowledge of the biology of these associations is scant, and for many mushrooms it is not known whether or not mycorrhizal relationships with tree roots exist.

The cap of the boletus mushroom (*Boletus edulis*) and the morels (class Ascomycetes) form a minor part of an international trade in mushrooms, being exported from India, France, and Italy after air drying. Several species of morels are edible, including *Morchella esculenta* (sponge mushroom), *M. angusticeps* (black morel), *M. crassipes* (thick-footed morel), and *M. hybrida* (spring mushroom). The egg yolk mushroom or chanterelle (*Cantherellus cibarius*) is a common vegetable in the rural areas of France, Italy, and many

Eastern European countries.

Wild mushrooms undoubtedly contributed to man's total diet when he relied solely on hunting animals and the products of plants in his immediate locality. Though modern man is also faced with many problems relating to his supplies of food, the common mushroom today is generally available throughout the world, irrespective of season, as a result of artificial techniques of culture. Also, the modern methods of culture provide a "guarantee" of edibility, a crucial feature which is not provided in the consumption of wild mushrooms unless, of course, the mushrooms are verified as being edible by experts trained in their identification. Some wild mushrooms contain toxic substances which can be fatal. It is imperative, therefore, that the consumption of wild mushrooms be discouraged, especially since artificial culture methods are now being applied to a wider range of species and types of mushroom. Recently, more emphasis has been given to the food value of mushrooms. Contrary to many popular beliefs, mushrooms, although low in calories, contain valuable protein, minerals, and vitamins. They can be viewed, therefore, as a useful source of supplementary food and are considered to be especially useful in predominantly vegetarian diets. *Agaricus bisporus*, the common or cultivated mushroom, and the most widely available, is unique, since it is now a relatively cheap vegetable for the home in Western countries and is highly regarded in sophisticated cuisine.

The materials on which mushrooms are grown are also relevant to their future status as a food. A variety of industrial, agricultural, and family wastes can be successfully used as media for growing various edible mushrooms. Even noxious wastes which pollute the environment can be converted by way of cultivation into a beneficial soil conditioner and fertilizer, which can complement the growing of other agricultural or horticultural crops.

In this chapter, the biology of mushrooms will be discussed and the methods used in the artificial culture of *A. bisporus*, the species most extensively consumed throughout the world, will be detailed. Some implications of recent research and knowledge on future prospects worldwide for the extension of cultivation and a more generalized availability of mushrooms will be discussed.

THE MUSHROOM FRUITBODY

The basic life cycle of a higher fungus is from spore to mycelium to fruitbody, the latter bearing spores again. Mycelia cultured from single spores are sometimes sterile and do not produce fruitbodies unless a mating occurs by fusion with another compatible mycelium generated from another spore. Multiple spore cultures are invariably fertile and are therefore used to establish mycelial cultures from which selections are made for manufacture of pure culture spawn— the first essential requirement for pure culture methods of production. In this highly specialized process of spawn manufacture, mycelial cultures are established on granular substrates such as wood chips, sawdust, tobacco stems, or rye or wheat grains.

A second transplant occurs at the stage of spawn inoculation onto the primary substrate base for culture. For most of the cultivated forms this base is wood or cereal straw, reflecting the enzymatic capacities of mushrooms to degrade lignin and cellulose. Mycelium from spawn soon colonizes an appropriate substrate, absorbing the required food materials for growth and accumulating food reserves. The manufacture of spawn and substrates is essentially a procedure in the establishment of vegetative growth.

After a given time, the typical macrostructure—the fruitbody— develops as an aerial structure and represents the reproductive stage. A fruitbody develops from an aggregation of hyphal cells, known as an initial, which then develops into a firm mass of tissue which becomes differentiated into a stalk (stipe) and cap (pileus). While this structure is still relatively small it is referred to as a pinhead. These eventually expand to form a fully differentiated fruitbody which, when fully mature, develops the spore-bearing tissue, ultimately releasing the spores into the atmosphere. The form and nature of the fruitbodies vary according to species (for detailed differences in structure and sexual mechanisms the reader should refer to standard mycological textbooks), but the general form and structure of the agarics are typified by the cultivated mushroom A. bisporus (Fig. 11.1). Fruits may be harvested for consumption in a relatively immature stage (button stage), a more advanced intermediate stage (cup stage), or when fully mature and the spore-bearing tissues or gills are fully exposed (flat stage).

This fundamental change in structure and form from mycelium to fruitbody is a complex biological process which for the cultivated species is not fully understood. It is known, however, that mycelia developed from some spores are infertile, while other spores from the same fruit generate mycelia which bear fruit and are therefore fertile. This establishes a genetic basis for the formation of fruitbodies, but many other factors related to environmental conditions directly affect the switch from vegetative to reproductive growth. There is a known requirement for light in some species, e.g., P. oestreatus. While there is no light involved in A. bisporus growth, other environmental factors

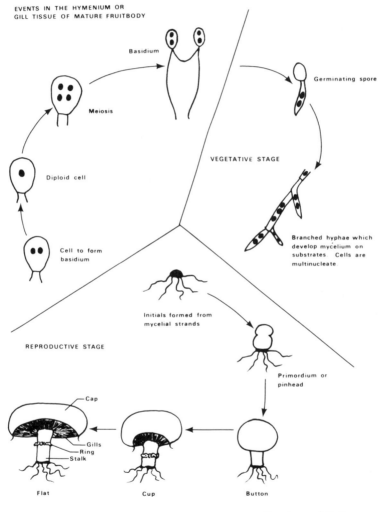

EVENTS IN THE HYMENIUM OR
GILL TISSUE OF MATURE FRUITBODY

Basidium

Meiosis

Germinating spore

Diploid cell

VEGETATIVE STAGE

Cell to form
basidium

Branched hyphae which
develop mycelium on
substrates. Cells are
multinucleate.

Initials formed from
mycelial strands

REPRODUCTIVE STAGE

Primordium or
pinhead

Cap

Gills
Ring
Stalk

Flat

Cup

Button

Courtesy of Z. Bano

FIG. 11.1. MAIN MORPHOLOGICAL FEATURES OF *AGARICUS BISPORUS*, THE CULTIVATED MUSHROOM

such as carbon dioxide concentration and temperature are linked to the formation of fruits. For many of the wild edible species the factors governing the formation of fruits have not been studied, but their seasonal occurrence suggests that environmental factors of the kind mentioned above are involved.

DEVELOPMENTS IN THE ARTIFICIAL CULTURE OF *A. BISPORUS*

Early Development

The first efforts to cultivate mushrooms are known to have occurred in France at about 1650. Apparently, horticulturists near Paris used compost after being collected from melon culture. Some 130 years later it was realized that this mushroom could grow without light. Its culture was then transferred to caves dug for the extraction of gypsum (calcium sulfate) and building stones. These caves in the Paris region are extensively used even in present times.

The natural self-heating properties of manure obtained from horse stables was found to be a good medium and was stacked into ridge beds in rows on the floors. These were inoculated with soil permeated with mycelia obtained from native soils adjacent to old manure heaps or in locations where horses congregated. Soil surrounding the walk-tracks of horses used for working mills was also a good source of mycelia.

Mushroom growing was known to be widespread in England in the 18th century. Abercrombie, a well known gardener, is known to have preferred cultivation in the open. However, in Abercrombie's *Practical Gardener* published in 1817, a description is given of the methods employed by other gardeners which indicates that in some cases mushroom beds were laid in "airy covered sheds" or sometimes under "a sort of awning of canvas." Some gardeners made the beds in glasshouses. This was the origin of the concept of protected mushroom culture in purpose-built structures now widely used in mushroom cultivation throughout the world.

Later writings of Callow (1831) describe a specially constructed house which was warmed by fire heat and adapted to the growth of mushrooms throughout the year. This house was modified internally and shelves were arranged one above the other on brackets attached to the wall. We can now see this development as being the basis for the shelf system of growing which is used today in Holland, Canada, and the United States. This report by Callow is the first recorded report of cropping throughout the year, irrespective of season.

Mushroom growing became established in the United States in the 19th century and was introduced by English, French, and Scandinavian gardeners employed by the wealthy citizens of New York and Philadelphia. Although many different kinds of buildings were used, greenhouses, usually thatched to provide insulation, were popular in the early part of the 20th century.

Structures and Systems

In about the second decade of this century the mushroom growing community of Pennsylvania in the United States adopted a standard house for growing. These were constructed of wood or hollow tiles to provide insulation and conformed to a standard size of approximately $5.5 \times 18.4 \times 4.6$ m high, sufficient to crop 465.5 m^2 of growing space in tiers of fixed shelves. These were called "standard singles" and had a sloping roof and a natural system of ventilation. Later, growers began building two of these units under one roof, termed "standard doubles." These constructions are widely used in the United States and Canada today, and their design provided the basic principles for protected and continuous mushroom growing.

Shortly after the Second World War, the tray system was developed. With the tray system, the bed is effectively split into movable portions contained in boxes which can be palletized and transported to the growing houses. This important development allows greater use of mechanization, a substantial increase in the potential size of growing unit, and a greater degree of capital investment. As a result, growing units cropping up to 20,000 m^2 of bed area are common and this system has directly contributed to the expansion of mushroom outputs in America and many European countries. Unlike the shelf method, containerized cropping in trays provides flexibility in the general system of growing and its application to commerce has tended to diversify general systems of growing. Today, a wide variety of houses of different construction and design exist to accommodate a variety of tray arrangements.

During the last decade a significant breakthrough in the design of mushroom houses has been achieved through the use of plastic film materials. These structures, consisting of polyethylene film as a vapor and weatherproof layer over and under an insulating layer of fiber glass or polystyrene, are inexpensive and durable even in extreme climates. Also, the replacement of the traditional wooden tray or shelf with inexpensive sacks of polyethylene to be disposed at the end of cropping minimizes the capital outlay required for mushroom growing and greatly assists in maintaining disease-free cultivation.

Pure Culture Inoculum—Spawn

While improvements in the construction and design of mushroom growing structures have advanced over the decades, considerable improvements in the techniques of growing have also contributed to the cultivation of mushrooms, now regarded as the most advanced, sophisticated, and predictable of horticultural crops.

The production of pure culture inoculum or spawn is a vital prerequisite for successful culture. In 1932, Sinden developed a method of pure culture spawn production by which pure culture mycelium was grown on grain. This method is now universally used as the means of vegetatively propagating mushroom mycelium. In this method, grain to which water and chalk (calcium carbonate) are added is used as the base medium. The production of grain spawn is now a highly mechanized and specialized process, and the application of modern aseptic techniques has eliminated sources of contamination which were common when manure was used as the base medium. Grain spawn can be safely distributed throughout the world from any one spawn manufacturer. It is claimed that the reservoir of nutrients available in the grain accelerates the growth of mycelia on the beds and that its granular nature allows the mycelia to be mixed throughout the growing medium instead of being planted on the surface.

Substrates and Pasteurization

For the successful growing of mushrooms, of equal importance to the inoculation with pure culture spawn is the preparation of the substrate which is to be inoculated. If incorrectly prepared, other organisms will dominate and the spawn will not establish. Even until about two decades ago, the preparation of horse manure-wheat straw substrates was largely a "hit or miss" affair and there is little doubt that these imperfections were responsible for the unpredictable nature of mushroom growing up to that time. Crop failures, however, were tolerable since the return from an occasional successful crop fully compensated for such losses.

A major breakthrough in the preparation of composts was achieved by the application of pasteurization procedures to composting— essentially the same process as is used in food manufacture and the pasteurization of milk and dairy products. In composting, pasteurization is achieved by subjecting the compost to a heat treatment with steam under controlled aerobic conditions, thus ensuring a substrate which is free from pests and pathogens which otherwise would compete with the inoculated spawn or cause disease.

More recently, a chemical fumigant, methyl bromide, has been introduced as a substitute to steam and this procedure is likely to gain preference over steam in view of the lower costs involved. In addition, it can serve adequately in situations where energy supplies for steam production are difficult to obtain.

Environment Control

Like green plants, mushrooms require the correct temperatures for growth and fruitbody formation. In the growth cycle of A. bisporus there are two optima, i.e., 25°C for growth in the compost and 14° to 18°C for the formation of the characteristic mushroom fruitbodies.

In the 1930s it was realized that carbon dioxide also influenced growth. At high concentrations, fruitbody growth is inhibited; even at levels above that in fresh air (up to 0.15%), growth of fruitbodies is distorted.

These two physiological responses have placed emphasis on control of environmental conditions in growing. Ventilation of the growing houses is an essential prerequisite and for most climatic conditions some degree of heating is necessary. Conversely, growing in extremes of heat requires a degree of air cooling. However, growing heat-tolerant strains or other species of Agaricus obviates this need.

Mechanization

Mechanical aids for use in mushroom growing have resulted from the seemingly innate abilities of growers to improvise. It is only in recent years that machinery designed specifically for the mushroom industry has been marketed. Up to about three decades ago the laborious task of turning compost stacks was done by hand, but mechanical turning equipment designed to mix and aerate compost is now standard.

The shelf or fixed-bed system of growing which predominates in the United States and Holland imposes major constraints on sophisticated mechanization, but the tray system provides a means by which units of compost can be readily transported by fork lifts over considerable distances. Large growing units are thus more flexible. The introduction of the tray system provided a stimulus for the phasing of culture procedures into peak-heating, spawning, and growing. This could in turn provide the means by which spawning and casing could be mechanized. In new modern plants, tray handling lines consisting of devices for tipping, spawn metering and mixing, compressing compost, casing, and automatic destacking are standard. These arrangements, together with sophisticated controls for growing, can be viewed as a preliminary to automated mushroom production.

Recently a number of farms have extended the principle of line operation to harvesting the crop. This involves transporting the crop from the growing area to a conveyor line with pickers on either side.

An alternative concept designed for the shelf system is to grow the mushrooms on a movable conveyor. At present this procedure serves to mechanize the tasks of filling and emptying but may lead to a system of mechanical harvesting—a logical next step in the development of mushroom growing systems.

Low Cost Systems

In sharp contrast to these sophistications are the relatively simple low cost systems of growing. With a seeming quality premium for the fresh product, small family units are common in most of the mushroom growing countries. These low cost systems are closely linked to the availability of what is known as "ready-mixed" compost. This is prepared in bulk quantities by compost specialists before delivery to the farm, with the grower taking the responsibility for pasteurizing the compost. The efficiency of these small production units is usually superior to the large sophisticated units now common in Europe. Although ready-mixed compost units are a postwar development, it is logical that they will develop further. Units which produce, pasteurize, and inoculate compost with spawn are already in existence in some countries, supplying compact units which require no further preparation before the growing stage.

COMMERCIAL GROWING PROCEDURES

Irrespective of the system adopted, there are procedures in growing *A. bisporus* which follow sequential stages. These are illustrated in Fig. 11.2 and can be summarized as follows.

(1) Spawn Manufacture

Pure cultures are derived from spores or tissues of selected mushrooms and germinated on agar media. Mycelium is then transferred to sterilized, moistened grain mixed with calcium carbonate or to tobacco stems mixed with humus and/or peat. Rye is the most commonly used grain but wheat, millet, and sorghum are also satisfactory. After about 3 weeks at 21°C the substrate has been colonized by mycelium and is referred to as spawn. It can be stored for up to 3 months in a cold storage room before using or dried and kept viable for over a year in a cool place. Spawn manufacture is a specialized activity requiring aseptic techniques.

(2) Compost Manufacture

Usually compost is prepared from a mixture of cereal straw and horse manure (stable bedding), with added chicken manure and activators. The straw/horse manure is stacked in windrows outdoors and turned at 2- or 3-day intervals by machines over a period of 10 to 15 days or more (Fig. 11.3 top). This phase of mushroom growing is least subject to control. Fermentation by the microflora natural to manures and straws proceeds at various rates during the composting period. Various groups of microorganisms are favored by the extent of

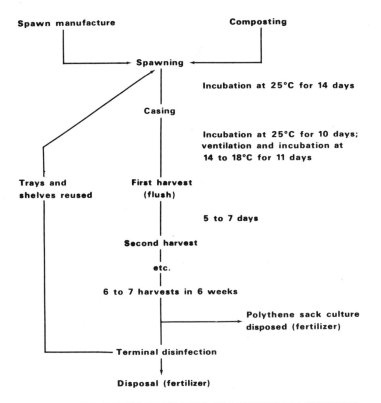

FIG. 11.2. SUMMARY OF STAGES IN MUSHROOM CULTURE

oxygen supply and by elevated temperature (Fig. 11.4). Thorough periodic mixing ensures a homogeneous breakdown of the compost.

Easily metabolized materials such as sugars and starches are degraded more rapidly during fermentation than are lignin and cellulose. Insoluble nitrogen compounds accumulate, favorably in the form of microbial biomass. The pH of the compost may rise to 9.0 during fermentation but declines to 7.5 to 7.0 at the completion of composting when most of the soluble nitrogen has been converted.

The compost is then filled into wooden trays at a depth of about 13 cm or onto shelf beds at a depth of 14 to 20 cm and pasteurized by the introduction of live steam or less often by methyl bromide. Aerobic conditions are maintained so as to complete the composting process. This normally takes five days, during which time the compost temperature rises and should be maintained at 55° to 63°C between the third and fifth day.

Almost any vegetable waste can be utilized in the preparation of

FIG. 11.3. PREPARATION OF MUSHROOM COMPOST IN A MODERN COM-
POSTING YARD IN ENGLAND (TOP) AND IN TAIWAN (BOTTOM)
Note Top Photograph, Courtesy of W. Darlington & Sons, Ltd.

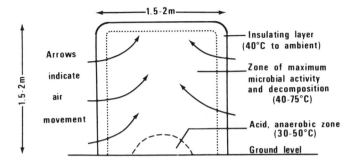

FIG. 11.4 CROSS SECTION OF STACK DURING COMPOSTING, SHOWING ZONA-
TION AND NATURAL AIR MOVEMENTS (CHIMNEY EFFECT)

compost. Most manures are suitable, provided appropriate modifications are made to the techniques of composting. Synthetic mixtures,
i.e., without animal manure, are also practical and commercial in most
locations, but again techniques are modified according to the
materials available.

(3) Spawning
    Spawn is added to compost by hand or machine, either over the
surface or mixed through. About 50 g of spawn is sufficient for 0.09 m^2
or 9.1 kg of fresh compost.

(4) Incubation or Spawn Running
    Once inoculated, compost is incubated at about 24°C. Accumulation
of carbon dioxide at this stage is stimulatory to growth of the
mycelium, and ventilation is not required. Circulation of air within the
incubation room, however, is practiced to maintain temperature
uniformity. Some heat is generated at this stage and care is required to
ensure that temperatures do not exceed 28°C in the compost,
otherwise damage is caused to the mycelium. Incubation time seldom
exceeds 14 days.

(5) Casing
    When the compost is colonized with mycelium, a layer of soil is
spread over the surface at a depth of 2 to 4 cm. This casing soil can be a
medium clay loam which is normally pasteurized or a peat/limestone
(calcium carbonate) mixture which requires no pasteurization.
    Incubation continues at 25°C for 10 days and carbon dioxide levels
in the air are maintained at relatively high levels to encourage growth
into the casing layer. Once mycelial growth reaches the surface, the
temperature is lowered to 14° to 18°C by the introduction of fresh air,
which also lowers the carbon dioxide content of the atmosphere. This

Courtesy of W. Darlington & Sons, Ltd.

FIG. 11.5. *AGARICUS BISPORUS* GROWING ON SHELVES

induces the formation of initials and primordia which develop into buttons and mature mushrooms (basidiocarps) (Fig. 11.5). Primordia form 14 to 18 days after casing and the first crop is harvested usually over a 3-day period commencing 21 days after the initial casing.

Mushrooms continue to develop in regular "flushes" or "breaks" at 5- to 7-day intervals. As cropping proceeds, yields decline and it is normal to crop for 5 to 6 weeks before disposing of the culture which is then marketed as a fertilizer and soil conditioner.

(6) Terminal Disinfection

In badly designed growing units or under conditions of bad husbandry, the monocropping system allows for the development of disease organisms. Although many disease organisms can be controlled by chemical treatment, it is common practice, especially on large growing units, to disinfect cultures *in situ* at the end of cropping by the application of heat (steam) or fumigation with methyl bromide. This is especially important if virus infections occur. A disinfection of the trays or shelves only is commonly practiced, but with poly-ethylene sack culture this terminal disinfection procedure is not required.

Mushroom growing can therefore be seen as a batch process and normally a growing structure yields over four crops per annum.

UNDERSTANDING THE BIOLOGY OF *A. BISPORUS*

Genetics and Breeding

Important contributions to the development of the mushroom industry have been made by science. The importance of carbon dioxide and temperature in the basic physiology of *A. bisporus* has already been discussed, as well as the implications of introducing pasteurization procedures in the control of disease. More recent discoveries on the genetics and nutritional physiology of *A. bisporus* are likely to have far-reaching implications on the extent and method of mushroom cultivation in the future.

Detailed studies by several researchers (Sinden 1937; Kligman 1943; Lambert 1960; Raper and Raper 1972; Miller and Kanaven 1972; Elliot 1972) confirm that *A. bisporus* (which bears two spores per basidium and not four, as does *A. campestris*) is heterothallic, possessing the unifactorial or bipolar system of sexuality. For mycelium to be fertile it must contain nuclei with different alleles of the incompatibility factor. The incompatibility factor is multiallelic.

This understanding is basic to the methodology of obtaining new strains for the manufacture of spawn, but of more significance is that these studies have given an impetus to plant breeders and mycologists to attempt breeding mushrooms for specific purposes. Improvements to strains, however, have been made by a continual process of selection from varieties obtained from single or multispore cultures. Now deliberate attempts at hybridization and combining desired characteristics may be possible. This greatly improves the prospects for improving strains and elevates mushroom breeding to a level comparable to that applied to higher crop plants. The techniques, however, are complex and arduous and the feasibility of deliberate breeding for commercial purposes has yet to be established.

Substrates

Compost.—Knowledge of sexuality only partially explains the mystique which has always been associated with mushroom growth and which presumably originates from the behavior of wild mushrooms appearing suddenly at given times of the year. Even in artificial culture there are many misconceptions and unknown factors which relate directly to mushroom growth on composts prepared from horse manure and to the necessity to case beds with a layer of soil in order to induce fruitbody formation.

Studies on the composting process have highlighted the active role of microorganisms in preparing straw, which forms the bulk of a compost, for subsequent enzymatic attack by the mycelium of *A. bisporus*. Chemical analyses (Gerrits *et al.* 1967), when linked to the ecological sequence of groups of microorganisms in composting (Hayes 1969, 1976; Stanek 1972), have shown that the major nutrients for mushroom growth are derived from lignin, cellulose and hemicellulose of the straw. The secondary nutrient, nitrogen, is derived from the body substrate of bacteria or biomass accumulated during composting. The predominant microorganisms are thermophilic bacteria and fungi, i.e., they grow well at temperatures above which most living organisms cannot exist. This feature strongly suggests that the protein and enzyme characteristics of these organisms are unique, which probably accounts for the specificity of growth of A. bisporus on compost and links growth with a heat-producing fermentation. Furthermore, productivity of cultures has been related directly to the population levels of the thermophilic bacteria in composting. A diagram indicating the types of microorganisms which sequentially predominate in the substrate during composting and on into the growth phase of A. bisporus is shown in Fig. 11.6.

As a result of these studies many new approaches are possible in the nutrition of the crop. This understanding has already changed the emphasis away from nitrogenous supplements in composting mixtures to less expensive supplements such as molasses and other wastes containing carbohydrate (Hayes and Randle 1969, 1970). Also, improvements in culture productivity have stemmed from this work by giving attention to the nutrient balance in composting mixtures, which favor specific microorganisms that are beneficial, rather than by providing nutrients to be consumed directly by the mushroom. Perhaps of greater impact is the fact that these studies have led to the extension of the range of materials which can be exploited for mushroom culture. Horse manure is not an essential prerequisite; virtually any vegetable waste can be utilized for composting purposes, including a variety of agricultural and industrial wastes. For example, a productive compost can be prepared entirely from the waste products of sugar cane growing and processing and from a number of obnoxious wastes from industrial processes.

Casing Soil.—The phenomenon of fruitbody formation and development in the casing layer has been associated with carbon dioxide levels in this soil layer, but now another factor has been highlighted. Fruitbody formation has been shown to require certain specific bacteria which are located in the casing layer. During the vegetative stage of growth, volatile by-products of A. bisporus metabolism accumulate in the casing layer and select for high population levels of the required bacteria (Hayes et al. 1969; Hayes 1972) (Fig. 11.7). Productivity of cultures can be related to the population of these bacteria and current studies are aimed at controlling their activity in order to produce more uniform cropping

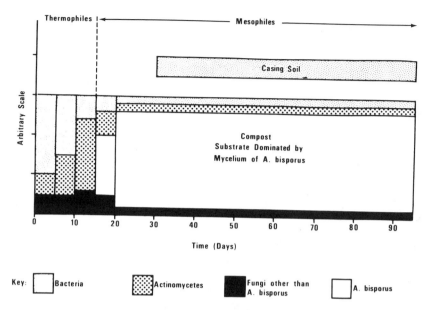

FIG. 11.6. ECOLOGICAL SEQUENCE OF MAJOR GROUPS OF MICROORGANISMS DURING STAGES OF *AGARICUS BISPORUS* CULTURE

VEGETATIVE
14 days at 25° C

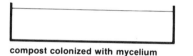

compost colonized with mycelium

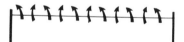

volatiles CO_2, ethanol, acetone

REPRODUCTIVE
18-21 days at 16°-18° C

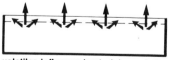

volatiles influence bacterial populations
in the soil layer

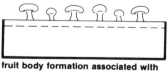

fruit body formation associated with
bacterial activity

FIG. 11.7. SUGGESTED RELATIONSHIP BETWEEN COMPOST AND CASING SOIL SUBSTRATES IN *AGARICUS BISPORUS* CULTURE. Volatile by-products of growth in compost accumulate and diffuse through the casing soil, creating an environment which selects for bacteria which are associated with fruitbody formation.

patterns so as to allow progress toward mechanical harvesting (Hayes 1974; Nair and Hayes 1974, 1975). Also, these studies define the requirements of soils to give maximum productivity and provide the basis for utilizing a wide range of materials for casing purposes.

STORAGE OF FRESH MUSHROOMS

Following harvest, a mushroom fruitbody continues to respire at a high rate and undergoes changes in morphology, appearance, and chemical composition. All of these changes are of considerable importance in the marketing, sale, and consumption of mushrooms. Respiration results in a loss of fresh and dry weight which is accelerated by increasing temperature (Fig. 11.8) Hammond and Nichols (1975) attribute this respiration in A. bisporus to the metabolism of mannitol and suggest that trehalose, glycogen, and probably amino acids in the harvested fruit may also function as postharvest respiratory substrates. While this loss of mannitol during postharvest respiration may not be very important from a nutritional point of view, according to Haddad and Hayes (1977) there is a considerable loss of protein, while total nitrogen remains relatively constant, indicating that during postharvest storage a shift occurs in the nitrogen components of fruitbodies which may affect food values significantly.

Mushrooms are harvested by hand and this process represents one of the major costs of production. The fruitbody of A. bisporus, once harvested, is a delicate structure which is susceptible to bruising. Bruising is caused by the rupture of the fragile hyphal strands which are exposed on the surface of the pileus or stipe and is accompanied by what is termed "browning." This browning is the result of phenoloxidase activity of the cellular phenolic compounds which are exposed as as result of the hyphal fracture.

The color changes from white to brown are also evident during prolonged storage. Analysis of the enzyme activity during storage led Goodenough (1976) to conclude that a "physiological aging" was distinct from the activity of polyphenoloxidase enzymes responsible for the color change.

Deterioration of fresh mushrooms is influenced by a combination of factors in addition to those causing color change. Opening and expansion of the pileus and stipe are important. Spoilage also results from growth of microorganisms which are part of the normal microflora of the mushroom. These processes are retarded by storage at low temperature, thus prolonging storage life.

In recent years, the use of punnets overwrapped with plastic film has been introduced to cater primarily to supermarket sales. One of

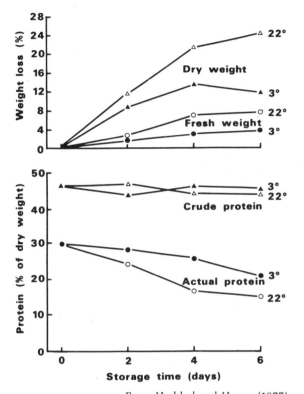

From Haddad and Hayes (1977)

FIG. 11.8. CHANGES IN THE COMPOSITION OF *AGARICUS BISPORUS* DURING STORAGE AT 3° AND 22°C

the benefits of this method is the lengthening of storage life of fresh mushrooms at room temperature. These benefits arise from the restricted permeability of the overwrap to oxygen, carbon dioxide, and water vapor (Nichols and Hammond 1973, 1975). Fresh weight loss is greatly retarded, as are the browning and opening of the pileus. Types of films vary in their effectiveness, and the integrity of the seal is an important factor in determining keeping quality at room temperature. Under refrigeration, perforated films are preferred in order to avoid carbon dioxide accumulation in the punnet, which may accelerate browning. *Clostridium botulinum* may also proliferate under anaerobic conditions. Vacuum cooling of mushroom prepacks is practiced by some modern mushroom-producing units, but because of the extra cost its use is not widespread.

PROCESSING MUSHROOMS

Canning

The most common method of processing mushrooms is by canning. The procedures adopted are common to most vegetables and for convenience may be considered in the following six stages.

(1) Preparation.—Normally mushrooms are harvested as buttons and canned as quickly as possible after harvest, but they may be cool-stored. Mushrooms are washed and graded, and blemished or diseased fruits are removed. In order to prevent browning, citric acid (0.1%) or sodium metabisulfite (0.3%) is added to the wash water.

(2) Blanching.—This is done in a 1% solution of sodium chloride by cooking in steam or hot water at a temperature of 95° to 100°C for 3 to 8 min.

(3) Filling of Cans.—Mushrooms are put into cans or jars containing brine (1.5% sodium chloride). Citric acid or ascorbic acid (0.1%) may also be added before the can is sealed.

(4) Sterilization.—Cans are sterilized in batch or continuous retorts.

(5) Cooling.—This is introduced to prevent overcooking and to help avoid the danger of stimulating the germination of heat-resistant bacterial spores.

(6) Labeling and Packaging.—The operation can proceed when the can temperature is below 35°C.

The size of cans varies according to country but usually conforms to standards of size, content, volume, and net drained weight. Mushrooms may be marketed as "Whole" (caps with attached stems), "Buttons" (whole mushrooms with short stems), "Sliced" (on average, 4 mm thick longitudinal slices), and "Pieces and Stems" (containing at least 50% caps, whole mushrooms, stalks, and other pieces of irregular shape and size).

Although the majority of canned preparations are in brine, numerous combinations exist to cater to local preferences in various countries. Soup preparations, for example, are universal and usually contain the equivalent of at least 30% whole mushrooms. More exotic preparations may also include cream, curry, butter or tomato sauce.

Dehydration

The process of drying mushrooms is slow and, if done artificially at an air temperature of 60° to 70°C, is also costly. For satisfactory drying, the moisture content should be reduced to about 12%. Dehydration is commonly practiced to preserve mushrooms collected from the wild, e.g., *Morchella esculenta* and other morels, shiitake in Japan, and the straw mushroom in Asia. Natural drying in the sun is eminently satisfactory for the latter species. Dehydration of *Agaricus* is, however, not widely practiced.

Freeze-Drying

This process involves freezing at –20°C in a closed vessel under vacuum. Freeze-dried *A. bisporus* readily reconstitutes on soaking in hot water for a few minutes and retains its flavor. For prolonged preservation, the freeze-dried product should be stored under a nitrogen atmosphere. Costs of freeze-drying are, however, prohibitive to widespread adoption of this method of processing.

Freezing

Mushrooms are frozen at –120°C with liquid nitrogen. Treatment times vary from 4 to 6 min, depending on size and whether sliced. Mushrooms frozen by this method can be stored for protracted lengths of time and are equal to fresh mushrooms in both flavor and appearance. Costs are, however, high, and freezing mushrooms by this method is only done exceptionally for the food processing trade.

Irradiation

Electron or gamma irradiation delays the post-harvest growth of mushrooms, thus retarding the opening of caps. The experimental work of Langerak (1972) has revealed that better natural color retention can be achieved by irradiation. This method, however, has not been commercialized, probably due to the high costs involved in applying treatments. Consumer acceptance of irradiated foodstuffs is not known and is likely to be an important factor in its future exploitation. In addition, irradiation as a method of food preservation is illegal in many countries.

Pickling

This method is traditional and popular in Italy and India and is becoming more widespread in England and other European countries as a means of taking maximum advantage of seasonal availability of mushrooms at a lower cost. Pickling is a relatively simple procedure and is conveniently done in the home. After blanching or tenderizing in a pan, a previously boiled solution of diluted acetic acid (vinegar) is added to the mushrooms in a sterilized jar and then sealed. Other ingredients may include, for example, peppercorns, chilies, garlic, or dill.

For further discussions of procedures and problems in processing mushrooms, the reader is directed to other reports (McArdle and Curwen 1962; Beelman *et al.* 1973; Beelman and McArdle 1975; Hanson 1975; Gormley and O'Riordain 1976).

TABLE 11.2
PROXIMATE COMPOSITION OF THE CULTIVATED MUSHROOM,
AGARICUS BISPORUS

Component (%)	Reference		
	Anderson and Fellers (1942)	McConnel and Esselen (1947)	Hayes and Haddad (1976)
Water	89.50	88.90	91.00
Crude protein (N X6.25)	3.94	3.95	3.50
Actual protein (amino acids)	—	—	3.15
Fat	0.19	0.26	0.40
Carbohydrates (includes mannitol)	4.01	4.75	—
Carbohydrates (monosaccharide)	—	—	2.45
Ash	1.26	1.14	0.90
Fiber	1.09	1.00	1.00

COMPOSITION AND FOOD VALUE

The value given to a foodstuff is based largely on its chemical composition. From the composition data for freshly cultivated A. bisporus, it can be seen that mushrooms contain carbohydrates, proteins, fat, and minerals (in ash), all of which contribute to food value (Table 11.2). Composition tables of this kind are useful in considering gross composition and give quantitative information on the amount of water and dry matter, but more detailed and specific analyses are required to describe the nature of the components so that deficiencies or an abundance of a particular nutrient can be assessed. Also, information is not given on other substances which may be of special significance in nutrition.

Carbohydrates

Many different carbohydrates have been identified in the cultivated mushroom but the most common components are reducing sugars, glycogen, hemicellulose, and relatively large amounts of mannitol (Hughes 1961; Holtz 1971). The nutritional significance of mannitol has yet to be demonstrated and is usually ignored in food value studies. According to Holtz (1971) mannitol is thought to fulfill an important physiological role in water uptake. Dommel (1964) showed that carbohydrates were the main constituents lost during canning.

Fat

Ten different fatty acids were monitored by Hughes (1962). The amounts of the essential fatty acid, linoleic, are exceptionally high compared with other vegetables.

Protein

There is considerable variation in values reported for the protein content of mushrooms. Even a zero value was recorded by Chatfield and Adams (1940). It is not possible to satisfactorily explain the differences in protein content given by various authorities, but there is good justification in not accepting crude protein (Nitrogen [N] × 6.25) as a true value, since some nitrogenous constituents of the fruitbody are not in the form of protein, e.g., nucleic acid, urea, chitin and chitosans.

Accurate estimations are possible using modern methods of hydrolysis followed by quantitation of amino acids. Using a typical sample of fresh A. bisporus fruitbodies grown according to commercial methods in our laboratory, a crude protein value of 38% of the dry matter was obtained. Amino acid analysis on the same sample gave a value of 35% protein. The difference, therefore, accounts for nonprotein nitrogen.

Digestibility values varying from 69 to 83% have been reported. Using in vitro enzymatic techniques, a digestibility value of 82% was obtained in the sample quoted above.

Data on the amino acid composition of the same sample, in addition to data from other species of mushrooms, are shown in Table 11.3. All of the essential amino acids required by an adult are present in A. bisporus, but the amount of methionine is relatively small, whereas tryptophan and lysine are present in relatively large amounts. This is of some significance since tryptophan and lysine are often deficient in vegetable proteins. Mushroom protein, like other fungal protein, is therefore intermediate in quality between vegetable and animal protein, and the supplementary value of mushroom protein in vegetarian diets is likely to assume more relevance in the future. A good account of various mushroom proteins is given by Kurtzman (1975).

Minerals

An analysis of the ash in A. bisporus has been given by Anderson and Fellers (1942) (Table 11.4). Mushrooms contain appreciable amounts of potassium, phosphorus, copper, and iron, but low levels of calcium. Except for iron, there is no knowledge on the availability of these minerals in the body.

Vitamins

Cultivated mushrooms do not contain vitamins A, D, or E. Anderson and Fellers (1942) reported the vitamin content on a 100 g fresh weight basis to be: 8.6 mg ascorbic acid, 0.12 mg thiamine, 0.52 mg riboflavin

TABLE 11.3

AMINO ACID COMPOSITION OF VARIOUS MUSHROOM FRUITBODIES

Amino Acid	Agaricus spp.		Morchella spp.[3]			Lentinus edodes[4]	Pleurotus ostreatus[5]
	bisporus[2]	campestris[3]	deliciosa	esculenta (caps)	crassipes		
Aspartic acid	3.14	4.2	5.2	5.5	4.9	6.5	6.4
Threonine[1]	1.48	2.8	3.0	3.0	3.6	4.4	3.2
Serine	1.89	2.7	2.9	2.8	3.3	4.3	3.5
Glutamic acid	7.06	9.1	10.0	8.8	8.3	22.5	11.7
Proline	2.50	2.8	2.2	3.8	2.5	3.6	3.2
Glycine	1.20	2.8	2.3	4.0	2.8	3.6	3.2
Alanine	2.40	4.7	2.2	4.6	4.2	5.1	4.5
Cystine	0.18	0.9	0.6	0.3	0.6	—	0.3
Methionine[1]	0.39	0.8	0.9	0.7	0.6	1.5	1.1
Valine[1]	1.63	3.4	2.5	4.1	2.1	4.4	3.6
Isoleucine[1]	1.28	2.4	2.7	2.9	2.9	3.6	3.0
Leucine[1]	2.16	3.9	3.6	4.3	4.8	5.8	4.8
Tyrosine	0.78	1.0	2.5	1.8	2.5	2.9	2.1
Phenylalanine[1]	1.55	3.1	3.2	2.6	3.2	4.4	2.6
Histidine	0.64	0	0	0	0	1.5	1.2
Lysine[1]	1.62	4.0	3.6	3.8	3.4	2.9	3.2
Arginine	1.90	3.0	3.2	3.2	3.0	5.8	3.7
Tryptophan[1]	3.94	—	—	—	—	—	1.0

[1] Essential amino acids.
[2] G per 100 g dry matter (Hayes and Haddad 1976).
[3] G per 16 g protein nitrogen (McKellar and Kohrman 1975).
[4] G per 16 g nitrogen (adapted from Sugimori et al. 1971.)
[5] G per 16 g nitrogen, gray type (adapted from Kalberer and Kunsch 1974).

TABLE 11.4

MINERAL CONTENT OF THE ASH OF
AGARICUS BISPORUS

Component	Content
Calcium, %	0.0024
Phosphorus, %	0.15
Potassium, %	0.50
Total iron, ppm	19.50
Available iron, ppm	5.95
Copper, ppm	1.35

Source: Anderson and Fellers (1942).

5.85 mg nicotinic acid (niacin), 2.38 mg pantothenic acid, and 0.018 mg biotin. These vitamins are not readily destroyed by cooking, canning, drying, or freezing (Williams and Esselen 1946). Mushrooms are usually considered to be rich in nicotinic acid and riboflavin.

Being a natural product, it can be expected that variations occur in the composition of fresh mushrooms. However, recent research in our laboratory designed to standardize the composition has revealed that composition can be affected by growing practice. The degree of watering during growing, for example, affects the dry matter content of the fruitbody. While some variations are inherent in mushrooms, buttons, cups, and flats, and mushrooms harvested from different flushes vary in composition, especially in the protein content (Table 11.5), perhaps this in part accounts for discrepancies in proximate composition recorded in the past. Current research should reveal

TABLE 11.5

PROTEIN CONTENT (%, DRY WT) IN BUTTONS, CUPS, AND
FLATS OF *AGARICUS BISPORUS* FROM FIRST
AND THIRD BREAK CROPS

Crop	Stage of Maturity		
	Buttons	Cups	Flats
First break (flush)			
Crude protein	38.52	34.30	31.80
Actual protein	35.72	31.58	29.00
Third break			
Crude protein	39.33	36.00	33.03
Actual protein	32.27	28.70	26.83

Source: Hayes and Haddad (1976).

variations that may be attributable to nutritional factors provided in substrates, a factor which to date has not been considered in commercial methods of culture.

MUSHROOM CONSUMPTION

In most countries there is a long-established consumer acceptance of white mushrooms, although this is not by any means universal. In some localities, e.g., California, there is a preference for cream or brown variants of A. bisporus, while in England there is an established preference for white mushrooms. Also, even within a country, consumer attitudes vary even on such matters as shape and size. In southern England mushrooms harvested at the button stage are preferred, while in the industrial towns of the north of England mature mushrooms at the flat stage are favored. There also exists a preference for fresh mushrooms in some countries, notably the United Kingdom and Italy, while in West Germany, Canada, and the United States canned mushrooms are dominant. Countries with a consumer preference for fresh mushrooms are not dependent on foreign imports to fulfill consumer demand, while countries with a high consumption rate of canned mushrooms import from countries in Asia where costs of production, especially the labor component, are relatively low.

Consumption of A. bisporus per capita is greatest in Western Europe and Canada, with West Germany having the largest per capita consumption in the world, being about 2.5 times that of the United States (Tables 11.6 and 11.7).

Most other cultivated mushrooms are sold in the fresh state close to centers of production. Shiitake, the dominant mushroom grown in Japan, is available as a fresh or dried product, but it is graded into five categories based primarily on size before being sold. The dried product is exported, primarily to Hong Kong, Singapore, and the United States. Pleurotus spp. (Fig. 11.9) and Volvariella volvacea are usually produced for local consumption and, in hot countries, are preserved by natural drying. There is a limited export of Volvariella in cans from Taiwan.

PRESENT AND FUTURE OUTLOOK

The location and systems of growing in the traditional mushroom-growing countries in Europe and in the United States have been influenced to a considerable degree by the nature and availability of the traditionally used substrate, horse manure, and by the acces-

TABLE 11.6
PRODUCTION OF *AGARICUS BISPORUS* (1974)

Country	Production	Consumption	Consumption as Fresh (%)
	(Metric Tons × 1000)		
United States	127	163	39
France	114	64	48
Great Britain	53.1	57	82
Holland	40	11	55
Italy	38	39	87
West Germany	29.2	125.5	26
Canada	19	28	32
Taiwan (Republic of China)	60	9	95
South Korea	28.3	1	98
China (People's Republic)	30	1	98

Source: Edwards (1976).

TABLE 11.7
MUSHROOM CONSUMPTION
(KG *AGARICUS BISPORUS* PER CAPITA)

Country	1969	1974
West Germany	1.12	2.02
France	0.9	1.22
Belgium	0.85	—
Holland	0.47	0.81
Italy	0.44	0.71
United States	—	0.80
United Kingdom	—	1.03
Canada	—	1.30
Taiwan (Republic of China)	—	0.58
South Korea	—	0.03
China (People's Republic)	—	0.01

Source: Edwards (1976).

Courtesy of Z. Bano
FIG. 11.9. OYSTER MUSHROOMS *(PLEUROTUS SP.)* GROWING ON STRAW

sibility of the required "know-how" involved in the methods of cultivation. In recent years, major industries have developed in countries which hitherto were not regarded as major producers. The small island of Taiwan, for example, has, within the decade 1960 to 1970, become the third largest producer of mushrooms in the world. Rice straw is used as a base material for composting, and growing structures of simple design made from native bamboo supports thatched with the leaves of native plants provide adequate protection. The availability of inexpensive labor was a key factor in the establishment phase of cultivation.

In Africa, mushroom cultivation is in its infancy, but many farms operate successfully in different countries. In South Africa, synthetic composts are prepared from wheat straw, while at some locations horse manure supplies are plentiful. Some growing units exploit the by-products of the sugar cane industry. Similarly, in Kenya local plant materials are available. The total production in Africa is consumed within Africa.

Of special interest are the methods employed in northern India. While some conventional farms produce mushrooms for local consumption, the concept of mother units which manufacture spawn and compost for delivery to a large number of satellite production units is being implemented. In this way the demanding and technically difficult operations of spawn manufacture and compost manufacture are centralized, while growing is done by villagers in structures of simple design or in outhouses. The mushrooms are consumed locally; better grade mushrooms are transported to other parts of India after canning. A range of locally available plant materials can be exploited for compost manufacture.

Of significance also in recent times is the introduction of new strains and species of *Agaricus* which grow optimally at relatively high temperatures and can therefore be cultivated in subtropical and tropical climates. *Agaricus bitorquis* is grown in Greece and in some African countries for this reason. Also, the basic methods of culture for *A. bisporus* can be applied to other species of mushrooms which are adapted to tropical environments. *Volvariella volvacea*, the Chinese straw mushroom, for example, is now grown in Taiwan in polyethylene structures on beds which are cased with a local soil, in much the same way as *A. bisporus* is cultivated (Ho 1972). This has resulted in a dramatic increase in the yield potential of this mushroom. Cotton wastes are known to be particularly suited to this fungus and its large-scale cultivation is now being considered in southern India and southeast Asia (Chang 1972).

Although it is a matter for research at present, techniques based on the culture of *A. bisporus* may well apply to other edible species. It is probable that with the range of substrates and with the development of new strains, the exploitation of this popular species will continue to increase.

The remarkable growth of mushroom cultivation worldwide reflects a changing attitude toward a foodstuff which for centuries has been associated with myth and superstition. The methods employed in the artificial culture of mushrooms provide a guarantee of edibility which cannot be provided by gathering mushrooms growing in the wild, and this, together with the ready availability of cultured mushrooms, has perhaps been the most significant factor in changing attitudes and increasing consumption. However, consumption rates vary from country to country. Clearly, many factors contribute to differences in food consumption, but it is apparent that considerable potential exists for this food in countries where consumption rates are presently at a low level. Also, there is little doubt that with a broader appreciation of the multiplicity of factors that control growth of mushrooms, there will in future years be an extension of the range of edible species that will be grown artificially, which may in turn extend the consumption of fungi generally.

With an increasing awareness of the importance of human nutrition, increasing attention should be given to the quality of the nutrients in mushrooms. The supplementary value of mushrooms in balancing largely vegetarian diets is of relevance to many countries. However, economic considerations must also be considered. In Western countries about 50% of the production costs are attributable to labor, while about 25% of costs are concerned with materials. In Asia, the reverse cost structure for labor and materials is evident. Thus it seems in those countries where labor is plentiful, low cost and extensive systems will be more appropriate for the future. In contrast, for Western countries to counteract these costs, increased mechanization and sophistication in growing will be required, ultimately leading to mechanical harvesting.

In view of our recent knowledge of the complex of factors which contribute to the normal developmental processes in *A. bisporus*, it would also seem necessary to further appraise the possibilities of applying conventional fermentation techniques to the production of mycelial biomass from appropriate mushroom strains. Although the conventional method of growing is a type of batch surface culture using solid substrates, excellent growth can be achieved using liquid substrates. However, in liquid culture growth does not proceed to the stage of fruitbody development.

Previous attempts at the production of mycelial biomass have not been exploited further, mainly because of the relatively low growth rate and the apparent failure of the mycelial product to emulate the flavor of the fruitbody. These attempts, however, were undertaken without consideration of the unique nutritional requirements of *A. bisporus* or of the associated biological requirements (bacteria) in the developmental processes. While at present the biochemical mechanisms involved in the development and production of flavor have not been determined, these recent findings suggest a new and fertile area for research and development. Other advances in the fermentation industry and in the general area of mushroom genetics and strain selection are also relevant to this concept. The widespread acceptance of mushrooms as a highly regarded and flavorful food offers prospects far beyond those for other fungi which are currently being exploited for the production of mycelial biomass.

The consequences of these foreseeable developments may well add a further dimension to the future prospects of this foodstuff which in the 19th century was only consumed by the aristocracy and until recently was considered a luxury even in Western countries. Nowadays, mushrooms are consumed regularly in almost every home.

REFERENCES

ABERCROMBIE, J. 1817. Abercrombie's Practical Gardener, or Improved System of Modern Horticulture, 2nd Edition (revised by J. Mean). Cadell and Davies, London.

ANDERSON, E. E., and FELLERS, C. R. 1942. The food value of mushroom, *Agaricus campestris*. Proc. Am. Soc. Hortic. Sci. *41*, 301-303.

ANGLE, R. Y., and TAMHANE, D. V. 1974. Mushrooms: an exotic source of nutritious and palatable food. Indian Food Packer. *28*, No. 5, 22-33.

BEELMAN, R. B., KUHN, G. C., and McARDLE, F. J. 1973. Influence of post-harvest and soaking treatments on yield and quality of canned mushrooms. J. Food Sci. *38*, 951-953.

BEELMAN, R. B., and McARDLE, F. J. 1975. Influence of post-harvest storage temperatures and soaking on yield and quality of canned mushrooms. J. Food Sci. *40*, 669-671.

BYRNE, P. F. S., and BRENNAN, P. J. 1975. The lipids of *Agaricus bisporus*. J. Gen. Microbiol. *89*, 245-255.

CALLOW, E. 1831. Observations on the Methods Now in Use for the Artificial Growth of Mushrooms, with a Full Explanation of an Improved Mode of Culture, by Which a Most Abundant Supply May Be Procured and Continued Throughout Every Month in the Year, with a Degree of Certainty That Has in No Instance Failed. Fellows, London.

CHANG, S. T. 1972. The Chinese Mushroom. Chinese Univ., Hong Kong.

CHATFIELD, C., and ADAMS, G. 1940. Proximate Composition of American Food Materials. U. S. Dep. Agric. Circ. No. *549*.

CHRISTENSEN, C. M. 1975. Molds, Mushrooms, and Mycotoxins. Univ. of Minnesota Press, Minneapolis, Minn.

COALE, C. W., and BUTZ, W. T. 1972. Impact of selected economic variables on the profitability of commercial mushroom processing operations. Mushroom Sci. *8*, 231-237.

DELCAIRE, J. R. 1976. Economics of edible mushrooms. *In* Biology and Cultivation of Edible Fungi. S. T. Change and W. A. Hayes (Editors). Academic Press, New York.

DOMMEL, R. M. 1964. Mushroom Shrinkage During Processing. M.S. Thesis. Pennsylvania State Univ., University Park.

EDWARDS, R. L. 1976. World mushroom production and consumption. Mushroom J. *45*, 282-285.

ELLIOT, T. J. 1972. Sex and the single spore. Mushroom Sci. *8*, 11-18.

GERRITS, J. P. G., MULLER, F. M., and BELS-KONING, H. C. 1967. Changes in compost constituents during composting, pasteurization, and cropping. Mushroom Sci. *6*, 225-243.

GOODENOUGH, P. W. 1976. How chilled storage affects the physiology of mushrooms. Mushroom J. *43*, 208-209.

GORMLEY, T. R., and O'RIORDAIN, F. 1976. Quality evaluation of fresh and processed oyster mushrooms *(Pleurotus ostreatus)*. Lebensm. Wiss. Technol. *9*, 75-79.

GRAY, W. D. 1970. The use of fungi as food and in food processing. CRC Crit. Rev. Food Technol. *1*, 225-329.

GRAY, W. D. 1972. The use of fungi as food and in food processing, II. CRC Crit. Rev. Food Technol. *3*, 121-215.

HADDAD, N., and HAYES, W. A. 1977. Factors influencing composition and food value of mushrooms. *In* Quality of Mushrooms. Proc. Aston Semin. in Mushroom Sci., Vol. 3. W. A. Hayes and F. C. Atkins (Editors). Birmingham, England.

HAMMOND, J. B. W., and NICHOLS, R. 1975. Changes in respiration and soluble carbohydrates during the post-harvest storage of mushrooms *(Agaricus bisporus)*. J. Sci. Food Agric. *26*, 835-842.

HANSON, H. P. 1975. Processing mushrooms. In Commercial Processing of Vegetables. Noyes Data Corp., Park Ridge, N. J.

HAYES, W. A. 1969. Microbiological changes in composting wheat straw/horse manure mixtures. Mushroom Sci. 7, 173-186.

HAYES, W. A. 1972. Nutritional factors in relation to mushroom production. Mushroom Sci. 8, 663-674.

HAYES, W. A. 1974. Microbiological activity in the casing layer and its relation to productivity and disease control. In The Casing Layer. Proc. Aston Semin. in Mushroom Sci., Vol. 1. W. A. Hayes (Editor). M. G. A., London.

HAYES, W. A. 1976. Mushroom nutrition and the role of microorganisms in composting. In Composting. Proc. Aston Semin. in Mushroom Sci., Vol. 2. W. A. Hayes (Editor). M. G. A., London.

HAYES, W. A., and HADDAD, N. 1976. The food value of the cultivated mushroom and its importance to the mushroom industry. Mushroom J. 40, 104-110.

HAYES, W. A., and RANDLE, P. E. 1969. Use of molasses as an ingredient of wheat straw mixtures used for the preparation of mushroom composts. Rep. Glasshouse Crops Res. Inst. (1968), 142.

HAYES, W. A., and RANDLE, P. E. 1970. An alternative method of preparing mushroom compost, using methyl bromide as a pasteurizing agent. Rep. Glasshouse Crops Res. Inst. (1969), 166.

HAYES, W. A., RANDLE, P. E., and LAST, F. T. 1969. The nature of the microbial stimulus affecting sporophore formation in Agaricus bisporus Lange (Sing.) Ann. Appl. Biol. 64, 177-186.

HO, M. 1972. Straw mushroom cultivation in plastic houses. Mushroom Sci. 8, 257-263.

HOLTZ, R. B. 1971. Qualitative and quantitative analysis of free neutral carbohydrates in mushroom tissue by gas-liquid chromatography and mass spectrometry. J. Agric. Food Chem. 19, 1272-1273.

HUGHES, D. H. 1962. Preliminary characterization of the lipid constituents of the cultivated mushroom Agaricus campestris. Mushroom Sci. 5, 540-546.

KALBERER, P., and KUNSCH, U. 1974. Amino acid composition of the mushroom oyster (Pleurotus ostreatus). Lebensm. Wiss. Technol. 7, 242-244.

KLIGMAN, A. M. 1943. Some cultural and genetic problems in the cultivation of the mushroom Agaricus campestris Fr. Am. J. Bot. 30, 745-763.

KURTZMAN, R. H. 1975. Mushrooms as a source of food protein. In Protein Nutritional Quality of Foods and Feeds. Part 2. Quality Factors — Plant Breeding, Composition, Processing, and Antinutrients. M. Friedman (Editor). Marcel Dekker, New York.

LAMBERT, E. G. 1960. Improving spawn cultures of cultivated mushrooms. Mushroom Sci. 4, 35-51.

LANGERAK, D. Is. 1972. The influence of irradiation and packaging upon the keeping quality of fresh mushrooms. Mushroom Sci. 8, 221-230.

LITCHFIELD, J. H., VELY, V. G., and OVERBECK, R. C. 1963. Nutrient content of morel mushroom mycelium: amino acid composition of the protein. J. Food Sci. 28, 741-743.

MALLOCH, D. 1976. Agaricus brunnescens: the cultivated mushroom. Mycologia 68, 910-919.

McARDLE, F. J., and CURWEN, D. 1962. Some factors influencing shrinkage of canned mushrooms. Mushroom Sci. 5, 547-551.

McARDLE, F. J., KUHN, G. D., and BEELMAN, R. B. 1974. Influence of vacuum soaking on yield and quality of canned mushrooms. J. Food Sci. 39, 1026-1028.

McCONNEL, J. E. W., and ESSELEN, W. B. 1947. Carbohydrates in cultivated mushrooms. Food Res. 12, 118-121.

McKELLAR, R. L., and KOHRMAN, R. E. 1975. Amino acid composition of the morel mushroom. J. Agric. Food Chem. 23, 464-467.

MILLER, O. K., JR. 1972. Mushrooms of North America. E. P. Dutton and Co., New York.

MILLER, R. E., and KANAVEN, D. L. 1972. Bipolar sexuality in the mushroom. Mushroom Sci. 8, 713-718.

NAIR, N. G., and HAYES, W. A. 1974. Suggested role of carbon dioxide and oxygen in casing soil. In The Casing Layer, Proc. Aston Semin. in Mushroom Sci., Vol. 1. W. A. Hayes (Editor). M. G. A., London.

NAIR, N. G., and HAYES, W. A. 1975. Some effects of casing soil amendments on mushroom cropping. Aust. J. Agric. Res. 26, 181-188.

NICHOLS, R., and HAMMOND, J. B. W. 1973. Storage of mushrooms in prepacks. The effect of changes in carbon dioxide and oxygen on quality. J. Sci. Food Agric. 24, 1371-1381.

NICHOLS, R. and HAMMOND, J. B. W. 1975. The relationship between respiration, atmosphere, and quality in intact and perforated mushroom pre-packs. J. Food Technol. 10, 427-435.

PARRISH, G. K., BEELMAN, R. B., McARDLE, F. J., and KUHN, G. D. 1974. Influence of post-harvest storage and canning on the solids and mannitol content of the cultivated mushroom and their relationship to canned product yield. J. Food Sci. 39, 1029-1031.

RAPER, J. R., and RAPER, C. A. 1972. Life cycle and prospects for interstrain breeding in Agaricus bisporus. Mushroom Sci. 8, 1-9.

SINDEN, J. W. 1932. Mushroom spawn and method of making same. U.S. Pat. 1,869,517. Dec. 11.

SINDEN, J. W. 1937. Mushroom experiments. Pa. Agric. Exp. Stn. Bull. 352.

SMITH, J. 1972. Commercial mushroom production—2. Process Biochem. 7, No. 3, 24-26.

SMITH, J. E., and BERRY, D. R. 1974. An Introduction to Biochemistry of Fungal Development. Academic Press, London.

STANEK, M. 1972. Microorganisms inhibiting compost during fermentation. Mushroom Sci. 8, 797-811.

SUGIMORI, T., OYAMA, Y., and OMICHI, T. 1971. Studies on Basidiomycetes. I. Production of mycelium and fruiting body from noncarbohydrate organic substances. J. Ferment. Technol. 49, 435-439. (Japanese)

TREVELYAN, W. E. 1975. Fungi. In Food Protein Sources. N. W. Pirie (Editor). Cambridge Univ. Press, Cambridge, England.

WEISS, B., and STILLER, R. L. 1972. Sphingo-lipids of mushrooms. Biochemistry 11, 4552-4557.

WILLIAMS, B., and ESSELEN, W. B. 1946. Mass. Agric. Exp. Stn. Bull. 434.

WU, L. C., and STAHMANN, M. A. 1975. Fungal protein. In Unconventional Sources of Protein. Coll. Agric. Life Sci., Univ. of Wisconsin, Madison.

12

Fungi as a Source of Protein

Anthony J. Sinskey

Fungi have been an indirect source of protein in man's diet for centuries. Records of the production of cheese, in which fungi are used to develop texture and flavor, go back a thousand years. Filamentous fungi and yeasts are widely used in the traditional food technologies of the Far East to modify various dietary staples. Modifications include improved taste, texture and digestibility. In these processes, however, there is characteristically no net increase in the protein content of the foodstuff.

Recent developments in the use of filamentous fungi, yeasts, algae and bacteria for the production of protein are usually referred to as single-cell protein research. The term (abbreviated SCP) is a relatively new one coined to denote the production of protein for animal or human consumption via microbial means. It is not a strict definition, since microorganisms such as filamentous fungi may be multicellular.

During the last decade much has been written about SCP. Three references which provide a wealth of information on both technological and nutritional aspects were edited by Mateles and Tannenbaum (1968), Davis (1974), and Tannenbaum and Wang (1975). The review by Gray (1970) gives an extensive list of references on filamentous fungi as sources of SCP. Litchfield (1977) has presented a review of the most recent developments in SCP research.

PRODUCTION OF PROTEIN FROM FUNGI

Fungi as a Source of Protein

An important advantage of protein derived from both filamentous and unicellular fungi (yeasts) is that production is largely independent of agriculture, and consequently of climatic factors. Therefore, SCP from fungi could be produced in arid lands, such as Africa, and has potential for closing the widening gap between food production and population.

The filamentous microfungi may have a number of advantages over yeasts or bacteria as sources of SCP. Nutritional value and sulfur-amino content of protein from filamentous fungi can be high compared to other sources of SCP. In addition, these fungi can be propagated on a wider variety of substrates generally discarded as wastes from food production, harvesting, and manufacturing operations. Renewable resources high in cellulose, such as straw, may also be utilized and thus converted into nutritionally useful biomass.

The fact that filamentous fungi can be propagated in a variety of physical forms, such as single cells, filaments, or pellets, offers another type of advantage over yeasts and bacteria. For example, the filamentous structure may simplify the process of harvesting and may also be useful in food texturization processes.

Before discussing any further the use of fungi as a source of protein, it is appropriate to review the basic principles involved in the bioengineering of fungal fermentations.

Basic Bioengineering Principles

Growth Rate.—The multiplication of single cells can be best described by the following equation (symbols are listed at the end of the chapter):

$$\mu = \frac{1}{x} \cdot \frac{dx}{dt}$$

(12.1)

where x is weight of cells (g/liter), t is time (hr), and μ is specific growth rate (hr^{-1}). Upon integration,

$$\frac{dx}{x} = \mu\,dt$$

is

$$\frac{\ln x_{final}}{x_{initial}} = \mu \cdot \Delta t$$

(12.2)

If doubling time td is defined as the time required for the mass to double, then

$$\ln 2 = \mu \cdot td \qquad (12.3)$$

and simplifying,

$$\mu = \frac{0.693}{td}$$

Under conditions where growth is not limited by the availability of nutrients, the actual growth rate is usually equal to maximum specific growth rate, termed μ_{max}, under a given set of environmental conditions.

For years many mycologists thought that filamentous fungi grew at a linear rather than at an exponential rate. Trinci (1969) demonstrated that exponential growth does in fact occur, even though fungi do not multiply by simple binary fission, but rather by extension of the ends of hyphae at branch points. Fungi normally grow exponentially until the mycelial mass is so large that diffusion of nutrients into the colony or pellet is limiting. Solomons (1975) reports true exponential growth up to 10 to 15 g dry weight/day $\cdot$ m^3.

The growth rate of fungi is dependent upon both the nutritional and environmental conditions of the growth menstruum. Maximum growth rates (μ_{max}) have been reported for a variety of fungi, as summarized in Table 12.1. Within limits, the higher the growth temperature, the higher the μ_{max}. Despite a common misconception, fungi exhibit a variety of μ_{max} which may approach that of bacteria, such as the value of 0.40 hr^{-1} for *Neurospora sitophila*.

Yield.—A fast-growing organism is not absolutely necessary for production of SCP. What is critical is the yield factor, or amount of cell biomass produced per g unit weight of carbon in the substrate. The maximum or theoretical yield to be expected from glucose, for example, is approximately 0.45 to 0.50 g of cells per g of glucose. This means that 45 to 50% of substrate carbon is converted into fungal cells under aerobic metabolism. Since approximately 50% dry weight of substrate cells is carbon, it is possible to calculate the amount of carbon which must be supplied in the growth medium in order to produce any particular weight of fungal cells.

For example, if 20 g/liter dry weight of cells are required, and metabolism is aerobic, then the carbon in the medium will be

$$\frac{20\text{g cells}}{\text{liter}} \times \frac{1\text{g carbon}}{0.5\text{g cells}} = \frac{40\text{g carbon}}{\text{liter}}$$

TABLE 12.1

MAXIMUM GROWTH RATES AND EFFECT OF TEMPERATURE
ON GROWTH RATE OF SELECTED FUNGI

Organism	Growth Rate, μ_{max} (hr^{-1})	Reference
Aspergillus nidulans	0.15 at 25°C	Trinci (1972)
	0.22 at 30°C	
	0.36 at 37°C	
A. niger	0.20	Carter and Bull (1969)
Neurospora sitophila	0.40	Solomons (1975)

If supplied as glucose, then

$$\frac{40g \text{ carbon}}{\text{liter}} \times \frac{180g \text{ glucose}}{72g \text{ carbon}} = \frac{100g \text{ glucose}}{\text{liter}}$$

are required. The above calculation assumes that all available carbon is utilized to produce cells, and energy required for maintenance is ignored. For a discussion of maintenance requirements of fungi the review paper by Righelato (1975) is recommended.

Achievement of maximum yield is dependent upon providing all nutrients in the proper concentration. Unfortunately, during fungal fermentation it becomes more and more difficult to supply adequate oxygen and thus maximal yields are not always achieved. Experimentally determined yields using glucose as the carbon source are summarized in Table 12.2.

For aerobic growth of organisms, oxygen must also be supplied along with the carbon source. Although air as a source of oxygen is readily available, it is difficult to incorporate into the fermentation system because of the low solubility of oxygen in aqueous fermentation broths. Therefore, considerable engineering efforts have been devoted to the development of effective and economical means of delivering oxygen to the fermentation menstruum.

Another important nutrient required in the production of protein from fungal cells is nitrogen, which is usually provided in the form of ammonia or ammonium salts. However, faster growth is usually observed with organic nitrogen, sources of which include some relatively inexpensive forms, such as soybean, peanut, and fish meals, and other enzymatic digests of protein-rich materials.

While nitrogen constitutes approximately 10% of the dry weight of yeasts, algae, and bacteria, it comprises only about 5% of the dry

TABLE 12.2

GROWTH YIELD CONSTANTS FOR SELECTED FUNGI

Organism	Yield $\left(\dfrac{\text{g cell}}{\text{g glucose}}\right)$	Reference
Absidia corymbifera	0.45	Graham et al. (1976)
Aspergillus nidulans	0.45	Righelato (1975)
Rhizopus oligosporus	0.38	Graham et al. (1976)
R. rhizopodiformis	0.37	Graham et al. (1976)
Tricholoma nudum	0.3 - 0.5	Solomons (1975)

weight of filamentous fungi. These organisms are structurally more complex and contain a lower proportion of their weight in the form of metabolically active proteins and nucleic acids in their cell cytoplasm, and a higher proportion as invert polysaccharide components in the cell wall.

Minimum nitrogen requirements in the fermentation medium can also be calculated for a desired cell yield. Normally it is assumed that 20 g of cell dry weight can be produced per g of nitrogen, an approximation which is useful in designing medium composition. The choice of nitrogen source will be dependent upon cost, availability, purpose of fermentation, and the organism to be grown.

Listed in Table 12.3 are the inorganic constituents of fungi which have to be supplied to the growth medium, in addition to nitrogen. Elements with concentrations of 0.01 g/100 g dry weight may have to be supplied as ingredients, but elements with lower concentrations are usually present in the growth medium as trace contaminants.

In determining yields, all environmental parameters should be properly documented. If the fermentation is conducted under batch conditions, the final pH should be noted. Sugar consumption should also be determined.

Monad Equation.—Equation (12.2) may seem to imply that exponential growth of fungi can occur for indefinite periods of time. However, experience indicates that the exponential growth rate decreases to a value known as the growth limiting concentration due to the decrease in concentration of the substrate. The Monad equation expresses this relationship between growth rate and substrate concentration:

$$\frac{dx}{dt} = \mu_{max} \left(\frac{S}{k_s + S}\right) x \qquad (12.4)$$

TABLE 12.3

INORGANIC CONSTITUENTS OF FUNGI

Element	Percentage of Dry Weight
Phosphorus	0.4 - 5.0
Sulfur	0.1 - 0.5
Potassium	0.2 - 2.5
Magnesium	0.1 - 0.3
Sodium	0.02 - 0.5
Calcium	0.1 - 1.4
Iron	0.1 - 0.2
Total Ash	2.0 - 8.0

Source: Adapted from Aiba *et al.* (1973).

where μ_{max} is again the maximum growth rate constant, S is the substrate concentration, and k_s is the half saturation constant for uptake of the growth limiting substrate. When growth rate μ is equal to 0.5 μ_{max}, k_s is equal to the substrate concentration S.

Equation (12.4) may be more accurately expressed in terms of substrate uptake (ds/dt) as a function of substrate concentration. Remembering that a yield constant Y (g cell dry weight per g substrate) allows one to convert growth rate (dx/dt) into changes of substrate concentration,

$$\frac{ds}{dt} = \frac{1}{Y} \cdot \frac{dx}{dt} = q_{max}\left(\frac{S}{k_s + S}\right)x \qquad (12.5)$$

where μ_{max} is assumed to be equal to $q_{max}Y$, and Y and q_{max} are constants. In batch cultures during exponential growth, k_s is very small compared with the substrate concentration; thus until exponential growth ceases, one can approximate x as $S_o \cdot Y$, where S_o is initial substrate concentration.

Continuous Culture.—It is often desirable to maintain exponential growth of a fungal culture. One method is the technique of continuous culture. In explaining this procedure the concept of a mass balance is employed.

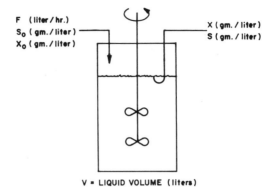

FIG. 12.1. SCHEMATIC DIAGRAM, INCLUDING DEFINITIONS, FOR A SINGLE-STAGE, WELL-MIXED CONTINUOUS CULTURE APPARATUS

Suppose a single fermentor, with volume V (liter) as diagrammed in Fig. 12.1, is inoculated with a fungal culture. Let

F = flow rate of medium (liter/hr)

x = cell mass concentration (g/liter)

S = limiting substrate concentration (g/liter)

$D = F/V$ = dilution rate (hr^{-1})

Cell accumulation is equal to the number of cells in the fermentor minus the number of cells going out with the effluent plus cell accumulation due to growth, which is expressed mathematically as follows:

$$V \cdot \frac{dx}{dt} = Fx_0 - Fx + V\left(\frac{dx}{dt}\right)_{growth} \qquad (12.6)$$

[Cell Accumulation = Cells In − Cells Out + Cell Growth]

Remembering that

$$\mu = \frac{1}{x} \cdot \frac{dx}{dt} \quad \text{and} \quad F/V = D$$

and then substituting into Equation (12.6),

$$\frac{dx}{dt} = D (x_0 - x) + \mu \cdot x \qquad (12.7)$$

Since no cells come into the fermentor, $x_o = 0$ and

$$\frac{dx}{dt} = \mu x - Dx \tag{12.8}$$

at steady state,

$$\frac{dx}{dt} = 0, \; \mu = D \tag{12.9}$$

A substrate mass balance can be similarly written

$$\frac{ds}{dt} = \frac{F}{V}(S_o - S) - \frac{\mu x}{Y} \quad \text{or} \quad \frac{ds}{dt} = D(S_o - S) - \frac{\mu x}{Y} \tag{12.10}$$

$$\left[\begin{array}{c} \text{Nutrient Accumulation} = \text{Nutrient In} - \text{Nutrient Out} - \text{Nutrients} \\ \text{Consumed Due to Cellular Growth} \end{array}\right]$$

and if at steady state

$$\frac{ds}{dt} = \frac{1}{Y} \cdot \frac{dx}{dt} = q_{max}\left(\frac{S}{k_s + S}\right)x = 0$$

then

$$D(S_o - S) = \frac{\mu \cdot x}{Y} \tag{12.11}$$

Since in continuous culture $D = \mu$, the concentration of x is then equal to

$$x = Y(S_o - S) \tag{12.12}$$

At a dilution rate less than μ_{max}, S is very small and consequently x can be approximated by $Y \cdot S_o$.

Another useful parameter is the substrate concentration S as a function of dilution rate D. At steady state conditions

$$\mu = D = \mu_{max}\left(\frac{S}{k_s + S}\right)$$

or in terms of substrate yields,

$$D = q_{max} \cdot Y \left(\frac{S}{k_s + S}\right)$$

Solving for S,

$$S = k_s \left(\frac{D}{q_{max} Y - D}\right)$$

Values of S can be determined experimentally. Also note that the maximum dilution rate at which a continuous culture system can operate is equal to $q_{max} Y$. The continuous culture system described here is also known as a chemostat, since growth rate is controlled by the concentration of the substrate. The chemostat gives a population of organisms in a steady state condition at any growth rate less than μ_{max}, and allows independent control of organism growth rate and cell concentration by manipulation of S_o and D.

Many variations on the basic approach of continuous culture are possible; e.g., multivessel continuous culture systems. Cells from the effluent of the fermentor vessel can also be recycled. For further information the classic articles by Herbert (1958) and Fencl et al. (1969) are recommended, as well as general references in biochemical engineering.

Pellet Growth Form.—It has been assumed in the above discussions that fungal growth is normally exponential, as it is in most single-cell systems. However, filamentous fungi may be grown under conditions where exponential growth is not observed, and for these situations other growth equations have been developed.

For example, several investigators have reported that filamentous fungal growth may be fitted to a cube root equation, such as

$$x^{1/3} = x_o^{1/3} + kt \qquad (12.13)$$

Pirt (1966) showed that cube root growth rates are a consequence of the pellet form of growth. When a fungal pellet exceeds a certain size, nutrients such as glucose or oxygen cannot diffuse into the pellet rapidly enough to maintain unrestricted growth of the whole pellet mass. This is because the pellet mass increases in proportion to the cube of its radius, while the surface through which substrate materials must diffuse increases in proportion to the square of the radius. Therefore, after reaching a certain size, the pellet can sustain exponential growth only at its periphery.

If it is assumed that the fungi are growing exponentially only at the periphery of the pellet, then

$$\frac{dx}{dt} = \mu \cdot x_p \qquad (12.14)$$

where x_p is the mass of peripheral growth zone.

In terms of radius r of the pellet, and the depth of the peripheral growth zone w, which is a constant that depends upon the growth-limiting substrate concentration, the rate of change of the radius is

$$\frac{dr}{dt} = \mu \cdot w \qquad (12.15)$$

and upon integration

$$r = \mu \cdot w \cdot t + r_o \tag{12.16}$$

The mass of the pellet with a density ρ is

$$x = {}^4\!/_3\, r^3\, \pi \cdot \rho \tag{12.17}$$

and substituting

$$x^{1/3} = \left(\frac{3}{4\,\pi\,\rho}\right)^{-1/3} \mu \cdot w \cdot t + x_o^{1/3} \tag{12.18}$$

Righelato (1975) reports that $({}^3\!/_4\pi\rho)^{-1/3}$ is a constant equal to 0.74, if the density of the pellet is 0.1 g dry weight per cm^3, and

$$x^{1/3} = 0.74 \cdot \mu \cdot w \cdot t + x_o^{1/3} \tag{12.19}$$

Note that this equation has the same form as that determined experimentally [equation (12.13)] by earlier investigators. If the radius of the pellet is smaller than w, then exponential growth occurs, and if greater than w, a cube root growth pattern of pellets is observed.

Surface Growth of Fungi.— Another way to grow fungi is on solid or semisolid surfaces. The most typical example is the surface of agar, but other solid-surface fermentations may be employed, such as the surface of cheese and grains. Difficulties arise in comparing fungal growth on solid surfaces with growth in broth cultures, since growth on agar is commonly linear whereas broth fermentations are many times exponential.

Pirt (1967) made an analysis of the surface growth of fungi which is similar to that of pellet growth. The analysis was refined by Trinci (1971) and confirmed for a variety of fungal strains. Experimental techniques for determining exponential growth rate μ from growth rates on solid surfaces are also described by Trinci (1971). The reasoning is as follows.

It is assumed that a colony grows as a disk with radius r, height H, and density ρ,

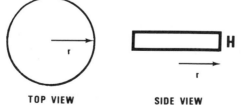

TOP VIEW   **SIDE VIEW**

and that growth is exponential, i.e.,

$$x_t = x_o\, e^{\mu t} = H\pi\rho\, r^2 = (H\pi\rho\, r_o^2)\, e^{\mu t} \tag{12.20}$$

Simplifying and taking logarithms of both sides,

$$\ln r_t = \ln r_o + \frac{\mu t}{2} \tag{12.21}$$

As in pellet growth, when a colony reaches a certain size, diffusion of nutrients into its center becomes limiting, and exponential growth at the center of the colony ceases. The growth of the colony then behaves as an annulus with exponential growth occurring in the outer ring of the annulus. If the annulus has a depth c, and mass x_c,

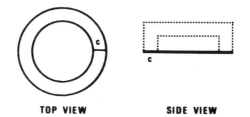

TOP VIEW   **SIDE VIEW**

$$\frac{dx}{dt} = \mu \cdot x_c \tag{12.22}$$

and x_c can be approximated by $2\pi Hc\rho$, where c is small compared with r and

$$\frac{dr}{dt} = \mu \cdot c \tag{12.23}$$

and

$$r_1 = c\mu t + r_o \tag{12.24}$$

Linear growth of the colony is expected since c and μ are constants.

The growth of colonies of *Aspergillus nidulans* was found by Trinci (1971) to be initially exponential and later linear at about 0.2 mm, as predicted by equation (12.21). In order to calculate the specific growth rate μ from this equation, depth c and change in colony diameter must be measured. Trinci (1971) described procedures for measuring c and calculated the specific growth rates of nine species of fungi, which agreed reasonably well with values obtained in submerged cultures. The value of c varied considerably even among strains having the same μ. Some data collected by Trinci (1971) is presented in Table 12.4.

The above procedure for determining μ by surface growth is useful in evaluating or screening fungi for SCP production. Growth rates can be determined for a number of fungi under a variety of nutritional and environmental conditions. Since c must be determined, the procedure provides insight into the degree of branching and hyphal growth.

TABLE 12.4

COMPARISON OF FUNGAL GROWTH RATE:
SUBMERGED VERSUS SURFACE CULTURE

| Organism | Specific Growth Rate (hr⁻¹) | | c¹ |
	Submerged	Surface	(μm)
Actinomucor repens	0.18	0.19	2500
Aspergillus niger	0.12	0.12	1100
A. wentii	0.15	0.12	1300
Geotrichum·lactis	0.35	0.43	420
Mucor racemosus	0.10	0.13	3400
Penicillium chrysogenum	0.16	0.15	500
Rhizopus stolonifer	0.14	0.11	8700

Source: Adapted from Trinci (1971).
[1]Depth of peripheral growth zone of a surface colony.

USE OF FUNGI FOR SCP

The overall objective of a fungal fermentation for SCP is to produce a nutritionally useful biomass as economically as possible. Furthermore, the product has to be nontoxic and suitable for food applications. Some factors influencing the choice of a fungal SCP process are summarized in Table 12.5. They can be divided into the three categories: (a) economy and technology; (b) nutrition and safety; and (c) application to foods. Most research to date has been in the areas listed under the first two categories. Research into applications to food is still in its infancy.

Fortunately, the Protein Advisory Group (PAG) of the United Nations has issued guidelines which provide valuable assistance in the planning of economic and technological, as well as nutritional and safety, aspects of SCP. The information is summarized in PAG Guidelines No. 12, 6 and 7.

PAG Guideline No. 12, entitled "Production of SCP for Human Consumption," deals in a general manner with suitability of organisms, raw materials (substrates), process variables, composition, quality control requirements, product safety and nutritional considerations.

Guideline No. 6 deals with preclinical testing of novel sources of protein and describes in detail the testing of new protein materials which is recommended prior to human feeding trials. These include evaluation of toxicity, microbiological contamination and sanitary

TABLE 12.5

FACTORS INFLUENCING A FUNGAL SCP PROCESS

Economic and Technological
  (1) Fungal strain
  (2) Protein yield per unit of substrate
  (3) Cost of substrate and pretreatment
  (4) Cost of other nutrients
  (5) Productivity (mass of protein per unit volume per day)
     and method of fermentation
  (6) Cost of sterilization of medium
  (7) Heat removal
  (8) Recovery operation and its cost

Nutritional and Safety
  (1) Protein content
  (2) Amino acid pattern
  (3) Protein digestibility
  (4) Effects of extraneous materials (e.g., nucleic acid)

Application to Foods
  (1) Flavor
  (2) Texture
  (3) Solubility
  (4) Color
  (5) Processing capabilities

---

testing, chemical tests, nutritional evaluation with test animals, reproductive and lactation studies, and teratogenic and mutagenic studies.

Guideline No. 7 is a companion document to No. 6 and deals with the human testing of SCP as a supplement to food mixtures. Both of these guidelines are a must for the serious reader and are available from the United Nations; they are also included as appendices in a book on SCP edited by Tannenbaum and Wang (1975).

Economic and Technical Considerations

Selection of Raw Materials.—Varieties of raw materials have been investigated as substrates for SCP production of fungi, including both industrial and agricultural wastes. Examples in industrial wastes and by-products which can be utilized as substrates are:

(1) sulfite waste liquor from the paper and pulp industry
(2) molasses from sugar refineries

(3) whey
(4) acetic acid
(5) hydrocarbons

Substrates derived from agricultural wastes are:

(1) citrus wastes
(2) sugar cane bagasse
(3) coffee by-products
(4) animal manure

Substrates can also be classified as renewable or nonrenewable resources. The best examples of renewable resources are starches derived from food processing, such as potato wastes, and cellulose from insoluble agricultural wastes, such as corn stalks, rice straw, and wheat hulls. Even municipal waste has been proposed as a source of cellulose for conversion into protein. For a general discussion of the use of fungi in waste control operations, the reader is referred to a book edited by Birch et al. (1976).

Before they are suitable for fermentation, substrates may require pretreatment. For example, much of the cellulose in waste plant materials is insoluble, and chemical, physical, or biological treatments are required to solubilize it. Unfortunately, such pretreatment usually increases the costs of fermentation.

A unique example of a biological pretreatment operation is the so-called "Symba" process, which is an acronym for the symbiotic growth of two yeasts. The yeasts Endomycopsis fibuligera (now Saccharomycopsis sp.) and Candida utilis (formerly Torulopsis utilis, also known as torula yeast) are cultivated together, using a starch substrate that can be derived from a variety of sources, such as potato wastes, cassava products and wheat (Skogman 1976). Endomycopsis fibuligera produces amylases which are needed to hydrolyze starch into sugars, while C. utilis, which cannot hydrolyze starch, consumes the sugar as quickly as it is released. The final product is 90% C. utilis by weight.

Conversion of whey and whey fractions from the dairy industry to SCP by lactose-assimilating Kluyveromyces fragilis has been studied in several countries. Production of SCP using C. utilis, C. krusei, Saccharomyces cerevisiae, and other yeast species was recently studied on a technical scale using acid whey, rennet whey, heat-deproteinized whey, ultrafiltration-deproteinized whey, dialysis-deproteinized whey, ultrafiltration-deproteinized skim milk, and lactose liquor residues (Lembke et al. 1975). The published Linzer, Waldhof, Wheast, Polyvit, and SAV procedures were used. References to and details of the procedures were presented. The use of whey as a substrate for conversion to protein is discussed in Chap. 6.

Production of SCP on hydrocarbons by yeasts has received a great amount of research attention. Pure linear alkanes (n-paraffins) and other petroleum distillates have been used successfully as carbon sources for culturing *C. utilis, C. tropicalis, C. lipolytica,* and *C. intermedia.* The use of n-paraffins has the advantage of eliminating the need for solvent purification of yeast cells (Lipinsky and Litchfield 1970). Although gas oil is a cheaper source of carbon, extensive solvent extraction of yeast cells is required to remove potentially toxic residues. Numerous reviews on the use of hydrocarbons as substrates for SCP production have been published (Johnson 1967; Humphrey 1970; Klug and Markovetz 1971; Shennan and Levi 1974; Shacklady 1975).

A wide range of chemical wastes and effluents can be utilized by yeasts. Edwards and Finn (1969) reviewed these substrates, analyzing the variables involved in SCP production, including the nature of the chemical effluents, yeast growth on synthetic organic compounds, and the economics and marketing problems involved. SCP production from methanol by *Hansenula polymorpha* has been studied by Cooney and Levine (1975). Ethanol also shows promise as a carbon source for SCP production by yeasts.

Selection of Fungal Strain.—Assuming a substrate has been chosen, a screening study is routinely employed to find nontoxic microfungi which are efficient utilizers of the substrate in question. Other criteria may include high optimal growth temperature and ability to grow at low pH, in addition to yield and growth rate.

In general the SCP fermentation process is exothermic and requires cooling. For example, in the following carbon balance as described by Imrie and Righelato (1976) (12.25)

$$\text{Carbohydrate} + \text{oxygen} \longrightarrow \text{SCP} + CO_2 + 27{,}000 \text{ kcal}$$
$$\text{(1 Kg C)} \quad \text{(1.3 Kg)} \quad \text{(0.5 Kg C)} \quad \text{(1.8 Kg)}$$

Fungi capable of growing at high temperatures are desirable since production costs of cooling may be substantially reduced. Seal and Eggins (1976) reported on the use of thermophilic fungi to upgrade agricultural waste. Temperatures between 50° and 60°C were employed, and several fungal species were able to grow at low pH. Von-Hoftsen (1976) also describes the use of a thermotolerant strain, *Sporotrichum pulverulentum,* which can grow at temperatures up to 40°C.

Fungi able to grow rapidly at low pH are also desirable, since pathogenic organisms are eliminated and sterilization requirements are reduced. An analogy is to be found in the sterilization requirements for high-acid (low pH) foods, where milder conditions may be employed by comparison with low-acid (high pH) foods.

TABLE 12.6

FUNGI EXAMINED FOR SCP PRODUCTION ON A VARIETY OF SUBSTRATES

Organism	Substrates	Comments	Reference
Filamentous fungi			
Aspergillus niger	Carob (Ceratonia sitiqua)	45% Total sugar is utilized	Imrie and Righelato (1976)
A. fumigatus	Cellulose	Cellulase is produced	Dunlap (1975)
Fusarium graminearum	Starch		Anderson et al. (1975)
Neurospora sitophila	Molasses		Anderson et al. (1975)
Myrothecium verrucaria	Cellulose (coffee waste)	Cellulase is produced	Dunlap (1975)
Rhizopus oligosporus	Mung bean whey plus 1% glucose	Carbon/nitrogen ratio affects yield	Graham et al. (1976)
Trichoderma viride	Cellulose, paper, bagasse	Reduction in chemical oxygen demand	Brown and Fitzpatrick (1976)
Yeast			
Candida utilis	Sulfite waste liquor	Utilizes pentoses	Lipinsky and Litchfield (1970)
Candida spp.	Refined petroleum fractions	Cleanup simplified if substrate is pure	Lipinsky and Litchfield (1970)
Hansenula polymorpha	Methanol	Grows well up to 42°C	Cooney and Levine (1975)
Kluyveromyces fragilis	Cheese whey	Utilizes lactose	Peppler (1968)

Table 12.6 lists some representative microfungi which have been investigated in detail on a variety of substrates and show promise for production of SCP.

Process Variables.—*Fermentors.*—After choosing a suitable fungus for the substrate to be utilized, the issue of how to propagate the organism needs to be considered. The approaches used for large-scale production processes can be either simple or complex.

Simple fermentation procedures are obviously desirable. The simplest fermentation scheme for fungi is a lagoon to which nutrients may be added as growth supplements. Several such schemes have been described by Church *et al.* (1972), and a schematic diagram is presented in Fig. 12.2. Such procedures are desirable when the substrate comes from a food plant, or other large-scale industrial operation, since a minimum of capital investment is required, and a waste disposal problem may be eliminated.

More sophisticated fermentation operations are also possible and can range from batchwise growth in simple tanks to continuous culture operations. The more sophisticated procedures are usually designed to increase the amount of oxygen supplied to the fermentation medium. A schematic diagram is presented in Fig. 12.3.

Aeration and Agitation.—Fermentors usually need to be aerated in order to supply a sufficient amount of oxygen to the fungus. In addition, it is usually desirable to agitate the fermentor in order to prevent solids and fungal cells from settling out. Frequently the aeration and agitation are combined as a single-unit process. Examples of a mechanically agitated fermentor and an airlift fermentor are schematically presented in Figs. 12.4 and 12.5. Both approaches are designed to maximize gas-liquid contact.

The transfer of oxygen to cells is a multiple step process in which oxygen passes from air bubbles across gas and liquid films and into the bulk liquid. The oxygen dissolved in the bulk liquid then diffuses to the liquid film surrounding the fungal cell surface and into the cell to the site of its utilization.

Biochemical engineers have determined that the rate-controlling step in this process is oxygen transfer across the liquid film around the bubble. The equation expressing amount of oxygen transferred is

$$N_a = k_L \cdot a \; (C^\alpha - C_L) \tag{12.26}$$

where N_a is the volumetric mass transfer rate (mMoles O_2/liter $\cdot$hr), k_L is the mass transfer coefficient (cm/hr), a is the specific interfacial area of mass transfer (cm^2/cm^3), C^α is the equilibrium oxygen concentration (mMoles/liter), and C_L is the dissolved oxygen concentration (mMoles/liter).

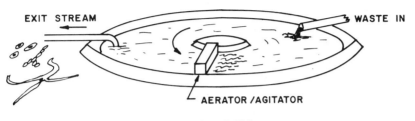

FERMENTATION DITCH

FIG. 12.2. SCHEMATIC DIAGRAM OF A SIMPLE FERMENTATION SCHEME FOR THE CONVERSION OF FOOD WASTES INTO FUNGAL PROTEIN

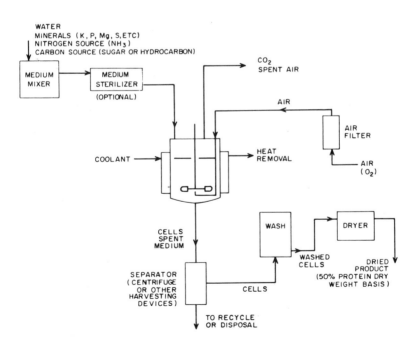

FIG. 12.3. FLOW SHEET FOR PRODUCTION OF FUNGAL SCP

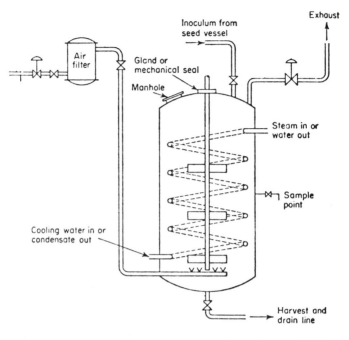

From Cooney (1976)

FIG. 12.4. SCHEMATIC DIAGRAM OF AN AGITATED FERMENTOR

Since it is practically impossible to determine a experimentally, k_L and a are usually combined into an overall mass transfer coefficient k_La.

Demand for oxygen is also expressed by the equation

$$N_a = \frac{\mu x}{Y_{O2}} \qquad (12.27)$$

where μ equals specific growth rate (hr^{-1}), x equals cell concentration (g/liter), and Y_{O2} equals oxygen yield (g cell/g O_2).

Since supply must equal demand

$$\frac{\mu \cdot x}{Y_{O2}} = K_La \ (C^{\alpha} - C_L) \qquad (12.28)$$

and since the dilution rate D (hr^{-1}) is equal to $\mu \cdot hr^{-1}$ in continuous culture,

$$\frac{D \cdot x}{Y_{O2}} = k_La \ (C^{\alpha} - C_L)$$

Considerable engineering effort has been directed at determining how the parameter k_La is influenced by fermentor design, type of

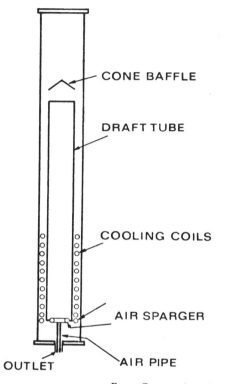

CONE BAFFLE

DRAFT TUBE

COOLING COILS

AIR SPARGER

OUTLET   AIR PIPE

From Cooney (1976)

FIG. 12.5. SCHEMATIC DIAGRAM OF AN AIRLIFT FERMENTOR

agitation, and flow rate of oxygen. Empirical equations have been developed, for example,

$$k_L a = \left(\frac{p}{V}\right)^\alpha (V_s)^\beta \qquad (12.29)$$

where p/V is power input in the liquid volume (kw/m³) and V_s is the superficial gas velocity (m/sec). Superficial gas velocity is the ratio of volumetric gas flow to the cross sectional area of the fermentor. In the laboratory α is usually around 0.95 and β is 0.7. In large-scale fermentors the exponents decrease to about half the laboratory values.

Methods for increasing oxygen transfer range from aerated tanks with baffles to agitation with a variety of stirrers. Much of the field of biochemical engineering is devoted to increasing the oxygen transfer rates in fermentation broths. For further information the reader is referred to Aiba *et al.* (1973).

TABLE 12.7

INTERRELATIONSHIP BETWEEN CELL YIELD, OXYGEN DEMAND,
AND HEAT LOAD

Substrate	Substrate Yield $\left(\dfrac{\text{g cell}}{\text{g substrate}}\right)$	Oxygen Yield[1] $\left(\dfrac{\text{g cell}}{\text{g O}_2}\right)$	Oxygen Demand $\left(\dfrac{\text{mMole}}{\text{liter} \cdot \text{hr}}\right)$	Heat Load $\left(\dfrac{\text{kcal}}{\text{liter} \cdot \text{hr}}\right)$
Glucose	0.5	1.3	80	10
Methanol	0.5	0.6	180	22
n-Alkanes	1.0	0.50	230	30

Source: Adapted from Cooney (1976).
[1]Based on productivity of 3.5 g cell/liter $\cdot$ hr.

Another way to meet the cellular demand for oxygen is by manipulating the driving force $(C^\alpha - C_L)$. C_L is fixed by solubility laws but C^α may be increased by using oxygen-enriched air or even pure oxygen. Therefore, in some fermentation operations the fermentor may be sparged with pure oxygen.

Interrelationship Between Cell Yield, Oxygen Demand, and Heat Load.—In SCP production the cell yield, oxygen demand, and heat load are interrelated, as shown in equation (12.25), with different substrates having different cell yields, oxygen demands, and cooling requirements (Table 12.7). Differences in oxygen demand among substrates are a function of molecular composition of the substrate and of the proportions of carbon and oxygen it contains. Heat wasted in production of a given quantity of cells is also an intrinsic feature of each substrate.

In addition, if the fermentation is operated under conditions which are not ideal, so that maximum cell yields are not produced, there may be large amounts of substrate which are unnecessarily converted into heat as shown in Fig. 12.6. Thus, cooling requirements are a function of substrate and efficiency of substrate utilization.

In SCP production, carbon source, oxygen demand, and cooling requirements are all significant cost components, accounting for approximately 65 to 70% of the total cost. The substrate itself accounts for 40 to 50% of the total cost.

Productivity of the Fermentation.—As mentioned previously high growth rates of fungal fermentations are not necessary for production of SCP. What is essential is high productivity, or amount of biomass produced per unit volume per hour.

Productivity P in a continuous culture operation is equal to the

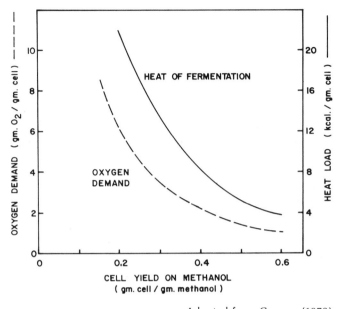

Adapted from Cooney (1976)
FIG. 12.6. EXAMPLE OF INTERRELATIONSHIP BETWEEN CELL YIELD, OXYGEN DEMAND, AND HEAT LOAD, WITH METHANOL AS SUBSTRATE

dilution rate D (hr^{-1}) times the cell concentration x (g/liter), or $P = Dx$. Continuous culture is more productive than batch fermentation because the organism is always growing at an exponential growth rate. This is contrary to batch systems, in which the organism exhibits a classic microbial growth curve—a lag phase and an exponential phase of growth followed by a stationary phase.

In a large industrial operation where fungi are grown batchwise, it may take up to 40 hr to complete a fermentation cycle of sterilizing, inoculating, growing, emptying and washing. On the other hand, in the same time period a continuous culture operation will have produced four times as much biomass at dilution rate 0.1 hr^{-1}. In addition, in continuous culture operations, time is saved in recycle.

Cell Harvesting and Recovery.—In SCP operations the cost of cell removal from the fermentor is dependent on cell concentration. The higher the cell concentration, the lower the cost per unit of protein recovered. Cooney (1976) has estimated that recovery from an effluent with a concentration of 2% solids will be approximately 50% more expensive than one with 3% solids.

Cell size also influences cost of cell separation procedures. An advantage of filamentous fungal SCP may be their larger size compared to bacterial or yeast SCP. Centrifugation is frequently used to remove bacterial and yeast cells in SCP operations, but the simpler process of filtration can be used with ferments containing large-celled filamentous fungi. Biomass is retained in the filter press and forms a filterable cake. Other even simpler filtration schemes are possible, including the use of screens or sieves.

Additional Processing Requirements.—SCP may be further processed, depending upon specific needs and requirements. For example, it may be washed to remove extraneous materials, solvents, and medium components. Specific processing treatments may be employed to reduce the RNA content or to release protein. The product may also be dried as a powder or as flakes.

Throughout the fermentation operations appropriate quality control procedures are absolutely necessary if the product is to be employed in animal feed or in human foods. Such quality control procedures include the Good Manufacturing Practices (GMP) applied in the food industry to ensure a safe and reproducible product.

Unique Aspects of Fungal SCP.—An important variable in growth of fungi is the morphological form of growth, which can be single-celled, filamentous, or pellet-like. A variety of factors—both nutritional and environmental—have been found to affect the morphological form. For example, the presence of high carbon dioxide tension can cause certain fungi to be converted from a yeast-like phase into a filamentous mycelium. Growth as pellets in submerged fermentation is influenced by the degree of shear or agitation, with low shear promoting pellet growth. Inoculum level and culture medium have also been reported to influence form of growth. Simple media require lower inoculum levels than more complex media in order to avoid pellet formation. The degree of hyphal branching may also be affected by pH.

A fruitful area for research may lie in investigating dimorphism in fungi and how it can be manipulated or controlled. For example, an advantage of growing fungi as pellets is the ease of recovery and processing. On the other hand, yeasts are easier to grow due to ease of nutrient delivery and agitation. Further understanding of the principles of dimorphism may lead to greater utilization of fungi for SCP production.

Nutrition and Safety

Protein Content.—Protein content of fungi is of fundamental importance. It is not only strain dependent but is also influenced by growth conditions. For example, fungi grown at a high growth rate normally have higher protein content as well as increased levels of nucleic acids.

Several studies have revealed that the protein content of fungi is equal to total nitrogen content times 6.25, using Kjeldahl nitrogen determinations. Caution has to be employed in using this technique, however, as some fungal SCP may be high in nonprotein sources of nitrogen, as can be seen from the data of Solomons (1975). Major sources of nonprotein nitrogen are chitin (in some fungi) and nucleic acids, primarily RNA, the levels of which vary with growth rate. Anderson *et al.* (1975) reported levels of nonprotein nitrogen in *Fusarium graminearum* ranging from 6% at low growth rates to 13% when organisms were grown at μ_{max}. However, in other fungi nonprotein nitrogen may constitute 20 to 50% of total nitrogen.

Solomons (1975) has described a method of nitrogen assessment for fungi which is based upon determination of total amino acid nitrogen and is more reliable than the Kjeldahl procedure. This method is a modification of the one described by Gehrke and Wall (1971), and results are considerably different from Kjeldahl determinations. For example, the total protein for *Trichoderma viride* as determined by the Kjeldahl procedure is 64% and by α-amino nitrogen determination 35%. Solomons (1975) reports that fungi contain from 3 to 7.5% α-amino nitrogen, or about 19 to 47% protein.

Net Protein Utilization.—The high total nitrogen value of fungal SCP may produce misleading results in biological studies. The rat assay, as described by Miller and Payne (1961), is an accepted method of measuring protein quality by determining net protein utilization (NPU). In this method a diet is used consisting of 10% protein or 1.6% N. If the value of 1.6% N is determined using a total nitrogen technique, such as Kjeldahl, the true value of fungal proteins will be seriously underestimated. Solomons (1975) reported that the NPU values of SCP from a variety of fungi ranged from 30 to 90 when protein was measured using the α-amino determination. The high value of 90 is associated with an exceptional source of SCP which is in the class of animal proteins.

Amino Acid and Vitamin Content.—Protein quality is normally assessed by biological means, as described above, and is dependent upon the type, availability, and amount of essential amino acid. In SCP sources, sulfur-containing amino acids (viz. methionine and cystine) are usually limiting. However, in some fungi essential amino acid patterns can be found to approximate hens' eggs and the 1957 Food and Agriculture Organization (FAO) provision pattern. The essential amino acid patterns of some fungi having potential as sources of SCP are shown in Table 12.8.

Summarized in Table 12.9 are additional nutritional data on the mycelium of various species of filamentous fungi. Note that the

TABLE 12.8
ESSENTIAL AMINO ACID PATTERNS OF SELECTED FUNGI

Amino Acid	1957 FAO Provisional Pattern[1]	Hen's Egg[1]	Filamentous fungi[1]			Candida utilis[1]	Yeasts	
			Penicillium chrysogenum	Fusarium graminearum	Neurospora sitophila		Kluyveromyces fragilis[4]	Saccharomyces cerevisiae[4]
Isoleucine	4.2	6.6	5.4	4.3	4.4	6.0[2] - 7.9[3]	5.5	5.5
Leucine	4.8	8.8	10.6	6.5	8.5	9.1 - 7.5	9.9	7.9
Lysine	4.2	6.4	3.8	7.5	5.9	7.1 - 8.7	8.8	8.2
Total aromatic amino acids	5.6	10	8.3	8.5	8.1	9.6 - 5.1	3.9	4.5
Total sulfur-containing amino acids	4.2	5.5	4.2	3.2	3.0	2.0 - 3.2	1.5	2.5
Cystine	2	2.4	1.3	1.3	1.1	0.4 - 1.4	—	—
Methionine	2.2	3.1	2.9	1.9	1.9	1.6 - 1.8	1.5	2.5
Threonine	2.8	5.1	5.3	5.1	6.0	6.1 - 5.5	5.5	4.8
Tryptophan	1.4	1.6	—	—	—	1.5 - 1.4	1.5	1.2
Valine	4.2	7.3	5.0	7.2	5.4	7.3 - 6.3	6.6	5.5
Total essential amino acids	31.4	51.3	42.6	42.3	41.3	50.3 - 45.6	44.7	42.6
Approximate chemical score	100	100	90	76	71	37 - 65	31	54
Limiting amino acid	None	None	Lysine	Sulfur-containing amino acids	Sulfur-containing amino acids	Sulfur-containing amino acids	Sulfur-containing amino acids	Sulfur-containing amino acids

[1] Grams of amino acid per 100 g of total amino acids; data from filamentous fungi adapted from Anderson et al. (1975).
[2] Adapted from Spicer (1971).
[3] Adapted from Inskeep et al. (1951).
[4] Grams of amino acid per 16 g N; adapted from Standard Brands, Inc. (1967); see also Table 6.3.

nucleic acid content is lower than that of faster growing organisms such as bacteria and yeasts. The digestibility of filamentous fungi is high, considering that the cell wall contains chitin. The usefulness of methionine supplementation should also be noted.

In addition to valuable protein in fungal biomass, substantial levels of vitamins, especially those in the B group, are present. Yeasts generally contain higher levels than do filamentous fungi. The gross composition and vitamin content of Candida utilis grown on sulfite waste liquor are listed in Table 12.10. These values may vary depending on the conditions of growth.

Ribonucleic Acid.—It has been reported by Sinskey and Tannenbaum (1975) that the nucleic acid content of SCP is high by comparison with conventional proteins, most of the nucleic acid content being in the form of RNA. The nucleic acid content of SCP varies, ranging from 8 to 25 g per 100 g protein, although filamentous fungi in general have lower levels than other SCP sources. An estimated 8 to 13% of yeast nitrogen is from purines and about 4% is from pyrimidines. Consumption of high amounts of these bases by humans can cause an increase in the amount of uric acid in the blood. Human feeding studies reported by the PAG have resulted in recommendations that humans limit their intake of single-cell nucleic acids to 2 g/day. Thus, either SCP protein intake has to be limited, or the nucleic acid content of SCP needs to be reduced.

Several techniques have been developed for controlling the RNA content in SCP, including:

(1) manipulation of growth rate
(2) based-catalyzed hydrolysis
(3) use of exogenous and endogenous enzymes
(4) chemical extraction of physical removal

Since the nucleic acid content of SCP is dependent upon growth rate, with slower growing organisms having less nucleic acid than faster growing ones, one can control nucleic acid content by manipulating rate of growth. One may also hydrolyze the nucleic acids by chemical or enzymatic means to produce low molecular weight materials which can then be removed from the cells by washing. Alkaline hydrolysis is one such chemical method. Enzymes, both endogenous and exogenous RNAases, can also be used for biological hydrolysis procedures. Each method has its advantages and disadvantages, and not every method is suitable for a given SCP operation. For a more detailed discussion the review article by Sinskey and Tannenbaum (1975) is recommended.

TABLE 12.9

NUTRITIONAL DATA ON SELECTED FUNGI

SPC Source	Nucleic Acid (g/100 g mycelium)	Lysine (g/16 g N)	Total Sulfur-containing Amino Acids (g/16 g N)	Digestibility (%)	NPU[1]
Aspergillus oryzae	3.4	5.2	2.2	71.0	67
Fusarium semitectum	3.2	5.0	2.4	83.0	92
Gliocladium deliquescens	3.9	5.3	2.3	78.0	80
Trichoderma viride	4.7	5.4	2.8	83.0	62

Source: Adapted from Worgan (1976).

[1]Based on true protein plus methionine supplementation ×100.

TABLE 12.10

PROXIMATE AND VITAMIN COMPOSITION OF *CANDIDA UTILIS* GROWN
ON SULFITE WASTE LIQUOR

Gross component	(g/100 g, dry wt)
Moisture	6
Ash	9
Phosphorus (as P)	2
Calcium (as Ca)	1
Crude protein (N × 6.25)	47
Crude fat	5
Carbohydrate (by difference)	27
Vitamins (μg/g)	
Biotin	2
Folic acid	21
Niacin	417
Pantothenic acid	37
Pyridoxine hydrochloride	33
Riboflavin	45
Thiamine	5

Source: Adapted from Inskeep *et al.* (1951).

Animal Feeding Studies.—Several animal feeding studies have been conducted with fungal SCP, including acute and chronic toxicity trials. Worgan (1976), for example, describes an acute toxicity study in which five rats were administered 40 g of *Fusarium semitectum* per kg body weight over a 24 hr period. No adverse symptoms were detected and all animals were normal in post-mortem examination. Chronic toxicity studies were also negative. Other feeding studies with pigs and poultry have not shown any adverse effects. Shacklady (1975) summarized toxicological experiments with hydrocarbon-grown yeast.

Duthrie (1975) reported on feeding trials in which animals were fed *Penicillium notatum-chrysogenum*. The protein was demonstrated to be of good quality, and no toxicity could be observed in rats, chicks, or mice. However, the organism is unsuitable for use in SCP production because production costs are too high. *Fusarium graminearum*, a more useful organism from the economic and technological viewpoint, was then investigated by means of long-term animal feeding and reproductive studies. No adverse findings have been reported. Human feeding tests are also being developed to determine acceptability, tolerance, and biological value of the products. Results of long-term human feeding studies are not yet available.

An excellent review article on the nutritional value of yeasts was written by Bressani (1968). A description of the benefits of supplementing corn, wheat flour, cottonseed, and sesame with SCP derived from yeast is given. Since yeasts are limiting in the sulfur-containing amino acids, they do not complement soybeans in the same way. This deficiency can be overcome by supplementing the diet with cystine and methionine.

Application to Foods

The most efficient way to utilize SCP is directly by human consumption, which results in reductions in labor, facilities, and animal waste. The utilization of SCP for human food has been reviewed by Rha (1975).

The use of fungal SCP as a functional ingredient in foods is an especially interesting possibility. Proposed uses include surface active agents, structure-orienting agents, and thickeners. Surface active properties of importance are foaming and whipping ability, while rheological properties of fungal SCP make it useful as a thickener. Its structure-orienting characteristics are of value for thermosetting, extrusion, and spinning.

Only limited information is available on the functional properties of fungal SCP. However Worgan (1976) reports that mycelium from *Fusarium semitectum* has a cohesive structure which can confer a meat-like texture to foods. The mycelium can also be produced with considerable variations in texture. Heating influences the shear forces that are required to break the mycelium. For example, the shear value for a heat treatment of 2 min at 80°C was 4.5 kg, while at 121°C for 30 min it was 5.0 kg.

The effect of drying conditions on functional properties of *Saccharomyces cerevisiae* has been investigated by Labuza *et al.* (1972). Functionality was shown to increase as drying temperatures increased. The foaming capacity and stability of *Saccharomyces (Kluyveromyces) fragilis* proteins are reported to be generally poorer than that of soy protein isolate; however, the emulsifying activity is better (Vananuvat and Kinsella 1975).

Torutein™, a SCP derived from *Candida utilis* grown on ethanol, is presently being tested for its functional characteristics in foods such as breads, cakes, cereals, pastas and ground meats (Schnell *et al.* 1976). It is reported to improve the flavor, texture, nutritional value and handling characteristics of certain food systems.

TABLE 12.11

WORLDWIDE PRODUCTION OF FUNGAL PROTEIN (1975-80)

Country	Organism	Substrate	Capacity (tons/year)
Czechoslovakia	Yeast	Ethanol	100,000
		n-Paraffins	100,000
Finland	Filamentous fungi	Sulfite liquor	10,000
France	Yeast	Gas oil	100,000
Great Britain	Yeast	n-Paraffins	100,000
	Filamentous fungi	Carbohydrate	4,000
Italy	Yeast	n-Paraffins	200,000
Japan	Yeast	n-Paraffins	100,000
Romania	Yeast	Methanol	60,000
USSR	Yeast	n-Paraffins	200,000
United States	Yeast	Ethanol	5,000

Source: Adapted from Wu and Stahmann (1975).

Summary

The use of fungi for human consumption and for improving protein-poor diets throughout the world is a real possibility, as is the use of fungal SCP for producing animal feeds from wastes. Estimates of the worldwide production of fungal protein (1975-1980) are given in Table 12.11. There is still much to be accomplished, however, in order to make fungal protein more attractive from the viewpoints of cost, toxicity, and palatability. The importance of the role played by fungal SCP in alleviating pending food problems depends on the willingness of industry and governmental agencies to support basic and applied research, and upon students' willingness to accept the challenge of engineering new foods for tomorrow.

LIST OF SYMBOLS

Growth Rate

x	cell concentration (g/liter)
t	time (hr)
μ	specific growth rate (hr^{-1})
td	doubling time
μ_{max}	maximum specific growth rate (hr^{-1})

Monod Equation

S	substrate concentration (g/liter)
k_s	half saturation constant for uptake of growth-limiting concentration
ds/dt	substrate uptake
Y	yield constant (g cell dry weight/g substrate)
dx/dt	growth rate
q_{max}	constant
S_o	initial substrate concentration (g/liter)

Continuous Culture

F	flow rate of medium (liter/hr)
$D = F/V$	dilution rate (hr^{-1})
V	volume (liter)

Pellet Growth

x_p	mass of peripheral growth zone
r	radius of pellet
w	depth of peripheral growth zone
ρ	density of pellet (g dry weight/cm^3)

Surface Growth

H	height
c	depth of annulus
x_c	mass of annulus

Aeration and Agitation

N_a	volumetric mass transfer rate (mMoles O$_2$/liter $\cdot$ hr)
k_L	mass transfer coefficient (cm/hr)
a	specific interfacial area of mass transfer (cm^2/cm^3)
C^α	equilibrium oxygen concentration (mMoles/liter)
C_L	dissolved oxygen concentration (mMoles/liter)
Y_{O2}	oxygen yield (g cell/g O$_2$)
p/V	power input in liquid volume (kw/m^3)
V_s	superficial gas velocity (m/sec)
α, β	exponents

Productivity

P	productivity (g/cm^3 $\cdot$ hr)

Thanks are due to M. Ludwig for her excellent editorial assistance, M. Crowley for unfailing secretarial assistance, and B. Larson for preparation of the drawings (Publ. No. 3071).

REFERENCES

AIBA, S., HUMPHREY, A. E., and MILLIS, N. F. 1973. Biochemical Engineering. 2nd Edition. Academic Press, New York.

ANDERSON, C., LONGTON, J., MADDIX, C., SCAMMELL, G. W., and SOLOMONS, G. L. 1975. The growth of micro-fungi on carbohydrates. In Single-Cell Protein II. S. R. Tannenbaum and D. I. C. Wang (Editors). M.I.T. Press, Cambridge, Mass.

BIRCH, C. G., PARKER, K. J., and WORGAN, J. T. 1976. Food from Waste. Applied Science Publishers, London.

BRESSANI, R. 1968. The use of yeast in human foods. In Single-Cell Protein. R. I. Mateles and S. R. Tannenbaum (Editors). M.I.T. Press, Cambridge, Mass.

BROWN, D. E., and FITZPATRICK, S. W. 1976. Food from waste paper. In Food from Waste. C. G. Birch, K. J. Parker, and J. T. Worgan (Editors). Applied Science Publishers, London.

CARTER, B. L. A., and BULL, A. T. 1969. Studies of fungal growth and intermediary carbon metabolism under steady and non-steady state conditions. Biotechnol. Bioeng. 11, 785-804.

CHURCH, B. D., NASH, H. A., and BROSZ, W. 1972. Use of Fungi Imperfecti in treating food processing wastes. Dev. Ind. Microbiol. 13, 30-46.

COONEY, C. L. 1976. Chemical sources of food: An approach to novel food sources. In Environmental Chemistry. J. O'M. Bockris (Editor). Plenum Publishing Corp., New York.

COONEY, C. L., and LEVINE, D. W. 1975. SCP production from methanol by yeast. In Single-Cell Protein II. S. R. Tannenbaum and D. I. C. Wang (Editors). M.I.T. Press, Cambridge, Mass.

DAVIS, P. 1974. Single Cell Protein. Academic Press, London.

DUNLAP, C. E. 1975. Production of single-cell protein from insoluble agricultural wastes by mesophiles. In Single-Cell Protein II. S. R. Tannenbaum and D. I. C. Wang (Editors). M.I.T. Press, Cambridge, Mass.

DUTHIE, I. F. 1975. Animal feeding trials with a microfungal protein. In Single-Cell Protein II. S. R. Tannenbaum and D. I. C. Wang (Editors). M.I.T. Press, Cambridge, Mass.

EDWARDS, V. H., and FINN, R. K. 1969. Fermentation media: chemical effluents. Process Biochem. 4, No. 6, 29-33, 39.

FAO Committee on Protein Requirements. 1957. FAO nutritional studies No. 16. FAO/ UN, Rome.

FENCL, Z., MACHEK, F., NOVAK, M., and SEICHERT, L. 1969. Control of culture activity as a function of growth rate in continuous cultivation. In Continuous Culture of Microorganisms. I. Malek, K. Beran, A. Fencl, J. Ricica, and H. Smrčková (Editors). Academic Press, New York.

FUJIMAKI, M., and MITSUDA, H. 1972. Conversion and Manufacture of Foodstuffs by Microorganisms. Saikon Publishing Co., Tokyo.

GEHRKE, C. W., and WALL, L. L. 1971. Automated trinitrobenzene sulfonic acid method for protein analysis in forages and grain. J. Assoc. Off. Anal. Chem. 54, 187-191.

GRAHAM, D. C. W., STEINKRAUS, K. H., and HACKLER, L. R. 1976. Factors affecting production of mold mycelium and protein in synthetic media. Appl. Environ. Microbiol. 32, 381-387.

GRAY, W. D. 1970. The use of fungi as food and in food processing. CRC Crit. Rev. Food Technol. 1, 225-329.

HERBERT, D. 1958. Some principles of continuous culture. In Recent Progress in Microbiology. VII Int. Congr. Microbiol., 381.

HUMPHREY, A. E. 1970. Microbial protein from petroleum. Process Biochem. 5, No. 6, 19-22.

IMRIE, F. K. E., and RIGHELATO, R. C. 1976. Production of microbial protein from carbohydrate wastes in developing countries. In Food from Waste. C. G. Birch, K. J. Parker, and J. T. Worgan (Editors). Applied Science Publishers, London.

INSKEEP, G. C., WILEY, A. J., HOLDERBY, J. M., and HUGHES, L. P. 1951. Food yeast from sulfite liquor. Ind. Eng. Chem. 43, 1702-1711.

JOHNSON, M. J. 1967. Growth of microbial cells on hydrocarbons. Science 155, 1515-1516.

KIHLBERG, R. 1972. The microbe as a source of food. Annu. Rev. Microbiol. 26, 427-466.

KLUG, M. J., and MARKOVETZ, A. J. 1971. Utilization of aliphatic hydrocarbons by microorganisms. Adv. Microb. Physiol. 5, 1-43.

LABUZA, T. P., JONES, K. A., SINSKEY, A. J., GOMEZ, R., WILSON, S., and MILLER B. 1972. Effect of drying conditions on cell viability and functional properties of single-cell protein. J. Food Sci. 37, 103-107.

LEMBKE, A., MOEBUS, O., GRASSHOFF, A., and REUTER, H. 1975. Technological and technical bases of production of single cell protein from dairy waste products. In Berichte uber Landwirtschaft, Sonderheft 192. Beundesanstalt fur Milchforschung, Kiel. (German)

LITCHFIELD, J. H. 1977. Single-cell proteins. Food Technol. 31, No. 5, 175-179.

LIPINSKY, E. S., and LITCHFIELD, J. H. 1970. Algae, bacteria, and yeasts as food or feed. CRC Crit. Rev. Food Technol. 1, 581-618.

MATELES, R. I., and TANNENBAUM, S. R. 1968. Single-Cell Protein. M.I.T. Press, Cambridge, Mass.

MILLER, D. S., and PAYNE, P. R. 1961. Problems in the prediction of protein values of diets. The influence of protein concentration. Br. J. Nutr. 15, 11-19.

OSER, B. L. 1972. Safety evaluation of single cell protein for human food use. In Conversion and Manufacture of Foodstuffs by Microorganisms. M. Fujimaki and H. Mitsuda (Editors). Saikon Publishing Co., Tokyo.

PEPPLER, H. J. 1968. Industrial production of single-cell protein from carbohydrates. In Single-Cell Protein. R. I. Mateles and S. R. Tannenbaum (Editors). M.I.T. Press, Cambridge, Mass.

PIRT, S. J. 1966. A theory of the mode of growth of fungi in the form of pellets in submerged culture. Proc. R. Soc. London, Ser. B, 166, 369-373.

PIRT, S. J. 1967. A kinetic study of the mode of growth of surface colonies of bacteria and fungi. J. Gen. Microbiol. 47, 181-197.

POMERANZ, Y. 1976. Single-cell protein from by-products of malting and brewing. Brew. Dig. 51, 49-55, 60.

RHA, C. 1975. Utilization of single-cell protein for human food. In Single-Cell Protein II. S. R. Tannenbaum and D. I. C. Wang (Editors). M.I.T. Press, Cambridge, Mass.

RIGHELATO, R. C. 1975. Growth kinetics of mycelial fungi. In The Filamentous Fungi. Vol. 1. Industrial Mycology. J. E. Smith and D. R. Berry (Editors). John Wiley and Sons, New York.

SCHNELL, P. G., AKIN, C., and FLANNERY, R. J. 1976. Properties of Torutein for food applications. Cereal Foods World 21, 311-315.

SEAL, K. J., and EGGINS, H. O. W. 1976. The upgrading of agricultural wastes by thermophilic fungi. In Food from Waste. C. G. Birch, K. J. Parker, and J. I. Worgan (Editors). Applied Science Publishers, London.

SHACKLADY, C. A. 1975. Yeasts grown on hydrocarbons. In Food Protein Sources. N. W. Pirie (Editor). Cambridge Univ. Press, Cambridge, England.

SHENNAN, J. L., and LEVI, J. D. 1974. The growth of yeasts on hydrocarbons. Prog. Ind. Microbiol. 13, 1-57.

SINSKEY, A. J., and TANNENBAUM, S. R. 1975. Removal of nucleic acids in SCP. In Single-Cell Protein II. S. R. Tannenbaum and D. I. C. Wang (Editors). M.I.T. Press, Cambridge, Mass.

SKOGMAN, H. 1976. Production of Symba-yeast from potato wastes. In Food from Waste. C. G. Birch, K. J. Parker, and J. T. Worgan (Editors). Applied Science Publishers, London.

SMITH, R. H., and PALMER, R. 1976. A chemical and nutritional evaluation of yeasts and bacteria as dietary protein sources for rats and pigs. J. Sci. Food Agric. 27, 763-770.

SNYDER, H. E. 1970. Microbial sources of protein. Adv. Food Res. 18, 85-140.

SOLOMONS, G. L. 1975. Submerged culture production of mycelial biomass. In The Filamentous Fungi. Vol. 1. Industrial Mycology. J. E. Smith and D. R. Berry (Editors). John Wiley and Sons, New York.

SPICER, A. 1971. Protein production by micro-fungi. Trop. Sci. 13, 239-250.

STANDARD BRANDS, INC. 1967. Debittered Brewer's Dried Yeast, Dried Fragilis Yeast. Standard Brands, Inc., New York.

TANNENBAUM, S. R., and D. I. C. WANG. 1975. Single-Cell Protein II. M.I.T. Press, Cambridge, Mass.

TERUI, G. 1972. Fermentation Technology Today. Soc. Ferment. Technol. Osaka, Japan.

TREVELYAN, W. E. 1976. Autolytic methods for the reduction of the purine content of baker's yeast, a form of single-cell protein. J. Sci. Food Agric. 27, 753-762.

TRINCI, A. P. J. 1969. A kinetic study of the growth of Aspergillus nidulans and other fungi. J. Gen. Microbiol. 57, 11-24.

TRINCI, A. P. J. 1971. Influence of the width of the peripheral growth zone on the radial growth rate of fungal colonies on solid media. J. Gen. Microbiol. 67, 325-344.

TRINCI, A. P. J. 1972. Culture turbidity as a measure of mould growth. Trans. Br. Mycol. Soc. 58, 467-473.

VANANUVAT, P., and KINSELLA, J. E. 1975. Some functional properties of protein isolates from yeast, Saccharomyces fragilis. J. Agric. Food Chem. 23, 613-616.

VONHOFSTEN, B. 1976. Cultivation of a thermotolerant Basidiomycete on various carbohydrates. In Food from Waste. C. G. Birch, K. J. Parker, and J. T. Worgan (Editors). Applied Science Publishers, London.

WORGAN, J. T. 1976. Wastes from crop plants as raw material for conversion by fungi to food or live-stock feed. In Food from Waste. C. G. Birch, K. J. Parker, and J. T. Worgan (Editors). Applied Science Publishers, London.

WU, L.-C., and STAHMANN, M. A. 1975. Fungal Protein. In Papers from a Workshop on Unconventional Sources of Protein. Coll. Agric. Life Sci., Univ. of Wisconsin, Madison.

ZEE, J. A., and SIMARD, R. E. 1975. Simple process for the reduction in nucleic acid content in yeast. Appl. Microbiol. 29, 59-62.

13

Metabolites of Fungi Used in Food Processing

R. J. Bothast and K. L. Smiley

ungal metabolites have traditionally been a part of food preparation. Fermented foods, based on the ancient koji process, are
staples in Oriental diets. The Western world is familiar with the
role of fungi in the ripening of cheeses such as Roquefort and Camembert, and fermented beverages have been consumed for centuries
throughout the world. No attempt will be made in this chapter to
review the historic role of fungi in food processing, since this topic is
covered rather extensively in Chap. 6, 9, and 10. Instead, an attempt
will be made to show the role of fungal metabolites in modern food
processing. The modern era was initiated in the 1920s after Thom and
Currie discovered in 1916 that Aspergillus niger formed large amounts
of citric acid from sucrose. About the same time, and as a result of the
work of Takamine, the fungal enzyme industry was established. From
this start the fermentation industry has expanded so that today
several organic acids and a variety of enzymes produced by fungi are
used in the manufacture of foods and beverages. In addition, fungal
fermentations are conducted to make amino acids, vitamins, flavoring
agents, mannitol, and fats and oils, all of which are of value in food
processing.

ORGANIC ACIDS

Organic acids function primarily as acidulants in modern food processing. In addition to rendering foods more palatable and stimulating to the consumer, Gardner (1968) has listed the following uses for acidulants.

(1)   Flavoring agents, where they may intensify certain tastes, blend unrelated taste characteristics, and mask undesirable aftertastes.

(2)   Buffers, in controlling the pH of food during various stages of processing, as well as of the finished product.

(3)   Preservatives, in preventing growth of microorganisms and the germination of spores which lead to the spoilage of food or cause food poisoning or disease.

(4)   Synergists to antioxidants, in preventing rancidity and browning.

(5)   Viscosity modifiers, in changing the rheological properties of dough and, consequently, the shape and texture of baked goods.

(6)   Melting modifiers, for such food products as cheese spreads and mixtures used in manufacturing hard candy.

(7)   Meat curing agents, together with other curing components in enhancing color, flavor, and preservative action.

Only the fungal-produced organic acids used in food processing will be covered in this chapter, and discussion will be confined largely to use and production. A list of these organic acids is shown in Table 13.1.

Citric Acid

Uses.—Citric acid is widely used in the preparation of soft drinks, cheeses and other dairy products, desserts, jams, jellies, candies, canned seafoods, wines, and frozen fruits. Citric acid is also used in gelatin food products and with artificial flavors of dry compounded materials such as soft drink tablets and powders. It functions as an antioxidant for inhibiting rancidity in fats and oils, and as an acidulant, buffer, emulsifier, and stabilizer in various food products.

Production by Molds.—Citric acid was first made commercially in England in 1860 from calcium citrate obtained from cull lemons in Italy and Sicily (Lockwood and Schweiger 1967). In 1922, Italy produced approximately 90% of the world's supply of citrates (Prescott and Dunn 1959). The introduction of mold-produced citric acid (Currie 1917) in the United States in 1923 broke this monopoly, and now most of the world's supply of citric acid is produced by carbohydrate fermentation.

The microorganisms most commonly used in the production of citric acid are selected strains of *Aspergillus niger* (Fig. 13.1). However, other molds such as *A. wentii*, *A. clavatus*, *Pencillium luteum*, *P.*

TABLE 13.1

FOOD ACIDS PRODUCED BY FUNGI

Acid	Formula	Produced By	Food and Beverage Use
Citric	$H_2C\text{-}COOH$ $HO\text{-}C\text{-}COOH$ $HO\text{-}C\text{-}COOH$ $H_2C\text{-}COOH$	Aspergillus niger, Candida lipolytica	Soft drinks, dairy products, desserts, jams, jellies, candies, canned seafoods, wines, frozen fruits, fats, and oils
Itaconic	$H_2C{=}C\text{-}COOH$ $H_2C\text{-}COOH$	Aspergillus itaconicus, A. terreus	Shortenings, resin coatings contacting food
Gluconic	$COOH$ $H\text{-}C\text{-}OH$ $HO\text{-}C\text{-}H$ $H\text{-}C\text{-}OH$ $H\text{-}C\text{-}OH$ $H_2C\text{-}OH$	A. niger, Penicillium chrysogenum	Bottle washing formulations, baking powder, bread mixes, emulsified meat products, desserts
Fumaric	$H\text{-}C\text{-}COOH$ $\parallel$ $H\text{-}C\text{-}COOH$	Species of Rhizopus and Mucor	Fruit drinks, desserts, wines, doughs, coatings, fats, dairy and meat products

Acid	Structure	Organism	Uses
Malic	H₂C-COOH / H-C-COOH / OH ($H_2C\text{-}COOH$, $H\text{-}C\text{-}COOH$, OH)	Species of *Aspergillus*, *Penicillium brevi-compactum*, yeasts	Beverages, jams, jellies, candy, syrups, sour dough, oils
Tartaric	COOH / H-C-OH / HO-C-H / COOH	*Penicillium notatum*, *A. niger*, *A. griseus*	Carbonated beverages, desserts, jellies
Succinic	H₂C-COOH / H₂C-COOH	Species of *Rhizopus*, *Mucor*, *Fusarium*	Flavorings
Oxalic	COOH / COOH	*A. niger*	Hydrolysis of starch to glucose
Lactic	CH₃ / H-C-OH / COOH	Species of *Rhizopus* and *Mucor*	Fruit juice, shortenings, mayonnaise, mincemeat, desserts, bakery, dairy, and meat products

FIG. 13.1. *ASPERGILLUS NIGER* NRRL-334 CULTURED ON CZAPEK AGAR FOR 10 DAYS AT ROOM TEMPERATURE

citrinum, Paecilomyces divaricatum, and *Mucor* spp. have been used in the laboratory

A surface culture process or shallow pan method and a submerged culture method are used in the commercial production of citric acid. Both fermentations rely on interruption of the Krebs citric acid cycle in a terminal stage of carbohydrate metabolism. In the surface culture method, a solution of suitable carbohydrate is inoculated with spores of *A. niger* and poured into shallow aluminum or stainless steel pans. The carbohydrate may be refined or crude sucrose, fructose, glucose, high-test cane syrup, or beet molasses. In general, sugar concentrations of 20 to 25% are required to produce high yields of citric acid. The high sugar concentrations are believed to inhibit the formation of acids other than citric, and the addition of ferricyanide (Clark and Lentz 1963) followed by filtration is reported to reduce the dissolved iron which interferes with citric acid accumulation (Schweiger 1961). Molliard (1922) showed that phosphate deficiency enhanced citric acid production, and undoubtedly the presence or absence of trace amounts of other elements in the medium may have a marked effect on the fermentation. Doelger and Prescott (1934) indicated that a pH of 1.6 to 2.2 is optimum for citric acid production. By balancing ammonium salts of mineral acids and alkali metal nitrates, the fermentation can be controlled to give only citric acid. Maximum acid production occurs when the rate of growth of mycelium is insignificant and there is no sporulation. During the first 5 to 7 days of fermentation, humidified air at 28° to 34°C is blown across the surface of the culture. After 8 to 10 days the solution is drained off and the mat is washed and pressed to remove any acid contained in it. Calcium citrate is then precipitated from a hot neutral solution. Next a sulfuric acid treatment is used to liberate citric acid. Usually about 70% of the weight of the sugar used in the medium can be recovered as citric acid (Schweiger 1961).

The submerged culture method of citric acid production has developed since 1930. The studies of Shu and Johnson (1947, 1948A, B) showed that the fermentation was quite sensitive to manganese and iron. Moyer (1953) demonstrated that methanol and ethanol stimulate production of citric acid by A. niger. However, it was not until (1) methods were developed for removal of metallic ions with ion exchange resins (Woodward et al. 1949), (2) copper was used as an antagonist for ions (Schweiger 1961), and (3) suitable strains of A. niger were selected that the present commercial process developed. Decationized solutions of high-test syrup, glucose, or sucrose are suitable for use as carbohydrate sources (Lockwood and Schweiger 1967). Except for ammonia, which is used to adjust the initial pH in the range of 2 to 4, nutrients are added similarly to the surface culture process. However, restriction of phosphate is unnecessary and air is bubbled through the fermentation solution at a rate of 0.5 to 1.5 (vol/vol/min) (Lockwood and Schweiger 1967). The fermentation is usually conducted over a 5- to 14-day period at 27° to 33°C. The acid is then harvested by a method similar to the one used in the surface process.

Production by Yeasts.—The commercial process for production of citric acid with A. niger has several disadvantages. For example, with time the citric acid-producing capability of the A. niger culture tends to degenerate, and more than 7 days are required for maximum acid production. Thus, it is obvious that the development of a rapid fermentation process is of considerable commercial importance. A typical process flow sheet is given in Fig. 13.2.

Over the last 10 years, several yeast strains of the genera Candida (perfect stage—Saccharomycopsis), Endomycopsis, Torulopsis, Hansenula, and Pichia have been studied that have the ability to accumulate substantial amounts of citric acid during the aerobic

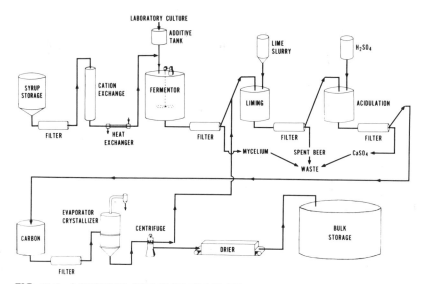

FIG. 13.2. A TYPICAL PROCESS FLOW SHEET FOR CITRIC ACID

fermentation of aqueous carbohydrate, hydrocarbon, and/or acetic acid-containing media (Roberts 1973; Hustede and Siebert 1974; Takayama and Tomiyama 1974; Furukawa and Kaneyuki 1975). Roberts (1973) found that yeasts will accumulate larger amounts of citric acid when halogen-containing agents (α-chloro- and α-fluoro-substituted lower alkanoic mono- and dicarboxylic acids) are included in the fermentation medium. A major problem with yeasts is that many strains gradually consume the citric acid produced during fermentation and for this reason the yield is reduced. Fukuda et al. (1974) emphasized the need for selecting strains of yeasts that are capable of utilizing hydrocarbons and incapable of using citric acid. Hustede and Siebert (1974) produced 128 g of citric acid/liter of hydrocarbon-containing medium in 4 days with a mutant strain of Candida oleophila. Furukawa and Kaneyuki (1975) reported yields of 150 g/liter in 7 days with Saccharomycopsis lipolytica on a hydrocarbon medium.

Ikeno et al. (1975) have recently studied citric acid production by mutant strains of Candida lipolytica on natural oils, fatty acids, glycerol, ethanol, and n-paraffin. They reported 102 g/liter of citric acid from palm oil in 94-hr fermentations and yields of citric acid against oil of 146%. On n-paraffin, yields of 170 g/liter were obtained. Although the potential for economical production of citric acid by yeasts exists, commercial production is still in the developmental stage.

Itaconic Acid

Uses.—Itaconic acid is principally used in resin coatings for paper. It offers superior properties in taking printing inks and in bonding. Food-related uses include itaconic acid-modified triglycerides as shortening compounds and the use of itaconic copolymers in coatings which come into contact with foods.

Production.—Itaconic acid was first reported by Kinoshita (1929) to be a fungal metabolite of *Aspergillus itaconicus*. Later Calam *et al.* (1939) found that certain strains of *Aspergillus terreus* produced significant amounts from glucose. Further studies (Lockwood and Reeves 1945; Moyer and Coghill 1945; Lockwood *et al.* 1945) led to the isolation of a superior strain of *A. terreus* (NRRL 1960) for the production of itaconic acid in surface and submerged culture. Subsequent developmental work by Lockwood and Nelson (1946), Pfeifer *et al.* (1953), and Batti and Schweiger (1963) led to the current process for itaconic acid production which is similar to the one described by Nubel and Ratajak (1962). In this process, spores of *A. terreus* are germinated 18 hr in a beet molasses medium. A 15% cane molasses medium containing zinc sulfate, magnesium sulfate, and cupric sulfate is then inoculated with a 20% (vol/vol) spore suspension. Air is bubbled through the solution, vigorous agitation is applied, and the temperature is maintained at 39° to 42°C. The pH drops from 5.1 to about 3.1 during the first 24 hr. Lime or ammonia is then used to readjust the pH to 3.8 and the fermentation continues for 2 more days. At the end of the fermentation, the mycelium is filtered off and washed, and the solution is concentrated by heating. The hot solution is decolorized with carbon, and on cooling the itaconic acid crystallizes. Alternatively, a calcium precipitation similar to one described for citric acid may be used. Yields of itaconic acid are about 85 g/liter. As with the citric acid fermentation, the itaconic acid fermentation is most efficient below pH 2 and when the mycelium growth rate is insignificant. The fermentation is also sensitive to metallic ions.

Gluconic Acid

Uses.—Many important applications have been found for gluconic acid and its sodium, calcium, and lactone derivatives in the food industry. The largest markets for gluconates are in alkaline bottle-washing formulations used in cleansing milk, beer, and soft drink bottles. Favorable properties of gluconates include ease of handling, low toxicity, low corrosion of metals, good cleaning efficiencies, compatability with aqueous systems, and prevention of calcium, magnesium, and iron deposits (Ward 1967). The latent acidity of δ-gluconolactone has led to its use as a component of baking powder, in which it is not reactive with sodium bicarbonate until water is added. This property along with edibility makes δ-gluconolactone desirable for use in chemically leavened bakery goods and bread mixes (Feldberg 1959). Also, because of its latent acidity this compound has produced excellent results in the controlled acidification and retention of red color in emulsified meat products (Sair 1961). This same latent acid property has been applied to cheesemaking by using the lactone as an acidogen before curd formation (Hammond and Deane 1961). Bland-flavored sherbets (Feldberg 1959) and cold milk puddings (Hunter and Rock 1960) have been prepared using δ-gluconolactone as a latent acidulant. Also the lactone has been used to reduce fat absorption in rolled waffle cones and doughnuts (Feldberg 1959).

Production.—Early attempts to produce gluconic acid by the surface cultivation of *Penicillium* and *Aspergillus* spp. were made by Herrick and May (1928) and May *et al.* (1929). May *et al.* (1934) and Gastrock and Porges (1938) later applied submerged cultivation techniques and calcium carbonate neutralization to gluconic acid production. Subsequent studies by Moyer *et al.* (1937), Wells *et al.* (1937), and Blom *et al.* (1952) on glucose concentration, agitation, aeration, and air pressure laid the groundwork for the production of gluconates by submerged growth of *Aspergillus niger* as it is practiced today. A glucose concentration of 30%, temperature maintained at 33° to 34°C, fermentors of several thousand liters capacity, aeration rates of 1.0 to 1.5 vol of air per min per vol of fermentation broth, air pressure on the fermentor of 30 psig (about 2900 g/cm^2), and a high degree of agitation are preferred for gluconic acid (sodium salt) production (Ward 1967). Theoretical yields of 95% are obtained. Fermentors can be inoculated with either spore suspensions or vegetative mycelium, or with mycelium separated from a previous fermentation. Calcium or sodium salts of gluconic acid are recovered by incorporating calcium carbonate or sodium hydroxide in the culture medium. Gluconic acid may be recovered from calcium gluconate broths or solutions by sulfuric acid treatment or from aqueous sodium gluconate solutions by ion exchange. The lactones are recovered by temperature-critical crystallization of oversaturated gluconic acid solutions. Below 30°C gluconic acid is recovered. From 30° to 70°C, the δ-gluconolactone is predominant, and above 70°C the γ-lactone is obtained (Pasternack and Giles 1934).

Fumaric Acid

Use.—Fumaric acid is used in fruit juice drinks, gelatin desserts, pie fillings, refrigerated biscuit doughs, maraschino cherries, and wines (Gardner 1968). Fumaric acid is useful in preparing edible coatings for candy, water-in-oil emulsifying agents, reconstituted fats, and dough conditioners (Bertram 1951). It also shows good antioxidant properties in lard, butter, cheese, powdered milk, sausage, bacon, nuts, and potato chips (Gardner 1968) and may be used for improving the whipping properties of gelatin and egg white (Conrad and Stiles 1954; Abbott et al. 1957). Fumaric acid is economical from the standpoint of cost and the quantities required for imparting acid tastes; however, its application is limited for some purposes because it goes into solution slowly and has a relatively low solubility in water (Gardner 1966).

Production.—Fumaric acid is produced principally by the fermentation of glucose or molasses with species of the genus *Rhizopus*. Foster (1949, 1954) has reviewed production of fumaric acid by these fungi. The level of various components of the fermentation medium such as carbohydrate, potassium, iron, magnesium, zinc, and copper can greatly influence the yield. Only under limited conditions do selected strains of *Rhizopus* produce primarily fumaric acid. Usually ethanol is produced with substantial proportions of other acids. Rhodes et al. (1959) reported fumaric acid yields of 60 to 70% in 3 to 8 days in shaken flasks containing 10 to 16% concentrations of glucose, sucrose, or the partially inverted sucrose of high-test molasses. Of the total acids produced, 75 to 80% was fumaric acid. Up to 25% greater yields of fumaric acid were obtained by the addition of methanol to fermentations conducted with molasses.

Even though fumaric acid can be produced in rather good yields by fermentation, the economics today are such that fumaric acid is produced commercially as a by-product in the manufacture of phthalic and malic anhydrides or by isomerization of malic acid with heat and a catalyst (Dmuchovsky and Franz 1967).

Malic Acid

Use.—Malic acid is a general purpose acidulant. It has unusual taste-blending characteristics and in some instances appears to have flavor-fixing qualities, as well as serving to overcome undesirable aftertastes (Gardner 1966). Apparently malic acid has a stronger acidic taste than does citric but not as strong as fumaric.

The United States Food and Drug Administration (1965) has included malic acid as a miscellaneous and/or general-purpose food additive in its list of GRAS substances. Malic acid has been used in the manufacture of apple jams, jellies, candy, and beverage products. However, the use of malic acid in food has been limited to date (Pederson 1971). Applications have been restricted primarily to specialty items such as a peach drink, confectionery products, apple-flavored hard candy, ice-cream syrups, and tobacco products (Irwin et al. 1967). Malic acid has been employed in extracting pectin, in the production of sour dough, and in the synthesis of emulsifying agents which inhibit the development of rancidity in oils (Gardner 1968). Pray and Powers (1966) used malic acid as an acidifying agent in canning tomatoes.

Production.—Malic acid has been found in cultures of a variety of fungi, including aspergilli, yeasts, and *Penicillium brevi-compactum* (May and Herrick 1932; Godin 1953). Abe *et al.* (1962) reported yields of levorotatory malic acid as high as 55 g/100 g of D-glucose for *Aspergillus flavus* and *A. parasiticus*. Iron, manganese, chromium, and aluminum ions also enhance malic acid production.

Despite the ability of fungi to produce malic acid, present commercial production consists of hydrating malic and fumaric acids in the presence of a suitable catalyst and separating the malic acid from the equilibrium product mixture (Irwin *et al.* 1967).

Tartaric, Succinic, Oxalic, and Lactic Acids

Other food acids which may be produced by fungi but are not produced commercially by fungal fermentation processes are tartaric, succinic, oxalic, and lactic. Tartaric acid is used in carbonated beverages. It especially has the ability to augment natural and synthetic grape flavors (Gardner 1968). Tartaric acid is also used in gelatin desserts, fruit jellies, and starch jelly candies (Chichester 1969). Succinic is used in flavoring and as a gelling agent for marmalades (Turi 1969). Oxalic acid is used in the hydrolysis of starch to glucose (Pederson 1971). Lactic acid is used in bakery products, liquid shortening, egg powder, cheese, dried food casein, fruit juice, frozen desserts, manufacture of beer, mincemeat, mayonnaise, and many other food products (Gardner 1968).

Tartaric acid may be produced by *Penicillium notatum, Aspergillus niger,* or *A. griseus,* and succinic acid can be produced in yeast fermentations (Pederson 1971). Many species of aspergilli, penicillia, and mucors produce oxalic acid (May and Herrick 1932). Lactic acid has been produced by many species of *Rhizopus, Mucor,* and at least one species of *Monilia. Rhizopus oryzae* has particularly outstanding ability to produce lactic acid (Prescott and Dunn 1959).

ENZYMES

Fungal enzymes have been used for hundreds of years, especially in the Orient. However, modern industrial enzyme technology probably started with Takamine (1894) and his work with *Aspergillus oryzae.* Today many industrial enzymes are of fungal origin. Those that are in commercial use are listed in Table 13:2.

α-Amylase

α-Amylase, although produced by many species of fungi, is made commercially by either *Aspergillus niger* or *A. oryzae.* Two general methods are used to produce α-amylase by molds. In one method, known as the Takamine process (Takamine 1894), the mold is grown

TABLE 13.2

COMMERCIAL FUNGAL ENZYMES USED IN FOOD PROCESSING

Enzyme	Produced By	Food and Beverage Uses
α-Amylase	Aspergillus oryzae, A. niger	Corn syrup, dextrose, bread and cracker baking, food dextrins, chocolate and licorice syrups
Glucoamylase	Aspergillus awamori, A. niger, Rhizopus niveus, R. delemar	Dextrose, dextrose syrup, baking, degrading gelatinized starch
Pectinases	A. niger, Coniothyrium diplodiella	Clarifying fruit juices and wine, coffee concentration
Naringinase	A. niger	Debittering grapefruit juice
Invertase (sucrase)	Saccharomyces cerevisiae, S. uvarum (formerly S. carlsbergensis)	Confectioneries, liqueurs, cordials, soft-centered chocolates, artificial honey
Lactase (β-D-galactosidase)	A. niger, Kluyveromyces lactis, K. (formerly Saccharomyces) fragilis	Dairy products, upgrading cheese whey
Protease	A. oryzae	Soy sauce, brewing, baking bread, crackers, tenderizing meat
Rennet (rennin, if pure)	Mucor pusillus, M. miehei, Endothia parasitica	Many types of cheese
Glucose oxidase, catalase	A. niger	Powdered egg products, brewing, wines, mayonnaise

on wheat bran in shallow trays. The bran is moistened with water, steam sterilized, and inoculated with spores of A. oryzae. After growth, the amylase is extracted from the bran with water. The extract may be concentrated below 50°C to a syrup or the α-amylase may be precipitated from the aqueous extract with alcohol and then dried at 55°C or less. In the other method the fungus, usually A. niger, is cultivated by submerged fermentation using a starch-salts medium. The α-amylase is generally recovered by filtering from the mycelium and concentrating the filtrate to a syrup under reduced pressure.

The largest use of fungal α-amylase in food processing is in the production of glucose syrups (Dale and Langlois 1940; Erenthal and Block 1962; Denault and Underkofler 1963; Underkofler et al. 1965). Normally starch is converted with acid until the dextrose equivalent (D.E.) is about 40 to 50. The dextrose equivalent value represents the quantity of reducing sugars formed, calculated as dextrose, and expressed as a percentage of the total solids. Following the acid conversion, fungal α-amylase is added to increase the D.E. to around 63. With the fungal α-amylase preparations available today, syrups with a high D.E. but low content of dextrose can be formed. These syrups are sweet and have low viscosity and good flavor.

Fungal amylase is also used extensively in the baking industry. Wheat flour does not naturally contain sufficient α-amylase for best baking quality. Fungal amylase from *A. oryzae* is normally added to flour to complement the α-amylase already present in cereals. The combined action of these two amylases supplies maltose for the panary fermentation. Maltose, or some form of fermentable sugar, is required for the yeast to generate carbon dioxide for raising the dough. Fungal α-amylase is preferred over either plant or bacterial α-amylase because it is heat labile and is inactivated early in the baking cycle before the starch gelatinizes (Rubenthaler et al. 1965). During baking, heat-stable amylases would act on the gelatinized starch and cause sticky, gummy loaves.

Glucoamylase

Most glucoamylase produced in the United States is from *Aspergillus awamori*. The mold is cultivated in a cornmeal medium by submerged fermentation (Smiley et al. 1964). The enzyme is extracellular and is recovered by filtration. The filtrate may be concentrated under a vacuum below 50°C to a thick syrup. For some purposes it is recovered by precipitation with alcohol. The precipitate is dried in the presence of an inert carrier.

In Japan glucoamylase is made from certain *Rhizopus* strains grown on wheat bran. The enzyme is extracted from the fermented bran with water. It has been claimed that the glucoamylase from *Rhizopus* contains less α-amylase and transferase than glucoamylase from *A. awamori*. Because *Rhizopus* glucoamylase is less stable on storage and cannot be produced by submerged fermentation, it has not been used extensively in the United States.

Glucoamylase is widely used in the corn wet milling industry to produce high-dextrose syrups from which crystalline corn sugar is prepared. Some glucoamylase is used in the baking industry. Generally, it accompanies the fungal α-amylase employed by bakers. However, it does have special benefit, since it forms glucose which is readily metabolized by yeast and probably accounts for the improved gassing and loaf volume. Pomeranz et al. (1964) and a group at Kansas State University have described the role of glucoamylase in bread manufacture.

Pectic Enzymes

Many fungi produce pectic enzymes, also referred to as pectinolytic or pectolytic enzymes. The enzyme as produced is actually a complicated mixture of enzymes. The predominant ones are (1) pectin methylesterase, also known as pectin esterase and pectase, which catalyzes the hydrolysis of the methyl ester groups in pectic substances with the formation of methyl alcohol and polygalacturonic acid; and (2) polygalacturonase, also known as pectinase, which degrades pectinic or pectic acid by hydrolysis of the glucosidic linkages of polygalacturonic acid to form smaller polygalacturonates which are water soluble (Endo 1964A,B). Commercial fungal pectinase consists predominantly of polygalacturonase.

Fungi used for commercial pectinase production are generally strains of the *Aspergillus niger* group. In Japan, *Sclerotinia libertiana* is used. Also, the Deuteromycete *Coniothyrium diplodiella* is a good source of pectic enzymes (Endo 1963).

Pectic enzymes are used for clarification of fruit juice, especially apple and grape juice (Aitken 1961; Fuleki and Hope 1964; Endo 1965). Both of these juices contain colloidal polygalacturonates which the enzymes degrade, causing the colloids to coalesce so they can be filtered out. The pectinase not only clarifies the juice but also increases the yield of juice from the crushed fruit.

Fungal pectic enzymes also have been found very useful in clarification of wines (Cruess *et al.* 1955). In modern practice the enzyme is added to fruit during crushing, which not only clarifies but also increases the yield of juice by as much as 10%. In making red wines, grapes are often crushed warm to extract more color. Unfortunately, this also extracts pectin and makes subsequent clarification more difficult unless fungal pectinases are employed to clarify the must.

Fungal pectic enzymes are also used in processing green coffee beans. The coffee cherry contains a mucilaginous layer surrounding the bean which must be removed before the bean can be dried. Natural fermentation to solubilize the mucilage requires 24 to 48 hr and is often accompanied by undesirable microbial growth which lowers the bean quality. By using fungal pectinases the "fermentation" time is reduced to about an hour and spoilage by microorganisms is forestalled.

Ishii and Yokotsuka (1972) have proposed using pectin transeliminase from *Aspergillus sojae* or *A. japonicus* to clarify fruit juice. Pectin transeliminase solubilizes the polygalacturonate without hydrolyzing the methyl ester groups. Consequently, the juice will contain little or no methanol.

Naringinase

Commercial preparations of naringinase are prepared from cultures of *A. niger*. This enzyme acts on the bitter principle in grapefruit, naringin, to form nonbitter derivatives (Griffiths and Lime 1959). The enzyme actually consists of a rhamnosidase and a β-glucosidase. Generally the glucosidase is inhibited by glucose naturally present in grapefruit, so that only splitting of rhamnose from naringin occurs. The resulting glucoside, prunin, is not bitter. In addition to naringinase, *A. niger* also produces pectinase which is undesirable in production of grapefruit juice because it destroys the cloudy appearance normally associated with this product. Therefore, the pectinase must be inactivated in some manner, such as by treatment of the enzyme preparation with urea (Thomas *et al.* 1958).

Invertase

Invertase production is widely distributed among the fungi. Commercial invertase is prepared from *Saccharomyces* yeast strains (Cochrane 1961). The enzyme is found in close association with the yeast cell; in order to free it, the plasmolyzed cells must be treated with a proteolytic enzyme such as papain. Invertase is used to make soft-centered candies by mixing it with the cast cream consisting of sucrose and other ingredients (Janssen 1962, 1963). After coating, the candies are held for 1 to 2 weeks to allow the invertase to reduce the sucrose to invert sugar, which dissolves in the available water and accounts for the liquid center.

Invertase is employed for making artificial honey, for manufacture of liqueurs, frozen desserts, and confections where high sucrose concentrations would tend to crystallize unless converted by invertase to the much more soluble invert sugar.

α-Galactosidase

Like invertase, α-galactosidase is widely distributed in fungi. There are two possible food applications for α-galactosidase. First, it has potential use in sugar refining. Sugar beet juice contains significant quantities of raffinose. Raffinose interferes with sucrose crystal formation from concentrated juice. By treating the juice with α-galactosidase the raffinose is degraded to sucrose and galactose, resulting in increased yield of sucrose crystals. For this purpose the enzyme must be free of invertase for obvious reasons. Fortunately, several fungi do produce significant levels of α-galactosidase free of invertase. *Mortiella vinacae* (Suzuki *et al.* 1969) and *Corticium rolfsii* (Kaji and Ichimi 1969) are examples of organisms which produce α-galactosidase with only minor amounts of invertase. The second possible application for α-galactosidase in foods is treatment of soybean "milk" to remove stachyose and raffinose (Sugimoto and VanBuren 1970). These two oligosaccharides have been implicated as causal factors of flatulence associated with consumption of soybeans (Rackis *et al.* 1970). In this instance the presence of invertase is not detrimental and fungal enzyme preparations from A. niger that are commonly used in food applications can be employed (Smiley et al. 1976).

In both of the above applications the α-galactosidase must be at least partially purified. Because this significantly increases enzyme cost, attempts have been made to conserve enzyme by immobilization on inert supports. Reynolds (1974) immobilized a bacterial α-galactosidase on nylon for the purpose of removing raffinose from sugar beet juice. Smiley et al. (1976) used ultrafiltration membranes in the form of hollow fibers to remove oligosaccharides from soybean milk. The advantage of this method is that the soy proteins are unable to diffuse to the enzyme site and are, therefore, unaffected by any contaminating protease that may be present.

Lactase

Lactose, though available in large quantities in cheese whey, has not found favor as a sugar for food use. It has low solubility which limits quantities that can be added to foods without crystallization. The presence of lactose crystals results in sandy or grainy products. Lactose is not very sweet; this also limits its usefulness. Enzymes have long been known which will effectively hydrolyze lactose to glucose and galactose. These sugars are considerably sweeter and much more soluble than lactose. The high cost of making lactase, coupled with the high cost of treatment, has effectively prevented its use. In the past, the most economical solution to whey disposal was simply to dump it into rivers and streams. With the present emphasis on environmental quality, this manner of disposal will no longer be tolerated. Consequently, new emphasis is being placed on utilizing whey for food use, and lactase can be expected to play a central role in this endeavor.

Lactase is produced by many kinds of microorganisms. Fungal lactase from A. oryzae and A. niger is well-suited for hydrolysis of whey because the optimum pH range is 4.5 to 5.5, near the normal pH of cheese whey. Pomeranz (1964) has shown that either spray-dried cheese whey or skim milk powder can be used to advantage in bread baking if a fungal lactase is used. The lactose content of these supplements replaces the need for sucrose added to the dough. The glucose resulting from lactose hydrolysis is utilized by the yeast to form carbon dioxide. The galactose, which is not fermented by bakers' yeast, reacts with the wheat proteins during baking to give an excellent crust color. Production of lactase by Kluyveromyces fragilis on cheese whey is described in Chap. 6.

Very acceptable dairy products can be made with lactase-hydrolyzed whole and skim milk (Thompson and Brower 1974; Thompson and Gyuricsek 1974). For this purpose a commercial lactase from *Saccharomyces (Kluyveromyces) lactis* is used because its optimum pH is near that of milk. Lactase-hydrolyzed milk has been used to make fluid whole milk, milk concentrates, ice cream, spray-dried whole and skim milk, yogurt, cottage cheese, Cheddar cheese and buttermilk. Not only are the lactase-hydrolyzed products acceptable organoleptically, but they can also be consumed by lactose-intolerant individuals (Guy *et al.* 1974). Cheddar cheese made from lactase-hydrolyzed milk ripens faster than cheese from normal milk. This is an important economic consideration for use of lactase-hydrolyzed milk (Thompson and Brower 1974).

Protease

Proteases, like amylases, have a variety of applications. Fungal proteases compete with plant and animal proteases for applications in food processing. In the Orient, proteolytic strains of *A. oryzae* are grown on rice or soybeans to produce koji; this koji is added to vats containing roasted soybeans, cereal grains, and salt brine to make soy sauce. *A. oryzae* protease is also used entensively in bread and cracker formulations where a particularly pliable, extensible dough is required. Addition of protease to doughs also permits sizable reduction in mix times (Coles 1952; Waldt 1965). This is possible because the doughs with protease can be slightly undermixed. The continued action of the enzyme beyond the mixing stage will result in doughs with comparable handling properties to fully mixed doughs without added protease.

Fungal protease is also useful as a chillproofing agent for beer, although normally papain is the enzyme of choice for this application. Fungal proteases have been tested for meat (beef) tenderization and have been found to be particularly effective on muscle fiber but show little action on connective tissue (collagen). Plant proteases (papain, bromelain, and ficin) readily attack connective tissue but show less action on muscle fiber. Therefore, meat tenderizer formulations are mixtures of proteases designed to tenderize particular cuts of beef (Underkofler 1959).

Rennin is traditionally a protease of calf stomach origin and is used universally in cheese manufacture. A growing shortage of calf stomachs has limited the supply of rennin and has spurred search for substitutes. Rennin is peculiar in that after the curd is formed, there is no further action on the casein. Many proteolytic enzymes will form acceptable curd but continuing proteolytic action is detrimental to the cheesemaking process. Some success has been obtained in duplicating calf rennin with enzymes from *Mucor* spp. Commercial rennin substitutes are now offered which are prepared from *Mucor pusillus*,

M. miehei, and Endothia parasitica. While enzymes from these fungi offer promise as a rennin substitute, they are still not completely satisfactory from the point of view of flavor and texture (Scott 1973). Acceptable cheese has been made by using mixtures of fungal and calf rennin. Perhaps the future of fungal rennins will be in extending rather than in substituting for calf rennet (mixture of rennins). Sardinas (1976) recently reviewed calf rennet substitutes, giving sources of higher plant, bacterial, and fungal milk-coagulating enzymes.

Glucose Oxidase

Glucose oxidase can be applied to food processing to remove traces of glucose or dissolved oxygen. For instance, if glucose is not removed from commercially dried egg whites, the product will darken on storage due to interaction between the sugar and protein. This nonenzymatic browning results in off-flavors and reduced protein solubility. Also there are losses in whipping properties and foam stability. These latter properties are most important in prepared cake mixes which depend on egg albumin as a functional ingredient.

To remove the glucose, glucose oxidase plus catalase is added to the egg whites (Baldwin et al. 1953). Since an excess of oxygen is required to remove all the glucose, hydrogen peroxide is also added. The catalase breaks down hydrogen peroxide, releasing oxygen. The glucose oxidase then utilizes the oxygen to oxidize glucose to gluconic acid and more hydrogen peroxide. Gluconic acid is not detrimental to the dried egg whites. The procedure can also be used for dried egg yolks as well as whole dried eggs.

Glucose oxidase can be used to remove oxygen from beverages such as white wines, beer, and fruit juices. It is necessary to have excess glucose present to remove all traces of oxygen so that with very dry wines it has been found necessary to add glucose to accomplish removal of oxygen (Ough 1960). It is desirable to remove oxygen from white wine to prevent later color formation and flavor deterioration (Yang 1955).

Most commercial glucose oxidase is isolated from A. niger fermentation broths (Swoboda and Massey 1965). One major supplier of glucose oxidase formerly isolated the enzyme from mother liquors of the gluconic acid fermentation. To be useful in food applications, the enzyme must be highly purified. For instance, the presence of protease in glucose oxidase preparations used for desugaring egg white would destroy the albumin. Likewise, in removing oxygen from mayonnaise (Bloom et al. 1956) the presence of amylase in the glucose oxidase would affect the starch that is used to give the product proper body.

Cellulase

Up to now, cellulase has had very limited application in food processing. Its relatively slow action on native cellulose makes it difficult to fit into a food processing operation. Cellulase is composed of at least two enzymes. One, called C_1, is required for the hydrolysis of insoluble and crystalline cellulose and is the rate-limiting reaction. The other component, known as C_x, is probably, in reality, two β-glucanases. One is an endo enzyme that randomly splits β-1,4 linkages in cellulose chains while the other is an exoenzyme that attacks the cellulose chain from the nonreducing end. Commercial cellulases are of fungal origin. Traditionally, culture filtrates of *A. niger* fermentations were used as their source; *A. niger* cellulase is low in the C_1 component, so its action on insoluble cellulose is very slow. It does work well on amorphous and soluble forms of cellulose such as carboxymethyl cellulose. A more practical source of cellulase is derived from mutant strains of *Trichoderma viride*. This organism gives rise to relatively large quantities of C_1 enzyme and is, therefore, more effective on crystalline cellulose.

Considerable interest has been manifested in recent years in using cellulase to make glucose from the huge amounts of cellulosic waste available annually. The glucose could be used as a source of food and feed as well as a chemical feedstock source for a variety of chemicals now obtained from petroleum. Many basic problems must be solved before cellulase technology will become a major factor as a source of either food or chemicals. A report by Spano *et al.* (1975) documents the present status of enzymatic hydrolysis of cellulosic waste to glucose.

Lipase

Fungal lipase is not generally produced and added *per se* to foods. Nevertheless, mention should be made of lipase because of its importance in cheese ripening. Italian cheeses such as provolone and Romano depend on lipase from kid rennet paste to provide the proper flavor. Lipase from other sources such as fungal lipase cannot substitute for kid (or sheep) rennet paste. Likewise, rennet extract will not substitute for the paste. Several other cheeses such as Stilton, Gorgonzola, Roquefort, and blue cheese depend on the lipase of *Penicillium roqueforti* for proper flavor development. This is, however, *in situ* action resulting from growth of the mold in the ripening cheese curd.

AMINO ACIDS

Many microorganisms for one reason or another may excrete free amino acids into their growth medium. Advantage is taken of this property to prepare amino acids for food, feed, and therapeutic purposes. Fungi have not generally been employed commercially for amino acid production, except for lysine. Lysine can be produced in good yields (2.0 g/liter) from a glucose-urea-salts medium by strains of *Ustilago maydis* and *Gliocladium* spp. (Dulaney 1957; Richards and Haskins 1957; Haskins and Spencer 1959). In addition to lysine, the fermentation broth contains glutamic acid and arginine.

Lysine can also be made from α-aminoadipic acid by *Candida utilis* (Sagisaka and Shimura 1957). Broquist *et al.* (1961) studied lysine production by *Saccharomyces cerevisiae* as well as by *C. utilis*. For highest yields a precursor had to be added to a corn steep liquor medium. Under the proper conditions about 70% of the precursor was converted to L-lysine on a molar basis. In addition, the lysine adheres tightly to the yeast cells so that a lysine-enriched yeast can be readily prepared.

Lysine is used to enrich wheat flour for baking purposes. It can also be added to polished rice to improve the nutritive value of this widely used grain.

In recent years immobilized amino-acylase from *A. oryzae* has been used to prepare L-amino acids from DL-acylamino acids (Tosa *et al.* 1967). The enzyme is ionically bound to DEAE-Sephadex. Acylamino acids are put through a column containing the enzyme complex where the L-acylamino acid is deacylated to the L-amino acid. The D-form, which is not affected by the enzyme, is racemized chemically and the resulting racemate is recycled. L-Methionine has been made extensively by this procedure in Japan. As an essential amino acid, methionine is used to reinforce vegetable proteins in foods and feeds. However, in the United States synthetic DL-methionine is used, even though twice as much DL as L must be added to give the same nutritional value.

PIGMENTS AND VITAMINS

Fungi are used for fermentative production of β-carotene and riboflavin. β-Carotene is produced by + and - strains of *Choanephora cucurbitarum*, *C. conjuncta*, *Blakeslea trispora*, and *B. circinans*. The medium is rather complicated and contains in addition to cornmeal and cottonseed meal such additives as deodorized kerosene, citrus molasses, and a vegetable oil (Ciegler *et al.* 1963). Yields of β-carotene of about 1.0 g/liter can be expected. β-Carotene is not being produced by fermentation since, at the present time, synthetic β-carotene is more economical. β-Carotene is widely used as a coloring agent in foods. It is also a source of vitamin A in foods and feeds.

Riboflavin has been produced commercially by the Ascomycetes, *Eremothecium ashbyii* (Rudert 1945) and *Ashbya gossypii* (Malzahn *et al.* 1959). Yields as high as 5.0 g/liter have been obtained with *A. gossypii* in a medium consisting of corn steep liquor, animal stick liquor, collagen peptone, and vegetable oil. Continuous feeding of glucose was also necessary. Most of the riboflavin produced by these fungi was not purified. The fermentation broth was concentrated, dried, and used as a supplement for animal feeds. The high concentration of riboflavin in the fermentation medium made it easy to isolate and use as a food supplement. Currently most riboflavin is made by chemical synthesis.

The plant hormone gibberellin is made by the mold *Gibberella fujikuroi* (Stodola 1958). The fermentation is quite long and involved and is usually conducted in two stages. The ratios of carbon:nitrogen must be carefully controlled. Yields of around 1.0 g/liter can be obtained in 450 or 500 hr (Borrow *et al.* 1959). Gibberellin is used as a plant growth regulator for α-amylase in brewery malt and for improving crop yields of cotton, grapes, and celery.

FLAVORINGS OF FUNGAL ORIGIN

Nucleotides

Use.—Flavor potentiators have been used extensively for more than 15 years to enhance the natural flavor of the foodstuff to which they are added. Monosodium glutamate (MSG) in particular has an established place in the potentiator market. However, 5'-inosinate (inosine-5'-monophosphate, IMP) and 5'-guanylate (guanosine-5'-monophosphate, GMP) are useful in improving the flavor of many foods (Nelson and Richardson 1967). Furthermore, Titus (1964) reported that these potentiators are beneficial in soups, gravies, and bouillons at levels of 25 to 100 ppm. They are also useful in improving the flavor of many products containing hydrolyzed proteins and can be used to replace beef extracts in many foods (Kuninaka 1972). Kuninaka *et al.* (1964) reported a strong synergistic action between MSG and flavor nucleotides. For example:

$$10 \text{ g MSG} + 1 \text{ g IMP} = 55 \text{ g MSG}$$
$$10 \text{ g MSG} + 1 \text{ g GMP} = 209 \text{ g MSG}$$

Consequently, combinations of MSG and the nucleotides are used for home cooking, especially in the Orient.

Production.—Monosodium glutamate is produced by neutralization of glutamic acid which can be produced by yeasts and molds.

However, glutamic acid is primarily a fermentation product of the bacterium, *Micrococcus glutamicus*. In contrast, the nucleotides are produced by treating ribonucleic acid (RNA), primarly from yeasts, with nuclease from *Penicillium citrinum* (Kuninaka 1972). The enzyme cleaves the 5′-phosphodiester linkages of the nucleosides to free the flavorful 5′-nucleotides. Also, nuclease from *A. oryzae* and *Streptomyces griseus*, an actinomycete, has been used (Nelson and Richardson 1967).

Yeast Cells

Yeast cells can serve as flavor enhancers or extenders in soups, gravies, sauces, snacks, and cereals. Reportedly, they are excellent carriers for flavorings such as cheese, smoke, barbecue, fish, and chicken. In some formulated meat products use of yeast permits a 20 to 50% reduction in pepper and other spices (Anon. 1974).

Mycelia of Mushrooms and Other Fleshy Fungi

The discovery in the late 1940s that mycelia from fleshy fungi could be efficiently produced in submerged culture led researchers to explore the growth characteristics of several mushroom genera. By manipulating medium composition, temperature, aeration, and other growth factors, investigators have produced mycelia with mushroom-like flavor from *Agaricus bisporus, A. blazei, Coprinus comatus, Tricholoma nudum,* and *Lepiota rachodes.* Mycelia of *Morchella hortensis, M. crassipes,* and *M. esculenta* have been found to have a desirable flavor and odor. The United States Food and Drug Administration has approved mycelia of *Morchella* spp. for sale as "morel mushroom flavoring" (Litchfield 1967). At one time, this flavoring was commercially available for use in dehydrated soup sauce and gravy formulations under the trade name Powdered Morel Mushroom Flavorings. Litchfield (1967) stated that this product was competing successfully with dried mushrooms imported from Europe. However, production has since been discontinued according to LeDuy *et al.* (1974).

Blue Cheese

Nelson (1970) described a submerged fermentation procedure for the production of blue cheese flavor by *Penicillium roqueforti* on a milk-based medium. The resulting flavor contained 7 to 12 times the ketone content of good quality commercial blue cheese and exhibited an efficacy four times that of blue cheese. The product is used commercially in salad dressing, snacks, and party dips where a blue cheese flavor is desired.

FATS AND OILS

Plant seeds are the principal source of fats and oils for edible and technical purposes. Animal fats and fish oil are subsidiary sources. However, as the demand and price increase, it may become feasible to use fungi for fat production. According to Whitworth and Ratledge (1974), *Rhodotorula* and *Lipomyces* spp. of yeast are most favored for fat production; lipid contents range up to 60% of the cell dry weight. Prescott and Dunn (1959) have listed *Penicillium*, *Aspergillus*, *Fusarium*, *Geotrichum*, and *Paecilomyces* spp. of molds as potential fat producers.

The fat from molds in general has a higher content of polyunsaturated fatty acids than that from yeasts and for this reason may have more commercial value (Whitworth and Ratledge 1974). In all cases, about 80% of the lipid content of the cell is triglycerides. The remaining lipids are usually phospholipids, sterols, and some sterol esters (Thorpe and Ratledge 1972). Moreover, different fatty acid compositions can be attained by using different organisms. It also has been found that, when nitrogen is limited, the microbial cell will continue to consume carbon and excessive lipid will accumulate. This flexibility of control over fungal fermentations allows the quality of the lipid to be improved. However, the successful production of fats by fungi will depend upon the abundance of a cheap and easily available substrate.

MANNITOL

Mannitol is generally prepared as a by-product of sorbitol manufacture. It can, however, be made by fermentation of glucose with *Aspergillus candidus* (Smiley *et al.* 1967). The organism requires sodium nitrate, an organic nitrogen source such as yeast extract, and daily feeding of glucose to obtain maximum mannitol yields. If a constant glucose source is not supplied, the organism utilizes the mannitol present and yields will be lowered.

Mannitol is used as a humectant and bulking agent in foods. Specifically, it is used in soybean chewing gums, sugarless hard candy, and sugarless chocolate.

REFERENCES

ABBOTT, A. J., JR., CRAWFORD, D. B., and KELLEY, K. M. 1957. Method of producing angel food batter and cake, U.S. Pat. 2,781,268. Feb. 12.

ABE, S., FURUYA, A., SAITO, T., and TAKAYAMA, K. 1962. Method of producing L-malic acid by fermentation. U.S. Pat. 3,063,910. Nov. 13.

AITKEN, H. C. 1961. Apple juice. *In* Fruit and Vegetable Juice Processing Technology, 2nd Edition. D. K. Tressler and M. A. Joslyn (Editors). AVI Publishing Co., Westport, Conn.

ANON. 1974. New torula yeast has synergistic effect. Food Process. (Chicago) *35*, No. 10, 29-31.

BALDWIN, R. R., CAMPBELL, H. A., THIESSEN, R., and LORANT, G. J. 1953. The use of glucose oxidase in the processing of foods with special emphasis on the desugaring of egg white. Food Technol. *7*, 275-282.

BATTI, M. A., and SCHWEIGER, L. B. 1963. Process for the production of itaconic acid. U.S. Pat. 3,078,217. Feb. 19.

BERTRAM, S. H. 1951. Emulsifying agent. U.S. Pat. 2,552,706. May 15.

BLOM, R. H., PFEIFER, V. F., MOYER, A. J., TRAUFLER, D. H., CONWAY, H. F., CROCKER, C. K., FARISON, R. E., and HANNIBAL, D. V. 1952. Sodium gluconate production. Ind. Eng. Chem. *44*, 435-440.

BLOOM, J., SCOFIELD, G., and SCOTT, D. 1956. Oxygen removal prevents rancidity in mayonnaise. Food Packer *37*, No. 13, 16-17.

BORROW, A., JEFFERYS, E. G., and NIXON, I. S. 1959. Process of producing gibberellic acid by cultivation of *Gibberella fujikuroi*. U.S. Pat. 2,906,671. Sept. 29.

BROQUIST, H. P., STIFFEY, A. V., and ALBRECHT, A. M. 1961. Biosynthesis of lysine from α-ketoadipic acid and α-amino-adipic acid in yeast. Appl. Microbiol. *9*, 1-5.

BRYCE, W. W. 1975. The production of microbial enzymes and some applications. Inst. Food Sci. Technol. Proc. *8*, 75-81.

CALAM, C. T., OXFORD, A. E., and RAISTRICK, H. 1939. The biochemistry of micro-organisms. LXIII. Itaconic acid, a metabolic product of a strain of *Aspergillus terreus* Thom. Biochem. J. *33*, 1488-1495.

CEIGLER, A., LOGODA, A. A., SOHNS, V. E., HALL, H. H., and JACKSON, R. W. 1963. Beta-carotene production in 20-liter fermentors. Biotechnol. Bioeng. *5*, 109-121.

CHICHESTER, D. F. 1969. Tartaric acid. *In* Kirk-Othmer Encyclopedia of Chemical Technology, Vol. 19, 2nd Edition. A. Standen (Editor). John Wiley and Sons, New York.

CLARK, D. L., and LENTZ, E. P. 1963. Submerged citric acid fermentation of beet molasses in tank-type fermentors. Biotechnol. Bioeng. *5*, 193-199.

COCHRANE, A. L. 1961. Invertase: Its manufacture and uses. Soc. Chem. Ind. (London) Monograph *11*, 25-31.

COLES, D. 1952. Fungal enzymes in bread baking. Proc. 28th Annu. Meet. Am. Soc. Bakery Eng., Chicago.

CONRAD, L. J., and STILES, H. S. 1954. Method of improving the whipping properties of gelatin and gelatin containing products and the resulting products. U.S. Pat. 2,692,201. Oct. 19.

CRUESS, W. V., QUACCHIA, R., and ERICSON, K. 1955. Pectic enzymes in wine-making. Food Technol. *9*, 601-607.

CURRIE, J. N. 1917. The citric acid fermentation of *Aspergillus niger*. J. Biol. Chem. *31*, 15-37.

DALE, J. K., and LANGLOIS, D. P. 1940. Syrup and method of making same. U.S. Pat. 2,201,609. May 21.

DENAULT, L. J., and UNDERKOFLER, L. A. 1963. Conversion of starch by microbial enzymes for production of syrups and sugars. Cereal Chem. *40*, 618-629.

DMUCHOVSKY, B., and FRANZ, J. E. 1967. Maleic anhydride, maleic acid, and fumaric acid. *In* Kirk-Othmer Encyclopedia of Chemical Technology, Vol. 12, 2nd Edition. A. Standen (Editor). John Wiley and Sons, New York.

DOELGER, W. P., and PRESCOTT, S. C. 1934. Citric acid fermentation. Ind. Eng. Chem. 26, 1142-1149.

DULANEY, E. L. 1957. Formation of extracellular lysine by Ustilago maydis and Gliocladium sp. Can. J. Microbiol. 3, 467-476.

ENDO, A. 1963. Pectic enzymes of molds. V and VI. The fractionation of pectolytic enzymes of Coniothyrium diplodiella. Agric. Biol. Chem. 27, 741-750, 751-757.

ENDO, A. 1964A. Pectic enzymes of molds. VIII, IX, and X. Purification and properties of endo-polygalacturonase I, II, and III. Agric. Biol. Chem. 28, 535-542, 543-550, 551-558.

ENDO, A. 1964B. Pectic enzymes of molds. XI. Purification and properties of exo-polygalaturonase. Agric. Biol. Chem. 28, 639-645. XII. Purification and properties of pectinesterase. Agric. Biol. Chem. 28, 757-764.

ENDO, A. 1965. Pectic enzymes of molds. XIII. Clarification of apple juice by the joint action of purified pectolytic enzymes. Agric. Biol. Chem. 29, 129-136.

ERENTHAL, I., and BLOCK, G. J. 1962. High fermentable noncrystallizing syrup and the process of making same. U.S. Pat. 3,067,066. Dec. 4.

FELDBERG, C. 1959. Glucono-δ lactone. Cereal Sci. Today 4, No. 4, 96-99.

FOSTER, J. W. 1949. Chemical Activities of Fungi. Academic Press, New York.

FOSTER, J. W. 1954. Fumaric acid. In Industrial Fermentations, Vol. I. L. A. Underkofler and R. J. Hickey (Editors). Chemical Publishing Co., New York.

FUJIMAKI, M., and MITSUDA, H. 1972. Proc. Int. Symp. Conversion and Manufacture of Foodstuffs by Microorganisms. Saikon Publishing Co., Tokyo.

FUKUDA, H., SUZUKI, T., AKIYAMA, S., and SUMINO, Y. 1974. Method for producing citric acid. U.S. Pat. 3,799,840. March 26.

FULEKI, T., and HOPE, G. W. 1964. Effect of various treatments on yield and composition of blueberry juice. Food Technol. 18, 568-570.

FURUKAWA, T., and KANEYUKI, H. 1975. Method for production of citric acid. U.S. Pat. 3,902,965. Sept. 2.

GARDNER, W. H. 1966. Choosing an acidulant. In Food Acidulants. Allied Chemical Corp., New York.

GARDNER, W. H. 1968. Acidulants in food processing. In Handbook of Food Additives. T. E. Furia (Editor). Chemical Rubber Co., Cleveland, Ohio.

GASTROCK, E. A., and PORGES, N. 1938. Gluconic acid production on pilot-plant scale—Effect of variables on production by submerged mold growths. Ind. Eng. Chem. 30, 782-789.

GODIN, P. 1953. Studies on the tertiary metabolism of Penicillium brevicompactum. Biochim. Biophys. Acta 11, 114-118. (French)

GRIFFITHS, F. P., and LIME, B. J. 1959. Debittering of grapefruit products with naringinase. Food Technol. 13, 430-433.

GUY, E. J., TAMSMA, A., KNOTSON, A., and HOLSINGER, V. H. 1974. Lactase-treated milk provides base to develop products for lactose-intolerant populations. Food Prod. Dev. 8, No. 8, 50, 54, 56, 58, 60, 74.

HAMMOND, E. G., and DEANE, D. D. 1961. Cheesemaking process. U.S. Pat. 2,982,654. May 2.

HASKINS, R. H., and SPENCER, J. F. T. 1959. Production of lysine, arginine and glutamic acids. U.S. Pat. 2,902,409. Sept. 1.

HERRICK, H. T., and MAY, O.E. 1928. The production of gluconic acid by the Penicillium luteum-purpurogenum group. II. Some optimal conditions for acid formation. J. Biol. Chem. 77, 185-195.

HUNTER, A. R., and ROCK, J. K. 1960. Cold milk puddings and method of producing the same. U.S. Pat. 2,949,366. Aug. 16.

HUSTEDE, H., and SIEBERT, D. 1974. Production of citric acid from hydrocarbons. U.S. Pat. 3,843,465. Oct. 22.

IKENO, Y., MASUDA, M., TANNO, K., OOMORI, I., and TAKAHASHI, N. 1975. Citric acid production from various raw materials by yeasts. J. Ferment. Technol. 53, 752-756.

IRWIN, W. E., LOCKWOOD, L. B., and ZIENTY, M. F. 1967. Malic acid. In Kirk-Othmer Encyclopedia of Chemical Technology, Vol. 12, 2nd Edition. A. Standen (Editor). John Wiley and Sons, New York.

ISHII, S., and YOKOTSUKA, T. 1972. Clarification of fruit juice by pectin trans-eliminase. In Proc. Int. Symp. on Conversion and Manufacture of Foodstuffs by Microorganisms, Kyoto, December 5-9. Saikon Publishing Co., Tokyo.

JANSSEN, F. 1962. Composition and production of cordial cherries. Confect. Prod. 28, No. 1, 54, 56, 58, 60, 75.

JANSSEN, F. 1963. Invertase. An important ingredient for cream centers. Candy Ind. Confect. J. 121, No. 6, 30, 33.

KAJI, A., and ICHIMI, T. 1969. Production and properties of galactosidases from Corticium rolfsii. Appl. Microbiol. 18, 1036-1040.

KIHLBERG, R. 1972. The microbe as a source of food. Annu. Rev. Microbiol. 26, 427-466.

KINOSHITA, K. 1929. Formation of itaconic acid and mannitol by a new filamentous fungus. J. Chem. Soc. Jpn. 50, 583-593. (Japanese)

KUNINAKA, A. 1972. Production of nucleotides by microorganisms. In Proc. Int. Symp. on Conversion and Manufacture of Foodstuffs by Microorganisms, Kyoto, December 5-9. Saikon Publishing Co., Tokyo.

LEDUY, A., KOSARIC, N., and ZAJIC, J. E. 1974. Morel mushroom mycelium growth in waste sulfite liquors as source of protein and flavoring. Can. Inst. Food Sci. Technol. J. 7, 44-50.

LITCHFIELD, J. H. 1967. Submerged culture of mushroom mycelium. In Microbial Technology. H. J. Peppler (Editor). Reinhold Publishing Corp., New York.

LOCKWOOD, L. B., and NELSON, G. E. N. 1946. Some factors affecting the production of itaconic acid by Aspergillus terreus in agitated cultures. Arch. Biochem. 10, 365-374.

LOCKWOOD, L. B., RAPER, K. B., MOYER, A. J., and COGHILL, R. D. 1945. The pro-duction and characterization of ultraviolet-induced mutations in Aspergillus terreus. III. Biochemical characteristics of the mutations. Am. J. Bot. 32, 214-217.

LOCKWOOD, L. B., and REEVES, M. D. 1945. Some factors affecting the production of itaconic acid by Aspergillus terreus. Arch. Biochem. 6, 455-469.

LOCKWOOD, L. B., and SCHWEIGER, L. B. 1967. Citric and itaconic acid fermentation. In Microbial Technology. H. J. Peppler (Editor). Reinhold Publishing Corp., New York.

MALZAHN, R. C., PHILLIPS, R. F., and HANSON, A. M. 1959. Riboflavin process. U.S. Pat. 2,876,169. March 3.

MAY, O. E., and HERRICK, H. T. 1932. Production of organic acids from carbohydrates by fermentation. U.S. Dep. Agric. Circ. 216.

MAY, O. E., HERRICK, H. T., MOYER, A. J., and HELLBACH, R. 1929. Semi-plant scale production of gluconic acid by mold fermentation. Ind. Eng. Chem. 21, 1198-1203.

MAY, O. E., HERRICK, H. T., MOYER, A. J., and WELLS, P. A. 1934. Gluconic acid. Production of submerged mold growths under increased air pressure. Ind. Eng. Chem. 26, 575-578.

MOLLIARD, M. 1922. A new acid fermentation product from Sterigmatocystis nigra. C. R. Acad. Sci. 174, 881-883. (French)

MOYER, A. J. 1953. Effect of alcohols on the mycological production of citric acid in surface and submerged culture. I. Nature of the alcohol effect. Appl. Microbiol. 1, 1-7.

MOYER, A. J., and COGHILL, R. D. 1945. The laboratory-scale production of itaconic acid by *Aspergillus terreus*. Arch. Biochem. 7, 167-183.

MOYER, A. J., WELLS, P. A., STUBBS, J. J., HERRICK, H. T., and MAY, O. E. 1937. Gluconic acid production. Development of inoculum and composition of fermentation solution for gluconic acid production by submerged mold growths under increased air pressure. Ind. Eng. Chem. 29, 777-781.

NELSON, J. H. 1970. Production of blue cheese flavor *via* submerged fermentation by *Penicillium roquefortii*. J. Agric. Food Chem. 18, 567-569.

NELSON, J. H., and RICHARDSON, C. H. 1967. Molds in flavor production. *In* Microbial Technology. H. J. Peppler (Editor). Reinhold Publishing Corp., New York.

NUBEL, R. C., and RATAJAK, E. J. 1962. Process for producing itaconic acid. U.S. Pat. 3,044,941. July 17.

OUGH, C. S. 1960. The use of glucose oxidase in dry wines. Wein Rebe 10, 14-23. (German)

PASTERNACK, R., and GILES, W. R. 1934. Process for the preparation of gluconic acid and its lactones. U.S. Pat. 1,942,660. Jan. 9.

PEDERSON, C. S. 1971. Microbiology of Food Fermentations. AVI Publishing Co., Westport, Conn.

PFEIFER, V. F., NELSON, G. E. N., VOJNOVICH, C., and LOCKWOOD, L. B. 1953. Production of itaconic acid. U.S. Pat. 2,657,173. Oct. 27.

POMERANZ, Y. 1964. Lactase (β-D-galactosidase). I. Occurrence and properties. Food Technol. 18, 682-687. II. Possibilities in the food industries. Food Technol. 18, 690-697.

POMERANZ, Y., RUBENTHALER, G. L., and FINNEY, K. F. 1964. Use of amyloglucosidase in breadmaking. Food Technol. 18, 138-140.

PRAY, L. W., and POWERS, J. J. 1966. Acidification of canned tomatoes. Food Technol. 20, 87-91.

PRESCOTT, S. C., and DUNN, C. G. 1959. Industrial Microbiology, 3rd Edition. McGraw-Hill Book Co., New York.

RACKIS, J. J., HONIG, D. H., SESSA, D. J., and STEGGERDA, F. R. 1970. Flavor and flatulence factors in soybean protein products. J. Agric. Food Chem. 18, 977-982.

REED, G. 1975. Enzymes in Food Processing, 2nd Edition. Academic Press, New York.

REYNOLDS, J. H. 1974. Immobilized α-galactosidase continuous flow reactor. Biotechnol. Bioeng. 16, 135-137.

RHODES, R. A., MOYER, A. J., SMITH, M. L., and KELLY, S. E. 1959. Production of fumaric acid by *Rhizopus arrhizus*. Appl. Microbiol. 7, 74-80.

RICHARDS, M., and HASKINS, R. J. 1957. Extracellular lysine production by various fungi. Can. J. Microbiol. 3, 543-546.

ROBERTS, F. F. 1973. Fermentation process for the production of citric acid. U.S. Pat. 3,717,549. Feb. 20.

RUBENTHALER, G., FINNEY, K. F., and POMERANZ, Y. 1965. Effects of loaf volume and bread characteristics of α-amylases from cereal, fungal, and bacterial sources. Food Technol. 19, 239-241.

RUDERT, F. J. 1945. Production of riboflavin by biochemical methods. U.S. Pat. 2,374,503. April 24.

SAGISAKA, S., and SHIMURA, K. 1957. Lysine biosynthesis. I. Lysine formation in yeast cells. J. Agric. Chem. Soc. Jpn. 31, 110-114. (Japanese)

SAIR, L. 1961. Production of meat emulsions. U.S. Pat. 2,992,116. July 11.

SARDINAS, J. L. 1976. Calf rennet substitutes. Process Biochem. 11, No. 4, 10, 12-14, 16, 17.

SCHWEIGER, L. B. 1961. Production of citric acid by fermentation. U.S. Pat. 2,970,084. Jan. 31.

SCOTT, R. 1973. Rennet and rennet substitutes. Process Biochem. 8, No. 12, 10-14.

SHU, P., and JOHNSON, M. J. 1947. Effect of the composition of the sporulation medium on citric acid production by Aspergillus niger in submerged culture. J. Bacteriol. 54, 161-167.

SHU, P., and JOHNSON, M. J. 1948A. The interdependence of medium constituents in citric acid production by submerged fermentation. J. Bacteriol. 56, 577-585.

SHU, P., and JOHNSON, M. J. 1948B. Citric acid production by submerged fermentation with Aspergillus niger. Ind. Eng. Chem. 40, 1202-1205.

SMILEY, K. L., CADMUS, M. C., HENSLEY, D. E., and LOGODA, A. A. 1964. High-potency amyloglucosidase-producing mold of the Aspergillus niger group. Appl. Microbiol. 12, 455.

SMILEY, K. L., CADMUS, M. C., and LIEPINS, P. 1967. Biosynthesis of D-mannitol from D-glucose by Aspergillus candidus. Biotechnol. Bioeng. 9, 365-374.

SMILEY, K. L., HENSLEY, D. E., and GASDORF, H. G. 1976. Alpha-galactosidase production and use in a hollow-fiber reactor. Appl. Microbiol. 31, 615-617.

SMITH, J. E., and BERRY, D. R. 1975. The Filamentous Fungi, Vol. I. Industrial Mycology. John Wiley and Sons, New York.

SPANO, L. A., MEDEIROS, J., and MANDELS, M. 1975. Enzymatic hydrolysis of cellulosic wastes to glucose. U. S. Army Natick Dev. Cent., Natick, Mass.

STODOLA, F. H. 1958. Source book on gibberellin, 1828-1957. U.S. Dep. Agric. Res. Serv. 71-11.

SUGIMOTO, H., and VANBUREN, J. P. 1970. Removal of oligosaccharides from soy milk by an enzyme from Aspergillus saitoi. J. Food Sci 35, 655-660.

SUZUKI, H. Y., OZAWA, Y., OOTA, H., and YASHIDA, H. 1969. Studies on the decomposition of raffinose by α-galactosidase of mold. I. α-Galactosidase formation and hydrolysis of raffinose by the enzyme preparation. Agric. Biol. Chem. (Japan) 33, 501-513.

SWOBODA, B. E. P., and MASSEY, V. 1965. Purification and properties of the glucose oxidase from Aspergillus niger. J. Biol. Chem. 240, 2209-2215.

TAKAMINE, J. 1894. Process of making diastatic enzyme. U.S. Pat. 525,820 and 525,823. Sept. 11.

TAKAYAMA, K., and TOMIYAMA, T. 1974. Process for producing citric acid. U.S. Pat. 3,809,611. May 7.

TERUI, G. 1972. Fermentation Technology Today. Proc. 4th Int. Ferm. Symp., Kyoto. Soc. Fermentation Technol., Japan, Osaka.

THOMAS, D. W., SMYTHE, C. V., and LABBEE, M. D. 1958. Enzymatic hydrolysis of naringinin, the bitter principle in grapefruit. Food Res. 23, 591-598.

THOMPSON, M.P., and BROWER, D. P. 1974. Manufacture of Cheddar cheese from hydrolyzed lactose milks. J. Dairy Sci. 56, 598. (Abstr.)

THOMPSON, M. P., and GYURICSEK, D. M. 1974. Manufacture of yogurt, buttermilk, and cottage cheese from hydrolyzed lactose milks. J. Dairy Sci. 57, 584. (Abstr.)

THORPE, R. F., and RATLEDGE, C. 1972. Fatty acid distribution in triglycerides of yeast grown on glucose or n-alkanes. J. Gen. Microbiol. 72, 151-163.

TITUS, D. S. 1964. Symposium on Flavor Potentiation. Arthur D. Little, Cambridge, Mass.

TOSA, T., MORI, T., FUSE, N., and CHIBATA, I. 1967. Studies on continuous enzyme reactions. IV. Preparation of a DEAE-Sephadex-aminoacylase column and continuous optical resolution of acyl-DL-amino acids. Biotechnol. Bioeng. 9, 603-615.

TURI, P. 1969. Succinic acid. *In* Kirk-Othmer Encyclopedia of Chemical Technology, Vol. 19, 2nd Edition. A. Standen (Editor). John Wiley and Sons, New York.

UNDERKOFLER, L. A. 1959. Meat tenderizer. U.S. Pat. 2,904,442. Sept. 15.

UNDERKOFLER, L. A., DENAULT, L. J., and HOU, E. F. 1965. Enzymes in the starch industry. Staerke *17*, 179-184.

U.S. FOOD AND DRUG ADMINISTRATION. 1965. Food additives. *In* Code of Federal Regulations. Title 21, 121.101. U.S. Gov. Printing Office, Washington, D.C.

WALDT, L. M. 1965. Fungal enzymes: Their role in continuous process bread. Cereal Sci. Today *10*, 447-450.

WALLEN, L. L., STODOLA, F. H., and JACKSON, R. W. 1959. Type Reactions in Fermentation Chemistry. U.S. Dep. Agric. Res. Serv. ARS-71-13.

WARD, G. E. 1967. Production of gluconic acid, glucose oxidase, fructose, and sorbose. *In* Microbial Technology. H. J. Peppler (Editor). Reinhold Publishing Corp., New York.

WELLS, P. A., MOYER, A. J., STUBBS, J. J., HERRICK, H. T., and MAY, O. E. 1937. Gluconic acid production. Effect of pressure, air flow, and agitation on gluconic acid production by submerged mold growths. Ind. Eng. Chem. *29*, 653-656.

WHITWORTH, D. A., and RATLEDGE, C. 1974. Microorganisms as a potential source of oils and fats. Process Biochem. *9*, No. 9, 14-22.

WOODWARD, J. S., SNELL, R. L., and NICHOLLS, R. S. 1949. Conditioning molasses and the like for production of citric acid by fermentation. U.S. Pat. 2,492,673. Dec. 27.

YANG, H. Y. 1955. Stabilizing apple wine with glucose oxidase. Food Res. *20*, 42-46.

14

Mycotoxins

N. D. Davis and U. L. Diener

F ungi benefit man in many ways. They play a part in the production of cheese, bread, antibiotics, vitamins, enzymes, glycerol, fats, livestock feeds from fermentation by-products, and facilitate alcohol and organic acid fermentations (Gray 1959). Mushrooms are a food source. Fungal activities that are detrimental to the food and beverage industry include plant diseases, food spoilage, and the production of mycotoxins, which is the subject of this chapter.

Certain fungi (molds) produce chemical substances that cause toxic symptoms when food containing them is ingested by man or animals. These chemicals are referred to as mycotoxins and their toxicity syndromes in animals are known as mycotoxicoses. It is inescapable that human and animal foods and feedstuffs are universally exposed to fungal invasion. This is true from planting of the crop through harvesting, transportation, and storage, even into the grocery store, restaurant, and home where it awaits final use by the consumer. There is a constant possibility that temperature and moisture may become favorable for fungal growth. Such growth is not always conspicuous or even visible to the untrained eye. It is also noteworthy that the very young are most susceptible to mycotoxins and that their diet for the first year or so consists of a very high concentration of cereals, grains, milk, etc.—the very foods in which mycotoxins most readily occur. The term mycotoxin as used in this chapter excludes alcohol, mushroom poisons, and various poisonous gases such as arsene, which are also known products of fungal metabolism. Mycotoxins discussed in this chapter are significant since they occur in rice, corn, wheat, barley, rye, other grains, peanuts, and cottonseed, which make up in excess of 90% of our food. Some mycotoxins are acutely toxic to laboratory animals with LD_{50}'s as low as 5 mg/kg of body weight. They cause a wide variety of disorders, depending upon the nature of the toxin, the dosage, the time span over which it is ingested, the age, species and condition of the animal, and the composition of the diet. A specific toxin may affect a wide range of experimental and domesticated animals including man, whereas another toxin is specifically toxic to only one or a few species. With respect to chronic toxicity syndromes, various mycotoxins have been found to be carcinogenic, teratogenic, tremorgenic, hemorrhagic, or dermatitic; most are hepatotoxins, nephrotoxins, or neurotoxins. More subtle chronic effects are merely the inhibition of protein synthesis and/or growth, with no conspicuous syndrome to enable doctors or veterinarians to diagnose the disease.

HISTORICAL

Ergotism or "holy fire" is the earliest and best known mycotoxicosis with respect to recorded effects on man. In the Middle Ages large-scale epidemics occurred among peasants in western and central Europe, where rye was used for bread. German and French scientists established that ergot poisoning resulted from the ingestion of bread prepared from rye parasitized by the fungus Claviceps purpurea(Groger 1972). However, it was not until the 1930s that the alkaloids responsible for the biological effects of ergot were isolated and identified. The last major incidence of ergotism in the United States involving humans was in 1825; outbreaks occurred in 1926 and 1927 in Russia and in 1928 in England.

Stachybotryotoxicosis is the result of ingestion of feeds (hay, straw, fodder, etc.) infected by the fungus Stachybotrys atra. This syndrome was first reported from Russia in 1931, where it severely affected horses causing a high degree of mortality (Forgacs 1972). Various domestic and experimental animals as well as humans were determined to be susceptible to this mycotoxin.

Alimentary toxic aleukia (ATA) appears to be the most recent and only mycotoxicosis in grains other than ergotism that has caused many human deaths during epidemics of the disease. An outbreak in 1932 resulted in mortality of 62% of those afflicted in one district in Russia. ATA was especially severe in Russia from 1942 to 1947, when in one district in 1944 up to 30% of the people were affected and mortality from 2 to 80% of those affected occurred in various districts.

ATA resulted from the ingestion of food processed from cereal grains, particularly proso millet, but also wheat and barley that had over-wintered in the field under snow and were infested with *Fusarium sporotrichioides* (Forgacs and Carll 1962; Joffe 1971).

Mycotoxicoses in rice have been suspected in Japan since 1891 (Kinosita and Shikata 1965), when it was shown that preparations from molds growing on stored rice were toxic to rabbits, guinea pigs, and frogs. The theory was advanced that beriberi was caused by a toxic substance elaborated by certain rice molds. Extensive research on rice mycotoxins has been conducted since 1940 by pioneering Japanese scientists. Several species of *Fusarium* have been associated with "red mold" or "scabby disease" of rice suspected to have caused various diseases and disorders in man as early as 1901. Toxigenic species of *Penicillium* associated with "yellowed rice toxins" have also been widely studied in Japan since 1954.

Early investigations of mycotoxins in the United States were con-cerned with outbreaks of moldy corn toxicosis in 1952 and 1957 that resulted in extensive losses of swine and cattle in Georgia and west Florida (Forgacs 1965). *Aspergillus flavus* and *Penicillium rubrum* were isolated from toxic corn and shown to be toxigenic to swine. Later it was shown that hepatitis-X syndrome in dogs was similar to that of moldy corn toxicosis in swine. Also, a hemorrhagic syndrome in poultry was recognized as being caused by moldy feeds (Forgacs and Carll 1962).

Facial eczema in sheep and cattle was observed in New Zealand prior to the turn of the century. Research impetus began following a severe outbreak in 1936. The mycotoxicosis was discovered to be caused by *Pithomyces chartarum (Sporodesmium bakeri)* growing on dead perennial ryegrass. Facial eczema in ruminants was later ob-served in Australia and in Florida.

Forgacs and Carll (1962) reviewed the status of mycotoxicoses at a time when few scientists were actively engaged in mycotoxin re-search. They concluded that "of the innumerable diseases to which man and beasts are heir, the mycotoxicoses are perhaps the most unfamiliar and the least investigated."

Recent interest in these neglected diseases evolved following the death of 100,000 turkey poults at 500 locations in England in 1960, which led to the discovery by British scientists of aflatoxin (a toxic metabolite of the fungus *A. flavus*) in the peanut meal fraction of the feed (Lancaster et al. 1961; Sargeant et al. 1961). *A. flavus* occurs worldwide in the soil and aflatoxin has been found on a wide variety of food crops grown in tropical and temperate climates. Since 1962, more than 2000 research papers on aflatoxin have been published, primarily due to the fact that aflatoxin was characterized as one of the most car-

cinogenic naturally occurring substances known. Research on aflatoxin has turned the spotlight on the total problem of mycotoxins in
human foods and animal feeds (Goldblatt 1969).

Alpert and Davidson (1969) suggested that fungal toxins are an
important cause of primary liver cancer in man in some parts of the
world. Kraybill and Shapiro (1969) concluded that the extensive data
on the pharmacological actions of mycotoxins in various animals is
"suggestive evidence from animal experimentation that provides a
strong case for cause and effect relationships between aflatoxins and
the incidence of diseases in man." Research by Joffe (1960), Scott
(1965), and Christensen et al. (1968) has shown that numerous fungi,
when grown on foodstuffs under favorable conditions, produce
metabolites toxic to one or more species of animals. Bamburg et al.
(1969), Hesseltine (1969), and Lillehoj et al. (1970), in independent
reviews on the chemical nature and biological effects of mycotoxins,
present summary tables of the chemical structures of over 30 toxic
metabolites, most of which have been elucidated since 1962. Ciegler et
al. (1971) and Kadis et al. (1971, 1972) published a comprehensive
review in three volumes, which is the most complete consideration of
mycotoxins to date. Purchase (1971), Campbell and Stoloff (1974), and
Frank (1974) have reviewed the implications of mycotoxins to human
health. Purchase (1974) and Rodricks (1976) have recently edited
books on mycotoxins and other fungal related food problems.

Martin et al. (1971), Saito et al. (1974), and Davis et al. (1975) have
found toxigenic fungi in foods intended for direct consumption by
humans, i.e., a variety of processed foods from grocery stores and
home refrigerators. Approximately one-third of the foods investigated
were contaminated with spores or mycelia of fungi that have been
demonstrated to be at least potentially toxigenic in the laboratory. Wu
et al. (1974), investigating aged cured meats, isolated numerous
Aspergillus and Penicillium spp. that have been shown to be toxigenic
under laboratory conditions. Although many fungi isolated from
foods have been shown to be toxigenic on media in the laboratory, in
very few instances has the fungus been grown on the kind of food from
which it was originally isolated and shown to form a mycotoxin in that
food. Also, in most cases the alleged toxins were not isolated and
identified. Thus, the toxicity of the foods to humans cannot even be
inferred. However, enough evidence has been accumulated to demonstrate that, until proved otherwise, moldy foods and beverages should
be considered dangerous and not consumed by man or animals. Also,
agricultural commodities that are to be processed as foods must be
grown, harvested, handled, stored, and processed by methods that
will prevent contamination and growth of fungi.

MYCOTOXINS OF CONTEMPORARY IMPORTANCE

Of the large number of toxic fungal metabolites that have been reported, this chapter describes those relatively few which in the judgment of the authors are or should be regarded as important in the United States with respect to production, harvest, storage, transportation, and consumption of food and beverages by humans and domesticated animals. Those mycotoxins included were judged to satisfy a set of criteria similar to those of Hesseltine (1976).

(1) The mycotoxin is judged capable of causing a disease situation in man.
(2) It can occur in agricultural commodities in nature.
(3) It is produced by fungi commonly found in the commodity in the field, in storage, etc.
(4) It is acutely or chronically toxic, i.e., it is carcinogenic, teratogenic, tremorgenic, estrogenic, hemorrhagic, etc.

The genera of fungi of greatest importance to humans and domesticated animals with respect to natural poisoning outbreaks are *Aspergillus, Penicillium,* and *Fusarium* spp. (Hesseltine 1976). A treatise on the taxonomy of toxigenic fungi will assist scientists in the identification of species of these and other genera (Booth and Morgan-Jones 1977). The general format followed here for each of the toxins discussed is, in order of presentation: name and description of the fungi involved; natural occurrence of fungi and mycotoxin; toxicology; chemistry; and bioproduction of the mycotoxin. Further descriptions of mycotoxigenic fungi and methods for detecting mycotoxins in foodstuffs are included in Chap. 1 and 15, respectively.

Aflatoxin

Of the known mycotoxins, the most important from the viewpoint of direct hazard to human health are the aflatoxins (Scott 1973). They are also the most extensively investigated mycotoxins. Goldblatt (1969) reviewed the history of turkey-X disease in England and the intriguing research that led to the discovery of the aflatoxins. Presently 18 aflatoxins are known, of which aflatoxins B_1, G_1 and M_1 are of primary importance in nature (Fig. 14.1).

The Fungi.— The aflatoxins are produced by *Aspergillus flavus* and the closely related *A. parasiticus* of the *A. flavus* group of aspergilli (Raper and Fennell 1965). The *A. flavus* group is a constituent of the microflora in air and soil and is associated with living and dead plants and animals throughout the world. *A. flavus* contributes extensively to the deterioration of stored grains and a wide variety of other agricultural commodities (Christensen 1974). *A. flavus* has conidial

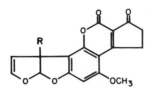

Aflatoxin B₁ R = H
Aflatoxin M₁ R = OH

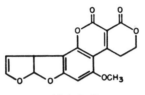

Aflatoxin G₁

FIG. 14.1. STRUCTURES OF THREE AFLATOXINS

heads that are pale to intense yellow or yellow-green when young, becoming deep green with age, not shifting to brown on Czapek's agar; sclerotia are produced in many strains; conidia are echinulate (ornamented, spiked); sterigmata are mostly double (two-tiered), and the heads are radiate or very loosely columnar. The conidial heads of *A. flavus* var. *columnaris* are columnar and bear sterigmata that are typically in a single series. *A. parasiticus* has radiate heads and uniseriate sterigmata. *A. flavus* generally retains its yellow-green or green color, whereas *A. parasiticus* gradually turns from grass green to ivy green to dark olive green, but not the brownish-green typical of other members of this group. *A. parasiticus* does not produce sclerotia or cleistothecia.

Natural Occurrence.—*A. flavus* will grow and produce aflatoxin on most agricultural commodities, many processed foods and beverages, and certain laboratory media. Aflatoxin can be a serious problem of peanuts, corn, cottonseed meal, rice, sorghum and other grains, Brazil nuts, pecans, and tree nuts in general (Stoloff 1976). Since aflatoxin and other mycotoxins occur in grains, beverages such as beer can also be contaminated with mycotoxins (Nip *et al.* 1975), although there is little documented evidence to this effect. *A. flavus* generally produces only aflatoxin B in the field, whereas *A. parasiticus* produces aflatoxins B and G.

Toxicology.—The effect of aflatoxins on animals is quite variable, depending on age, sex, species, nutritional condition of the animal, dosage level, frequency, and composition of the diet. Aflatoxins are mutagenic, carcinogenic, teratogenic, and acutely toxic to most experimental and domesticated animals and man. Sensitivity to the toxins varies greatly from species to species (Butler 1974). The organ primarily affected is the liver, but changes can be seen in most other organs (Butler 1969). Aflatoxins are acutely toxic with B_1 LD_{50} values of 5 to 7 mg/kg in rats, 1.4 mg/kg in guinea pigs, and 364 μg/kg in day-old ducklings. It is carcinogenic at 15 ppb in the diet of rats, whereas 4 to 8 ppb daily induced tumors in rainbow trout in 12 months and 0.8 ppb caused tumors in 20 months. Currently, 20 ppb is the maximum permissible level in agricultural commodities in the United States, but 15 ppb may become the limit in foods with the possibility of allowing higher levels in certain animal feeds.

Chemical Properties.—Aflatoxin B_1 (Fig. 14.1) has a molecular weight of 312 and an empirical formula of $C_{17}H_{12}O_6$. It decomposes without melting at 268° to 269°C. It is optically active $(\alpha)_D - 558$, fluoresces bright blue under long wave ultraviolet (UV) light and exhibits UV absorption maxima at 223, 265, and 362 nm ($\epsilon = 25,600$, 13,400 and 21,800). It chromatographs on silica gel, thin-layer chromatography (TLC) plates at approximately R_f 0.5 in chloroform-methanol (97:3). The compound is slightly soluble in water, soluble in most solvents of intermediate polarity, including alcohols and chloroform, and is insoluble in hexane. It can be extracted from most foodstuffs using chloroform-methanol (97:3) or 70% aqueous acetone (Assoc. Off. Anal. Chem. 1975). Detailed physical properties are described by Pons and Goldblatt (1969) and by Ciegler et al. (1971). Qualitative and quantitative analytical standards are available commercially and official methodology for analysis has been developed by the AOAC (1975) for a variety of commodities (see Chap. 15).

Bioproduction.—A. flavus will grow and produce aflatoxins on a wide variety of laboratory media and natural substances (Diener and Davis 1969). Yields of 2 mg/g have been obtained on natural substrates such as rice, shredded wheat, peanuts, corn, soybeans, and coconut. Optima for fungal growth and aflatoxin production are 2 to 3 weeks at 25° to 30°C and 88 to 95% relative humidity. Optimal substrate moisture and incubation time vary with the substrate. High surface area to volume ratios with good aeration enhance aflatoxin production where the substrate is in equilibrium with an appropriate relative humidity so that drying is prevented.

Aflatoxin yields of 200 to 300 mg/liter have been observed in stationary surface cultures on semisynthetic liquid media such as YES, which consists of 2% yeast extract and 20% sucrose (Davis et al. 1966). Adequate zinc (0.4 to 2 ppm) is essential for high yields and a mixture of organic nitrogen sources is preferred. Methods for large-scale production and purification of aflatoxins were reviewed by

Diener and Davis (1969).

Aflatoxins other than Aflatoxin B₁.—Of the 18 known aflatoxins, aflatoxin B_1 is the most important in terms of occurrence and toxicity. Aflatoxins G_1 and M_1 rank next in importance (Fig. 14.1). Aflatoxin G_1 appears greenish under UV light when compared to B_1 and appears on TLC plates at a slightly lower R_f value than B_1. G_1 also has slightly different chemical properties. Properties of G_1 and other aflatoxins have been reviewed (Detroy et al. 1971; Ciegler 1975). The LD_{50}'s of aflatoxins B_1, B_2, G_1, and G_2 for day-old ducklings are reported as 0.36, 0.78, 1.69 and 2.45 mg/kg, respectively (Carnaghan et al. 1963). Aflatoxin M_1 is 4-hydroxyaflatoxin B_1. It is found in the milk and urine of cows and other mammals fed toxic peanut meal. M_1 has approximately the same order of toxicity as B_1 and causes the same general effects in animals. In addition, M_1 appears to induce renal tubular necrosis, a lesion not commonly found following aflatoxin B_1 administration to nonruminants. Other important hydroxylated metabolites of aflatoxin B_1 are aflatoxins P and Q. These metabolites may be conjugated with various moieties and occur in water soluble portions of waste products and thus require modified techniques for extraction and analysis. For further discussion of aflatoxins in dairy products, see Chap. 6.

Ochratoxin A

Of several ochratoxins that have been identified, ochratoxin A (Fig. 14.2) appears to be the important toxic metabolite of *Aspergillus ochraceus*. It was first described by Van der Merwe et al. (1965A,B) and shown to be toxigenic to ducklings, mice, and rats when produced on a wide variety of foodstuffs. Ochratoxin has received much attention by mycotoxicologists, since it was discovered to occur in nature in a variety of products. Steyn (1971), Harwig (1974), and Nesheim (1976) have recently reviewed the literature on ochratoxin.

The Fungi.—The principal ochratoxin-producing fungi are *A. ochraceus, A. melleus, A. sulphureus, Penicillium viridicatum*, and *P. cyclopium*. The most important of the group are *A. ochraceus* and *P. viridicatum*. The first three fungi are members of the *A. ochraceus* group and will be briefly described here, while the two penicillia are described elsewhere. Detailed descriptions of the aspergilli are given by Raper and Fennell (1965).

Conidial heads of *A. ochraceus* are globose when young. Sterigmata occur in two tiers or series. Sclerotia are typically but not regularly produced. Conidial heads of *A. sulphureus* are pale pure yellow and sclerotia are cream to pale yellow, whereas conidia of *A. ochraceus* and *A. melleus* are dull yellow cream, buff, or ochraceous. Sclerotia are abundant in most strains of *A. melleus*, 400 to 500 μ in diameter, pure yellow when young, then brown. Those of *A. ochraceus* are

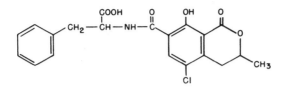

FIG. 14.2. STRUCTURE OF OCHRATOXIN A

vary from pink to purplish when mature. Other species of the group have mature sclerotia that are black, orange, white, or buff, while one does not produce sclerotia.

Natural Occurrence.—A. ochraceus is of fairly common occurrence in soil and agricultural commodities. It has been isolated from most cereal grains, white beans, peanuts, Brazil nuts, cottonseed, citrus fruits, and tobacco. Ochratoxin A has been found in barley, corn, oats, rye, wheat, white beans, mixed feed grains, peanuts, and coffee beans (Harwig 1974; Stoloff 1976).

Toxicology.—LD$_{50}$'s vary from 0.5 to 54 mg/kg of body weight, depending on species of animal. Oral LD$_{50}$ in day-old ducklings has been reported to be 150 μg/duckling and 116 to 135 μg/chick in 7-day-old chicks. Detailed effects of various levels of ochratoxin in swine have been reported by Krogh et al. (1974). Effects are often non-specific, but chronic symptoms include hepatic and renal lesions, failure or impairment. Also, necrosis of renal tubules and periportal liver cells and, in some species, fatty infiltration of the liver is common. Enteritis may also occur along with visceral gout.

Chemical Properties.—Ochratoxin A has a molecular weight of approximately 404 and an empirical formula of $C_{20}H_{18}ClNO_6$ (Fig. 14.2). It is commercially available as colorless crystals that have a melting point of 94° to 96°C. When crystallized from xylene the melting point is 169°C (no solvation). Methodology for ochratoxin analysis was recently reviewed by Nesheim (1976). Ochratoxin A fluoresces greenish-blue under long wave UV on silica gel TLC plates and changes to deep blue on exposure to ammonia fumes. It has an R$_f$ value of approximately 0.7 in toluene-ethyl acetate-90% formic acid (6:3:1). It exhibits UV absorption maxima at 214 and 334 nm with corresponding extinction coefficients of 36,800 and 6500, respectively. Mass, infrared, and nuclear magnetic resonance spectra and other physical characteristics were reviewed by Steyn (1971). Ochratoxins B, C, and other ochratoxin esters have been investigated relative to various properties, but there is little evidence that any of these toxins other than ochratoxin A occur naturally.

Bioproduction.—*A. ochraceus* will grow and produce ochratoxin on a variety of natural and synthetic media. The liquid or solid media are extracted with solvents such as chloroform or chloroform-methanol. The extract is partitioned by column chromatography and the ochratoxin crystallized from benzene. Ochratoxin has been produced on corn, shredded wheat, winter wheat, soybeans, and rice (Steyn 1971). Production averaged 239 mg/100 g shredded wheat with water levels of 40 to 70 ml/100 g of wheat, incubated 17 to 25 days at 22°C (Schindler and Nesheim 1970), using a strain of *A. ochraceus* designated M-298. Ochratoxin A has also been produced on numerous artificial media, including mycological broth plus yeast extract (Van Walbeek *et al.* 1968), 2% yeast extract and 4% sucrose (YES) (Davis *et al.* 1969, 1972), and a synthetic medium containing glutamic acid as the nitrogen source (Ferreira 1968). *A. ochraceus* NRRL 3174 has been used frequently for bioproduction purposes.

Sterigmatocystin

This mycotoxin (Fig. 14.3) has been found in natural products in relatively few instances. However, its similarity to aflatoxin in chemical and biological properties places it among the potentially important mycotoxins in foods.

The Fungi.—The most common fungus producing sterigmatocystin in agricultural commodities and foods is *Aspergillus versicolor*, described in detail by Raper and Fennell (1965). *Bipolaris* sp., *A. nidulans*, *Penicillium luteum*, and *A. flavus* may also produce this mycotoxin or closely related compounds such as O-methyl-sterigmatocystin. Recently, production has been reported for *A. chevalieri*, *A. ruber*, and *A. amstelodami* (Schroeder and Kelton 1975). *A. versicolor* has conidial heads that, although greatly variable in color, are most often light yellow, sometimes buff to orange, and occasionally flesh colored, exuding a deep red pigment. Compacted mycelium and sclerotia are absent, sterigmata are two-tiered, and mature conidia do not exceed 4μ.

Natural Occurrence.—*A. versicolor* is widely distributed in nature in soil, grains, feedstuffs, bread, cereal products, dried meats, and cheese. Sterigmatocystin has been found in grain, green coffee, and food samples from Mozambique (Van der Watt 1974).

Toxicology.—Sterigmatocystin generally exhibits biological activity similar to aflatoxin, although it is not as potent as aflatoxin B_1. LD_{50}'s vary from 60 to 166 mg/kg in rats, depending on the route of administration and other factors, to 32 mg/kg in monkeys (Hesseltine 1976). Details of carcinogenicity and toxicology of sterigmatocystin were reviewed by Van der Watt (1974).

Chemical Properties.—Sterigmatocystin (Fig. 14.3) is a yellow crystalline compound having a molecular weight of 338 and an

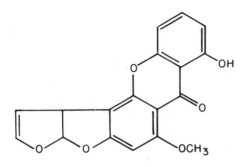

FIG. 14.3. STRUCTURE OF STERIGMATOCYSTIN

empirical formula of $C_{19}H_{14}O_6$. Structurally, it is related to aflatoxin and consists of a xanthone nucleus attached to a bifuran ring. Official methods of analysis have been published for its determination in barley and wheat (Assoc. Off. Anal. Chem. 1975). The compound exhibits a dull brick red fluorescence under UV and has an R_f of about 0.8 in chloroform-methanol (98:2) on silica gel TLC plates. Fluorescence changes to yellow on spraying with aluminum chloride solution. It is soluble in chloroform and pyridine, sparingly soluble in many other organic solvents, and insoluble in water, sodium hydroxide, and sodium carbonate solutions. Concentrated solutions give a green color with ferric chloride. It is optically active, $(\alpha)_D-387$, has maximal absorptivity in UV at 205, 233, 264, and 325 nm, and decomposes at 265°C. Other physical and chemical properties have been discussed by Davis et al. (1960).

Bioproduction.—Sterigmatocystin can be produced on natural and synthetic media in much the same manner as aflatoxin. Yields of 0.75 to 1.2 g/kg of sterigmatocystin have been produced on maize meal in 21 days at 27°C, while A. versicolor gave 13 g/kg of sterigmatocystin from dried mycelium produced in glucose-salts medium in 21 days at 30°C. The toxin can be extracted with acetone, partitioned by silica gel column or TLC procedures, crystallized from acetone, and sublimed. Methods and factors affecting bioproduction of sterigmatocystin have been reported by Schroeder and Kelton (1975) and Halls and Ayres (1975).

Rubratoxin B
The fungus associated with rubratoxin was first implicated in moldy corn toxicoses in pigs, cattle, and poultry (Burnside et al. 1957; Forgacs 1965). Reviews of this toxin were published by Moss (1971) and Newberne (1974B).

The Fungi.—*Penicillium rubrum* is the principal rubratoxin-producing fungus. *P. purpurogenum* may also produce the toxin (Natori *et al.* 1970). Both species are in the *P. purpurogenum* series of Raper and Thom (1949). In this series, the colonies are generally velvety, growth is restricted on Czapek's agar, on which variously colored aerial hyphae from yellow to orange or red shades, and massed conidial structures that are deep yellow to green and gray-green are produced. Reverse sides of plates are typically deep cherry red or purplish red with pigment, which diffuses away from the colony. Other features include penicilli that are typically biverticillate and symmetrical, elliptical conidia, and a pronounced odor that is often aromatic or fragrant, commonly suggesting apples or walnuts when cultured on malt agar. *P. purpurogenum* generally produces an intense red or purple pigmentation on the reverse side of the colony, has rough conidia, and yellow or orange-red aerial hyphae. *P. rubrum* also develops deep red pigmentation on reverse of colony, but has smooth conidia and aerial hyphae in yellow-green to gray-green shades.

Natural Occurrence.—*P. rubrum* is widely distributed in nature in soils and organic matter. It has been repeatedly isolated from moist paper and weathered fabrics. Substrates are often discolored by red or reddish pigments. Rubratoxin has been frequently associated with moldy corn toxicosis in swine, cattle, and poultry, and hepatitis-X in dogs. How much of the toxicity has been due to rubratoxin alone and how much was due to other mycotoxins is not clear. The extent of occurrence of rubratoxin in natural products is not known, which is probably due to the difficulty of quantitative analysis for this toxin.

Toxicology.—Corn invaded by *P. rubrum* has been shown to be toxic to pigs, mice, horses, goats, poultry, dogs, rabbits, cattle, and guinea pigs. Oral LD_{50}'s were 100 to 200 mg/kg in rats, 120 mg/kg in mice, 83 mg/kg in day-old chicks, and 60 mg/kg in ducklings. Intravenous (iv) LD_{50}'s are reported to be 6.5 mg/kg in mice, while intraperitoneal (ip) injections give 3.75 mg/kg in mice and 5 mg/kg in ducklings. The liver is the primary organ damaged. Weight losses in poultry are caused by 500 mg/kg of rubratoxin in the diet.

Chemical Properties.—Rubratoxin B (Fig. 14.4) is a white crystalline compound with a molecular weight of 518 and an empirical formula of $C_{26}H_{30}O_{11}$. It decomposes at 185° to 186°C. It is optically active, $(\alpha)_D$ + 67. It can be partially extracted from natural products such as corn with ethyl acetate. The toxin exhibits UV absorption maxima at 204 (ϵ = 30,000) and 251 nm (ϵ = 9683). The TLC R_f value is approximately 0.56 in chloroform-methanol-acetic acid (80:20:2). It appears as a dark spot against a fluorescent background on fluorescing silica gel TLC sheets and may also be visualized by production of a fluorescent derivative on nonfluorescent silica TLC

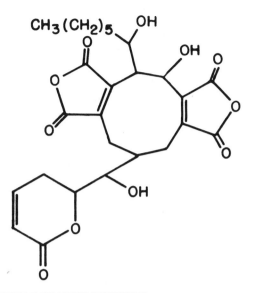

FIG. 14.4. STRUCTURE OF RUBRATOXIN B

plates following heating of the plates at 200°C for 10 min. Additional information on the properties of rubratoxin is presented by Moss (1971).

Bioproduction.—Although rubratoxin B can be produced on natural products, it is more easily obtained and purified from nutrient solution of stationary cultures. The toxin is excreted into the medium and little is retained in the mycelium. Hayes and Wilson (1968) described a convenient method for bioproduction. Yields on the order of 0.60 to 0.94 g/liter of nutrient solution have been reported. Cultures not maintained on natural media may lose their ability to produce rubratoxin.

Penicillic Acid

Penicillic acid has long been recognized as a toxic fungal metabolite, possibly as early as 1896. It was isolated from corn on which *P. puberulum* had grown (Alsberg and Black 1913). The carcinogenicity of penicillic acid and the fact that it is produced by a large number of fungi commonly associated with foodstuffs makes it one of the more important mycotoxins. Reviews on this toxin have been prepared by Ciegler *et al.* (1971) and Wilson (1976).

The Fungi.—A large number of the fungi have been found to produce at least small amounts of penicillic acid. The majority of these fungi are *Penicillium* spp. belonging to the Asymmetrica Fasciculata group of the penicillia (Hesseltine 1976). Members of the *A. ochraceus* series (see ochratoxin) also produce penicillic acid, viz, A. ochraceus, A. alliaceus, A. melleus, A. sclerotiorum, and A. sulphureus. Penicillium spp. that produce the toxin include aurantio-virens, baarnense, cyclopium, fennelliae, janthinellum, lividum, martensii, palitans, puberulum, roqueforti, simplicissimum, stoloniferum, and viridicatum. Detailed descriptions of these fungi are found in Raper and Thom (1949). Species of the P. cyclopium series are perhaps the most important penicillic acid producers. These are characterized by blue-green colonies in which the conidiophores occur in clusters giving a tufted appearance in marginal areas. Reverse sides of colonies are often orange-brown to deep red or purple. Strong moldy or earthy odors are often prominent. P. martensii has colonies that are usually bright blue-green and narrowly but distinctly zonate, whereas P. cyclopium and P. puberulum colonies are dull blue-green and mostly indistinctly zonate or nonzonate. The surface of P. cyclopium colonies is usually tufted or granular with definite fascicles or bundles in marginal areas, compared with P. puberulum colonies which are less fasciculate and generally appear velvety.

Natural Occurrence.—Fungi of the P. cyclopium group are abundant and ubiquitous in nature. They do not parasitize fruit, but are often present as secondary invaders. Penicillic acid has been demonstrated to occur in corn, beans, and tobacco. P. martensii, which is associated with blue eye disease of corn, has been shown to be toxigenic probably due to elaboration of penicillic acid.

Toxicology.—Penicillic acid has been described as an antibiotic with some antihelminthic and antiviral activity as well as antibacterial activity. It is also highly toxic to a variety of animals, microbes, and cell cultures. It proved to be too toxic for use in therapy. The toxin has an LD_{50} value of 100 mg/kg for mice supplied subcutaneously and 250 mg/kg by iv injection. A level of 10 μg/ml caused nuclear abnormalities in HeLa cells (Wilson 1976). This toxin is significant in its carcinogenicity, producing malignant tumors in rats on subcutaneous injection (Dickens and Jones 1961). One mg produced tumors in each of four rats in 48 weeks. Its biological activity has been reviewed by Hesseltine (1976).

Chemical Properties.—Penicillic acid (Fig. 14.5) is a crystalline material with a molecular weight of 170 and an empirical formula of $C_8H_{10}O_4$. It is soluble in hot water, alcohol, chloroform, and many other solvents, but it is insoluble in petroleum ether and hexane. It melts at 83° to 85°C, but is stable in various solvents and foodstuffs to 100°C or more. The toxin can be extracted from solid or liquid media with ether, sublimed at 80° to 90°C, and crystallized from pentane. Maximum UV absorption occurs at 226 to 227 nm (ϵ = 10,471). The most satisfactory analytical method involves silica gel TLC procedures. The R_f value is

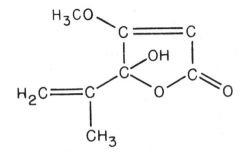

FIG. 14.5. STRUCTURE OF PENICILLIC ACID

about 0.45 in chloroform-ethyl acetate-formic acid (6:4:1). Upon exposure to ammonia fumes it produces a stable complex that fluoresces blue under UV light.

Bioproduction.—Penicillic acid can be produced on solid or liquid media using *P. martensii, P. cyclopium,* or *A. ochraceus.* Up to 3.2 g/liter have been reported for *A. ochraceus* on yeast extract solution and up to 2.89 g/kg on corn for *P. martensii* at 20°C (Ciegler 1972).

Patulin

This mycotoxin is important in the United States because of its potent biological effects and its likely widespread distribution in nature. Toxic outbreaks in livestock have been attributed to patulin. It can conceivably gain entrance to the human food chain via fruits and fruit juices. Patulin has been known at one time or another as clavicin, clavitan, claviformin, and expansin. Scott (1974) and Stott and Bullerman (1975) have reviewed the research and significance of patulin.

The Fungi.—A large number of fungi have been found that produce patulin, most of which are either penicillia or aspergilli. The *Penicillium* spp. are largely in the Asymmetria Fasciculata group and include *claviforme, cyclopium, divergens, equinum, expansum, granulatum, griseofulvum, lanosum, lapidosum, leucopus, melinii, novae-zeelandiae,* and *urticae (patulum)* (Hesseltine 1976). Other patulin-producing fungi include *Aspergillus clavatus, A. giganteus, A. terreus,* and *Byssochlamys nivea.* These species are described by Raper and Thom (1949) and Raper and Fennell (1965). *Byssochlamys* is described by Brown and Smith (1957). The *P. expansum* series is characterized by colonies that are typically dull yellow-green, gray-green, or glaucus shades. Conidiophores are long, up to 500 μ or more, and smooth or finely roughened. Conidia are abundant and the walls are conspicuously roughened. In *P. crustosum* the conidia often form definite crusts. *P. urticae*, which is in a separate series, has colonies that are typically pale to dull gray-green or gray shades, rather than yellow-green, and it has conidiophores that are typically smooth-walled. *P. urticae* is distinguished from the *P. italicum* series in that the latter has sterigmata 8 to 12 μ long and the colonies tend to spread on corn steep and malt agars, whereas *P. urticae* colonies are restricted on these agars and its sterigmata are only 4.5 to 6.0 μ in length.

Natural Occurrence.—*P. expansum* occurs commonly in soil and organic debris and frequently upon pomaceous (pome) fruits. It is easily recogizable as the blue mold rot of apples in storage. This is the most destructive species of *Penicillium* to apples, cherries, grapes, quinces, pears, butter, and meats. Much of the rot once ascribed to *P. glaucum* should have been attributed to *P. expansum*. *P. urticae* is also widely distributed in nature but it is not particularly abundant. It is generally found in soil and decaying vegetation. On citrus it produces a pronounced purplish discoloration but does not cause conspicuous tissue degeneration.

Patulin has been found in commercial apple juice and cider where moldy apples were used. Whether the toxin is present in other processed fruits is not known, but it is suspected. Much additional research on patulin is needed, since *P. expansum* is so commonly associated with noncitrus fruits.

Toxicology.—Patulin is both carcinogenic and mutagenic as well as acutely and chronically toxic. It is antibiotic toward various micro-organisms and is toxic to some higher plants. Detailed data on biological properties are supplied by Ciegler *et al.* (1971). The toxin is carcinogenic to male rats with repeated subcutaneous injections of 2 mg/kg of body weight. The LD_{50} to chicks is 170 mg/kg per os (po). Sublethal dosages of 10 mg in 150 g birds produce extensive hemorrhages in the digestive tract in 6 to 48 hr. Animal poisoning leads to convulsions and death.

Chemical Properties.—Patulin (Fig. 14.6) is a crystalline material with a molecular weight of 154 and an empirical formula of $C_7H_6O_4$. It absorbs in the UV at 276 nm. The compound melts at 111°C. It is soluble in water, very soluble in ethyl acetate, and soluble in most other organic solvents, except petroleum ether. Pohland and Allen (1970) described a method for analysis of patulin in corn. Stoloff *et al.* (1971) included patulin in their mycotoxin multidetection system. The toxin is observed on TLC plates as a dark spot on a light background at R_f 0.41 in toluene-ethyl acetate-90% formic acid (6:3:1). It can be visualized as a yellow fluorescent spot on TLC plates after spraying with p-anisaldehyde reagent (Scott 1974).

Bioproduction.—Patulin can be produced on Czapek-Dox and Raulin-Thom media with various isolates of *P. urticae* or *P. expansum*.

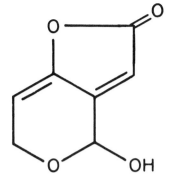

FIG. 14.6. STRUCTURE OF PATULIN

P. urticae grown in potato-dextrose broth gave 1.2 to 2.7 g/liter of patulin at 25°C in 14 days. Wilson and Nuovo (1973) produced up to 120 to 150 mg/liter of apple juice inoculated with a culture of *P. expansum*, which they originally isolated from apples decayed in refrigerated storage at O°C.

Citrinin

Citrinin was first isolated in 1931 as a pure compound from broth fermented by *P. citrinum* and later isolated from a variety of *Aspergillus* and *Penicillium* spp. When tested as an antibiotic, it caused marked renal toxicity (Carlton *et al.* 1974). The presence of citrinin in agricultural commodities and foods has not been adequately investigated and it may prove to be an important naturally occurring mycotoxin in the United States.

The Fungi.—*P. citrinum* and *P. viridicatum* appear to be the most important citrinin-producing fungi in nature, but others produce citrinin under laboratory conditions. These are *P. citreo-viride, P. expansum, P. fellutanum, P. implicatum, P. jensenii, P. lividum, P. notatum, P. palitans, Aspergillus candidus, A. niveus,* and *A. terreus.* The *P. citrinum* series includes *P. steckii* and *P. corylophilum* in addition to *P. citrinum* (Raper and Thom 1949). *P. citrinum* colonies grow restrictedly on Czapek's agar, producing a surface that appears velvety with medium to heavy sporulation. Colonies are typically furrowed radially, bluish-green to yellowish-green with dark centers surrounded by lighter shades and usually show bright yellow to orange-pink shades in reverse. Beads of yellow exudate rich in citrinin are typically produced on the surface of older colonies. Conidiophores are erect, less than 200μ in length, smooth-walled and mostly unbranched. Conidia are 2.5 to 3.2 μ in diameter, globose to elliptical, and have walls either smooth or very finely roughened.

The *P. viridicatum* series consists of *P. palitans, P. olivino-viride,* and *P. viridicatum. P. viridicatum* also grows restrictedly on Czapek's agar. It is principally characterized by colonies that are yellow-green when young and in which the conidiophores are markedly fasciculate. Older colonies are generally darker green. Penicilli are up to 65 to 70 μ in length and conidia occur in tangled chains up to 100 to 150 μ long with conidiophores that are commonly 150 to 250 μ by 3.5 to 4.5 μ. Conidia are mostly 3 to 4.5 μ in diameter with walls that are delicately roughened.

Natural Occurrence.—*P. citrinum* is widespread and abundant in nature, particularly in warmer climates. It is frequently associated with deteriorating fabrics and agricultural commodities such as corn and rice, bakery products, and various foods and feeds. The fungus is common in rice and in the rice-producing areas of the world. It has been isolated from toxic yellow rice (Saito *et al.* 1971). *P. viridicatum* occurs on humus or vegetation in contact with soil. Some strains are weak pathogens or secondary invaders of pomaceous fruit; others have been isolated from animal feeds. The species is not considered especially abundant in nature.

Citrinin has been found as a natural contaminant of barley that was nephrotoxic to swine. A citrinin-producing strain of *P. viridicatum* that also produced ochratoxin A was isolated from the barley. There is much interest in the separate and combined effects of ochratoxin and citrinin on animals and the fate of these mycotoxins during malting and brewing (Krogh *et al.* 1974; Nip *et al.* 1975). In our laboratory we have twice isolated *P. citrinum* from moldy bread that contained citrinin. Citrinin has also been found in wheat, oats, and rye at levels of 70 to 80,000 ppb and in yellowed rice.

Toxicology.—The acute toxicity of citrinin has been determined in a number of experimental animals (Saito *et al.* 1971). The LD$_{50}$ was 67 mg/kg in rats, 35 mg/kg in mice, and 37 mg/kg in guinea pigs. The LD$_{50}$ for rabbits was 19 mg/kg. Ambrose and DeEds (1945) have reported extensively on the toxicity of this chemical. Carlton *et al.* (1974) found that beagle dogs given 20 or 40 mg pure citrinin per kg of body weight developed renal disease. Renal lesions consisted of degeneration and necrosis of the tubular epithelium plus a variety of abnormal clinico-pathological symptoms. In rats, swine, and other animals, the kidney is the primary target organ, showing swelling, increased urinary output, and other physiological disorders. Swine given oral doses of 20 to 40 mg/kg showed growth depression, glucosuria, proteinurea, and increased blood urea nitrogen. Krogh (1974) has recently reviewed mycotoxic nephropathy in swine.

Chemical Properties.—Citrinin (Fig. 14.7) is a yellow crystalline compound with a molecular weight of 250 and an empirical formula of

Only **Spector** has **SENTRY**™

pp. 414-514

JH

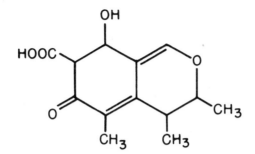

FIG. 14.7. STRUCTURE OF CITRININ

$C_{13}H_{14}O_5$. It has a melting point of 172°C and is optically active $(\alpha)_D$ – 37.4. It fluoresces lemon yellow under long wave UV light. On TLC plates it has an R_f of about 0.56 in toluene-ethyl acetate-formic acid (5:4:1) and spots exhibit a characteristic tailing or comet effect at most concentrations. Maximal UV absorption is at 319, 253, and 222 nm in ethanol with $\log_e$ at 3.67, 3.92 and 4.34, respectively (Neely et al. 1972). The toxin is soluble in most organic solvents and water, depending on pH. It is easily crystallized from cold ethanol solutions and is precipitated from aqueous solutions on acidification to pH 1.5. Additional chemico-physical properties are discussed by Saito et al. (1971).

Bioproduction.—Citrinin can be produced in natural media such as rice. However, extraction and purification of citrinin is greatly simplified when it is produced in nutrient solution. Timonin and Rouatt (1944) obtained 500 to 700 mg of citrinin per 200 ml of mineral salts medium with Aspergillus candidus incubated 20 days at 26°C. Medium-scale production of citrinin has been described by Davis et al. (1975). They obtained up to 1.75 g of citrinin per liter with P. citrinum in stationary culture in a 5-gal (19-liter) carboy containing 4 liters of 4% sucrose and 2% yeast extract. After 25 days, citrinin was precipitated by lowering the pH of the filtrate to 1.5 with hydrochloric acid and recovered by filtration, and crystallized several times from ethanol.

Zearalenone

This fungal metabolite functions as a hormone-like chemical rather than a direct toxin. However, it is generally considered to be a mycotoxin when present in moldy feeds as it causes serious estrogenic disorders in swine and other animals. Farmers and veterinarians in the Midwestern and Southeastern United States have long observed estrogenic disorders as a sporadic disease syndrome in swine. Reports in the scientific literature as early as 1928 associated estrogenism with consumption of moldy corn (McNutt *et al.* 1928; Koen and Smith 1945). Urry *et al.* (1966) completed isolation and structural characterization of the causal agent zearalenone, also known as F-2 toxin. Certain derivatives of the compound have since been utilized to stimulate animal growth when used at carefully controlled levels. The beneficial aspects of zearalenone derivatives are briefly discussed by Ciegler (1975). Recent reviews were compiled by Mirocha and Christensen (1974) and Pathre and Mirocha (1976).

The Fungi.—Fungi that produce zearalenone are members of the genus *Fusarium*, some species of which have a perfect (sexual) stage that is classified in the genus *Gibberella*. Fusaria that are known to produce zearalenone include *F. graminearum* (formerly *F. roseum*), *F. moniliforme*, *F. sporotrichioides*, *F. oxysporum*, and *F. tricinctum*. *F. graminearum* is most often implicated in field outbreaks of estrogenism and will be described here. The reader is referred to the monograph of the genus *Fusarium* (Booth 1971) and to a treatise on toxigenic fungi (Booth and Morgan-Jones 1977) for full descriptions of toxigenic *Fusarium* spp.

F. graminearum colonies on potato-sucrose agar are usually pigmented, but some lack pigment. Most colonies are pink to crimson, but occasionally are white or with sectors that are yellow or white. Colony appearance varies with pH and medium composition (Booth 1971). Macroconidia are formed from single, lateral, globose phialides or from multibranched conidiophores. Conidophores may aggregate into sporodochia. Conidia may be falcate to sickle-shaped, with or without an elongated apical cell, or they may be markedly dorsiventral and have a well marked foot cell. Septation is fine with from 3 to 7 septa per conidium. Spore size ranges from 2.5 to 5μ by 35 to 62μ. Mature conidia tend to be 3-septate and 3 to 5μ by 20 to 30μ. Chlamydospores are intercalary, single, or in chains or clumps. They are globose, thick-walled, colorless or pale brown, have a smooth outer wall, and generally measure 10 to 12μ. Chlamydospores are absent on many media. Also, various strains lose the ability to produce chlamydospores on repeated subcultures. Perithecia occur on a wide range of graminaceous hosts. Cultures may produce perithecia on wheat straw in flasks and on weak media such as potato-carrot agar (Booth 1971). Perithecia are ovoid with a rough wall. They measure 140 to 250μ in diameter. Asci are clavate and measure 60 to 85μ by 8 to 11μ, and generally contain 8 ascospores. Ascospores are generally colorless, curved, and eventually 3-septate, and 17 to 25.5μ by 3.5μ in size.

Natural Occurrence.—*F. graminearum* occurs mostly on cereals, but also on corn and tomato. It is a plant pathogen capable of causing seedling and kernel blight, root rot, and scab on wheat and other grains, and seedling blight, stalk rot, and ear rot on corn, to mention just a few of many disorders attributed to the fungus. Numerous reports have attributed poisoning of swine, cattle, or poultry to grain infested with *F. graminearum* (perfect stage *Gibberella zeae*). Infection with the fungus and subsequent zearalenone production in grain is apparently enhanced by periods of low temperature (10° to 12°C) not uncommonly associated with harvesting and storage of high-moisture corn in the Midwestern United States. Christensen *et al.* (1965) and Mirocha *et al.* (1967) have reported on various aspects of zearalenone production by *F. graminearum*. Natural outbreaks of zearalenone mycotoxicosis have occurred in Europe as well as in the Midwestern and Southeastern United States. The mycotoxin can apparently be produced on grains in the field, during harvest, storage, and processing of the grain, or subsequently during storage of any food or feed containing the grain.

Toxicology.— Estrogenism attributed to zearalenone is characterized by infertility of both sexes of swine, rats, mice, and guinea pigs. Other possibly susceptible animals include cattle, chickens, and turkeys. In swine, the effects are pronounced in females and include swollen vulva, enlarged nipples, prolapsed rectums, and atrophied ovaries collectively referred to as vulvovaginitis. Males frequently exhibit enlarged mammary glands and reduced testes. Levels of 1 to 5 ppm in feed are considered physiologically significant in swine. In the Midwest, surveys have revealed samples of corn containing 2909 ppm (Mirocha and Christensen 1974) and 450 to 750 ppb (Shotwell *et al.* 1971).

Zearalenone has a low order of acute toxicity, with LD_{50}'s of approximately 1, 2, and 20 g/kg in mice, rats and dogs, respectively. Thus, the importance of this mycotoxin lies in its effect on reproductive systems of animals fed moldy corn containing in excess of 1 to 5 ppm of zearalenone. Mirocha *et al.* (1971) discuss additional physiological effects of zearalenone.

Chemical Properties.— Zearalenone is a white crystalline substance with a molecular weight of 318, an empirical formula of $C_{18}H_{22}O_5$, and the structure illustrated by Fig. 14.8. It is classified as a β-resorcyclic acid lactone and is related chemically to other fungus metabolites such as the pathotoxin curvularin produced by *Curvularia* spp. (Pathre and Mirocha 1976). Zearalenone is soluble in a wide variety of solvents, including dilute alkali, acetone, alcohols, chloroform, and ether. It is slightly soluble in petroleum ether and insoluble in water. It melts at 164° to 165°C. UV absorptivity maxima are at 236, 274, and 316 nm with extinction coefficients of 29,700, 13,909 and 6020, respectively. The compound is optically active, $(\alpha)_D$–170.5. Zearalenone fluoresces blue-green under long wave UV light (360 nm) and more greenish

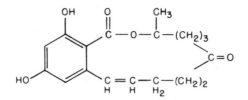

FIG. 14.8. STRUCTURE OF ZEARALENONE

under short wave UV light (260 nm). It has an R_f of about 0.7 on TLC plates developed in toluene-ethyl acetate-formic acid (6:3:1). It is ferric chloride and 2,4-dinitrophenylhydrazine positive and gives a green spot with 50% sulfuric acid that quickly turns to yellow. Various derivatives of zearalenone have been prepared, many of which are biologically active (Shipchandler 1975). Present analytical methods for zearalenone are based on that of Eppley (1968) for assay of zearalenone in corn or on that of Roberts and Patterson (1975).

Bioproduction.—Mirocha and Christensen (1974) reported on zearalenone production on corn, wheat and rice in which moisture levels were adjusted to 45 to 60%. Cultures were grown at 24° to 27°C for 1 to 2 weeks and then 12° to 14°C for 4 to 6 weeks. Yields up to 60 mg/kg of substrate have been obtained.

Trichothecenes

Trichothecenes are a group of potent, biologically active compounds. Some of these toxins appear to have been the causative agent of several severe outbreaks of mycotoxicoses, including alimentary toxic aleukia (ATA) and stachybotryotoxicosis in Russia, red mold poisoning in Japan, and moldy corn toxicoses in the United States (Eppley 1975). Of the 27 known naturally occurring trichothecenes isolated to date, T-2 toxin and vomitoxin appear to be the best known relative to natural disease outbreaks in the United States. The possible presence of trichothecenes in corn and grains in the United States poses one of the more serious mycotoxin problems in this country, the extent of which is not known due to the lack of specific and sensitive analytical procedures for these compounds. The trichothecenes have been reviewed by Smalley and Strong (1974), Saito and Ohtsubo (1974), and Bamburg (1976). Emetic factors are discussed by Prentice and Dickson (1968), Ueno et al. (1974), and Vesonder et al. (1973).

The Fungi.—Trichothecenes are produced by a variety of species belonging to the genera *Gibberella, Fusarium, Myrothecium, Tricho-*

derma, Cladosporium, Cephalosporium, Trichothecium, and perhaps others. T-2 toxin has been reportedly produced by *Gibberella zeae (F. graminearum, F. roseum), F. equiseti, F. lateritium, F. oxysporum, F. poae, F. solani, F. sporotrichioides,* and *F. tricinctum.* Details of morphology and taxonomy of these fungi are given by Booth (1971) and Booth and Morgan-Jones (1977). The most important T-2 producer is *F. tricinctum.* Vomitoxin is produced primarily by *F. graminearum,* which has already been treated in this chapter in the section on zearalenone. Other fungi reported to produce the emetic and rejection factors are *F. avenaceum, F. moniliforme, F. equiseti,* and *F. culmorum* (Hesseltine 1976).

F. tricinctum produces colonies that become deep red to purple on the surface of the agar. Aerial mycelium is sparse to floccose (Booth 1971). Microconidia are one-septate and 4.5 to 7.5 μ long. Macroconidia are sickle-shaped, have a well formed foot cell, are three- to five-septate, and measure 26 to 53μ by 3 to 4.8μ. Pale to orange sporodochia may be formed. Chlamydospores are intercalary, globose, and 10 to 16μ long.

Natural Occurrence.—Records of *F. tricinctum* are unreliable according to Booth (1971) due to misidentification and oversimplification of the species concept. The closely related *F. poae,* with which it has often been confused, is reported to have a wide host and geographical range. It is most common in temperate regions where it is found on woody seedlings, herbaceous and gramineous hosts, and presumably, as in the case of most fusaria, it is found in soil as mycelium and chlamydospores. *F. tricinctum* has been isolated from fescue, moldy corn, sorghum, peas, cranberries, rice, and bean hulls.

Distribution of *F. graminearum,* a producer of vomitoxin, has already been discussed. T-2 has been isolated from cattle feed containing moldy corn in Wisconsin (Hsu *et al.* 1972). According to Hesseltine (1976) that is the only confirmed instance in which T-2 has been isolated from naturally contaminated feed. However, this or closely related trichothecenes have been suspected in a number of natural outbreaks. According to Smalley and Strong (1974), these compounds are probably responsible for a variety of disorders, including low-temperature moldy corn toxicoses, swine rejection (Kotsonis *et al.* 1975), and emesis associated with moldy corn, and a variety of other naturally occurring mycotoxicoses. The trichothecene vomitoxin has been isolated from heavily scabbed, naturally molded corn in a number of instances. Trichothecenes appear to be the cause of ATA in Russia, which caused many human deaths in outbreaks in 1932 and 1942 to 1947.

Toxicology.—Biological activity of T-2 toxin and other trichothecenes has been discussed in detail by Bamburg and Strong (1971). The compound is active against some plants and fungi, brine shrimp, insects, and a broad spectrum of animals, including mice, rabbits, rats, trout, cattle, swine, and man. The LD_{50} was 5.2 mg/kg in male mice (ip) and 6.1 mg/kg (po) in rainbow trout. In various animals, including cattle and swine, effects of T-2 toxin are inflammation and hemorrhaging of the gastrointestinal tract, edema, leucopenia, degeneration of the bone marrow, and death. In turkeys, encrustations of the beak were characteristic. A general effect of T-2 is to inhibit protein synthesis with as little as 0.03 μg/ml of reaction mixture reducing ^{14}C-leucine incorporation by 50% in rabbit reticulocytes. The effect is so characteristic that Ueno et al. (1973) consider a mycotoxin to be a trichothecene if it inhibits leucine uptake and causes radio-mimetic injury. T-2 and other trichothecenes also are potent inducers of dermal necroses, making the compound hazardous to laboratory personnel concentrating and analyzing these substances from samples. Mouse and rabbit skin tests are among the more sensitive tests for T-2 type toxins. Additional symptoms of trichothecenes are reviewed by Joffe (1971) in his discussion of ATA in Russia.

A number of trichothecenes have been found (or inferred) to cause emesis in swine, ducklings, dogs, horses and man. These compounds were mostly associated with heavily scabbed grain in Europe and Japan. Vesonder et al. (1973) isolated a new trichothecene, vomitoxin, from corn infected with F. graminearum. Apparently vomitoxin is also responsible for the well known refusal of swine to eat certain lots of moldy corn. Vomitoxin is not as potent a dermal toxin as T-2 toxin.

Chemical Properties.—Both T-2 toxin and vomitoxin are 12,13-epoxy-Δ^9-trichothecenes. T-2 toxin (Fig. 14.9) is a colorless crystalline material with a molecular weight of 466 and an empirical formula of $C_{24}H_{34}O_9$. The compound does not absorb appreciably in the UV range, except for end absorption, but is optically active,$(\alpha)_D$+ 15. It melts at 151° to 152°C. It has a TLC R_f of approximately 0.36 in toluene-ethyl acetate-formic acid (6:3:1) and 0.28 in benzene-methanol-acetic acid (24:2:1). It is not visible on the TLC plate under visible or UV light, but may be visualized by spraying with various reagents, such as treatment with p-anisaldehyde reagent followed by heating (Scott et al. 1970). In this case it becomes pinkish under visible light and blue fluorescent under UV light. T-2 is soluble in alcohols, acetone, ethyl acetate, benzene, and chloroform, but is not soluble in hexane or water.

Vomitoxin has a molecular weight of 296, an empirical formula of $C_{15}H_{20}O_6$, and the structure shown in Fig. 14.10. Other properties are similar to that of T-2 toxin. It has an R_f of 0.45 on TLC plates developed in chloroform-methanol-water (80:20:0.1) and is visualized as a yellow-brown spot after spraying with p-anisaldehyde reagent and

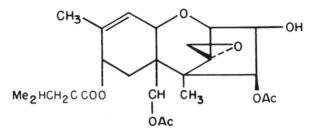

FIG. 14.9. STRUCTURE OF T-2 TOXIN

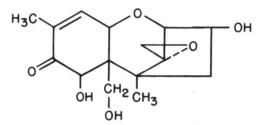

FIG. 14.10. STRUCTURE OF VOMITOXIN

heating at 110°C. It is soluble in alcohols, toluene, and pyridine, insoluble in hexane and butanol, and only slightly soluble in chloroform and acetone. It is readily extractable in water. These and other properties are described by Vesonder et al. (1973). The compound is inactivated by autoclaving, but is stable in water to 92°C. Additional data are presented by Yoshizawa and Morooka (1973).

Bioproduction.—T-2 toxin has been produced by growing F. tricinctum on soymeal, corn and peptone, and in corn steep liquor-supplemented nutrient solution. Burmeister (1971) reported yields of 7.5 g/kg on white corn grits. T-2 was extracted from solid substrates with ethyl acetate and from nutrient solutions on charcoal from which it was eluted with methanol (Ueno et al. 1973). The toxin was purified on silica gel columns and crystallized from benzene-hexane. T-2 was produced at temperatures from below freezing to about 28°C with the most toxin produced at 15°C.

Vomitoxin has been extracted from naturally infected corn using 40% aqueous methanol following extraction with butanol (Vesonder et al. 1973). Crude vomitoxin was precipitated by adding cold absolute ethanol to concentrated aqueous methanol extracts and subsequently purified using column chromatography and preparative TLC. Presumably vomitoxin could be produced in the laboratory on most grains by appropriate strains of *F. graminearum*. A low temperature treatment such as that described for T-2 toxin might be expected to enhance production.

Tremorgens

Various mycotoxins characterized as tremorgens have been reviewed by Ciegler (1975), Ciegler *et al.* (1976) and Binder and Tamm (1973). Although drugs with such action are apparently uncommon in nature (Everett *et al.* 1956), tremorgenic mycotoxins have been isolated from a diverse group of fungi. Wilson and Wilson (1964) reported an unnamed tremorgen from *A. flavus*. Later Wilson *et al.* (1968) isolated a toxin (now called penitrem A) from cultures of *P. cyclopium*. A similar compound was isolated from *P. palitans*, and later two closely related tremorgens were isolated and designated penitrem (tremortin) B and C (Ciegler 1975). Cole *et al.* (1972) reported the production of verruculogen by *P. verruculosum* and later Cole *et al.* (1974) isolated another tremorgen from *P. paxilli*. Related compounds known as cytochalasins (Aldridge *et al.* 1972) have been isolated from a variety of fungi; cytochalasin E has been shown to be a naturally occurring toxin of *A. clavatus* (Buchi *et al.* 1973).

How important and widespread the tremorgenic mycotoxins are in nature is not known. However, some of them have been involved in the illness and death of humans and domesticated animals.

The Fungi.—Tremorgenic mycotoxins have been reported from a diverse group of fungi. They are *A. clavatus*, *A. flavus*, *A. fumigatus*, *P. cyclopium*, *P. martensii*, *P. palitans*, *P. paxilli*, *P. puberulum*, *P. verruculosum*, and *Rosellinia necatrix*. Of those listed, *A. clavatus*, *P. paxilli*, and *P. verruculosum* are described here. *P. palitans* is very similar to *P. viridicatum*, which is described elsewhere in this chapter, except that the conidial heads quickly become dark yellow-green.

P. verruculosum is classified in the Biverticillata Symmetrica group (Raper and Thom 1949). Its conidiophores produce a simple terminal symmetrical whorl of several closely packed sterigmata. Colonies are yellow-green and do not produce perithecia, sclerotia, or coremia. The *P. funiculosum* series (*P. funiculosum*, *P. verruculosum*, *P. islandicum*, *P. varians*, and *P. piceum*) is characterized by colonies that have a more or less cottony or tufted appearance. The conidiophores arise primarily from aerial hyphae or strands of hyphae. *P. funiculosum* and *P. verruculosum*, unlike the other two species in the series, have colonies that spread broadly on plates of most media rather than being restricted in culture. *P. verruculosum* has conidia that are globose and conspicuously roughened or ornamented compared with conidia of *P. funiculosum*, which are elliptical and more or less smooth. Also, *P.*

verruculosum colonies are uncolored or greenish or brownish on the reverse sides, whereas reverse sides of *P. funiculosum* colonies are pink or red to orange-brown or even black.

P. paxilli is classified in the *P. brevi-compactum* series of the Asymmetrica Velutina section of *Penicillium,* which also includes *P. stoloniferum* (Raper and Thom 1949). In this series, colonies are usually yellow-green to gray-green, velvety, conspicuously furrowed, and heavy-sporing. Conidia are subglobose to elliptical, smooth or nearly so, and borne in long tangled chains. In *P. paxilli* the penicillus is unbranched and terminally crowded with 5 to 8 metulae.

A. clavatus colonies are bluish-green, conidiophores are colorless and smooth, conidial heads and vesicles are clavate (club-shaped), vesicles are fertile over the entire surface, sterigmata are uniseriate, and conidia are elliptical, smooth, and somewhat thick-walled (Raper and Fennell 1965).

Natural Occurrence.—*P. verruculosum* is worldwide in distribution, but less abundant than its close relative *P. funiculosum,* which is one of the most common of all soil fungi (Raper and Thom 1949). Both fungi have been isolated from decaying vegetation, fabrics, and wood. *P. islandicum* is a closely related, important mycotoxin-producing fungus, which occurs in rice in Japan. *P. verruculosum* has been isolated from peanuts and found to produce the tremorgenic mycotoxin verruculogen (Cole *et al.* 1972).

P. palitans occurs commonly upon decaying vegetation in contact with soil and occasionally on pomaceous fruits. Ciegler (1969) first reported the production of a tremorgenic mycotoxin produced by this fungus, which was isolated from a sample of moldy commercial feed and implicated in the death of dairy cows.

P. paxilli and other members of the *P. brevi-compactum* series are widespread in nature but not abundant. *P. paxilli* has been isolated from soil, decaying plants, vegetables, and cereal grains. It has also been isolated from insect-damaged pecans and shown to produce a tremorgenic mycotoxin (Cole *et al.* 1974). The closely related *P. stoloniferum* produces the mycotoxin mycophenolic acid. This fungus has occasionally been isolated from bread. Also, Wilson *et al.* (1968) isolated a tremorgen from *P. cyclopium* that was found to be the principal fungus present in feed stuffs causing death in sheep.

A. clavatus is common throughout the world in soil and on decaying vegetation and is notable in that it can tolerate strongly alkaline solutions (Raper and Fennell 1965). The fungus was isolated from toxic feed pellets as early as 1953. Certain strains of the fungus elaborate kojic acid, a mycotoxin also produced by a large number of *A. flavus* isolates. More recently Glinsukon *et al.* (1974) isolated *A. clavatus* from mold-infested rice in Thailand, which was implicated in the death of a child. This isolate was shown to produce several tremorgens, including cytochalasin E.

Toxicology.—Very little confirmed information is available from which generalizations can be made regarding the toxicology of the tremorgens (Ciegler 1975). The LD_{50}'s in mice (mg/kg) are: verruculogen (*P. verruculosum*), 2.4; penitrem A (also known as tremortin A) (*P. palitans*), 1.1; penitrem B (*P. cyclopium*), 5.8; and fumitremorgen A (*A. fumigatus*),< 5. An insufficient quantity of purified toxin was available for LD_{50} determinations of the tremorgen from *A. flavus*. Cytochalasin E has been reported to kill rats within a few hours after dosing, the LD_{50} value being 2.6 and 9.1 mg/kg by ip and po, respectively (Buchi *et al.* 1973). Death was attributed to circulatory collapse caused by massive extravascular effusion of plasma.

Most laboratory animals and several farm animals are susceptible to neurotoxicity of the tremorgens (Ciegler 1975). Sublethal symptoms include tremors, ataxia, loss of grasping ability, and general irritability. Larger dosages result in convulsions and death. Surviving animals appear to recover fully. Ciegler (1975) discusses a variety of biochemical and clinical changes accompanying tremorgen toxicosis.

Chemical Properties—The known tremorgens are high molecular weight compounds with basic structures similar to that of fumitremorgen A (Fig. 14.11) (Eickman *et al.* 1975). This toxin has a molecular weight of 579 and an empirical formula of $C_{33}H_{45}N_3O_6$. The known tremorgens melt from 211° to 239°C and have UV absorptions typical of various indole derivatives with at least two absorption peaks in the vicinity of 224 to 233 nm and 275 to 295 nm. Verruculogen has the empirical formula of $C_{30}H_{37}N_3O_7$; penitrem A, $C_{37}H_{44}NO_6Cl$; and penitrem B, $C_{37}H_{45}NO_5$. These compounds are mostly soluble in polar organic solvents and mostly insoluble in water. Some of them may decompose in chloroform. Penitrem A has an R_f of about 0.4 in chloroform-methanol (95:5) and is visualized as a green spot on spraying with 2% ferric chloride in butanol. Verruculogen has an R_f of approximately 0.3 in chloroform-acetone (93:7) and is visualized as a mustard colored spot under UV after spraying with ethanol-sulfuric acid (1:1).

Cytochalasin E has a molecular weight of 495 and an empirical formula of $C_{28}H_{33}NO_7$. It decomposes at 206° to 208°C and has the structure illustrated in Fig. 14.12 (Buchi *et al.* 1973).

Bioproduction.— The tremorgens have been produced on a variety of natural media. Cole *et al.* (1972) grew *P. verruculosum* on nutrient-supplemented shredded wheat for 14 days at 28°C and extracted verruculogen with hot chloroform. The toxic fraction was purified by column chromatographic procedures using 1-day-old cockerels for

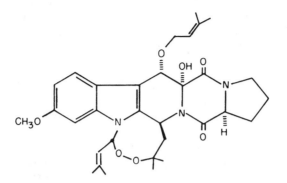

FIG. 14.11. STRUCTURE OF FUMITREMORGEN A

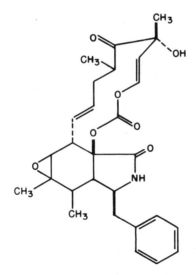

FIG. 14.12. STRUCTURE OF CYTOCHALASIN E

bioassays. Yield data were not given. The *P. paxilli* tremorgen was produced on the same medium in 5 to 7 days at 25°C and purified in much the same manner as verruculogen. *P. palitans* (Ciegler 1969) was grown in nutrient solution consisting of 0.5% dried commercial potato flakes, 2% sucrose, and 4% corn steep liquor. Cultures were incubated statically at 28°C for 1 week. Penitrem A was extracted from mycelial mats using chloroform-methanol and purified using preparative TLC procedures. Dried residues from chloroform extracts were bioassayed by injection into mice. Earlier Wilson et al. (1968) produced penitrem A by growing P. cyclopium for 1 week at 25°C on a variety of media consisting of aqueous food suspensions. The mycelium and medium were extracted with ethyl ether in a soxhlet apparatus, purified by column chromatography procedures, and bioassayed for toxicity to mice and rats. Yield data were not provided. Demain et al. (1976) produced enhanced yields of cytochalasin E by growing A. clavatus on pearled barley in agitated solid state fermentations for 12 days at 30°C. Toxicity was determined at various stages of purification using rat lethality as a guide. The toxin was extracted with methylene chloride and precipitated with petroleum ether. Approximately 35 mg of cytochalasin E (80 mg of combined tremorgens) were isolated in pure form from each kg of barley.

OTHER MYCOTOXINS

Because of space limitation it is not possible to include a detailed description of all of the important known mycotoxins in this chapter. Those without a direct or relatively easy path of entry into the human food chain are omitted as are those not known at this time to occur in the United States. Some others are not included because they are largely of historical interest. Recent reviews discuss the nature and importance of a number of these mycotoxins (Purchase 1974; Diener et al. 1976; Rodricks 1976).

Ergot Alkaloids

Ergotism is the earliest and probably the best known mycotoxicosis. Ergot alkaloids, primarily of Claviceps purpurea, produced in rye and other grains and compounded into bread, were responsible for periodic outbreaks of St. Anthony's Fire, Holy Fire, or ergotism in Europe throughout the Middle Ages. This disease has been recently reviewed by Van Rensburg and Altenkirk (1974). The last major incidence in the United States involving humans occurred in 1825. Outbreaks occurred in 1926 and 1927 in Russia and in 1928 in England. In severe cases of ergotism loss of limbs and death have resulted from gangrene. A second type of effect is the production of hallucinogenic effects. The ergot alkaloids are also important medicinally and are produced and used beneficially in a variety of ways. Although ergotism in humans is now rare, a type of ergotism occurs today in the Southeastern United States involving animals grazing pasture grasses such as dallis grass.

Sporodesmins

Sporodesmins are related mycotoxins produced by *Pithomyces chartarum* and other fungi in various agricultural commodities. This group of mycotoxins has been reviewed by Atherton *et al.* (1974). Of particular importance in field outbreaks of disease are sporodesmins produced by *P. chartarum* in pasture grasses in New Zealand, Australia, and possibly in Florida in the United States. Facial eczema and edema of grazing animals is induced on exposure of infected animals to light.

Alimentary Toxic Aleukia

Alimentary toxic aleukia (ATA) is a mycotoxicosis thought to be caused by a trichothecene and has recently been reviewed by Joffe (1974). It has been responsible for deaths of many humans and animals in Russia, mostly in outbreaks in 1932 and 1942 to 1947. Mortality from 2 to 80% has been recorded for those affected in various outbreaks. ATA results from the ingestion of cereal grains that have overwintered in the fields under snow and which were infected with various species of *Fusarium* and *Cladosporium*. Toxin production follows freezing conditions (Joffe 1965; 1971). Symptoms include fever, hemorrhagic rash, bleeding from the nose, throat and gums, necrotic angina, anemia, and leukopenia (Forgacs and Carll 1962). Symptoms may gradually grow worse, terminating in severe anemia and total exhaustion of bone marrow. ATA has not been reported to occur in the United States, although the trichothecenes T-2 and vomitoxin occur as previously discussed.

Another trichothecene that occurs in the Southern United States may be the cause of the disease known as fescue foot. Evidence is far from conclusive, but diacetoxyscirpenol, which causes cattle to become lame on ingesting pasture grass such as fescue infested with *Fusarium* spp., may be implicated in this disease. Fescue that is highly fertilized with poultry manure seems more susceptible. Another *Fusarium* metabolite, butenolide, has also been implicated in fescue foot disease. Disease symptoms include lameness in hind quarters, gangrene of the tail and hoof, fever, and eventually death.

Stachybotryotoxicosis

Stachybotryotoxicosis is a disease reported primarily in horses in Russia. Severe hemorrhaging and necrosis of many tissues may occur. The cause of the disease is a group of metabolites produced primarily in hay by *Stachybotrys chartarum* (*S. alternans*). These metabolites may also produce a rash in humans coming in contact with the hay. This mycotoxicosis has been reviewed by Forgacs (1972) and Rodricks and Eppley (1974).

Bermudagrass Tremors

Bermudagrass tremors is a disease of cattle first reported in Oklahoma in 1951, but currently occurs in Louisiana, Mississippi, Alabama, and Georgia. Nervous disorders, tremors, and staggers are produced in cattle grazing toxic bermudagrass or fed the toxic hay. The disease usually occurs in the fall following alternately dry and wet weather. It is not known for certain that the disease is caused by a mycotoxin.

Stress Metabolites (Paramycotoxins or Phytoalexins) of Host Plants

Pink rot dermatitis is a skin disorder of celery workers that results during the harvesting and handling of celery infected with pink rot disease caused by *Sclerotinia sclerotiorum.* Photoreactive furanocoumarins known as psoralens are responsible for the disorder. Blistering lesions form on exposure of skin to sunlight following contact with the psoralens. Apparently the furanocoumarins are produced by the celery in response to invasion by the fungus (Perone 1972). Similarly, a stress-produced toxin, ipomeamarone, may be produced by sweet potato roots in response to invasion by fungi, insects, or other environmental factors (Wilson and Boyd 1974; Burka and Wilson 1976). Another example of diseases of this type is the formation of the anticoagulant dicoumarol in moldy sweet clover. Apparently the hydroxycoumarin is produced naturally in sweet clover, but on spoilage by fungi the compound is converted to dicoumarol, an anticoagulant that causes fatal hemorrhaging in animals. Toxic coumarins are discussed in detail by Perone (1972). For additional discussion of stress metabolite accumulation in fungi-infected plants, see Chap. 4.

Slaframine and Cyclopiazonic Acid

Forage such as red clover infected with *Rhizoctonia leguminicola* (black patch) may cause excessive salivation in grazing animals. Animals may also become bloated and die. In this case, the implicated toxic compounds slaframine and cyclopiazonic acid are produced by the fungus rather than the host plant. *Rhizoctonia* toxins have been reviewed by Broquist and Snyder (1971) and by Aust (1974).

Akakabi-byo

As early as 1901 Japanese scientists began to associate moldy rice and grains with various diseases of man and animals. Much of our knowledge of mycotoxins today is due to decades of exacting and thorough research in Japan. Akakabi-byo (red mold or scab disease) of wheat, barley, rice, corn, and oats produced or stored in unusually wet and cold seasons is associated with toxicity to horses, cows, and pigs, and with headaches, vomiting, and diarrhea in humans. *Fusarium* spp. and trichothecenes were ultimately determined to be the cause of many of these disorders. Toxicological effects of trichothecenes have been previously discussed. Saito and Tatsuno (1971) and Saito and Ohtsubo (1974) have reviewed akakabi-byo and related mycotoxins in Japan.

Yellowed Rice Toxins

Another important group of mycotoxins and mycotoxicoses in Japan has been reviewed by Saito *et al.* (1971) under the name of yellowed rice toxins. They have been the subject of research since 1891. These are largely mycotoxins produced by *Penicillium* spp. growing in rice in Japan, including some of the toxins previously discussed in this chapter, e.g., citrinin. Citreoviridin, produced by *Penicillium citreoviride,* causes prominent nervous symptoms in animals, including excitation, ascending paralysis, respiratory paralysis, and death (Ueno 1974). Cyclochlorotine produced by *P. islandicum* primarily affects the liver, causing fibrosis and cirrhosis. Luteoskyrin and rugulosin, produced by *P. islandicum,* are highly toxic compounds that produce liver necrosis, affect DNA polymerase, inhibit energy transfer, and uncouple oxidative phosphorylation in mitochondria. Erythroskyrine, also produced by *P. islandicum,* causes paralysis and hepatic damage with nephrotic changes and injury to the lymphatic system. These and related mycotoxins are reviewed by Enomoto and Ueno (1974).

Finally, it should be mentioned that there are many fungi that have been shown in the laboratory to produce mycotoxins, but which have not been adequately studied to determine the extent to which their toxins are produced in food or feeds under natural field or storage conditions. For example, Hesseltine (1976) discusses toxins of *Alternaria* spp. fungi currently associated with field crops such as sorghum where the fruit is exposed to the weather. He strongly recommends that this genus and its metabolites be given more attention. Table 14.1 summarizes representative fungi and mycotoxins associated with various agricultural commodities.

MYCOTOXIN CONTROL

Research concerned with the control of mycotoxin-producing fungi and their metabolites in human and animal foods has been reviewed by Goldblatt (1969; 1970), Diener (1973), and Hesseltine (1973). Practical approaches to the control of mycotoxins include prevention of fungal growth, removal of toxin(s), and inactivation of toxin(s). Research that provides guidelines for control of mycotoxins and toxin-producing fungi in agricultural commodities is that concerned with aflatoxin in peanuts (Goldblatt 1970). Very little is known concerning control of mycotoxins other than aflatoxin in crops other

TABLE 14.1

REPRESENTATIVE FUNGI AND MYCOTOXINS ASSOCIATED WITH VARIOUS AGRICULTURAL COMMODITIES

Mycotoxin (Disease)	Fungus	Natural Occurrence
Aflatoxins	Aspergillus flavus[1] A. parasiticus[1]	Peanuts, corn, cottonseed meal, Brazil nuts, pecans, rice, sorghum
Akakabi-byo (red mold or scab disease)	Fusarium spp.	Rice, barley, wheat, corn, oats
Alimentary toxic aleukia (ATA)	Fusarium spp. Cladosporium spp.	Grains
Bermudagrass tremors	Unidentified fungi	Bermudagrass pastures
Citrin	Penicillium citrinum[1] P. viridicatum[1] P. citreo-viride P. expansum P. fellutanum P. implicatum P. jensenii P. lividum	Wheat, oats, rye, rice
	P. notatum P. palitans P. steckii P. corylophilum Aspergillus candidus A. niveus A. terreus	
Ergot alkaloids	Claviceps purpurea	Rye and other grains, dallis grass and other pasture grasses
Fescue foot	Fusarium spp.	Fescue pastures
Ochratoxins	Aspergillus ochraceus[1] A. melleus A. sulphureus Pencillium viridicatum[1] P. cyclopium	Corn, oats, barley, rye, wheat, white beans, peanuts, cottonseed meal, Brazil nuts, citrus fruits, tobacco, coffee

Patulin	Penicillium expansum[1]	Apple juice, various processed fruits
	P. cyclopium	
	P. claviforme	
	P. divergens	
	P. equinum	
	P. granulatum	
	P. griseofulvum	
	P. lanosum	
	P. lapidosum	
	P. leucopus	
	P. melinii	
	P. novae-zeelandiae	
	P. urticae (P. patulum)[1]	
	Aspergillus clavatus	
	A. giganteus	
	A. terreus	
	Byssochlamys nivea	
Penicillic acid	Penicillium cyclopium[1]	Corn, beans, tobacco
	P. aurantio-virens	
	P. baarnense	
	P. fennelliae	
	P. janthinellum	
	P. lividum	
	P. martensii[1]	
	P. palitans	
	P. puberulum	
	P. roqueforti	
	P. simplicissimum	
	P. stoloniferum	
	P. viridicatum	
	Aspergillus ochraceus	
	A. alliaceus	
	A. melleus	
	A. sclerotiorum	
	A. sulphureus	
Psoralens	Sclerotinia sclerotiorum	Celery
Rubratoxin	Penicillium rubrum[1]	Corn, various grains
	P. purpurogenum	
Slaframine	Rhizoctonia leguminicola	Red clover
Sporodesmins	Pithomyces chartarum	Pasture grasses
Stachybotryotoxicosis	Stachybotrys chartarum (S. alternans)	Hay
Sterigmatocystin and derivatives	Aspergillus amstelŏdami	Grains, green coffee, miscellaneous foodstuffs
	A. chevalieri	
	A. flavus	
	A. nidulans	
	A. versicolor[1]	
	Bipolaris sorokiniana	
	Penicillium luteum	

TABLE 14.1 (Continued)

Mycotoxin (Disease)	Fungus	Natural Occurrence
Tremorgens	Aspergillus clavatus A. flavus A. fumigatus Penicillium crustosum P. cyclopium P. martensii P. palitans P. paxilli P. puberulum P. verruculosum Rosellinia necatrix	Peanuts, various moldy commercial feeds, rice
Trichothecenes	Fusarium avenaceum F. culmorum F. equiseti F. graminearum (F. roseum, Gibberella zeae)[1] F. lateritium F. moniliforme F. oxysporum F. poae F. solani F. sporotrichioides F. tricinctum[1]	Corn and various grains, contaminated feed
Yellowed rice disease	Penicillium citreo-viride P. islandicum P. rugulosum	Rice
Zearalenone	Fusarium graminearum (F. roseum)[1] F. moniliforme F. oxysporum F. sporotrichioides F. tricinctum	Corn and various grains

[1]The most important toxin producing species.

than peanuts and possibly corn. Peanuts will therefore receive the greatest attention here.

Prevention

Fungi attack peanut pods and kernels mostly while they are developing in the soil before harvest and after lifting during curing and storage, whenever environmental conditions are favorable for their growth and development. The moisture content of the pod and kernel, the physical condition of the pod, the quality, viability, and physiological condition of the kernel, the ambient temperature, the period of time peanut pods remain in the soil beyond maturity, the conditions for drying and curing in windrow or stack, the length of time and aeration in storage, and the activity of insects are the principal factors that determine the degree of proliferation of fungal growth in peanuts. Many of the same factors also affect the extent of proliferation of fungi in grains, legumes and other oilseeds. The specific fungal species that develop in a given environment depends on moisture, temperature, presence of competing microorganisms, and the nature and physiological state of the peanut pod and/or kernel. These and other factors govern the metabolism of fungi and their capacity for utilization of peanut pod and kernel for growth and the production of metabolites.

Much of the deterioration of developing peanuts caused by the invasion of soil fungi has occurred by digging time. After digging, control over fungal growth may be exercised by reducing the moisture in pods and kernels rapidly in such a manner as to minimize undesirable changes in sensory qualities. Artificial drying with forced air and supplemented heat is used when moisture levels of kernels are above safe storage moisture levels (9 to 10%). Subsequent storage at kernel moisture contents in equilibrium with 70% relative humidity prevents any further fungal deterioration (Diener 1973).

Research utilizing natural genetic resistance of peanuts to invasion by A. flavus is a practical approach to control that is in progress. However, production of an acceptable variety is expected to take years.

Removal

An important method of control of aflatoxins is the elimination of contaminated commodities from food and food marketing channels (Goldblatt 1970). For example, peanuts contaminated by fungi, particularly A. flavus, are mostly diverted at peanut marketing points and used for oil stock and nonedible uses. It has been demonstrated that the amount of aflatoxin in peanuts is correlated with the proportion of loose shelled kernels and the number of shrivelled, rancid, or discolored kernels. When these are discarded, the remaining peanuts are relatively free of aflatoxin. Most of the aflatoxin resides in a relatively small number of contaminated kernels. Culling of these kernels is accomplished by screening for size at shelling plants, by removing discolored kernels manually on picking tables, and by utilizing various mechanical or electronic sorting devices that pass or reject kernels on the basis of color when scanned by a photoelectric cell. Another procedure of potential value is "air classification" (Goldblatt 1970). This is based on the principle that less dense or lighter nuts are more likely to have been invaded by *A. flavus* and thus possibly contaminated with aflatoxin. Experimental results indicate that air separation could be a useful tool for reducing aflatoxin contamination in peanuts.

Removal of aflatoxin from peanut oil is readily accomplished with sodium hydroxide and bleaching earth. Also, extraction with solvents offers good possibilities for removing aflatoxins from oilseed meals. After extraction, the solvent can be removed and recovered for reuse without appreciable reduction in nutritional quality of the peanut meal protein, although some of the meal or feed carbohydrates are removed with the aflatoxins.

Inactivation

Treatments for degrading, destroying, or inactivating aflatoxins by heat, chemical, or biological methods must not impair nutritive value of the material or leave any deleterious residues. Moisture and pressure increase the amount of aflatoxin destroyed by heat, but cooking alone does not reduce the aflatoxin content of contaminated peanut and cottonseed meals to levels safe for feed use. However, pilot plant roasting of contaminated peanuts reduced aflatoxin B_1 levels by 40 to 70% (Goldblatt 1970). Gamma rays do not significantly degrade aflatoxins in dry peanuts. Chemical inactivation using various acids, bases, etc., has met with some measure of success (Beckwith *et al.* 1976). Ammonia, methylamine, sodium hydroxide, and formaldehyde may be used to reduce aflatoxin levels and appear practical for large-scale treatments. Grain preservatives such as 2% ammonia or 1% propionic acid for preventing aflatoxin and ochratoxin formation in corn has shown potential in recent experiments (Vandegraft *et al.* 1975). Gardner *et al.*(1971) found that ammoniation of aflatoxin-contaminated peanut and cottonseed meals reduced aflatoxins to undetectable levels using large-scale equipment.

It is feasible to reduce the aflatoxin content of contaminated peanut products below the 20 ppb limit of the Food and Drug Administration. Separation by hand picking, electronic sorting, and/or air classification will remove the vast majority of aflatoxin-contaminated kernels. Extraction with certain solvents to achieve essentially complete removal of aflatoxins is technically feasible. Heat alone is relatively ineffective, although simple roasting such as that given to peanuts in the manufacture of peanut butter does result in a significant reduction in aflatoxin content and affords a certain margin of safety for such products.

Treatment with the bacterium *Flavobacterium aurantiacum* removes aflatoxin from solutions and, along with hydrogen peroxide treatment, may be useful for the elimination of aflatoxins from certain beverages (Goldblatt 1970). Additional research is expected to identify optimal conditions for nearly complete elimination of aflatoxins by ammoniation with minimal damage to protein quality.

In the final analysis, control of fungal deterioration and mycotoxins in peanuts, corn, wheat, rice, and other foodstuffs is both a pre- and post-harvest problem of controlling the factors influencing fungal growth. The grower must harvest the food crop at maturity and follow recommended curing and drying techniques to rapidly reduce seed moisture to safe levels for storage. Food processors should determine the presence of mycotoxins by chemical or other analytical procedures before processing. Removal by separation of toxin-contaminated materials is essential. Mycotoxin extraction from toxic seeds or food products with solvents is possible where economically feasible. Also, inactivation by heat, ammonia, methylamine, sodium hydroxide, and other methods is possible.

It is hoped that additional research will eventually permit complete elimination of aflatoxin and other mycotoxins from agricultural commodities with minimal damage to food quality. Much research is aimed at achieving this goal.

REFERENCES

ALDERMAN, G. G., and MARTH, E. H. 1974. Experimental production of aflatoxin on intact citrus fruit. J. Milk Food Technol. *37*, 451-456.

ALDRIDGE, D. C., BURROWS, B. F., and TURNER, W. B. 1972. The structures of the fungal metabolites cytochalasins E and F. J. Chem. Soc. Chem. Commun. 148-149.

ALPERT, M. E., and DAVIDSON, C. S. 1969. Mycotoxins, a possible cause of primary carcinoma of the liver. Am. J. Med. *46*, 325-329.

ALSBERG, C. L., and BLACK, O. F. 1913. Contributions to the study of maize deterioration. Biochemical and toxicological investigations of *Penicillium puberulum* and *Penicillium stoloniferum*. U.S. Dep. Agric. Bull. Bur. Plant Ind. *270*, 5-48.

AMBROSE, A. M., and DEEDS, F. 1945. Acute and subacute toxicity of pure citrinin. Proc. Soc. Exp. Biol. Med. *59*, 289-291.

ASSOC. OFF. ANAL. CHEM. 1975. Natural poisons. *In* Official Methods of Analysis, 12th Edition. Assoc. Off. Anal. Chem., Washington, D. C.

ATHERTON, L. G., BREWER, D., and TAYLOR, A. 1974. *Pithomyces chartarum. In* Mycotoxins. I. F. H. Purchase (Editor). Elsevier Scientific Publishing Co., New York.

AUST, S. D. 1974. *Rhizoctonia leguminicola*-Slaframine. *In* Mycotoxins. I. F. H. Purchase (Editor). Elsevier Scientific Publishing Co., New York.

AUSTWICK, P. K. C. 1975. Mycotoxins. Br. Med. Bull. *31*, 222-229.

BALDWIN, J. E., BARTON, D. H. R., and SUTHERLAND, J. K. 1965. The nonadrides. Part IV. The constitution and stereochemistry of byssochlamic acid. J. Chem. Soc. *1965*, 1787-1798.

BAMBURG, J. R. 1976. Chemical and biochemical studies of the trichothecene mycotoxins. *In* Mycotoxins and Other Fungal Related Food Problems. J. V. Rodricks (Editor). Adv. Chem. Ser. *149*, 114-162. Am. Chem. Soc., Washington, D. C.

BAMBURG, J. R., and STRONG, F. M. 1971. 12,13-Epoxy trichothecenes. *In* Microbial Toxins, Vol. 7. S. Kadis, A. Ciegler, and S. J. Ajl (Editors). Academic Press, New York.

BAMBURG, J. R., STRONG, F. M., and SMALLEY, E. B. 1969. Toxins from moldy cereals. J. Agric. Food Chem. *17*, 443-450.

BEAN, G. A., SCHILLINGER, J. A., and KLARMAN, W. L. 1972. Occurrence of aflatoxins and aflatoxin-producing strains of *Aspergillus* spp. in soybeans. Appl. Microbiol. **24**, **437-439.**

BECKWITH, A. C., VESONDER, R. F., and CIEGLER, A. 1976. Chemical methods investigated for detoxifying aflatoxins in foods and feeds. *In* Mycotoxins and Other Fungal Related Food Problems. J. V. Rodricks (Editor). Adv. Chem. Ser. *149*, 58-67. Am. Chem. Soc., Washington, D. C.

BINDER, M., and TAMM, C. 1973. The cytochalasins: A new class of biologically active microbial metabolites. Angew. Chem. Int. Ed. Engl. *12*, 370-380.

BOOTH, C. 1971. The Genus *Fusarium.* Commonw. Mycol. Inst. Kew, Surrey, England.

BOOTH, C., and MORGAN-JONES, G. 1977. Toxic Fungi. Academic Press, New York.

BROQUIST, H. P., and SNYDER, J. J. 1971. *Rhizoctonia* toxin. *In* Microbial Toxins, Vol. 7. S. Kadis, A. Ciegler, and S. J. Ajl (Editors). Academic Press, New York.

BROWN, A. H. S., and SMITH, G. M. 1957. The genus *Paecilomyces* Banier and its perfect stage *Byssochlamys* Westling. Trans. Br. Mycol. Soc. *40*, 17-89.

BUCHANAN, J. R., SOMMER, N. F., and FORTLAGE, R. J. 1975. *Aspergillus flavus* infection and aflatoxin production in fig fruits. Appl. Microbiol. *30*, 238-241.

BUCHANAN, J. R., SOMMER, N. F., FORTLAGE, R. J., MAXIE, E. C., MITCHELL, F. G., and HSIEH, D. P. H. 1974. Patulin from *Penicillium expansum* in stone fruits and pears. J. Am. Soc. Hortic. Sci. *99*, 262-265.

BUCHI, G., KITAURA, Y., YUAN, S. S., WRIGHT, H. E., CLARDY, J., DEMAIN, A. L., GLINSUKON, T., HUNT, N., and WOGAN, G. N. 1973. Structure of cytochalasin E, a toxic metabolite of *Aspergillus clavatus.* J. Am. Chem. Soc. *95*, 5423-5425.

BULLERMAN, L. B. 1974. Inhibition of aflatoxin production by cinnamon. J. Food Sci. *39*, 1163-1165.

BULLERMAN, L. B. 1976. Examination of Swiss cheese for incidence of mycotoxin-producing molds. J. Food Sci. *41*, 26-28.

BULLERMAN, L. B., HARTMAN, P. A., and AYRES, J. C. 1969A. Aflatoxin production in meats. I. Stored meats. Appl. Microbiol. *18*, 714-717.

BULLERMAN, L. B., HARTMAN, P. A., and AYRES, J. C. 1969B. Aflatoxin production in meats. II. Aged salamis and aged country-cured hams. Appl. Microbiol. *18*, 718-722.

BULLERMAN, L. B., and OLIVIGNI, F.J. 1974. Mycotoxin producing potential of molds isolated from Cheddar cheese. J. Food Sci. *39*, 1166-1168.

BURKA, L. T., and WILSON, B. J. 1976. Toxic furanosesquiterpenoids from mold-damaged sweet potatoes *(Ipomoea batatas).* In Mycotoxins and Other Fungal Related Food Problems. J. V. Rodricks (Editor). Adv. Chem. Ser. *149*, 387-399. Am. Chem. Soc., Washington, D. C.

BURMEISTER, H. R. 1971. T-2 toxin production by *Fusarium tricinctum* on solid substrate. Appl. Microbiol. *21*, 739-742.

BURNSIDE, J. E., SIPPEL, W. L., FORGACS, J., CARLL, W. J., ATWOOD, M. B., and DOLL, E. R. 1957. A disease of swine and cattle caused by eating moldy corn. II. Experimental production with pure cultures of molds. Am. J. Vet. Res. 18, 817-824.

BUTLER, W. H. 1969. Aflatoxicosis in laboratory animals. In Aflatoxin. L. A. Goldblatt (Editor). Academic Press, New York.

BUTLER, W. H. 1974. Aflatoxin. In Mycotoxins. I. F. H. Purchase (Editor). Elsevier Scientific Publishing Co., New York.

CAMPBELL, T. C., and STOLOFF, L. 1974. Implications of mycotoxins for human health. J. Agric. Food Chem. 22, 1006-1016.

CARLTON, W. W., SANSING, G., SZCZECH, G. M., and TUITE, J. 1974. Citrinin mycotoxicosis in beagle dogs. Food. Cosmet. Toxicol. 12, 479-490.

CARNAGHAN, R. B. A., HARTLEY, R. D., and O'KELLEY, J. 1963. Toxicity and fluorescence properties of the aflatoxins. Nature (London) 200, 1101.

CHRISTENSEN, C. M. 1974. Storage of Cereal Grains and Their Products. Am. Assoc. Cereal Chem., St. Paul, Minn.

CHRISTENSEN, C. M., NELSON, G. H., and MIROCHA, C. J. 1965. Effect on the white rat uterus of a toxic substance isolated from Fusarium. Appl. Microbiol. 13, 653-659.

CHRISTENSEN, C. M., NELSON, G. H., MIROCHA, C. J., and BATES, F. 1968. Toxicity to experimental animals of 943 isolates of fungi. Cancer Res. 28, 2293-2295.

CHU, F. S., CHANG, C. C., ASHOOR, S. H., and PRENTICE, N. 1975. Stability of aflatoxin B1 and ochratoxin A in brewing. Appl. Microbiol. 29, 313-316.

CIEGLER, A. 1969. Tremorgenic toxin from Penicillium palitans. Appl. Microbiol. 18, 128-129.

CIEGLER, A. 1972. Bioproduction of ochratoxin A and penicillic acid by members of the Aspergillus ochraceus group. Can. J. Microbiol. 18, 631-636.

CIEGLER, A. 1975. Mycotoxins: Occurrence, chemistry, biological activity. Lloydia 38, 21-35.

CIEGLER, A., KADIS, S., and AJL, S. J. 1971. Microbial Toxins, Vol. 6. Academic Press, New York.

CIEGLER, A., MITZLAFF, H. J., WEISLEDER, D., and LEISTNER, L. 1972. Potential production and detoxification of penicillic acid in mold-fermented sausage (salami). Appl. Microbiol. 24, 114-119.

CIEGLER, A., VESONDER, R. F., and COLE, R. J. 1976. Tremorgenic Mycotoxins. In Mycotoxins and Other Fungal Related Food Problems. J. V. Rodricks (Editor). Adv. Chem. Ser. 149, 163-177. Am. Chem. Soc., Washington, D. C.

COLE, R. J., KIRKSEY, J. W., MOORE, J. H., BLANKENSHIP, B. R., DIENER, U. L., and DAVIS, N. D. 1972. Tremorgenic toxin from Penicillium verruculosum. Appl. Microbiol. 24, 248-250.

COLE, R. J., KIRKSEY, J. W., and WELLS, J. M. 1974. A new tremorgenic metabolite from Penicillium paxilli. Can. J. Microbiol. 20, 1159-1162.

DAVIES, J. E., KIRKALDY, D., and ROBERTS, J. C. 1960. Studies on mycological chemistry. Part VII. Sterigmatocystin, a metabolite of Aspergillus versicolor (Vuillemin) Tiraboschi. J. Chem. Soc. 1960, 2169-2178.

DAVIS, N. D., DALBY, D. K., DIENER, U. L., and SANSING, G. A. 1975. Medium-scale production of citrinin by Penicillium citrinum in a semi-synthetic medium. Appl. Microbiol. 29, 118-120.

DAVIS, N. D., DIENER, U. L., and ELDRIDGE, D. W. 1966. Production of aflatoxins B1 and G1 by Aspergillus flavus in a semisynthetic medium. Appl. Microbiol. 14, 378-380.

DAVIS, N. D., SANSING, G. A., ELLENBURG, T. V., and DIENER, U. L. 1972. Medium-scale production and purification of ochratoxin A, a metabolite of Aspergillus ochraceus. Appl. Microbiol. 23, 433-435.

DAVIS, N. D., SEARCY, J. W., and DIENER, U. L. 1969. Production of ochratoxin A by *Aspergillus ochraceus* in a semisynthetic medium. Appl. Microbiol. *17*, 742-744.

DAVIS, N. D., WAGENER, R. E., DALBY, D. K., MORGAN-JONES, G., and DIENER, U. L. 1975. Toxigenic fungi in food. Appl. Microbiol. *30,* 159-161.

DEMAIN, A. L., HUNT, N. A., MALIK, V., KOBBE, B., HAWKINS, H., MATSUO, K., and WOGAN, G. N. 1976. Improved procedure for production of cytochalasin E and tremorgenic mycotoxins by *Aspergillus clavatus.* Appl. Environ. Microbiol. *31,* 138-140.

DETROY, R. W., LILLEHOJ, E. B., and CIEGLER, A. 1971. Aflatoxin and related compounds. *In* Microbial Toxins, Vol. 6. A. Ciegler, S. Kadis, and S. J. Ajl (Editors). Academic Press, New York.

DICKENS, F., and JONES, H. E. H. 1961. Carcinogenic activity of a series of reactive lactones and related substances. Br. J. Cancer *15,* 85-100.

DIENER, U. L. 1973. Deterioration of peanut quality caused by fungi. *In* Peanuts—Culture and Uses. Stone Printing Co., Roanoke, Va.

DIENER, U. L., and DAVIS, N. D. 1969. Aflatoxin formation by *Aspergillus flavus. In* Aflatoxins. L. A. Goldblatt (Editor). Academic Press, New York.

DIENER, U. L., DAVIS, N. D., and MORGAN-JONES, G. 1976. Nature and importance of mycotoxins in grains. Biodeterioration of Materials, Vol. 3. J. M. Sharpley and A. M. Kaplan (Editors). Applied Science Publishers, Ltd., Essex, England.

EICKMAN, N., CLARDY, J., COLE, R. J., and KIRKSEY, J. W. 1975. The structure of fumitremorgin A. Tetrahedron Lett. No. *12,* 1051-1054.

EMEH, C. O., and MARTH, E. H. 1976A. Production of rubratoxin by *Penicillium rubrum* in a soy whey-malt extract medium. J. Milk Food Technol. *39,* 95-100.

EMEH, C. O., and MARTH, E. H. 1976B. Cultural and nutritional factors that control rubratoxin formation. J. Milk Food Technol. *39,* 184-190.

EMEH, C. O., and MARTH, E. H. 1976C. Rubratoxin formation in soybean substrates. J. Milk Food Technol. *39,* 253-257.

ENOMOTO, M., and UENO, I. 1974. *Penicillium islandicum* —(Toxic yellowed rice)— luteoskyrin-islanditoxin-cyclochlorotine. *In* Mycotoxins. I. F. H. Purchase (Editor). Elsevier Scientific Publishing Co., New York.

EPPLEY, R. M. 1968. Screening method for zearalenone, aflatoxin, and ochratoxin. J. Assoc. Off. Agric. Chem. *51,* 74-78.

EPPLEY, R. M. 1975. Methods for the detection of tricothecenes. J. Assoc. Off. Agric. Chem. *58,* 906-908.

ESCHER, F. E., KOEHLER, P. E., and AYRES, J. C. 1973A. Effect of roasting on aflatoxin content of artificially contaminated pecans. J. Food Sci. *38,* 889-892.

ESCHER, F. E., KOEHLER, P. E., and AYRES, J. C. 1973B. Production of ochratoxins A and B on country-cured ham. Appl. Microbiol. *26,* 27-30.

EVERETT, G. M., BLOCKUS, L. E., and SHEPPARD, I. M. 1956. Tremor induced by tremorine and its antagonism by anti-parkinson drugs. Science *124,* 79.

FERREIRA, N. P. 1968. The effect of amino acids on the production of ochratoxin A in chemically defined media. Antonie van Leeuwenhoek, J. Microbiol. Serol. *34,* 433-440.

FORGACS, J. 1965. Stachybotryotoxicosis and moldy corn toxicosis. *In* Mycotoxins in Foodstuffs. G. N. Wogan (Editor). M.I.T. Press, Cambridge, Mass.

FORGACS, J. 1972. Stachybotryotoxicosis. *In* Microbial Toxins, Vol. 8. S. Kadis, A. Ciegler, and S. J. Ajl (Editors). Academic Press, New York.

FORGACS, J., and CARLL, W. T. 1962. Mycotoxicoses. Adv. Vet. Sci. *7,* 273-382.

FRANK, H. K. 1974. Mycotoxin problems in foods and beverages. Zentrabbl. Bakteriol. Parasitenkd., Infektionskr. Hyg. Abt. 1; Orig, Reihe B *159,* 242-434. (German)

GARDNER, H. L., KOLTUN, S. P., DOLLEAR, F. G., and RAYNER, E. T. 1971. Inactivation of aflatoxins in peanut and cottonseed meals by ammoniation. J. Am. Oil Chem. Soc. 48, 70-73.

GLINSUKON, T., YUAN, S. S., WIGHTMAN, R., KITAURA, Y., BUCHI, G., SHANK, R. C., WOGAN, G. N., and CHRISTENSEN, C. M. 1974. Isolation and purification of cytochalasin E and two tremorgens from Aspergillus clavatus. Plant Foods for Man 1, 113-119.

GOLDBLATT, L. A. 1969. Aflatoxin. Academic Press, New York.

GOLDBLATT, L. A. 1970. Chemistry and control of aflatoxin. Pure Appl. Chem. 21, 331-353.

GRAY, W. D. 1959. The Relation of Fungi to Human Affairs. Henry Holt, New York.

GROGER, D. 1972. Ergot. In Microbial Toxins, Vol. 8. S. Kadis, A. Ciegler, and S. J. Ajl (Editors). Academic Press, New York.

HALLS, N. A., and AYRES, J. C. 1973. Potential production of sterigmatocystin on country-cured ham. Appl. Microbiol. 26, 636-637.

HALLS, N. A., and AYRES, J. C. 1975. Factors affecting the production of sterigmatocystin in semisynthetic media. Appl. Microbiol. 30, 702-703.

HARWIG, J. 1974. Ochratoxin A and related metabolites. In Mycotoxins. I. F. H. Purchase (Editor). Elsevier Scientific Publishing Co., New York.

HARWIG, J., CHEN, Y. L., and COLLINS-THOMPSON, D. L. 1974. Stability of ochratoxin A in beans during canning. Can. Inst. Food Sci. Technol. J. 7, 288-289.

HARWIG, J., CHEN, Y. K., KENNEDY, B. P. C., and SCOTT, P. M. 1973. Occurrence of patulin and patulin-producing strains of Penicillium expansum in natural rots of apple in Canada. Can. Inst. Food Sci. Technol. J. 6, 22-25.

HAYES, A. W., and WILSON, B. J. 1968. Bioproduction and purification of rubratoxin B. Appl. Microbiol. 16, 1163-1167.

HESSELTINE, C. W. 1969. Mycotoxins. Mycopathol. Mycol. Appl. 39, 371-383.

HESSELTINE, C. W. 1973. Recent research for the control of mycotoxins in cereal. Pure Appl. Chem. 35, 251-257.

HESSELTINE, C. W. 1974. Natural occurrence of mycotoxins in cereals. Mycopathol. Mycol. Appl. 53, 141-153.

HESSELTINE, C. W. 1976. Mycotoxins other than aflatoxins. In Biodeterioration of Materials, Vol. 3. J. M. Sharpley and A. M. Kaplan (Editors). Applied Science Publishers, Ltd., Essex, England.

HSU, I. C., SMALLEY, E. B., STRONG, F. M., and RIBELIN, W. E. 1972. Identification of T-2 toxin in moldy corn associated with a lethal toxicosis in dairy cattle. Appl. Microbiol. 24, 684-690.

JARVIS, B. 1971. Factors affecting the production of mycotoxins. J. Appl. Bacteriol. 34, 199-213.

JOFFE, A. Z. 1960. The microflora of overwintered cereals and its toxicity. Bull. Res. Counc. Isr., Sect. D, 90, 101-126.

JOFFE, A. Z. 1965. Toxin production by cereal fungi causing toxic alimentary aleukia in man. In Mycotoxins in Foodstuffs. G. N. Wogan (Editor). M.I.T. Press, Cambridge, Mass.

JOFFE, A. Z. 1971. Alimentary toxic aleukia. In Microbial Toxins, Vol. 7. S. Kadis, A. Ciegler, and S. J. Ajl (Editors). Academic Press, New York.

JOFFE, A. Z. 1974. Toxicity of Fusarium poae and F. sporotrichioides and its relation to alimentary toxic aleukia. In Mycotoxins. I. F. H. Purchase (Editor). Elsevier Scientific Publishing Co., New York.

KADIS, S., CIEGLER, A., and AJL, S. J. 1971. Microbial Toxins, Vol. 7. Academic Press, New York.

KADIS, S., CIEGLER, A., and AJL, S. J. 1972. Microbial Toxins, Vol. 8. Academic Press, New York.

KING, A. D., JR., BOOTH, A. N., STAFFORD, A. E., and WAISS, A. C., JR. 1972. Byssochlamys fulva metabolite toxicity in laboratory animals. J. Food Sci. 37, 86-89.

KINOSITA, R., and SHIKATA, T. 1965. On toxic moldy rice. In Mycotoxins in Foodstuffs. G. N. Wogan (Editor). M.I.T. Press, Cambridge, Mass.

KOEHLER, P. E., HANLIN, R. J., and BERAHA, L. 1975. Production of aflatoxins B1 and G1 by Aspergillus flavus and Aspergillus parasiticus isolated from market pecans. Appl. Microbiol. 30, 581-583.

KOEN, J. S., and SMITH, H. C. 1945. An unusual cause of genital involvement in swine associated with eating moldy corn. Vet. Med. 40, 131-133.

KOTSONIS, F. N., SMALLEY, E. B., ELLISON, R. A., and GALE, C. M. 1975. Feed refusal factors in pure cultures of Fusarium roseum 'graminearum'. Appl. Microbiol. 30, 362-368.

KRAYBILL, H. F., and SHAPIRO, R. E. 1969. Implications of fungal toxicity to human health. In Aflatoxin. L. A. Goldblatt (Editor). Academic Press, New York.

KROGH, P. 1974. Mycotoxic nephropathy. In Mycotoxins. I. F. H. Purchase (Editor). Elsevier Scientific Publishing Co., New York.

KROGH, P., HALD, B., GJERSTEN, P., and MYKEN, F. 1974. Fate of ochratoxin A and citrinin during malting and brewing experiments. Appl. Microbiol. 28, 31-34.

KROGH, P., HALD, B., and PEDERSEN, E. J. 1973. Occurrence of ochratoxin A and citrinin in cereals associated with mycotoxic porcine nephropathy. Acta Pathol. Microbiol. Scand. Sect. B, 81, 689-695.

KUC, J. 1972. Compounds accumulating in plants after infection. In Microbial Toxins, Vol. 8. S. Kadis, A. Ciegler, and S. J. Ajl (Editors). Academic Press, New York.

LANCASTER, M. C., JENKINS, F. P., and PHILP, J. M. 1961. Toxicity associated with certain samples of groundnuts. Nature (London) 192, 1095-1096.

LILLEHOJ, E. B., CIEGLER, A., and DETROY, R. W. 1970. Fungal toxins. In Essays in Toxicology. F. R. Blood (Editor). Academic Press, New York.

MARTIN, P. M. D., GILMAN, G. A., and KEEN, P. 1971. The incidence of fungi in foodstuffs and their significance based on a survey in the Eastern Transvaal and Swaziland. In Symposium on Mycotoxins in Human Health. I. F. H. Purchase (Editor). Macmillan Press, London.

McNUTT, S. H., PURWIN, P., and MURRAY, C. 1928. Vulvovaginitis in swine. Preliminary report. J. Am. Vet. Med. Assoc. 73, 484-492.

MIROCHA, C. J., and CHRISTENSEN, C. M. 1974. Oestrogenic mycotoxins synthesized by Fusarium. In Mycotoxins. I. F. H. Purchase (Editor). Elsevier Scientific Publishing Co., New York.

MIROCHA, C. J., CHRISTENSEN, C. M., and NELSON, G. H. 1967. Estrogenic metabolite produced by Fusarium graminearum in stored corn. Appl. Microbiol. 15, 497-503.

MIROCHA, C. J., CHRISTENSEN, C. M., and NELSON, G. H. 1971. F-2 (Zearalenone) estrogenic mycotoxin from Fusarium. In Microbial Toxins, Vol. 7. S. Kadis, A. Ciegler, and S. J. Ajl (Editors). Academic Press, New York.

MISLIVEC, P. B., DIETER, C. T., and BRUCE, V. R. 1975. Mycotoxin-producing potential of mold flora of dried beans. Appl. Microbiol. 29, 522-526.

MOSS, M. O. 1971. The rubratoxins, toxic metabolites of Penicillium rubrum Stoll. In Microbial Toxins, Vol. 6. A. Ciegler, S. Kadis, and S. J. Ajl (Editors). Academic Press, New York.

NATORI, S., SAKAKI, S., KURATA, H., UDAGAWA, S. I., ICHINOE, M., SAITO, M., UMEDA, M., and OHTSUBO, K. 1970. Production of rubratoxin B by Penicillium purpurogenum Stoll. Appl. Microbiol. 19, 613-617.

NEELY, W. C., ELLIS, S. P., DAVIS, N. D., and DIENER, U. L. 1972. Spectroanalytical parameters of fungal metabolites. 1. Citrinin. J. Assoc. Off. Anal. Chem. 55, 112-1127.

NESHEIM, S. 1976. The ochratoxins and other related compounds. In Mycotoxins and Other Fungal Related Food Problems. J. V. Rodricks (Editor). Adv. Chem. Ser. 149, 276-295. Am. Chem. Soc., Washington, D.C.

NEWBERNE, P. M., 1974A. The new world of mycotoxins — animals and human health. Clin. Toxicol. 7, 161-177.

NEWBERNE, P. M. 1974B. Penicillium rubrum-Rubratoxins. In Mycotoxins. I. F. H. Purchase (Editor). Elsevier Scientific Publishing Co., New York.

NIP, W. K., CHANG, F. C., CHU, F. S., and PRENTICE, N. 1975. Fate of ochratoxin A in brewing. Appl. Microbiol. 30, 1048-1049.

OHMOMO, S., SATO, T., UTAGAWA, T., and ABE, M. 1975. Isolation of festuclavine and three new indole alkaloids, roquefortine A, B and C, from cultures of Penicillium roqueforti (production of alkaloids and related substances by fungi). Agric. Biol. Chem. 39, 1333-1334.

PATHRE, S. V., and MIROCHA, C. J. 1976. Zearalenone and related compounds. In Mycotoxins and Other Fungal Related Food Problems. J. V. Rodricks (Editor). Adv. Chem. Ser. 149, 178-227. Am. Chem. Soc., Washington, D.C.

PERCEBOIS, G., BASILE, A. M., and SCHWERTZ, A. 1975. The existence on strawberries of ascospores of strains of Byssochlamys nivea capable of producing patulin. Mycopathologia 57, 109-111. (French)

PERONE, V. B. 1972. The natural occurrence and uses of the toxic coumarins. In Microbial Toxins, Vol. 8. S. Kadis, A. Ciegler, and S. J. Ajl (Editors). Academic Press, New York.

POHLAND, A. E., and ALLEN, R. 1970. Analysis and chemical confirmation of patulin in grains. J. Assoc. Off. Anal. Chem. 53, 686-687.

PONS, W. A., JR., and GOLDBLATT, L. A. 1969. Physicochemical assay of aflatoxins. In Aflatoxin. L. A. Goldblatt (Editor). Academic Press, New York.

PRENTICE, N., and DICKSON, A. D. 1968. Emetic material associated with Fusarium species in cereal grains and artificial media. Biotechnol. Bioeng. 10, 413-427.

PURCHASE, I. F. H. 1971. Symposium on Mycotoxins in Human Health. Macmillan Press, London.

PURCHASE, I. F. H. 1974. Mycotoxins. Elsevier Scientific Publishing Co., New York.

RAPER, K. B., and FENNELL, D. I. 1965. The Genus Aspergillus. Williams and Wilkins Co., Baltimore.

RAPER, K. B., and THOM, C. 1949. A Manual of the Penicillia. Williams and Wilkins Co., Baltimore.

ROBERTS, B. A., and PATTERSON, D. S. P. 1975. Detection of twelve mycotoxins in mixed animal feedstuffs, using a novel membrane cleanup procedure. J. Assoc. Off. Anal. Chem. 58, 1178-1181.

RODRICKS, J. V. 1976. Mycotoxins and Other Fungal Related Food Problems. Adv. Chem. Ser. 149. Am. Chem. Soc., Washington, D.C.

RODRICKS, J. V., and EPPLEY, R. M. 1974. Stachybotrys and stachybotryotoxicosis. In Mycotoxins. I. F. H. Purchase (Editor). Elsevier Scientific Publishing Co., New York.

SAITO, M., ENOMOTO, M., and TATSUNO, T. 1971. Yellowed rice toxins. In Microbial Toxins, Vol. 6. A. Ciegler, S. Kadis, and S. J. Ajl (Editors). Academic Press, New York.

SAITO, M., ISHIKO, T., ENOMOTO, M., OHTSUBO, K., UMEDA, M., KURATA, H., UDAGAWA, S., TANIGUCHI, S., and SEKITA, S. 1974. Screening test using HeLa cells and mice for detection of mycotoxin-producing fungi isolated from foodstuffs. An additional report on fungi collected in 1968 and 1969. Jpn. J. Exp. Med. 44, 63-82.

SAITO, M., and OHTSUBO, K. 1974. Trichothecene toxins of *Fusarium* species. *In* Mycotoxins. I. F. H. Purchase (Editor). Elsevier Scientific Publishing Co., New York.

SAITO, M., and TATSUNO, T. 1971. Toxins of *Fusarium nivale*. *In* Microbial Toxins, Vol. 7. S. Kadis, A. Ciegler, and S. J. Ajl (Editors). Academic Press, New York.

SARGEANT, K., SHERIDAN, A., O'KELLEY, J., and CARNAGHAN, R. B. A. 1961. Toxicity associated with certain samples of groundnuts. Nature (London) *192*, 1096-1097.

SCHINDLER, A. F., and NESHEIM, S. 1970. Effect of moisture and incubation time on ochratoxin A production by an isolate of. *Aspergillus ochraceus*. J. Assoc. Off. Anal. Chem. *53*, 89-91.

SCHROEDER, H. W., and KELTON, W. H. 1975. Production of sterigmatocystin by some species of the genus *Aspergillus* and its toxicity to chicken embryos. Appl. Microbiol. *30*, 589-591.

SCOTT, DeB. 1965. Toxigenic fungi isolated from cereal and legume products. Mycopathol. Mycol. Appl. *25*, 213-222.

SCOTT, P. M. 1973. Mycotoxins in stored grains, feeds, and other cereal products. *In* Grain Storage. R. N. Sinha and W. E. Muir (Editors). AVI Publishing Co., Westport, Conn.

SCOTT, P. M. 1974. Patulin. *In* Mycotoxins. I. F. H. Purchase (Editor). Elsevier Scientific Publishing Co., New York.

SCOTT, P. M., and KENNEDY, B. P. C. 1975. The analysis of spices and herbs for aflatoxins. Can. Inst. Food Sci. Technol. J. *8*, 124-125.

SCOTT, P.'M., KENNEDY, B. P. C., HARWIG, J., and CHEN, Y. K. 1974. Formation of diketopiperazines by *Penicillium italicum* isolated from oranges. Appl. Microbiol. *28*, 892-894.

SCOTT, P. M., LAWRENCE, J. W., and VAN WALBEEK, W. 1970. Detection of mycotoxins by thin-layer chromatography: Application to screening of fungal extracts. Appl. Microbiol. *20*, 839-842.

SCOTT, P. M., MILES, W. F., TOFT, P., and DUBE, J. G. 1972. Occurrence of patulin in apple juice. J. Agric. Food Chem. *20*, 450-451.

SCOTT, P. M., and SOMERS, E. 1968. Stability of patulin and penicillic acid in fruit juices and flour. J. Agric. Food Chem. *16*, 483-485.

SHERWOOD, R. F., and PEBERDY, J. F. 1974. Production of the mycotoxin, zearalenone, by *Fusarium graminearum* growing on stored grain. I. Grain storage at reduced temperature. J. Sci. Food Agric. *25*, 1081-1087.

SHIH, C. N., and MARTH, E. H. 1972. Experimental production of aflatoxin on brick cheese. J. Milk Food Technol. *35*, 585-587.

SHIPCHANDLER, M. T. 1975. Chemistry of zearalenone and some of its derivatives. Heterocycles *3*, 471-520.

SHOTWELL, O. L., GOULDEN, M. L., BOTHAST, R. J., and HESSELTINE, C. W. 1975. Mycotoxins in hot spots in grains. I. Aflatoxin and zearalenone occurrence in stored corn. Cereal Chem. *52*, 687-697.

SHOTWELL, O. L., HESSELTINE, C. W., VANDEGRAFT, E. E., and GOULDEN, M. L. 1971. Survey of corn from different regions for aflatoxin, ochratoxin, and zearalenone. Cereal Sci. Today *16*, 266-273.

SMALLEY, E. B., and STRONG, F. M. 1974. Toxic trichothecenes. *In* Mycotoxins. I. F. H. Purchase (Editor). Elsevier Scientific Publishing Co., New York.

STEYN, P. S. 1971. Ochratoxin and other dihydroisocoumarins. *In* Microbial Toxins, Vol. 6. A. Ciegler, S. Kadis, and S. J. Ajl (Editors). Academic Press, New York.

STOLOFF, L. 1976. Occurrence of mycotoxins in foods and feeds. *In* Mycotoxins and Other Fungal Related Food Problems. J..V. Rodricks (Editor). Adv. Chem. Ser. *149*, 23-50. Am. Chem. Soc., Washington, D.C.

STOLOFF, L., NESHEIM, S., YIN, L., RODRICKS, J. V., STACK, M., and CAMPBELL, A. D. 1971. A multimycotoxin detection method for aflatoxin, ochratoxins, zearalenone, sterigmatocystin, and patulin. J. Assoc. Off. Anal. Chem. 54, 91-97.

STOTT, W. T., and BULLERMAN, L. B. 1975. Patulin: A mycotoxin of potential concern in foods. J. Milk Food Technol. 38, 695-705.

STOTT, W. T., and BULLERMAN, L. B. 1976. Instability of patulin in Cheddar cheese. J. Food Sci. 41, 201-203.

TAYLOR, A. 1971. The toxicology of sporidesmins and other epipolythiadioxopiperazines. In Microbial Toxins, Vol. 7. S. Kadis, A. Ciegler, and S. J. Ajl (Editors). Academic Press, New York.

TIMONIN, M. I., and ROUATT, J. W. 1944. Production of citrinin by Aspergillus sp. of the candidus group. Can. J. Public Health 35, 80-88.

UENO, Y. 1974. Citreoviridin from Penicillium citreo-viride Biourge. In Mycotoxins. I. F. H. Purchase (Editor). Elsevier Scientific Publishing Co., New York.

UENO, Y., ISHII, K., SATO, N., and OHTSUBO, K. 1974. Toxicological approaches to the metabolites of Fusarium. VI. Vomiting factor from moldy corn infected with Fusarium spp. Jpn. J. Exp. Med. 44, 123-127.

UENO, Y., NAKAJIMA, M., SAKAI, K., ISHII, K., SATO, N., and SHIMADA, N. 1973. Comparative toxicology of trichothec mycotoxins: Inhibition of protein synthesis in animal cells. J. Biochem. (Tokyo) 74, 285-296.

URRY, W. H., WEHRMEISTER, H. L., HODGE, E. B., and HIDY, P. H. 1966. The structure of zearalenone (from Gibberella zeae, Fusarium graminearum). Tetrahedron Lett. 27, 3109-3114.

VANDEGRAFT, E. E., HESSELTINE, C. W., and SHOTWELL, O. L. 1975. Grain preservatives: Effect on aflatoxin and ochratoxin production. Cereal Chem. 52, 79-84.

VAN DER MERWE, K. J., STEYN, P. S., and FOURIE, L. 1965A. Mycotoxins. Part 2. The constitution of ochratoxins A, B, and C, metabolites of Aspergillus ochraceus Wilh. J. Chem. Soc. 7083-7088.

VAN DER MERWE, K. J., STEYN, P. S., FOURIE, L., SCOTT, DeB. and THERON, J. J. 1965B. Ochratoxin A, a toxic metabolite produced by Aspergillus ochraceus Wilh. Nature (London) 205, 1112-1113.

VAN DER WATT, J. J. 1974. Sterigmatocystin. In Mycotoxins. I. F. H. Purchase (Editor). Elsevier Scientific Publishing Co., New York.

VAN RENSBURG, S. J., and ALTENKIRK, B. 1974. Claviceps purpurea—Ergotism. In Mycotoxins. I. F. H. Purchase (Editor). Elsevier Scientific Publishing Co., New York.

VAN WALBEEK, W. 1973. Fungal toxins in foods. Can. Inst. Food Sci. Technol. J. 6, 96-105.

VAN WALBEEK, W., SCOTT, P. M., and THATCHER, F. S. 1968. Mycotoxins from food-borne fungi. Can. J. Microbiol. 14, 131-137.

VESONDER, R. F., CIEGLER, A., and JENSEN, A. H. 1973. Isolation of the emetic principle from Fusarium-infected corn. Appl. Microbiol. 26, 1008-1010.

WELLS, J. M., and PAYNE, J. A. 1976. Toxigenic species of Penicillium, Fusarium, and Aspergillus from weevil-damaged pecans. Can. J. Microbiol. 22, 281-285.

WILSON, B. J., and BOYD, M. R. 1974. Toxins produced by sweet potato roots infected with Ceratocystis fimbriata and Fusarium solani. In Mycotoxins. I. F. H. Purchase (Editor). Elsevier Scientific Publishing Co., New York.

WILSON, B. J., and WILSON, C. H. 1964. Toxin from Aspergillus flavus: Production on food materials of a substance causing tremors in mice. Science 144, 177-178.

WILSON, B. J., WILSON, C. H., and HAYES, A. W. 1968. Tremorgenic toxin from Penicillium cyclopium grown on food materials. Nature (London) 220, 77-78.

WILSON, D. M. 1976. Patulin and penicillic acid. *In* Mycotoxins and Other Fungal Related Food Problems. J. V. Rodricks (Editor). Adv. Chem. Ser. *149*, 90-109. Am. Chem. Soc., Washington, D.C.

WILSON, D. M., and JAY, E. 1975. Influence of modified atmosphere storage on aflatoxin production in high moisture corn. Appl. Microbiol. *29*, 224-228.

WILSON, D. M., and NUOVO, G. J. 1973. Patulin production in apples decayed by *Penicillium expansum.* Appl. Microbiol. *26*, 124-125.

WU, C. M., KOEHLER, P. E., and AYRES, J. C. 1972. Isolation and identification of xanthotoxin (8-methoxypsoralen) and bergapten (5-methoxypsoralen) from celery infected with *Sclerotina sclerotiorum.* Appl. Microbiol. *23*, 852-856.

WU, M. T., AYRES, J. C., and KOEHLER, P. E. 1974. Toxigenic aspergilli and penicillia isolated from aged, cured meats. Appl. Microbiol. *28*, 1094-1096.

YOSHIZAWA, T., and MOROOKA, N. 1973. Biological modification of trichothecene mycotoxins: Acetylation and deacetylation of deoxynivalenols by *Fusarium* spp. Appl. Microbiol. *29*, 54-58.

15

Methods for Detecting Mycotoxins in Foods and Beverages

L. B. Bullerman

D etection of mycotoxins in foods, feeds, and beverages is a very broad and complex subject. This complexity is caused by many factors inherent in the diverse nature of the various commodities that may be contaminated with mycotoxins. A detailed review of the subject is beyond the scope of this chapter and will not be presented. Rather, this chapter will deal with the broad, general principles involved in mycotoxin analysis, and the reader will be directed to the selected bibliography at the end of the chapter for references to specific procedures and detailed review articles. On the basis of background given in this chapter, it is hoped that the reader will be able to use the literature to select the procedures for toxin analyses best suited to a specific commodity.

Since the occurrence of events which led to the discovery of aflatoxins in the United Kingdom in 1960, considerable research has been devoted to the development of methods for the detection and assay of mycotoxins in foods, feeds, and beverages. Such efforts would have been largely uncoordinated were it not for the efforts of certain societies and agencies throughout the world. Agencies such as the Canadian Food and Drug Directorate, the United States Food and Drug Administration working through the Association of Official Analytical Chemists (AOAC), and other societies like the American

Association of Cereal Chemists (AACC), the American Oil Chemists' Society (AOCS), and the International Union for Pure and Applied Chemistry (IUPAC) have cooperated in directing and coordinating the development of official methods of analysis of various commodities for mycotoxins. For example, before a method is adopted it is evaluated through collaborative studies in which skilled analysts in different laboratories attempt to obtain identical results using the same method on samples of the same food, feed, or beverage. Each society then forms its own judgments on the value of a method and makes recommendations to its constituency. These methods are then published in books or manuals of official methods of each society. Such methods are available for specific toxins such as aflatoxins and for specific products such as peanut, cottonseed, corn, copra, green coffee, mixed feeds, etc.

Most of the work on development of methods to date has concentrated on methods for quantitating aflatoxins in various commodities. However, work on analytical methods for aflatoxins has stimulated the development of methods for other mycotoxins. The full significance of these other toxins as potential hazards to human and animal health is not fully known at this time, nor is the extent of contamination of commodities with them completely clear. As better methods are developed, these questions will be answered and analytical methods for other mycotoxins will become more routine.

Since most of the work done on the development of analytical methods for mycotoxins has been concerned with aflatoxins, this chapter will deal with the principles inherent in these methods. However, all of the analytical methods for other mycotoxins basically involve the same steps, including sampling, extraction, cleanup, separation, and quantitation. The general method used is diagrammed in Fig. 15.1. The majority of the methods available are physico-chemical methods, but since mycotoxins are toxic to many life forms, biological assays have also been developed. This chapter will deal with both physico-chemical and biological methods of assay.

SAMPLING

The first problem encountered with mycotoxin analysis is the selection of a representative sample. This problem is most severe with grain, nut, and oilseed commodities. Within a given lot of such products, mycotoxin contamination may be concentrated in a relatively small percentage (i.e., less than 0.5%) of the kernels. However, these kernels may contain very high levels of a mycotoxin. For example, individual peanut kernels have been reported to contain aflatoxin B_1 at levels as high as 1,100,000 ppb (1.1 mg/g) (Cucullu et al. 1966). The

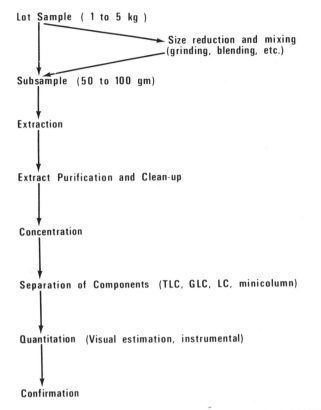

FIG. 15.1. DIAGRAM OF GENERAL STEPS INVOLVED IN ANALYSIS AND QUANTITATION OF MYCOTOXINS FROM AGRICULTURAL COMMODITIES

presence of one such kernel with an extremely high concentration would be sufficient to contaminate 10,000 sound kernels with a very high level of aflatoxin. Further processing of such a lot of peanuts into peanut butter or of a lot of corn containing one highly contaminated kernel into meal would result in a finished product high in aflatoxin, yet the initial sample would have been negative for aflatoxin if it had not included the contaminated kernel. Thus, the importance of adequate sample selection for mycotoxin assay cannot be overemphasized. Sampling plans for obtaining representative lot samples have been published by several workers (Tiemstra 1969; Whitaker and Wiser 1969; Whitaker et al. 1970).

Techniques for sample preparation have also been published (Stoloff et al. 1969, 1972). In the case of seeds and nuts, these procedures involve reduction in particle size by grinding or milling for efficient extraction, along with good comminution and mixing of the entire sample to obtain a representative portion for analysis. Stoloff (1972) recommends taking a large number of randomly scattered subsamples that have been finely ground and well mixed as the best approach to obtaining representative samples. The Official Methods of Analysis of the AOAC (1975) recommends that the entire laboratory lot sample be ground and thoroughly mixed, and that the subsamples be taken from this. In the case of free flowing powders, such as flours, and in the case of liquids or pastes, further size reduction may not be needed, although thorough mixing before removing the subsamples is still necessary. The equipment used for size reduction and mixing of samples is also discussed in AOAC's Official Methods of Analysis (1975). After thorough mixing, the subsamples should be taken with the same effort to be representative as was applied to taking the lot sample. Whenever possible, the sample should be subdivided by a random dividing procedure, such as riffling or similar means, until the desired sample size is achieved. When this type of subdivision is not possible, a composite number of small randomly taken portions should be used. In the case of liquids, any particulate matter should be resuspended before the samples are taken.

Sample size is an important consideration in obtaining representative samples. The uneven distribution of mycotoxins in a commodity makes it desirable to test an entire batch or lot of suspect material, but this of course is impractical. Usually, lot samples are limited to sizes of 1 to 5 kg. From this sample the subsample is taken.

The size of subsamples also varies, depending upon the method of analysis, and has ranged from 20 to 100 g. A subsample size of 50 g is used by most methods and appears to be the best size to obtain both solvent economy and a representative sample.

PHYSICO-CHEMICAL METHODS

Early methods for aflatoxin analysis employed defatting of the sample prior to extraction. Later it was found that aflatoxin recovery was not hampered by the presence of the lipids, and that fat and pigment removal from the extract was simpler and faster than removal from the sample. Subsequent methods, therefore, employed this latter approach. This approach also has been found to be effective for other mycotoxins. The main operations now employed in physico-chemical assay of mycotoxins involve (a) extraction, (b) extract purification and cleanup, (c) separation of extract components by thin-layer chromatography (TLC), and (d) estimation of the quantity of mycotoxin on the thin-layer plates. Other methods of separation, such as gas-liquid chromatography (GLC), high pressure (performance) liquid chromatography (HPLC), also referred to as high speed liquid chromatography (HSLC) or simply liquid chromatography (LC), and spectrophotometry, are being developed for many of the mycotoxins and may replace TLC quantitation in certain situations.

Extraction

One factor that complicates the chemical analysis of mycotoxins is the problem of extraction. No single extraction procedure is adequate for all commodities. This is due to the diverse nature of the commodities that may contain mycotoxins. Commodities with high contents of lipids and pigments require a different treatment than products that are low in these components. These interfering materials and mycotoxins often are soluble in the same solvents. Selective extraction of the toxins or extensive purification procedures are required to produce clean extracts in these situations. Also, toxins such as patulin and penicillic acid may bind to substrate material and not be recoverable by methods that are sufficient for other toxins. Similarly, solid substrates may require different extraction solvents and procedures than liquid substrates. The choice of solvent also depends on the properties of the toxin being extracted. Mycotoxins are soluble in slightly polar solvents and usually insoluble in completely nonpolar solvents. The degree of solubility in water varies with the toxin. Often, combinations of solvents combined in various ratios are found to be most efficient.

Early methods of extracting aflatoxins from peanut products employed exhaustive techniques in which the sample was extracted for several hours in a Soxhlet extractor using methanol as the solvent. Extraction time with these methods ranged from 4 to 6 hr. Defatting and lipid removal were necessary before extraction. During the long extraction times about 1 to 2% of the aflatoxin B_1 was lost per hour of extraction. Even with prior lipid removal, most exhaustive methanol extracts still contained large amounts of polar lipids, pigments, and carbohydrates. Because of these limitations, equilibrium extraction systems which were more rapid and which produced cleaner extracts were explored. These methods were based on the recognition that aqueous solvents were needed to penetrate hydrophilic tissues in order to extract the toxins. Various combinations of solvents with water were studied, including methanol-water, acetone-water, and chloroform-water. While the solubility of aflatoxins in water is low, with the proper proportion of water to chloroform, the partitioning of aflatoxins from water to chloroform occurs rapidly. Neutral fats and lipids are rather insoluble in these slightly polar solvents, resulting in cleaner extracts. By adding fat solvents such as hexane to the extraction solvent, many of the fats and lipids can be partitioned into the hexane portion of the solvent which can then be discarded. Again this results in cleaner extracts.

Two official methods for extraction of aflatoxins from peanut products are recognized by the AOAC. The first of these, known as the CB (Contaminants Branch) method or AOAC Method 1, employs an extracting solvent of chloroform and water (91:9), followed by purification using a silica gel column. The second method, known as the BF (Best Foods) method or AOAC Method 2, uses an extraction solvent of methanol and water (55:45) plus hexane (39:32 + 29). The aflatoxins are extracted and partitioned into the methanol phase, and fats and lipids are partitioned into the hexane phase. The aflatoxins are then transferred from methanol-water to chloroform and concentrated for analysis.

Extraction solvents used in official AOAC methods for aflatoxins in other commodities include acetone-water (85:15) for cottonseed and corn; chloroform-water (91:9) for soybeans, green coffee beans, and pistachio nuts; and chloroform-water (71:29) for coconut, copra, and copra meal, preceded by hexane defatting for cocoa beans. Dairy products such as fluid milk, powdered milk, butter and cheese, and other foods such as meats can be analyzed for aflatoxins using similar methods or slight modifications of the methods used for peanuts and grains. These solvents are summarized in Table 15.1.

Official methods of analysis for several mycotoxins other than aflatoxins are also available. Ochratoxins and their esters can be extracted with a mixture of chloroform and 0.1 M phosphoric acid, chloroform and sodium bicarbonate, methanol-water, chloroform-methanol, and acetonitrile-water using procedures similar to those described for aflatoxins. Extraction of sterigmatocystin from grains can be accomplished with chloroform, or mixtures of chloroform-methanol, acetonitrile-water, and water-chloroform. Patulin is very soluble in ethyl acetate and may be extracted from apple and other fruit juices by a simple liquid-liquid extraction using ethyl acetate. Extraction of patulin from flours may also employ ethyl acetate, while extraction from grains employs acetonitrile-water or acetonitrile-hexane. The latter solvent may also be used to extract patulin from sausage and cheese. Penicillic acid is soluble in chloroform and can be extracted from foods using methods similar to those used for aflatoxins and patulin. Acidification of the extraction solvent improves the recovery of penicillic acid. Extraction of zearalenone from grain and hay can be achieved using solvent pairs such as ethanol-water, acetonitrile-water, and water-chloroform. Tricothecenes can be extracted with ethyl acetate. Other experimental methods for extraction of mycotoxins are discussed in the literature, but they are not routine

TABLE 15.1

SUMMARY OF EXTRACTION SOLVENTS USED IN OFFICIAL AOAC METHODS
OF ANALYSIS FOR SEVERAL MYCOTOXINS

Toxin	Commodity	Extraction Solvent
Aflatoxins	Corn, cottonseed	Acetone:water (85:15)
	Green coffee beans, soybeans, coconut, copra, copra meal	Chloroform:water (91:9)
	Cocoa beans	Defat with hexane then chloroform
	Peanut products, pistachio nuts	Chloroform:water (91:9) or methanol:water (55:45) plus hexane (39:32 + 29)
	Powdered milk	Acetone:water (70:30)
Ochratoxins	Barley	Chloroform + 0.1 M phosphoric acid
Patulin	Apple juice	Ethyl acetate
Sterigmatocystin	Barley, wheat	Acetonitrile: 4% potassium chloride (9:1)

and must be adapted to individual situations. Stoloff (1972) has done an excellent job of summarizing the details of various methods. Methods for the simultaneous extraction of several mycotoxins are also being studied. These involve the use of similar principles and combinations of solvents as previously described. In most cases, the extraction of mycotoxins from substrates other than those discussed here can be accomplished by adapting one of the existing solvent systems and methods to the particular product.

Purification and Cleanup

Purification and cleanup of mycotoxin extracts can be accomplished by liquid-liquid partitioning, followed by precipitation of impurities and then removal of impurities using column chromatography or preparative TLC. As previously mentioned, early methods for quantitating aflatoxins involved Soxhlet extraction of fats with a non-polar solvent prior to aflatoxin extraction. This method may still be employed in certain situations, but for the most part other methods are now used.

Partitioning between solvents can occur during extraction as is the case with the solvent mixture of chloroform and water. When other aqueous solvents such as methanol-water and acetone-water are used, the toxins are partitioned into the chloroform layer after extraction. In these situations some prior concentration of the aqueous phase may be needed. Also, a prior cleanup step involving precipitation of interfering materials using lead acetate may be necessary. Lead acetate precipitation removes plant pigments, lipids, fatty acids, and other unknown materials which may cause streaking on TLC plates. When the solvents used in the partitioning cleanup are immiscible, the partitioning can be done in a separatory funnel. When mycotoxins are partitioned from aqueous solutions into chloroform, emulsions are sometimes formed. These emulsions can usually be broken by adding salt solutions, anhydrous sodium sulfate, or celite, or by warming or centrifugation.

Column chromatography may also be used to effect partitioning of mycotoxins from one solvent to another and thereby purify the extract. Columns packed with silica gel, cellulose, acidic alumina, or Florisil may be used to clean and purify an extract. The sample extract is usually added to the column in chloroform or another appropriate solvent, and then washed with one or more solvents in which the toxins are insoluble or less soluble than the impurities. After removal of impurities, the toxins are eluted from the column using a solvent in which the toxin is soluble. The toxin solution can then be collected, concentrated, and examined for quantity of toxin present. Some loss of toxin on the column due to incomplete elution can occur, and the analyst needs to be aware of this possibility. In certain situations, the cleanup and purification step may be omitted. This may occur if the extracts are very clean to start with or if only qualitative screening results are required.

Thin-layer Chromatography

After extraction and purification, the sample extract normally must be concentrated before analysis by thin-layer chromatography. This concentration step may be accomplished by evaporating the solvent in a rotary evaporator using reduced pressure, or on a steam bath or enclosed hot plate. If a steam bath or hot plate is used, it is normally recommended that the sample be concentrated under a stream of nitrogen to prevent loss of some of the toxin due to oxidation. If extreme accuracy is not required, this may be unnecessary. Once the sample is concentrated, it is quantitatively transferred to a small vial or other sealable container and again dried under a stream of nitrogen. After dissolving the residue in a known volume of solvent, the sample is ready for TLC assay.

Thin-layer chromatography can be used to identify and estimate the quantity of mycotoxins in sample extracts. Thin-layer chromatography plates are made by coating glass plates with a thin layer

(0.25 mm) of silica gel. Specific, highly purified grades of silica gel are required to achieve adequate resolution of mycotoxins, especially aflatoxins. The Official Methods of Analysis recommends the grades best suited for specific toxin separations. After the TLC plates are coated, they are allowed to air dry for a few minutes and are then activated by heating at a prescribed temperature for a specific length of time. Specific conditions of activation vary depending on individual laboratory conditions. After activation, the TLC plates should be stored in a desiccator to prevent readsorption of moisture. Pre-coated glass and plastic plates are available that have been perfected to the point where they are now satisfactory for use in quantitating mycotoxins.

To do an analysis using the TLC plate, microliter quantities of the extract are spotted on the plate along an imaginary line approximately 4 cm from the bottom edge of the plate. Either a microliter syringe or disposable capillary tubing pipettes are used to apply the sample extract to the plate. It is important that the extract be applied to the plate in as small a spot a possible. This can be accomplished by applying the extract slowly while gently blowing or directing a stream of air at the spot to speed evaporation of the solvent. This may, however, affect the separation and resolution of spots on the developed plate. If this practice causes problems, a stream of dry nitrogen gas can be used to speed evaporation. This procedure will result in small, tight spots that separate better when the plate is developed. After application of the sample extract and appropriate standards, the plates are developed in a solvent, usually by ascending chromatography. With a given developing solvent, each mycotoxin will have a characteristic migration and separation pattern which can be expressed as the R_f value. The R_f value is defined as the distance the compound has migrated divided by the distance the solvent front has migrated from the origin of the spot:

$$R_f = \frac{\text{distance compound traveled (cm)}}{\text{distance solvent front traveled (cm)}}$$

Thus, R_f values are always expressed as decimal numbers less than 1.0. In most solvent systems, the four main aflatoxins have a separation pattern in which B_1 has the highest R_f, followed in order by B_2, then G_1, and finally G_2. However, some developing solvent systems may change the R_f of aflatoxins to give a sequence of B_1, G_1, B_2, G_2. The R_f value of a compound may vary from one laboratory to another, depending on conditions, and is best expressed as an approximate value. Table 15.2 summarizes approximate R_f values for several

TABLE 15.2

APPROXIMATE Rf VALUES OF SEVERAL MYCOTOXINS ON SILICA GEL CHROMATOGRAPHY PLATES IN SEVERAL SOLVENT SYSTEMS[1]

Toxin	Fluorescent Color in UV Light (365 nm)	Color in Visible Light	Approximate Rf Value	Solvent System	Solvent Ratio	Reference
Aflatoxin B₁	Blue		0.56	C:M	97:3	Wilson and Hayes (1973)
B₂	Blue		0.53	C:M	97:3	Wilson and Hayes (1973)
G₁	Green		0.48	C:M	97:3	Wilson and Hayes (1973)
G₂	Green		0.46	C:M	97:3	Wilson and Hayes (1973)
M₁	Blue		0.40	C:M	97:3	Wilson and Hayes (1973)
M₂	Blue		0.30	C:M	97:3	Wilson and Hayes (1973)
Ochratoxin A	Greenish-blue		0.55	T:Ea:F	6:3:1	Scott et al. (1970)
Sterigmatocystin	Red-brown		0.85	T:Ea:F	6:3:1	Scott et al. (1970)
Patulin	Pale blue[2]	Yellow[4]	0.41	T:Ea:F	6:3:1	Scott et al. (1970)
Penicillic acid	Bright blue[2]	Yellow[4]	0.47	T:Ea:F	6:3:1	Scott et al. (1970)
Citrinin	Yellow		0.16–0.48 (streak)	T:Ea:F	6:3:1	Scott et al. (1970)
Rubratoxin B	Dark spot[3]		0.54–0.59	Ga:M:C	2:20:80	Moss (1971)
Penitrem A (tremorgen A)		Green[5]	0.60	E:Cy	3:1	Purchase (1974)
Zearalenone	Faint blue		0.78	T:Ea:F	6:3:1	Scott et al. (1970)
T-2 toxin		Gray-pink[6]	0.36	T:Ea:F	6:3:1	Scott et al. (1970)
Diacetoxyscirpenol		Pink[6]	0.33	T:Ea:F	6:3:1	Scott et al. (1970)
Vomitoxin		Yellowish-brown[6]	0.45	C:M:W	80:20:0.1	Vesonder et al. (1973)
Verruculogen	Mustard yellow[7]	Slate-gray[7]	0.30	C:A	93:7	Cole et al. (1972)

[1] A = acetone; C = chloroform; Cy = cyclohexane; E = ethyl ether; Ea = ethyl acetate; F = formic acid (90%); Ga = glacial acetic acid; M = methanol; T = toluene; W = water.

[2] After exposure to ammonia fumes (ochratoxin A also fluoresces bright blue after exposure to ammonia).

[3] Quenches fluorescence of fluorescing Silica Gel (HF₂₅₄).

[4] After spray with phenylhydrazine.

[5] After spray with 2% ferric chloride in butanol.

[6] After spray with anisaldehyde.

[7] After spray with 50% ethanolic sulfuric acid.

mycotoxins.

Conditions which give best results in the TLC analysis of mycotoxins have been studied extensively. Most of this work has been done with aflatoxins and has resulted in the availability of many solvent systems for the development and separation of mycotoxins on TLC plates. The solvents recommended in Official Methods of the AOAC are acetone-chloroform (1:9) for aflatoxins; benzene-methanol-acetic acid (18:1:1) for ochratoxins; benzene-methanol-acetic acid (90:5:5) for sterigmatocystin; and toluene-ethyl acetate-90% formic acid (5:4:1) for patulin, penicillic acid, and zearalenone. Many other solvent systems have been reported to be useful for TLC of mycotoxins. One such frequently used solvent system is benzene-ethanol-water (45:35:19), upper phase. The TLC plates may be developed in either unlined, unequilibrated tanks, or lined, completely equilibrated tanks, depending on the particular solvent used.

In addition to the developing solvent, many other factors affect the separation and resolution of mycotoxins, especially aflatoxins. Nesheim (1969) studied the conditions and techniques for the TLC separation of aflatoxins. He found that the most frequent cause of poor resolution and tailing of the aflatoxin spots was the variability in the commercial silica gel adsorbents. Variations in quality were observed within single lots of silica gel from one container to the next. Other factors which affected the separation and resolution of the toxins were adsorbent particle size, concentration and nature of the calcium sulfate binder, thickness of the silica gel layer, moisture content of the silica gel layer, and the vapor phase composition in the developing chamber. Most of these conditions can be adequately controlled so that resolution of mycotoxins, especially aflatoxins, is now fairly routine.

After removal from the developing chambers and evaporation of solvents, the plates are ready for examination and quantitation. Mycotoxins may be quantitated on TLC plates by visual comparison to standards or instrumentally using a fluorodensitometer. The visual techniques involve comparison of R_f values and the color and intensity of fluorescence of the sample to a range of concentrations of standards, with a judgment regarding the concentration of the standard which matches the sample. Dilution of the sample to barely detectable levels is necessary to achieve greatest accuracy by this method. The accuracy of the visual estimation method is said to be in the neighborhood of ± 20%. However, a skilled analyst who is experienced in reading TLC plates could most likely improve on this accuracy. The densitometric determination of concentration is said to have a precision of ± 3%. However, fluorodensitometry, while a good research tool, is expensive, time consuming, and not always practical for routine sample analysis. These techniques work well for mycotoxins that are fluorescent, such as aflatoxins, of which the B and M toxins fluoresce blue and the G toxins fluoresce green, ochratoxin A which fluoresces greenish-blue, and sterigmatocystin which fluoresces a dull brick red when exposed to long-wave ultraviolet (UV) light. Zearalenone which fluoresces a bluish-green in short-wave UV light can also be detected in this way. Patulin and penicillic acid, while not intensely fluorescent in long-wave UV light, can be made to fluoresce by exposure to ammonia fumes or by spraying with 3 to 4% aqueous ammonium hydroxide. After this treatment, patulin fluoresces a pale blue and penicillic acid fluoresces a bright intense blue when exposed to long-wave UV light. Ochratoxin can also be made to fluoresce an intense blue on exposure to ammonia. Patulin and penicillic acid can also be detected by spraying with phenylhydrazine and heating to form yellow-colored phenylhydrazones, which are visible in natural light and which fluoresce in long-wave UV light.

Gas-liquid Chromatography

The application of gas-liquid chromatography (GLC) to analysis of aflatoxins has not been highly successful. However, the technique has been applied successfully to the analysis of patulin, penicillic acid, and zearalenone. Patulin can be identified using GLC of silyl ether, acetate, and chloroacetate derivatives, and penicillic acid can be detected as a trifluoroacetate derivative. GLC of patulin without derivatization can also be employed. GLC can be applied to the detection of patulin in apple juice, corn, and rice, and penicillic acid in apple juice, corn, and dried beans. Further application of GLC to the detection and analysis of certain mycotoxins is anticipated.

Liquid Chromatography

Liquid chromatography (LC) or high pressure liquid chromatography (HPLC) is a technique that is also being applied to analysis of mycotoxins. Techniques have been developed for aflatoxins, patulin, and penicillic acid. Since LC is a modification of column chromatography, it is likely that methods will be developed for additional mycotoxins. The main drawback to LC at this time is that the equipment is relatively expensive, making it somewhat impractical for routine methodology.

Spectrophotometry

A spectrophotometric method for quantitating aflatoxin solutions based on UV absorption at 363 nm has been developed. Using the appropriate absorption maxima and extinction coefficients, this basic method may be applied to other mycotoxins. These methods are based on the principles of Beer's Law and the linearity of absorption as related to concentration. This method, however, has been used primarily as a research technique rather than for routine analyses.

Confirmatory Tests

Many compounds may have fluorescent and chromatographic characteristics similar to mycotoxins. These compounds may be other mold metabolites or natural constituents of certain commodities. Therefore, a spot on a TLC plate is only presumptive evidence of toxin identity, even though the R_f and fluorescent properties may match those of the toxin standard. Several confirmatory tests involving the formation of derivatives with chromatographic properties different from the original compound have been developed. These may involve formation of derivatives followed by TLC to determine changes in chromatographic patterns, or formation of derivatives by direct application of spray reagents to developed TLC plates. Changes in fluorescence or formation of colored derivatives are changes which may be observed and compared to similar changes in standards. Derivative formation has been applied to aflatoxins, sterigmatocystin, patulin, penicillic acid, and zearalenone. Probably the best confirmatory method, however, is a combination of mass spectrometry with TLC and GLC techniques.

Rapid Methods

Most methods for extraction and analysis of mycotoxins involving TLC are somewhat time consuming, and in many situations rapid methods of analyses are needed. For example, buyers of agricultural commodities have needs for screening tests that are capable of detecting aflatoxins in a matter of minutes rather than hours. Methods have been developed that now permit detection of aflatoxins in several agricultural commodities in 15 to 30 min. These methods have been applied primarily to aflatoxins and to a lesser extent other mycotoxins.

A screening method for aflatoxins in cottonseed and corn based on fluorescence of the seeds on exposure to long-wave UV light has been used with some success. Reports from several laboratories have shown a correlation between the presence of a bright greenish-yellow (BGY) fluorescence in cotton fibers and the occurrence of *Aspergillus flavus* in these fluorescing fibers. The BGY fluorescence is thought to be due to kojic acid produced by the fungus which is converted to the fluorescing substance by plant tissue peroxidases. The fluorescence also occurs in corn and other grains in which *A. flavus* has grown. The correlation with aflatoxin is based on the fact that most aflatoxin producing strains of *A. flavus* and *Aspergillus parasiticus* also generally produce kojic acid. The test is known as the "black light" test because of the use of UV light. The kernels are exposed to the UV light and observed for the characteristic type of BGY fluorescence. The kernels may have to be broken or sectioned since the fluorescence is on the interior of the kernel. If the characteristic fluorescence is observed, it is presumptive evidence for the presence of aflatoxins. It is very important that the test be read properly, since most biological materials contain fluorescent compounds. Only the specific yellow-green fluorescence can be counted as positive. If just any fluorescence is counted as positive, many false positives can result. The test is best utilized as a screening technique, since there is not a 100% correlation between BGY fluorescence and the presence of aflatoxins. If used as a presumptive test coupled with chemical tests, the black light test can be useful in detecting lots of grains or oilseeds contaminated with aflatoxins.

Other rapid methods for the detection of aflatoxins have involved chemical extraction of the toxins and chromatography in short columns of silica gel, or silica gel in combination with Florisil and alumina in layers. This test is known as the "minicolumn" procedure. A diagram of the construction of a minicolumn is shown in Fig. 15.2. The sample is extracted in a high speed blender with aqueous methanol or some other appropriate solvent, and the extract is applied to the minicolumn. The extract is forced through the column by applying a slight pressure. Columns containing standard aflatoxin solutions are developed and compared to columns containing the test samples. Comparable fluorescent bands indicate the presence of aflatoxins (Fig. 15.3). The procedure is essentially qualitative, since the exact amount of aflatoxins cannot be determined, but levels of aflatoxins as low as 5 to 15 ppb for mixed feeds and 2 to 5 ppb for seeds such as corn, cottonseed, and peanuts can be detected. Since the current Food and Drug Administration guideline for aflatoxin is 20 ppb, this test is useful in detecting commodities that would be in violation of federal guidelines. Even if the guidelines are lowered to 15 ppb as suggested by some, this test will still be useful.

By combining the black light test and the minicolumn test, a relatively simple and rapid procedure for screening commodities for aflatoxin contamination can be implemented. If more specific determinations of toxin levels are needed, these tests can be supplemented by the traditional extraction and TLC methods. The minicolumn can be prepared in the laboratory or can be obtained from commercial sources (Myco-Lab Company, P.O. Box 321, Chesterfield, Mo. 63017, and Tudor Scientific Glass Company, 555 Edgefield, Belvedere, S.C. 29841).

BIOLOGICAL ASSAY METHODS

Numerous biological systems have been studied for their potential

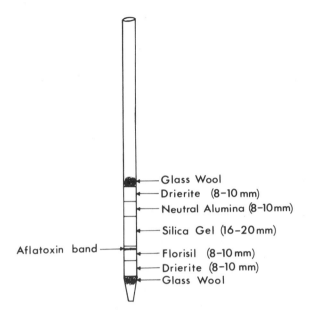

FIG. 15.2 DIAGRAM OF THE CONSTRUCTION OF ONE TYPE OF MINICOLUMN (ADAPTED FROM ROMER 1975)

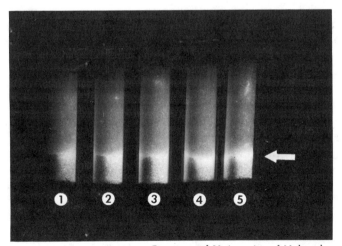

Courtesy of University of Nebraska

FIG. 15.3. DEVELOPED MINICOLUMNS SHOWING, FROM LEFT TO RIGHT, (1) SOLVENT CONTROL COLUMN AND COLUMNS CONTAINING (2) 10, (3) 20, (4) 50, AND (5) 100 PPB OF AFLATOXINS. Arrow indicates the position of the aflatoxin band.

use as bioassays for detecting and quantitating mycotoxins. These systems include cell and tissue cultures, animals, and microorganisms.

Cell and Tissue Cultures

The only accepted bioassay method given in the AOAC Manual of Methods is the chicken embryo bioassay for aflatoxin B_1. However, this system can also be used to test for toxicity of other mycotoxins. The chicken embryo bioassay is actually classed as a tissue culture system, since the test system is an embryo.

The AOAC manual recommends that eggs from fertile inbred single-comb White Leghorns be used, although other strains suitable for research or bioassay can also be used. After the eggs are candled, the air cell is marked and the surface is sterilized by swabbing with alcohol (or other suitable disinfectant) prior to drilling a hole into the shell covering the center of the air cell. The hole can best be made using a small, high-speed electric drill, although a sharpened hypodermic needle can also be used. The material to be tested can be dissolved in one of several solvents such as water, propylene glycol, ethanol, or vegetable oils. Organic solvents such as ethyl acetate and chloroform can also be used if they are highly purified and if used in very small amounts (20 μl or less). The eggs can be injected in either the air cell or the yolk sac. However, the air cell method in which the extract is placed on the inner egg membrane is much easier to perform and requires less skill. The eggs can be injected prior to incubation or several days after the embryo has started to develop. By injecting several days after incubation, the presence of a live embryo can be assured, but the time of injection is critical. After about 4 days of incubation the embryo is at the weakest point in its development. This can be an advantage when detecting minute quantities of toxins but can also be a disadvantage in that high background mortality may occur, making the test difficult to interpret. If the embryo is too old, it may be resistant to some toxins and give an inaccurate response. For routine work, injection prior to incubation is probably most desirable, especially if the eggs have a high fertility rate. Best results are obtained if the eggs are incubated in an automatic incubator set at 37.4°C and 60% RH. If the incubator can be adjusted to automatically turn the eggs, turning should be done every 2 hr. Otherwise, the eggs should be hand turned twice daily. The eggs should be candled beginning on the fourth day and every day thereafter until hatching.

The number of eggs to be used in a test may vary somewhat, but the recommended numbers are 30 noninjected controls and 20 for each sample and injected control. Injected controls consist of eggs injected with solvent containing no toxin. Sometimes as few as 10 eggs per sample and 20 for the noninjected control are used. More accurate results are obtained if more eggs are used. Death of embryos is counted as a positive test. However, background mortality must also be taken into account, and if the percentage mortality is equal to or less than the background mortality, the extract is considered to be nontoxic, or the test is considered invalid. Dead embryos should be removed from the shell and examined for abnormalities which may indicate teratogenic effects of toxic extracts. Abnormalities usually involve the beak, wings, legs, or feet (Fig. 15.4).

The egg embryo test has decided advantages and disadvantages. It is relatively inexpensive as bioassays go, rather simple to run, and very sensitive. The biggest disadvantage is that the embryos, while being very sensitive, are not specific in their response and may be killed by numerous compounds other than mycotoxins. Also, sometimes high background mortality is experienced which complicates the interpretation of the test. This occurs most often when eggs suitable for research and bioassay are not available and ordinary fertile hatching eggs must be used. In spite of its disadvantages, however, the egg embryo test remains a good tool for detecting toxicity of mold culture extracts to higher animals, and it is inexpensive enough to be used as a toxicity screening test in certain situations. It also has great value in confirming toxicity of suspected compounds in culture extracts.

A number of other tissue and cell cultures have been used to detect mycotoxins. Among these are tracheal organ cultures from day-old chicks, catfish cell cultures, HeLa cells, liver cells, and human embryonic lung cells. While these systems are sensitive to mycotoxins, they have not found routine use in the detection and toxicity testing of mycotoxins.

Animals

Other animal test organisms that are sensitive to mycotoxins and especially aflatoxins are brine shrimp, zebra fish larvae, brown planaria, the crustacean *Cyclops fuscus*, eggs and larvae of the amphibian *Triturus alpestris*, and fertilized eggs of the mollusk *Bankia setacea*, the cleavage of which is inhibited by low levels of aflatoxin B_1. Of these systems, brine shrimp probably offer the greatest potential for routine detection and testing of mycotoxins. This organism is sensitive to many mycotoxins and is easy to work with because eggs, not live animals, are kept on hand. Brine shrimp eggs are readily available in most pet shops and stores that carry tropical fish supplies.

Among larger animals, the trout and the duckling are the most sensitive to aflatoxins. Both of these organisms have been used in bioassays for aflatoxins and both exhibit rather characteristic pathological changes in the liver which are characteristic of aflatoxin

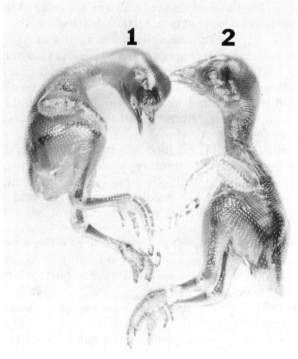

Courtesy of Dr. J. Verrett. U.S. Food and Drug Administration
FIG. 15.4. MALFORMED CHICK EMBRYO TREATED WITH OCHRATOXIN (1) AND
NONTREATED EMBRYO (2)

intoxication. However, trout and ducklings are harder to work with
than are the simpler organisms because the observer must be trained
in pathology to detect the changes that occur in the liver. Also,
obtaining and maintaining these animals is more difficult and
expensive. These animals have value in confirming aflatoxins but are
more useful as research tools than for routine use. Other animals that
may be used to detect mycotoxins include mice, rats, and day-old
chicks. A number of other larger animals have been used to study the
physiological and pathological effects of mycotoxins but have not
been used as assay systems.

Microorganisms

Many bacteria have been tested for sensitivity to mycotoxins for the
purpose of developing microbiological assay techniques. Those bacteria that show the most promise for detecting and assaying mycotoxins are *Bacillus megaterium*, *B. subtilis*, and *B. stearothermophilus*. Of

these organisms, *B. stearothermophilus* seems to be most sensitive, being capable of detecting as low as 0.01 μg aflatoxin (Reiss 1975B). *B. megaterium* and *B. subtilis* have similar sensitivities, around 1 to 2 μg of aflatoxins and patulin (Reiss 1975A; Stott and Bullerman 1975A). Other bacteria that have been studied include *Escherichia coli*, *Staphylococcus aureus*, and a *Flavobacterium* sp. The protozoan *Tetrahymena pyriformis* has also been examined. The *Bacillus* spp. appear to have the greatest promise for sensitive bioassays of mycotoxins. With bacterial assay systems a standard curve may be constructed, similar to those used for antibiotic assays, which can be used to obtain quantitative results (Fig. 15.5).

MYCOTOXIN STANDARDS

In order to recognize and quantitate mycotoxins using either physico-chemical or biological methods, it is necessary to employ authentic reference standards. Standards are available from a number of commercial sources. A list of these sources is given in the AOAC Official Manual of Methods in the section dealing with aflatoxins. Commercial sources of mycotoxin standards in the United States are summarized in Table 15.3. In addition to the U.S. sources of mycotoxin standards, they can also be obtained from Makor Chemicals, Ltd., P.O. Box 6570, Jerusalem, Israel, which produces aflatoxins, ochratoxin A, patulin, penicillic acid, rubratoxin B, sterigmatocystin, and diacetoxyscirpenol. Sources of mycotoxin standards in Europe are Rijksinstitut voor de Volksgezonheid, P.O. Box 1, Bilthoven, The Netherlands, and Senn Chemicals, CH-8157 Dielsdorf, Switzerland. In addition, mycotoxins that are not commercially available can sometimes be obtained from scientists who have worked with the specific compound in question, as evidenced by the scientific literature.

Any mycotoxin standard used for analytical purposes should be checked for purity by TLC and by determining the molar absorption and UV spectra. Procedures for doing this with aflatoxins are given in the AOAC Manual of Methods. This is especially important if the standards are to be used for quantitative work. Aflatoxins may adsorb to glass and may not completely redissolve when attempts are made to put them back into solution. Therefore, the resulting solution should be recalibrated to determine exact concentration. If the standards are to be used for qualitative screening only, the question of concentration becomes less critical and a certain amount of impurity can be tolerated. Again, while the information given in the AOAC Manual applies primarily to aflatoxins, most of the principles apply similarly to other mycotoxins.

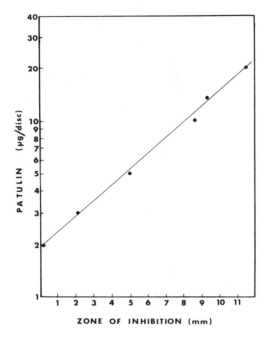

FIG. 15.5. A TYPICAL STANDARD CURVE FOR BIOASSAY OF PATULIN USING *BACILLUS MEGATERIUM*

Mycotoxin standards should be protected from light, particularly light containing UV rays, when possible, because this may cause deterioration and degradation. Standard solutions should be stored in the dark and in a freezer if possible, or in a refrigerator at the very least. Also, toxin standards should be used rapidly and the container should be quickly reclosed to prevent evaporation of the solvent and inadvertent concentration of the standard. Mycotoxin standards should also be checked frequently to redetermine concentration, since deterioration of the toxins may accompany prolonged storage and evaporation during normal use may cause concentration of the standard to occur.

The relative R_f positions of several mycotoxin standards on a TLC plate are shown in Fig. 15.6.

LABORATORY SAFETY

The extent to which mycotoxins can affect human health is not fully known or understood. In light of this lack of knowledge, all mycotoxins and extract preparations should be considered to be very toxic

TABLE 15.3

COMMERCIAL SOURCES OF MYCOTOXIN STANDARDS
IN THE UNITED STATES

Source	Address	Toxin Standards Available
Aldrich Chemical Company, Inc.	940 West St. Paul Ave. Milwaukee, WI 53233	Aflatoxins Diacetoxyscirpenol Ochratoxin A Rubratoxin B Sterigmatocystin
Applied Science Laboratories, Inc.	State College, PA 16801	Aflatoxins
Calbiochem	P.O. Box 12087 San Diego, CA 92112	Aflatoxins Diacetoxyscirpenol Ochratoxin A Patulin Rubratoxin B Sterigmatocystin
Sigma Chemical Company	P.O. Box 14508 St. Louis, MO 63178	Aflatoxins Sterigmatocystin
Supelco, Inc.	Supelco Park, Bellefonte, PA 16801	Aflatoxins
Wale Corporation	1975 Ocean Avenue P.O. Box 27174 San Francisco, CA 94127	Aflatoxins Citrinin Ochratoxin A Penicillic acid Patulin P. R. toxin Sterigmatocystin T-2 toxin Zearalenone

Courtesy of University of Nebraska

FIG. 15.6. RELATIVE R$_f$ POSITIONS OF SEVERAL MYCOTOXINS ON A DEVEL-
OPED SILICA GEL G-HR (BRINKMAN INST., INC.) TLC PLATE USING TOLUENE
ETHYL ACETATE: FORMIC ACID (90%) (6:3:1). From left to right the toxins are ochra-
toxin A, patulin, penicillic acid, aflatoxins (top to bottom = B$_1$, B$_2$, G$_1$, G$_2$), and citrinin.

substances and handled accordingly. In addition, the aflatoxins are
known carcinogens which have been found to be very potent when
tested in animals. It is not known if mycotoxins can enter the body by
portals other than the alimentary tract, viz., the lungs or by skin
absorption. The aflatoxins are very hydrostatic and if handled in the
dry state could easily become airborne and inhaled. T-2 toxin is sufficiently volatile that skin irritation can result from careless handling
of solutions of this toxin. In addition, a number of other mycotoxins
can also be very irritating to the skin and can cause dermal lesions in
test animals. Besides the mycotoxins themselves, some of the molds
which produce mycotoxins are also potential pathogens. *A. flavus*, for
example, can cause a lung disease known as aspergillosis, and all mold
spores are potentially allergenic. The spores can affect certain individuals by causing severe sinus and respiratory allergenic responses.

Most of the mycotoxins are generally not heat labile and are not
destroyed by heating to ordinary food and beverage sterilization
temperatures. In order to achieve destruction of aflatoxins, for
example, temperatures in excess of 300°C are required.

Recommendations for routine safety procedures have been made by
Stoloff and Trager (1965). Sodium hypochlorite is a strong oxidizing
agent and is known to destroy aflatoxins and render them nontoxic.
The effect on other mycotoxins is thought to be equally destructive.
Ordinary liquid household bleach containing 5 to 6% active sodium

hypochlorite is very effective in destroying mycotoxins as well as fungal spores and mycelia. Bleach solution can be used therefore as an effective detoxification and antifungal agent in a program of laboratory safety. Liquid bleach, either full strength or diluted tenfold, can be used to decontaminate glassware and other apparatus. Contact time should be at least 30 sec. If full strength bleach is used, care should be taken that the contact time is not too long, since etching of glassware and destruction of metals and plastics will occur. Strong acid cleaning solutions will also destroy the toxins. Bleach also can be used to clean up areas where toxin solutions or extracts have been spilled. It may be necessary to cover these areas with paper towels before treatment with bleach to ensure adequate contact of the surface with the bleach and to take up excess solution. Surfaces of work areas should also be wiped down with bleach solutions diluted tenfold. Chemical fume hoods can periodically be subjected to chlorine gas by mixing equal volumes of undiluted bleach and 6 N hydrochloric acid and immediately closing the door without the exhaust fan turned on. After the gas has diffused through the hood for several minutes, the fan can be turned on to exhaust the chlorine gas.

Several precautions should be taken to protect workers who must handle toxic solutions, cultures, or commodities. Workers should routinely wear protective gear such as disposable or rubber gloves when working with toxins. Workers should avoid breathing dust or spores that may be associated with contaminated food or toxic cultures. This can be accomplished by working under a fume or exhaust hood when handling toxic materials. When dry aflatoxins or other powdery materials are handled, the worker should wear a respirator or disposable mask to prevent inhalation of potentially harmful toxin. Whenever possible, aflatoxins should be in solutions and handled as such to avoid working with the aflatoxin powder which is very electrostatic. Disposable laboratory coats are also helpful in protecting the worker's clothing from contamination with mycotoxins. Other disposable materials, such as pipettes, can greatly aid in preventing exposure to toxins during cleaning and washing. Such disposable items should be treated with bleach before being discarded. Workers should use pipetting bulbs rather than the mouth when pipetting toxin solutions to prevent any oral exposure to toxins. If mycotoxins, toxic organisms, or culture material come in contact with the skin, the area should be immediately washed with full-strength bleach if possible or otherwise with diluted bleach solutions, followed by washing with germicidal soap or detergent and rinsing with copious amounts of water. Of course, food, beverages, and smoking should not be allowed in the laboratory in order to prevent possible contamination and exposure.

Toxic cultures, commodities, and samples to be disposed of should first be sterilized by autoclaving if possible. Such materials should then be either incinerated or soaked in full strength bleach for 30 to 60 min before disposal. The treated material should then be disposed of in tightly sealed containers.

In general, if a few simple safety precautions are routinely followed, there should be minimum hazard associated with the analysis of commodities for mycotoxins. Each laboratory should establish and strictly adhere to a definite safety program.

Published with the approval of the Director as paper No. 5175, Nebraska Agricultural Experiment Station, Lincoln.

REFERENCES

ALISAUSKAS, V. A. 1974. Determination of aflatoxin in grain and feedstuffs. Food Technol. Aust. 233-235,237.

ARCHER, M. 1974. Detection of mycotoxins in foodstuffs by use of chicken embryos. Mycopathol. Mycolog. Appl. 54, 453-467.

ASSOC. OFF. ANAL. CHEM. 1975. Official Methods of Analysis of the AOAC, 12th Edition. Assoc. Off. Anal. Chemists, Washington, D. C. Sections 26.001-26.115.

AYRES, J. C., LILLARD, H. S., and LILLARD, D. A. 1970. Mycotoxins: methods for detection in foods. Food Technol. 24, 161-164, 166.

BARABOLAK, R., COLBURN, C. R., and SMITH, R. J. 1974. Rapid screening method for examining corn and corn-derived products for possible aflatoxin contamination. J. Assoc. Off. Anal. Chem. 57, 764-766.

CHU, F. S. 1974. Studies on ochratoxins. CRC Crit. Rev. Toxicol. 2, 499-524.

CIEGLER, A., DETROY, R. W., and LILLEHOJ, E. B. 1971. Patulin, penicillic acid and other carcinogenic lactones. In Microbial Toxins, Vol. VI. A. Ciegler, S. Kadis, and S. J. Ajl (Editors). Academic Press, New York.

CIEGLER, A., and LILLEHOJ, E. B. 1968. Mycotoxins. Adv. Appl. Microbiol. 10, 155-219.

COLE, R. J., KIRKSEY, J. W., MOORE, J. H., BLANKENSHIP, B. R., DIENER, U. L., and DAVIS, N. D. 1972. Tremorgenic toxin from Penicillium verruculosum. Appl. Microbiol. 24, 248-250.

CUCULLU, A. F., LEE, L. S., MAYNE, R. Y., and GOLDBLATT, L. A. 1966. Determination of aflatoxins in individual peanuts and peanut sections. J. Am. Oil Chem. Soc. 43, 89-92.

CUCULLU, A. F., PONS, W. A., and GOLDBLATT, L. A. 1972. Fast screening method for detection of aflatoxin contamination in cottonseed products. J. Assoc. Off. Anal. Chem. 55, 1114-1119.

DETROY, R. W., LILLEHOJ, E. B., and CIEGLER, A. 1971. Aflatoxin and related compounds. In Microbial Toxins, Vol. VI. A. Ciegler, S. Kadis, and S. J. Ajl (Editors). Academic Press, New York.

EPPLEY, R. M. 1975. Methods for the detection of trichothecenes. J. Assoc. Off. Anal. Chem. 58, 906-908.

FENNELL, D. I., BOTHAST, R. J., LILLEHOJ, E. B., and PETERSON, R. E. 1973. Bright greenish-yellow fluorescence and associated fungi in white corn naturally contaminated with aflatoxin. Cereal Chem. 50, 404-414.

HALD, B., and KROGH, P. 1975. Detection of ochratoxin A in barley, using silica gel minicolumns. J. Assoc. Off. Anal. Chem. 58, 156-158.

HOLADAY, C. E., and BARNES, P. C., JR. 1973. Sensitive procedure for aflatoxin detection in peanuts, peanut butter, peanut meal and other commodities. J. Agric. Food Chem. 21, 650-652.

HOLADAY, C. E., and LANSDEN, J. 1975. Rapid screening method for aflatoxin in a number of products. J. Agric. Food Chem. 23, 1134-1136.

LEGATOR, M. S. 1969. Biological assay for aflatoxins. In Aflatoxin: Scientific Background, Control and Implications. L. A. Goldblatt (Editor). Academic Press, New York.

MARTH, E. H. 1967. Aflatoxins and other mycotoxins in agriculture products. J. Milk Food Technol. 30, 192-198.

MOSS, M. O. 1971. The rubratoxins, toxic metabolites of Penicillium rubrum Stoll. In Microbial Toxins, Vol. VI. A. Ciegler, S. Kadis, and S. J. Ajl (Editors). Academic Press New York.

NESHEIM, S. 1969. Conditions and techniques for thin layer chromatography of aflatoxins. J. Am. Oil Chem. Soc. 46, 335-338.

NESHEIM, S. 1976. The ochratoxins and other related compounds. In Mycotoxins and Other Fungal Related Food Problems. J. V. Rodricks (Editor). Adv. in Chem. Ser. No. 149. Am. Chem. Soc., Washington, D. C.

PONS, W. A., CUCULLU, A. F., FRANZ, A. O., LEE, L. S., and GOLDBLATT, L. A. 1973. Rapid detection of aflatoxin contamination in agricultural products. J. Assoc. Off. Anal. Chem. 56, 803-807.

PONS, W. A., and GOLDBLATT, L. A. 1969. Physico chemical assay of aflatoxins. In Aflatoxin: Scientific Background, Control and Implications. L. A. Goldblatt (Editor). Academic Press, New York.

PURCHASE, I. F. H. 1974. Penicillium cyclopium. In Mycotoxins. I. F. H. Purchase (Editor). Elsevier Scientific Publishing Co., New York.

REISS, J. 1975A. Bacillus subtilis: A sensitive bioassay for patulin. Bull. Environ. Contam. Toxicol. 13, 689-691.

REISS, J. 1975B. Mycotoxin bioassay using Bacillus stearothermophilus. J. Assoc. Off. Anal. Chem. 58, 624-625.

ROMER, T. R. 1975. Screening method for the detection of aflatoxins in mixed feeds and other agricultural commodities with subsequent confirmation and quantitative measurement of aflatoxins in positive samples. J. Assoc. Off. Anal. Chem. 58, 500-506.

ROSEN, J. D., and PARELES, S. R. 1975. Quantitative analysis of patulin in apple juice. J. Agric. Food Chem. 22, 1024-1026.

SCOTT, P. M. 1969. The analysis of foods for aflatoxins and other fungal toxins—a review. Can. Inst. Food Technol. J. 2, 173-177.

SCOTT, P. M., LAWRENCE, J. W., and VAN WALBEEK, W. 1970. Detection of mycotoxins by thin-layer chromatography: Application to screening fungal extracts. Appl. Microbiol. 20, 839.

STEYN, P. S. 1971. Ochratoxin and other dihydroisocoumarins. In Microbial Toxins, Vol. VI. A. Ciegler, S. Kadis, and S. J. Ajl (Editors). Academic Press, New York.

STOLOFF, L. 1972. Analytical methods for mycotoxins. Clin. Toxicol. 4, 465-494.

STOLOFF, L., CAMPBELL, A. D., BECKWITH, A. C., NESHEIM, S., WINBUSH, J. S., and FORDHAM, O. M., JR. 1969. Sample preparation for aflatoxin assay: The nature of the problem and approaches to a solution. J. Am. Oil Chem. Soc. 46, 678-684.

STOLOFF, L., DANTZMAN, J. G., and WEGENER, J. 1972. Preparation of lot samples of nut meats for mycotoxin assay. J. Am. Oil Chem. Soc. 49, 264-266.

STOLOFF, L., NESHEIM, S., YIN, L., RODRICKS, J. V., STACK, M., and CAMPBELL, A. D. 1971. A multimycotoxin detection method for aflatoxins, ochratoxins, zearalenone, sterigmatocystin and patulin. J. Assoc. Off. Anal. Chem. 54, 91-97.

STOLOFF, L., and TRAGER, W. 1965. Recommended decontamination procedures for aflatoxins. J. Assoc. Off. Anal. Chem. 48, 681-682.

STOTT, W. T., and BULLERMAN, L. B. 1975A. Microbiological assay of patulin using Bacillus megaterium. J. Assoc. Off. Anal. Chem. 58, 497-499.

STOTT, W. T., and BULLERMAN, L. B. 1975B. Patulin: A mycotoxin of potential concern in foods. J. Milk Food Technol. 38, 695-706.

THORPE, C. W., and JOHNSON, R. L. 1974. Analysis of penicillic acid by gas-liquid chromatography. J. Assoc. Off. Anal. Chem. 57, 861-865.

TIEMSTRA, P. J. 1969. A study of the variability associated with sampling peanuts for aflatoxin. J. Am. Oil Chem. Soc. 46, 667-672.

VESONDER, R. F., CIEGLER, A., and JENSEN, A. H. 1973. Isolation of the emetic principle from Fusarium-infected corn. Appl. Microbiol. 24, 248-250.

WARE, G. M., THORPE, C. W., and POHLAND, A. E. 1974. Liquid chromatographic method for determination of patulin in apple juice. J. Assoc. Off. Anal. Chem. 57, 1111-1113.

WHITAKER, T. B., DICKENS, J. W., and WISER, E. H. 1970. Design and analysis of sampling plans to estimate aflatoxin concentrations in shelled peanuts. J. Am. Oil Chem. Soc. 47, 501-504.

WHITAKER, T. B., and WISER, E. H. 1969. Theoretical investigations into the accuracy of sampling shelled peanuts for aflatoxin. J. Am. Oil Chem. Soc. 46, 377-379.

WILSON, B. J., and HAYES, A. W. 1973. Microbial toxins. In Toxicants Occurring Naturally in Foods, 2nd Edition. N.A.S.-N.R.C., Washington, D. C.

16

Methods for Detecting Fungi in Foods and Beverages

B. Jarvis

B oth filamentous and unicellular forms of microfungi are widely distributed in soil, air, and water and occur as important contaminants of a wide range of agricultural commodities and manufactured food and beverage products. In certain sections of the food and allied industries, molds and yeasts have long been used to produce desirable changes in foods, e.g., mold ripening of cheeses and meats, alcoholic fermentations, etc. However, in different circumstances molds and yeasts may be responsible for spoilage of foods, and the consumption of mold-spoiled materials is associated with human disease.

It is the growing awareness of the significance of mycotoxins during the past decade which has led to enhanced interest in food mycology, for too long the neglected branch of food microbiology (Jarvis 1975; Mossel *et al.* 1975). This interest has been strengthened by an increasing desire among legislators to introduce both national and international quantitative standards for the microbiological quality of foods. Although most proposed standards are concerned primarily with protection of public health, others (e.g., certain draft European Economic Community Directives on Food Quality) relate more closely to establishment of quality criteria, a practice long pursued by reputable food and beverage manufacturers.

Since several different approaches may be used for mycological analyses, it is essential to define the objectives of any investigation and to take note of the environmental and microenvironmental conditions affecting microorganisms on or in any food or beverage substrate before analyses are initiated. The choice of method to be used will depend on whether quantitative data are required, the nature and condition of the sample, whether the analysis is of a routine nature, and whether standard methods are prescribed by regulatory authorities.

The results obtained in any mycological analysis will be influenced by many factors including method of sampling, sample preparation, the culture medium used, and the incubation conditions. This chapter will consider the implications of methodological aspects of the isolation and enumeration of fungi from foods and beverages.

SAMPLING

Mycological analyses can be undertaken on either a quantitative or a qualitative basis, but a meaningful sampling program is fundamental to both. Present day attitudes to quantitative microbiology are based on the use of statistical sampling plans which attempt to relate the degree of potential hazard to the intensity of sampling (ICMSF 1974; Mossel 1975). However, few sampling plans have been derived specifically for mycological analysis. Two types of plans exist: two-class and three-class plans.

For those situations where high counts of mold may indicate a possible health hazard from mycotoxins (e.g., for infant, geriatric, and other dietetic foods) a two-class sampling plan should be used. Such plans permit the occurrence of particular organisms only below a critical level. A scheme of this type has been recommended for UNICEF dried weaning formulae and requires that not more than two samples of ten tested may contain 100 mold spores per g of food (Mossel et al. 1974).

In a three-class plan, tolerances are set for "acceptable levels" (m) and "unacceptable levels" (M) of organisms; a certain number of samples (c) of those tested (n) are permitted to contain levels of organisms within the marginally acceptable area (i.e., between m and M). However, for most foods a problem arises in establishment of appropriate values for m, M, c, and n. The values of c and n are derived from the acceptable degree of risk which the manufacturer (or regulatory authority) is prepared to accept and can be derived statistically (ICMSF 1974). Values for m and M must be based on data obtained from adequate surveys to assess what is technologically achievable under Good Manufacturing Conditions and must take into account the confidence limits of the methodology and the natural distribution of specific groups of organisms in a particular type of food. For a more detailed and erudite discussion of this topic the reader is referred to Mossel (1975).

Because of the variation in contamination level of many foods, especially bulk-stored materials, sampling should ideally be based on a statistical program, but in routine practice the level of sampling which can be applied is often limited by the available laboratory facilities and the costs of extensive sampling. A frequently used compromise is to composite multiple samples prior to laboratory analysis.

Bulk consignments of products such as grain, oilseeds, and nuts can be sampled using apparatus which permits the withdrawal of representative samples. Similarly, liquid bulks can be sampled using a flow diverter or by means of a vacuum sampling tube connected to an appropriate sterile sample receiver. Such apparatus has been used successfully for sampling viscous solutions of food stabilizers contained in large barrels. Whenever possible, liquids should be mixed before sampling.

Sampling of manufactured food and beverage products at the factory, warehouse, or retail outlet involves drawing an appropriate number of packs or units for analysis. When retail packs are taken for official analysis, it is essential to mix the contents well before subsampling to obtain the three replicate official samples often required by law. In any sampling of retail items all appropriate data should be recorded, including manufacturers' codes, "sell by" or "manufacturing" dates, pack sizes, etc. When a sample is required for both chemical and microbiological analyses, it is essential that the microbiologist be able to draw a subsample aseptically before allowing the chemist access to the sample.

Sampling Solid Foods

For quantitative analysis a known weight of sample is examined after an appropriate treatment to recover the microorganisms present; qualitative samples are frequently handled in a similar manner. Details of procedures for drawing samples are well documented (see, for example, Thatcher and Clark 1968; Board and Lovelock 1973; ICMSF 1974; AOAC 1975).

Frozen foods are generally best sampled from the frozen state to prevent multiplication of yeasts and molds, but can be sampled after a limited period (e.g., 12 hr) in a refrigerator. Samples of frozen foods can be taken using a sterile saw, auger, hammer and chisel, or other suitable instrument.

Nonfrozen foods such as chilled meats and meat products, bakers' confectionery, candies, dried foods, etc., can be sampled using a sterile knife, scissors, trier, or other implement. For composite foods (e.g., meat or fruit pies) it is essential that the laboratory sample be either representative of the whole product (i.e., pastry case and filling in appropriate proportions), or, preferably, that the component parts be tested separately. Dried foods should be ground before subsampling.

When specific sampling procedures are prescribed by law, it is essential that laboratory staff abide by the required procedures even though this may involve a departure from normal laboratory practice.

Total Mycoflora.—Dried foods, including nuts, oilseeds, and cereals, are best ground to a fine powder before blending with an appropriate diluent. A small domestic coffee grinder, sterilized with alcohol between samples, provides an adequate method of grinding. If heavy mold contamination is anticipated, grinding should be carried out in a pathogen cabinet to protect the operator and laboratory from airborne mold spores. Other foods are normally blended with diluent without preliminary grinding, although preliminary portioning may be advantageous for hard fruits and similar commodities.

Two traditional forms of blending are commonly used (Fig. 16.1). Bottom-drive macerators, such as the Waring Blendor or Atomix, are used with autoclavable jars of glass or stainless steel fitted with clamp-on lids. Top-drive homogenizers, such as the MSE blender or the Silverson mixer, have separate sterilizable sample containers but the blade units are normally sterilized in situ by alcohol and/or flaming. A Silverson mixer fitted with a sealed sample jar and blade unit is particularly suitable for homogenization of samples containing potentially pathogenic fungi, since the design eliminates aerosol dispersal.

Top-drive homogenizers frequently operate at higher rpm than do bottom-drive blenders, but transfer less heat to the sample. They also have the advantage that the sample container can be immersed in an ice bath during homogenization. The problem of temperature rise during blending can be overcome partly by the use of chilled diluents. If adequate maceration and blending are not achieved in the first 2 to 3 min, the sample container can be refrigerated before reblending.

A more recent approach to sample preparation is to pummel the sample in a sterile plastic bag. The technique of "stomaching" (Sharpe 1973) has found wide acceptance in Europe and elsewhere and uses an apparatus known as the Colworth Stomacher (Fig. 16.2). The advantages of stomaching are that there is no increase in sample temperature, the laboratory does not have to purchase, wash, or sterilize large numbers of homogenizer jars or lids, and the blending action forces microorganisms out of foods without producing a homogeneous slurry. Comparative tests in several laboratories have shown equal or higher recoveries of microorganisms from foods treated by stom-

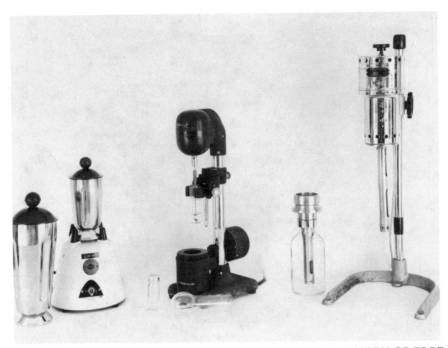

FIG. 16.1. MECHANICAL BLENDERS USED FOR HOMOGENIZATION OF FOOD SAMPLES. From left: Bottom-drive homogenizer (Atomix) fitted with a 250ml stainless steel container and with a 1 liter container to the side; top-drive homogenizer (MSE) fitted with a micro-sample attachment, showing spare square-bottomed microcontainer (10ml) and a macrocontainer(150 ml)in front: (far right) sealed unit for use with Silverson top-drive homogenizer.

aching than by conventional blending methods (Sharpe 1973;Tuttlebee 1975).

Surface Mycoflora.—The numbers of fungi on the surface of a food, or on work surfaces and process equipment, are normally expressed in terms of unit area sampled. Approximate surface areas of spherical or other bodies can be derived mathematically (Marshall and Walkley 1951).

Shaking samples with diluent has been used for many years to remove surface fungi but repeated shaking in several portions of diluent is necessary to remove the majority of propagules (Laszilo and Nyiredy 1974). Removal of surface microorganisms can frequently be improved by the use of abrasives such as sand or ballotini; recent studies in the author's laboratory have shown the usefulness of agitation with sterile sand to remove organisms from the surfaces of fruits (Blood 1976).

FIG. 16.2. COLWORTH STOMACHER 400. This stomacher is in open position to show the paddles. The sample, contained in a plastic bag, is placed between the paddles and the front plate which is locked in the vertical position by the top clamp. Several sizes of Stomachers are available.

Surface fungi can be removed by swabbing with either cotton or alginate wool swabs, using a template to delineate the area, but recovery of organisms even from alginate swabs dissolved in Calgon is rarely complete (Walter 1967). Impression samples can be taken using agar sausages (Cate 1965), or with agar in Rodac dishes (Angelotti et al. 1964) or mounted on strong gauze tape (Foster 1960). Velvet and other forms of replica pads can be used in the examination of irregular surfaces. The Sellotape impression technique of Endo (1966) provides a useful method for removing organisms from surfaces, and can be used for both total and viable counts of yeasts (Davenport 1967; Beech and Davenport 1971).

Direct coating of the sample surface with agar followed by incubation in a sterile container permits cultivation of organisms in situ and is particularly useful for demonstrating ecological associations. However, as with other replica plating techniques, colonies can be formed from both individual cells and clones of cells on the sample surface.

Probably the best method for removing all surface microorganisms is to macerate a thin layer of surface tissue removed with a sterile vegetable peeler. The method is not suitable, however, for routine control purposes since the exceptionally careful aseptic technique required makes sampling too time-consuming.

Internal Mycoflora.—It is frequently not appreciated that many foods, especially fruits, vegetables, and tree nuts, can have an internal mycoflora. This may result from airborne contamination of fruits at the time of pollination, or from fungal invasion of deep tissues.

The normal procedure for detection of internal contamination is to chemically sterilize the surface before maceration or grinding of the sample. Booth (1971A) has summarized the various methods of surface sterilization regularly used by mycologists. The author's preference is to immerse the sample briefly in 75% (vol/vol) ethyl alcohol and then to transfer the sample to a solution of sodium hypochlorite containing about 1000 ppm available chlorine for 2 min. After thorough rinsing in sterile distilled water the sample is dried on sterile filter paper before maceration.

An alternative procedure is to remove the surface layers aseptically, but this suffers the same disadvantages described for surface sampling. Recommended microscopic procedures for internal contamination of samples sometimes recommend a similar procedure [see, for instance, the APHA (1967) method for Howard Mold Count of butter].

Sampling Liquids

After thorough mixing, liquid subsamples are drawn using a pipet, or a wide-bore tube for viscous materials. The sample may be evaluated directly by examining membranes through which it has been filtered or following serial dilution.

In sampling from bulk liquids where both surface and deep spoilage fungi might occur (e.g., pickle stock brines), it is advisable to draw separate representative samples from appropriate areas, the deep sample being taken in a manner which avoids contamination by surface yeast or mold films. Mixed solid-liquid packs (e.g., brined or acid-packed pickles) should be sampled to permit separate analysis of the component parts.

Sampling Gases

Determination of the level of fungal contaminants in the air of a processing plant or laboratory is often of considerable importance. The numerous procedures developed for sampling gases have been reviewed in detail by Davies (1971). Probably the most commonly

TABLE 16.1

DILUENTS FOR YEAST AND MOLD COUNTS

Diluent	Used For	Reference
Phosphate-buffered water	General purpose	Am. Public Health Assoc. (1966, 1967)
Distilled water	General purpose	Am. Public Health Assoc. (1967)
0.85% (wt/vol) sodium chloride solution	General purpose	
Tween 80 in water	Salad emulsions	
¼ Strength Ringer's salts solution	General purpose	Ministry of Agriculture, Fisheries, and Food (1968)
¼ Strength Ringer's salts solution +0.15% (wt/vol) agar	Butter	
0.05 to 0.10% (wt/vol) Tween 80 in water	General purpose	Jarvis (1973)
0.05% (wt/vol) Tween 80 in ¼ strength Ringer's solution		LeBars (1972)
0.10% (wt/vol) dimethylsulfoxide solution	General purpose	LeBars (1972)
10° Brix sucrose solution	Osmophiles	Lloyd (1975)
20° Brix sucrose solution		Corry (1976)

used method in industry is the "air exposure" or "settle" plate method, but newer methods such as liquid impingement and membrane filtration (Millipore 1967) are finding increasing use.

PREPARATION OF SAMPLES FOR ANALYSIS

Dilution Series and Diluents

The primary dilution resulting from blending of the sample is normally further diluted in a sequential manner, usually by tenfold serial dilutions, to cover the expected range of sample contamination.

The choice of diluent for both primary and subsequent dilutions is dictated by the requirements of any "standard method," by the nature of the sample, and by personal preference. Some diluents recommended for general purpose use are summarized in Table 16.1. The author's preference is 0.05% (wt/vol) Tween 80 in ¼ strength Ringer's solution.

Yeasts and molds in sugar syrups, fruit juices, and high-salt solutions are in a microenvironment of high osmotic pressure. Consequently, it is essential to avoid osmotic shock which might result from the use of general purpose diluents. Corry (1976) used 20% sucrose solution as diluent when studying the effects of high sugar concentrations on the heat resistance of yeasts, while Lloyd (1975) used 10% sucrose for isolation of osmophilic yeasts from preserved ginger. Halophilic organisms are best diluted in a high-salt [18% (wt/vol) sodium chloride + 0.1% (wt/vol) peptone] diluent.

Although the pH of diluents is commonly near neutrality, it may occasionally be desirable to use a low-pH diluent for isolation of acidophilic yeasts from products such as pickles and tomato or fruit sauces. For high-fat foods such as butter and processed cheeses it may be necessary both to use a diluent containing a stabilizing agent (e.g., 0.15% agar) (M.A.F.F. 1968) and to warm the sample and diluent in order to obtain good emulsion of the sample.

Resuscitation and Activation Techniques

Comparatively little work has been carried out on the need to resuscitate sublethally injured yeasts and molds from foods. However, it is widely recognized that even uninjured cells are sensitive to the presence in culture media of trace impurities originating from the use of impure agars (Booth 1971B) or from metallic contamination of distilled water (Booth 1971A).

Activation of spores to stimulate germination has frequently been achieved by the use of heat, cold, or chemical "germinants." The most commonly used method has been to heat the primary dilution at 50° to 75°C for periods up to 30 min. Splittstoesser et al. (1970) found that activation of ascospores of Byssochlamys spp. and other organisms was optimal when the fruit homogenate was heated for 60 min at 70°C. They recommended blending the sample with 16° Brix Concord grape juice followed by immersion of the blender jar (contained in a plastic bag) in a 70°C water bath for 2 hr before plating in potato dextrose agar. Similar results were achieved by blending with 5% (wt/vol) yeast extract solution at pH 3.6.

QUANTITATIVE AND QUALITATIVE CULTURAL METHODS

Culture Media

Media for the isolation of fungi are of three types: general purpose, selective, and differential (diagnostic). General purpose media include those which will permit growth of most types of microorganisms encountered (including bacteria), although no single medium will support growth of all organisms. Selective media permit the growth of some fungi while inhibiting others and may be derived by the use of chemical inhibitors, by selection of pH value, or by the omission of specific nutrients. Differential media are used to isolate groups of fungi with particular characteristics, e.g., lipolytic or pectolytic organisms.

General Purpose Media.— The choice of a general purpose medium should be based, whenever possible, on the natural habitat of the fungi under investigation. For this reason many workers have used natural vegetable or fruit decoctions for primary isolation of yeasts and molds from foods. Such media include fish broth, apple juice agar, pea agar, wort agar, and many others. Formulae of many standard media are given by Booth (1971B).

When natural media are used the ingredients should be available throughout the year in a consistent form. Beech and Davenport (1971) described the production and storage of a pectin-free apple juice suitable for media production. Other workers have resorted to the use of canned or frozen vegetables and juices (e.g., V-8 juice) to provide a consistent supply of ingredients. However, in factory quality control laboratories time rarely permits preparation of media from natural ingredients.

General purpose media based on malt and/or yeast extracts fortified with glucose or other carbohydrate and solidified with 1.5 to 2% (wt/vol) agar permit the growth of the majority of fungi commonly encountered in food spoilage. Many basic recipes exist (Booth 1971B) and modifications of several commonly used media are available in dehydrated form from commercial suppliers of culture media (Table 16.2). In most instances the various commercial formulations compare well in performance and they are frequently available in both agar and broth forms. Special modifications of some media are also available for use with membrane filters. In most instances the user is recommended to acidify the media to pH 3.5 to 4.5 to control growth of bacteria.

Chemically defined or semidefined media are frequently used for morphological identification of mold isolates. Various modifications of traditional media, such as Czapek Dox solution, can be prepared in the laboratory (Booth 1971B) or purchased in dehydrated form. When using commercial dehydrated media for identification purposes, it is essential to ensure that known cultures give typical reactions on the medium, since use of some commercial formulations results in abnormal morphological characteristics, especially of aspergilli and penicillia.

Medium formulations which enhance sporulation (Sloan et al. 1961) or perithecium production (Acha and Villanueva 1961) are frequently useful in identification studies. Media for identification of yeasts are given by Lodder (1970) and Davenport (1973).

Selective Media.— Almost any of the general purpose media can be

TABLE 16.2

COMMERCIALLY AVAILABLE DEHYDRATED GENERAL PURPOSE
CULTURE MEDIA FOR YEASTS AND MOLDS

Medium	Source[1]	Comments
Buffered yeast agar	O	Modification of yeast-salt agar of Davis (1931)
Czapek solution agar	D, O, B	Oxoid medium has modified formula
Malt agar	D, B	—
Malt extract agar	D, O, B	—
Mycological agar	D, B	For culture maintenance
Potato dextrose agar	D, O, B	—
Sabouraud dextrose agar	D, O, B	BBL medium contains polypeptone and additional dextrose
Wort agar	D, O, B	BBL medium is adjusted to pH 7.0
Whey agar	O, B	For molds and yeasts in butter (Am. Public Health Assoc. 1967)

[1]D—Difco, Detroit, Michigan, U.S.A.;O—Oxoid Ltd.,Wade Road, Basingstoke, England
(distributed in the U.S.A.by K.C. Biological Ltd., Lenexa, Kansas);B—BBL,Division of
Bioquest, Cockeysville, Maryland, U.S.A.

made more selective for yeasts and molds by incorporating antibiotics
or other chemical inhibitors, and/or by reducing the pH of the medium.

Use of Acidified Media. —The time-honored method of making
media more selective for yeasts and molds is by acidification to a pH
within the range of 3.5 to 4.5. Various standard methods recommend
the use of media such as potato dextrose or malt agars acidified with
tartaric, lactic, citric, phosphoric, or hydrochloric acids (Table 16.3).

Sterilization of media to a pH below 5.0 results in partial hydrolysis
of the agar, with a consequent reduction in its setting power. Normal
practice is therefore to acidify media after sterilization and before
pouring plates. However, in some instances it is possible to retain a
satisfactory gel by increasing the agar concentration (Carr 1956).

Acidification of media restricts growth of all but acid tolerant
species of bacteria while permitting growth of yeasts and molds, many
of which are capable of growth at pH values down to about 2.5. There
are sound ecological reasons for the use of low-pH media when
isolating fungi from acid foods, but some fungi (e.g., certain *Mucor*

TABLE 16.3

MEDIA FOR MYCOLOGICAL ANALYSIS OF VARIOUS FOODS

Food Type	Medium[1]	Reference
Liquid sugars, soft drinks, spices, dried fruits and vegetables, eggs, egg products, bottle rinses	PDA, MA	
Dry sugars, syrups, molasses	PDA, DA	Am. Public Health Assoc. (1966)
Fruit juices and concentrates	PDA, MA; PCA + 5% (wt/vol) juice	
	WA + 45° Brix sugars[2]	Scarr (1959)
Frozen and fermented foods	PDA, MA, DA	
Meat and meat products	PDA	
Mayonnaise, salad creams	PDA, MDA, TMEA	Am. Public Health Assoc. (1966)
Brined and pickled vegetables	DA, DSB	
Cereals, flour and cereal products	PDA, MA, MSA	
Butter, cream and dairy products	PDA	Am. Public Health Assoc. (1967)
	MA, YSA	Ministry of Agriculture, Fisheries, and Food (1968)
Gelatin	RBCA	BSI[3]
Casein, caseinates, potato products, chocolate and confectionery	OGY	EEC[4]

[1]PDA—potato dextrose agar, acidified; MA—malt agar, acidified; DA—dextrose agar, acidified; PCA—plate count agar; WA—wort agar; MDA—malt dextrose agar, acidified; TMEA—Trommer's malt extract agar, acidified; DSB—dextrose broth plus salt; MSA—malt salt agar, acidified; YSA—yeast salt agar (Davis 1958); RBCA—rose bengal chloramphenicol agar (modification of Jarvis 1973); OGY—oxytetracycline glucose yeast extract agar (Mossel et al. 1962).
[2]For osmophiles.
[3]Draft British Standard Method, not yet published.
[4]Various draft European Economic Community directives.

spp. and *Schizosaccharomyces pombe*) are incapable of growth at low pH. Several workers have demonstrated lower counts of yeast and mold propagules on acid media than on neutral pH media containing antibiotics (Mossel et al. 1962; Overcast and Weakley 1969; Koburger 1970, 1971, 1972B, 1973; Hup and Stadhouders 1972; Jarvis 1973). By contrast Ladiges et al. (1974) obtained significantly higher counts on Sabouraud dextrose agar (SDA) at pH 3.5 than on SDA containing antibiotics. However, they also obtained lower counts on potato dextrose agar at pH 3.5, thus indicating that the composition of the medium affects response in acid conditions.

The inhibitory effect of reduced pH towards fungi is dependent also in part on the acid used, phosphoric and hydrochloric acids being more inhibitory than lactic, citric, or tartaric acids (Koburger 1971). Mossel et al. (1962) suggested that the reduced counts at low pH may be the result of formation of antimicrobials by reaction between the acid and certain constituents of the medium.

Use of Chemical Inhibitors. —Since many fungi grow better at neutral or slightly alkaline pH values, many studies have been carried out to compare low pH media with media containing antibiotics to inhibit bacterial growth. Goldberg (1959) defined the ideal characteristics an antibiotic should have to be incorporated in selective culture media. For fungal media the primary characteristic is that the antibiotic should inhibit bacteria without inhibiting any fungi. The antibiotics most frequently used for fungal media are detailed in Table 16.4. Of the commonly used antibiotics, gentamicin and chloramphenicol are probably the inhibitors of choice since they are stable in the medium, inhibit a wide spectrum of bacteria, and do not affect growth of yeasts or molds.

Media relying on a single antibiotic to provide bacteriostasis may fail when challenged with particular groups of bacteria. The oxytetracycline glucose yeast extract medium of Mossel et al. (1962) permits growth of many Bacillus spp. (Put et al. 1968; Put and Conway 1974), although the significance of this is disputed by Mossel et al. (1970). Media containing chlortetracycline (Jarvis 1973) permit growth of tetracycline-resistant Enterobacteriaceae frequently found in raw meats (Jarvis and Williams 1975). These can be overcome by replacing the tetracycline with chloramphenicol or gentamicin (Jarvis and Williams 1975; Mossel et al. 1975).

The addition of rose bengal to media for isolation of fungi has been used for many years (Martin 1950; Miller and Webb 1954). Overcast and Weakley (1969), Jarvis (1973), and Mossel et al. (1975) compared media with and without rose bengal and concluded that rose bengal prevents formation of excessive aerial mycelium from filamentous fungi. However, Mossel et al. (1975) reported inhibition of certain strains of yeast by rose bengal.

Isolation of yeasts from mold-contaminated samples is aided by the incorporation of sodium or calcium propionate in culture media (Lund 1956; Bowen and Beech 1967). The required level of propionate will depend largely on the pH of the medium, since more undissociated propionic acid will occur at lower pH values. Another commonly used inhibitor, diphenyl, is less effective than propionate (Beech and Carr 1955).

TABLE 16.4

COMBINATIONS OF ANTIBACTERIAL AGENTS USED IN
YEAST AND MOLD CULTURE MEDIA

Concentration (ppm) of[1]							
CH	CTC	G	K	OTC	RB	S	Reference
—	—	—	—	100	—	—	Mossel et al. (1962, 1970)
—	—	50	—	100	—	—	Mossel et al. (1970, 1975)
—	—	50	—	100	35	—	Mossel et al. (1975)
—	20	—	—	—	20	—	Overcast and Weakley (1969)
—	10	—	—	—	50	—	Jarvis (1973)
100	—	—	—	—	50	—	Jarvis and Williams (1975)
100	100	—	—	—	—	—	⎰ Koburger (1972A, B) ⎱ Koburger and Farhat (1975)
100	—	—	100	—	—	—	Ladiges et al. (1974)
50	—	—	—	—	—	30	⎱ Hup and Stadhouders (1972)
50	—	50	—	—	—	30	
—	—	—	—	—	33	30	Martin (1950)

[1]Antibacterial agents coded: CH, Chloramphenicol; CTC, Chlortetracycline; G, Gentamicin; K, Kanamycin; OTC, Oxytetracycline; RB, Rose bengal; S, Streptomycin.

Differential inhibitors can be used with advantage to restrict growth to certain groups of fungi. Actidione (cycloheximide) has long been used in association with antibiotics to selectively isolate "wild yeasts" from brewery plants and alcoholic beverages (Green and Gray 1950; van der Walt and van Kerken 1961). A fuchsin-sulfite mixture is used in Schwarz diagnostic medium to suppress growth of ale and beer yeasts, while permitting growth of most strains of wild yeast associated with brewing (Seidel 1973). Lin (1973) reported that although many culture yeasts are able to grow on Schwarz diagnostic medium, differences in colonial morphology permitted differentiation between culture yeasts and wild yeasts. Botran (2,6-dichloro-4-nitroaniline) has been used in media for the selective isolation of *Aspergillus flavus* from foods (Bell and Crawford 1967; Reiss 1972). Reiss's medium, which contained both botran and streptomycin, restricted growth or inhibited many other species of mold while permitting growth of *A. flavus*.

Xerophilic and xerotolerant fungi (Pitt 1975) can be isolated on media of reduced water activity, but they differ in their tolerance to salts and sugars (see Chap. 2). Brine yeasts (e.g., *Debaryomyces* spp.) tolerate sodium chloride (salt) concentrations up to 18 to 20% (wt/vol) and halophilic fungi such as *Wallemia sebi* (*Sporendonema sebi*) can grow in almost saturated salt solutions. By contrast, other fungi are inhibited by salt concentrations of 3 to 11%. Consequently, use of media with high salt concentrations permits isolation of halophiles, but the nitrogen source and pH of the media are often critical (Beech and Davenport 1971). Schoop's medium (containing sodium chloride, 10 g; glycerol, 0.5 g; glucose, 1.0 g; peptone, 1.5 g; fresh orange juice, 1 to 3 ml; and distilled water to 100 ml; pH 4.8) provides an excellent medium for growth of halophilic molds such as W. sebi. Etchells and Jones (1946) recommended the use of dextrose broth containing 5 or 10% added salt for isolation of film-forming yeasts from pickles (APHA 1966). Christensen (1946) used malt salt agar (APHA 1966) to isolate storage molds from cereal flours; this medium should not be confused with the yeast salt agar of Davis (1958) which does not contain sufficient salt to reduce the water activity of the medium significantly.

Osmophilic and osmotolerant yeasts are isolated readily on media containing high concentrations of carbohydrate. Scarr's (1959) medium contains Difco wort agar dissolved in a 45° Brix syrup containing 35 parts sucrose and 10 parts glucose. Van der Walt (1970) recommended the use of agars with 50 and 60% (wt/wt) glucose for isolation of osmotolerant and osmophilic yeasts, respectively, while Mossel (1951) preferred a medium containing 60% (wt/wt) fructose because of its higher solubility. Pitt (1975) provides recipes for several media containing up to 60% glucose for the isolation and identification of xerotolerant yeasts and molds. For detection of yeasts by membrane filtration of sugar syrups and fruit juices, the nutritionally rich medium of Devillers (1957) with 45° to 50° Brix sugar concentration is recommended.

Nutritionally Limited Media.—Media based on single carbon, nitrogen, or vitamin sources have been used widely for isolation of specific groups of yeasts. Lysine (Walters and Thistleton 1953) and ethylamine (van der Walt 1962) have been used for isolation of wild yeasts in brewery cultures. Similarly lactose can be used as a sole carbon source for isolation of *Kluyveromyces (Saccharomyces) lactis* when combined with incubation at 45°C. Nitrite has been proposed as a single nitrogen source for isolation of *Debaryomyces* spp. (Wickerham 1957), although some *Brettanomyces* spp. can also utilize nitrite (van der Walt 1963).

Differential Media.—Relatively few differential media are used in isolating and identifying fungi. Media which stimulate pigment production in yeasts are described by Beech and Davenport (1971). A differential medium containing ferric citrate permits rapid presumptive identification of *A. flavus* and *A. parasiticus* strains, all of which produce an intense yellow-orange coloration on this medium (Bothast and Fennell 1974).

It is sometimes useful to be able to isolate and/or enumerate fungi with particular metabolic characteristics, e.g., production of specific enzymes or toxins. Koburger (1972A) detected proteolytic activity in fungi using ¾-strength plate count agar reinforced with 0.4% gelatin and containing both chlortetracycline and chloramphenicol to inhibit bacteria. In the same study he detected lipolytic activity using a medium containing calcium chloride and Tween 40; lipolysis resulted in precipitation of calcium palmitate. Bours and Mossel (1973) compared tributyrin agar with a fat emulsion agar and Eijkman lipolysis plates. Of 346 spoilage yeasts isolated from margarine, only 70 (20%) hydrolyzed tributyrin, whereas 319 (92.2%) and 338 (97.7%) produced lipases on Eijkman and fat emulsion plates, respectively. Care must therefore be taken to select appropriate substrates for estimation of enzymic activities. Hankin and Anagnostakis (1975) and Anagnostakis and Hankin (1975) used a variety of media for detection of enzymic activities of fungi (Table 16.5).

Production of fluorescent pigments on a modified Czapek solution agar has been shown to correlate well with ability of A. flavus strains to produce aflatoxins (Hara et al. 1974). A similar approach has been used for some years by the author; the medium contains a thin layer of kieselgel beneath the agar on which fluorescent metabolites are adsorbed and concentrated. In tests of several hundred isolates about 20% false positive reactions and no false negatives were obtained when compared with results of aflatoxin analyses.

Quantitative Enumeration of Fungi

Counts of Fungal Propagules on Solid Media.—Enumeration of yeasts and molds on solid media utilizes the same basic techniques as are used by bacteriologists, i.e., pour plate and surface spread method and, for yeasts, the drop counts method of Miles and Misra (1938). Details of these standard procedures are given by Thatcher and Clark (1968).

A more recent approach to colony counts, which permits large savings in both materials and technician time, is the spiral plate method of Gilchrist et al. (1973) in which 7 μl of inoculum are spread in the form of an Archimedean spiral across the surface of an agar medium. The method has been shown to be extremely useful for enumerating both bacteria and yeasts, but it cannot be used for mold-infected samples, since the mold colonies spread rapidly across the inoculum lines. Since the lower limit of detection by this method is 500 to 1000 colony-forming units (cfu) per g of original sample (based on use of the primary 1 in 10 dilution), it cannot be applied to low-count samples.

TABLE 16.5

SUBSTRATES USED FOR DETECTING FUNGAL ENZYMES

Enzyme	Substrate	Detected on Solid Media by
Amylase	Soluble starch	Iodine reaction with residual starch in medium
Deoxyribonuclease	DNA	Toluidine blue reaction with, or acid precipitation of, residual DNA
	Tween 20[1]	Precipitation of calcium laurate
	Tween 80[1]	Precipitation of calcium palmitate
Lipase	Thin layer of fat[1,2]	Precipitation of calcium salts of fatty acids
	Emulsified fat	Zone of clearance in medium around colony
	Tributyrin	
Pectintranseliminase	Pectin at pH 7.0	Precipitation of pectin with hexadecyltrimethyl ammonium bromide
Polygalacturonase	Pectin at pH 5.0	
Protease	Gelatin	Precipitation of residual protein with acid or ammonium sulfate
	Casein	Zone of clearance in medium around colony

[1]It is essential to use a medium rich in calcium, e.g., whey agar.
[2]Poured into Petri dish, chilled, and overlayered with whey agar (Bours and Mossel 1973).

Opinions vary regarding the advantages of pour plate or surface plate methods, but Koburger and Norden (1975) found no significant differences between the methods in tests on a range of foods. The pour plate permits detection of fungi at relatively low levels of contamination since larger sample volumes can be used than with any surface plate method. Indeed, by using double strength agar it is possible to detect as few as one or two propagules in a 10g sample by mixing equal volumes of liquid sample and medium before pouring into one large, or several standard, Petri dishes. This procedure has been used successfully for assessing the level of fungal contamination of fruit sauces. The major disadvantages of pour plate techniques are that propagules may be thermally damaged by mixing with molten agar at 42° to 45°C and colonies are produced at various depths throughout the medium, making enumeration difficult.

Surface plating techniques permit all propagules to develop in a single plane under identical atmospheric conditions. However, the method requires plates with dry surfaces, yet they must not become case-hardened. Problems can arise also by uneven distribution of sample across the surface and by confluence of mold colonies.

For any quantitative analysis, at least duplicate plates should be prepared for each dilution level tested and for each culture medium and incubation condition. When both yeasts and molds are expected in a sample, plates should be examined at frequent intervals to prevent the possibility of colonies becoming obscured by overgrowth of molds. The need for this is alleviated to some extent by using media containing rose bengal (Jarvis 1973).

Enumeration of colonies of fungal propagules is best done manually; although instruments for automatic counting are commercially available, they lack the power of discrimination needed to differentiate between colonies of differing size and are unable to distinguish between large colonies and nearly confluent colonies.

Membrane Filtration Techniques.—The technique of membrane filtration is widely used in the food and beverage·industry and is particularly useful for detecting low-level contamination of samples. Liquid samples, or supernatant fluids of food homogenates clarified by low-speed centrifugation (not greater than 500 G), are filtered under negative or positive pressure through a membrane of cellulose acetate, nylon, or other material which is incubated on a pad soaked with culture medium. Details of various procedures for membrane filtration are given in the trade literature (Millipore 1967, 1973).

Membranes printed with a grid to aid colony counting are usually white but black membranes are sometimes used for counting translucent colonies of osmotolerant yeasts. For routine work a membrane pore size of 0.80 or 1.20μ m is normal, but smaller pore sizes $(0.45 \mu$m) may be used for very small yeasts. Halden *et al.* (1960) used membranes with 0.80 and 0.45 μm pore sizes to distinguish between two types of yeast.

The flow rate during filtration will depend upon the pore size of the membrane and the viscosity of the sample being filtered. Viscous samples such as sugar syrups are normally diluted before filtration but care must be taken to avoid osmotic shock to osmophilic species.

A new concept in membrane filtration (Millipore 1974) uses a plastic dip sampler containing an absorbent pad impregnated with dehydrated culture medium and covered with a 0.45 μm pore membrane. When the sampler is immersed in a liquid, a volume of 1 ml is drawn into the pad, the liquid rehydrating the medium and the organisms being retained on the membrane. The medium contains an unspecified bacterial inhibitor and is reported to be suitable for growth of a wide range of fungi. For detecting osmotolerant yeasts, it is essential to dilute the sample in a high (45° Brix) sugar diluent and to allow adequate time for penetration of the sample into the pad.

Counting of yeast and mold colonies on membranes, particularly after short incubation periods, is aided by use of a hand lens or, preferably, a stereoscopic microscope.

Most Probable Number Technique.—Liquid culture of yeasts and molds is normally undertaken on a qualitative basis, e.g., detection of film yeasts in brines (Etchells and Jones 1946). However, the level of fungal contamination can be determined by multiple inoculation methods in liquid media, most probable number (MPN) techniques being applied. Koburger and Norden (1975), using media containing antibiotics, demonstrated that MPN techniques gave consistently higher counts on a range of food samples than did either surface or pour plating techniques, and in several instances recovered fungi from samples which did not yield any fungal propagules on solid media. This may be related to resuscitation of sublethally damaged propagules, which often grow more readily at glass-liquid interfaces.

The major objection to MPN methods is the large quantity of medium required for any individual test and the low level of statistical significance which can be applied to quantitative results. However, the method permits estimation of much lower propagule numbers than can be achieved by plating procedures.

Qualitative Procedures

Fungi can be isolated from foods by pour plate, surface plate, or liquid culture systems, using a variety of media. Primary liquid cultures require plating onto agar to isolate the organisms. This may sometimes be useful for enrichment of damaged cells or fungi with specific physiological characteristics which represent only a small proportion of the total mycoflora.

Procedures which permit cultivation of spoilage fungi *in situ* are often useful in studies of ecological associations. Incubation of a sample on a bed of sterile filter paper pulp impregnated with sterile water (Keyworth 1951) or humectant solution permits direct observation of growth on the sample. Various modifications of this procedure exist (Booth 1971A), including the use of humidity chambers. The latter method was used by Denizel *et al.* (1976) to study the interactions of naturally occurring molds in relation to aflatoxin formation on pistachio nuts.

These methods can also be applied after surface sterilization of samples to remove gross contaminants, followed by dissection or grinding of the sample.

Incubation Conditions

Various authorities recommend a range of temperatures and incubation times for routine detection of molds and yeasts in foods (Table 16.6). Most food-associated fungi grow readily over the range 17° to 27°C (Koburger 1973), although higher temperatures must be used for thermophiles. Studies of fungi from frozen or refrigerated foods are best carried out at an incubation temperature within the range of 4° to 15°C.

In initial isolation studies it may be advantageous to incubate replicate cultures at, for example, 15°, 25°, and 32° to 37°C, in order to determine whether groups of fungi with different temperature requirements occur in the sample. Incubation at temperatures of 5° to 25°C will probably reflect more accurately the nature of the spoilage mycoflora and the potential quality of the food than will incubation at higher temperatures.

The duration of incubation is related to the types of organisms present and to the incubation temperature, 4 to 5 weeks being needed at 4°C but only 5 to 7 days at 25°C. Osmophiles may require incubation for several weeks before significant growth occurs in some media (Tilbury 1976). Under such circumstances it is essential to prevent dehydration of the media during incubation.

Plates are normally incubated in darkness, but brief exposure to ultraviolet or visible light encourages sporulation and pigment production by fungi. The effects of light on growth of fungi have been reviewed in detail by Leach (1971).

MICROSCOPIC METHOD OF ANALYSIS

Low Power and Stereoscopic Microscopy

Direct Microscopic Examination.—Examination of foods by stereoscopic microscopy at magnifications up to 400X can provide data on

TABLE 16.6

INCUBATION CONDITIONS FOR YEAST AND MOLD COLONY COUNTS

Food Type	Incubation Temperature (°C)	Time (Days)	Reference
Meats and meat products, fruit juices and concentrates	21	5	Am. Public Health Assoc. (1966)
Dairy products	21 or 25	5	Am. Public Health Assoc. (1967)
	22	3 or 5	Ministry of Agriculture, Fisheries, and Food (1968)
Other foods	32	3 to 5	Am. Public Health Assoc. (1966)

mold colonization and, where growth is extensive, can permit rapid generic identification of many molds. The method provides a rapid screen for the overt quality of stored agricultural products and foods.

Microscopic examination of sections of plant materials can provide information on fungal invasion of deep tissues but it is often necessary to stain the sections using multiple stain systems, such as Pianese IIIb [Martius yellow, 0.01 g; malachite green, 0.50 g; acid fuchsin, 0.10 g; 95% (vol/vol) ethyl alcohol, 50 ml; distilled water, 150 ml] which stains the plant tissue green and mycelium deep pink.

Low power (100X, 400X) microscopy of food homogenates prepared as wet-mount slides or in a hemocytometer provides an indication of the numbers of yeast cells and/or the amount of fungal mycelium. However, microscopic examination gives little information regarding the viability of the organisms since so-called "vital staining" rarely correlates well. Direct microscopic examination of Sellotape impressions from surfaces, or of Sellotape-agar slide cultures, provides a rapid indication of relative contamination levels and the latter method permits rapid assessment of viable cells after a brief period of incubation (Beech and Davenport 1971).

Microscopic examination of membrane filters after staining and clearing provides a rapid and convenient method for direct counting of cells. After filtration of the sample, organisms on the membrane are stained with a dye such as Ponceau S and then fixed with acetic acid. The membrane is placed onto a glass microscope slide and dried by passing through a smokeless burner flame. The membrane is cleared with immersion oil and examined under a cover slip. Counts of cells are made for a proportion of the membrane grid squares. Viable counts can be made by this procedure after incubation of the membrane on a suitable culture medium for sufficient time to permit microcolony development. Even after incubation for 2 to 3 days it may be advantageous to examine membranes by stereoscopic microscope for enumeration of small colonies.

Fluorescent Microscopy.—The use of fluorescent materials as microscopic tracers has been a subject of growing interest during recent years but has still not been fully explored by mycologists. Cultivation of organisms on a medium containing a fluorescent brightener (e.g., Tinopal 4BMT; Geigy) results in uptake of sufficient brightener for the organisms to be detected readily using a microscope equipped for ultraviolet light microscopy. Details of procedures for staining with optical brighteners and of suitable optical filter combinations are given by Preece (1971A).

Fluorescent stains have been used to differentiate viable from nonviable yeasts following membrane filtration (Cranston and Calver 1974). After washing the membrane in dilute ammonium hydroxide, the yeasts are stained sequentially with Clayton yellow, acridine hydrochloride, and primulline and, after further washing, the membrane is cleared with xylene for microscopic analysis. Good correlation is reported between microscopic and cultural estimates of yeast cell numbers.

Although serological techniques are finding increasing use in medical mycology (Preece 1971B), their application to food mycology is still in its infancy. A major problem in fungal serology is the occurrence of "common antigens" which make many serological techniques relatively nonspecific. However, cross-reactions could potentially be used to provide rapid indications of general mold contamination of foods and beverages.

The area of serology which has shown greatest promise to date for food mycology is the use of immunofluorescent techniques (IMF) (Preece 1971A). Direct and indirect staining techniques have been applied for the detection of a wide range of strains, including *Absidia, Alternaria, Aspergillus, Botrytis, Candida, Cladosporium, Fusarium, Nocardia, Penicillium, Saccharomyces, Sporotrichum,* and *Torulopsis* (Preece and Cooper 1969; Preece 1971A; Warnock 1973). Details of slide culture (Booth 1971A) and staining procedures for IMF (Preece 1971A) have been reported.

Warnock (1973) used IMF to study the fungal invasion of barley grains by *Alternaria, Aspergillus,* and *Penicillium* spp. Denizel (1974) used indirect IMF to demonstrate the presence of *A. flavus* mycelium in the vascular bundle of pistachio nut shells which contained afla-

toxin-contaminated kernels. IMF permits the tracing of fungal penetration into food matrixes and confirmation that microscopic structures observed are fungal hyphae. However, care must be taken to
check the extent of autofluorescence in both food materials and fungal
mycelium, otherwise misleading results may be obtained.

Howard Mold Count

The Howard Mold Count (HMC) was developed in 1910 by Howard
(1911) for two purposes: (1) to provide manufacturers of tomato
products with a method for checking product quality, and (2) to enable
the United States Federal Food Law Enforcement Authority to prevent
interstate commerce of tomato products manufactured from moldy or
decomposed fruits. Since that time the method has been refined and
has been used as the basis of assessing the amount of moldy tomato
tissue for both trade and legislative control purposes. Details of the
Official Method are given by AOAC (1975) and an amplification of the
laboratory technique is given by Troy (1968).

Legislative standards and advisory specification (usually incorporated in a Code of Practice) based on the HMC have been adopted in
many countries over the past 60 years. There is no doubt that enforcement of HMC limits has resulted in improvements in the quality of
tomato and other fruit products. It is noteworthy that the method has
been recommended also for assessing mold levels in jams, syrups,
juices, pulps, spices, herbs, infant foods, butter, and other commodities. Nonetheless, the method is frequently criticized as being of low
precision and it is a matter of debate whether the method is suitable
for legislative enforcement on single samples taken at retail outlets.

The variance in count obtained by competent analysis is reported to
be of the order of ± 7.4% (Vas et al. 1959; Dakin 1964), a level which
permits a wide range of "counts" for replicate analyses on a single
sample. Such variations reflect the distribution of mold mycelium in
tomato solids, as well as the problems of microscopic interpretation
sometimes experienced even by skilled analysts. Furthermore, the
HMC method is sensitive to the effects of processes to which the
product may have been exposed, e.g., homogenization and concentration. Recent studies by the author have confirmed earlier results by
Eisenberg (1968) and by other workers that comminution of tomato
products renders the HMC procedure invalid since mold hyphae are
fragmented and spread throughout the product. For this reason the
HMC should not be applied to comminuted fruit products. A more
objective analytical technique is required, such as estimation of
fungal chitin.

A procedure sometimes used as an adjunct to the HMC on fruit is the
rot fragment count, full details of which are given by AOAC (1975).

Such techniques are also subject to wide laboratory and interpretative errors.

Machinery Mold Count

A method for the quantitation of *Geotrichum candidum* in canned fruits and vegetables and their beverages and juices has been adopted as official first action by the United States Food and Drug Administration (Cichowicz and Eisenberg 1974; AOAC 1975). The method is applicable to products from which *Geotrichum*, referred to as machinery mold, can be recovered without being masked by large amounts of plant tissue. The organism will grow over a wide range of temperature and pH values. Counts of *Geotrichum* hyphal fragments in processed foods have shown good correlation with conditions of sanitation in the processing plant. Differentially stained machinery mold clumps have a characteristically feathered appearance as a result of regularly spaced, 45° angle branches. Walls of uniform length cells in septate hyphae are parallel, except for those of terminal cells which taper gradually or abruptly (Fig. 16.3). The machinery mold method for assessing in-plant sanitation is not without serious problems. Other molds can interfere with *Geotrichum* counts, plant materials can mask machinery mold fragments, and there is presently a lack of trained and experienced analysts to perform the technique.

INDIRECT QUANTITATIVE METHODS

Estimation of Cell Wall Chitin

The cell walls of many molds differ from those of plants, vertebrate animals, bacteria, and yeasts since they contain a high proportion of chitin (poly-β-1:4-N-acetyl-glucosamine) in association with other structural polysaccharides such as mannan, glucan, and cellulose (Bartnicki-Garcia 1968). The chitin content of various mold mycelia ranges from about 20 to 80 μg per mg dry weight (measured as glucosamine) (Ride and Drysdale 1971, 1972; Swift 1973; Jarvis 1976). The level of chitin varies somewhat between mold species, with the age of the mycelium, and with the analytical method applied. In the author's experience the analytical technique proposed by Ride and Drysdale (1972) is the method of choice; chitin is hydrolyzed by alkali to chitosan, which is precipitated with alcohol, and residual glucosamine is determined colorimetrically. The method has been applied successfully to the assessment of mold in various fruit products. It has also been shown to be a very sensitive tool for the estimation of mold in cereal grains (Austwick 1976).

The greatest potential disadvantages for estimation of mold by chitin assay are (1) variation in intrinsic glucosamine levels in the

Courtesy of Dr. W. V. Eisenberg, U.S. Food and Drug Administration
FIG. 16.3. TYPICAL FRAGMENTS OF *GEOTRICHUM CANDIDUM* (MACHINERY MOLD). An unstained specimen is shown in Fig. 1.21.

food, (2) contamination by invertebrates which contain chitin in their exoskeletons, and (3) deliberate use of higher fungi as sources of protein. Allowances can be made for average intrinsic glucosamine levels and, provided that insect infestation can be eliminated, there is no reason why chitin analysis could not be used to estimate levels of mold contamination of many food commodities. The method might also find application in controlling levels of added fungal protein in foodstuffs.

Estimation of Metabolic Activity

Carbon Dioxide Production.—Ingram (1960) developed a simple method for detecting yeasts in fruit juices and beverages which is sensitive to a single viable yeast cell per 100 ml of sample. The test material is diluted with an equal volume of 4% (wt/vol) dibasic sodium phosphate solution contained in a 500 ml medical flat bottle which is incubated on its side to expose a large surface area. A fermentation lock, attached to the bottle by means of a plugged glass tube, contains a borax-thymol blue indicator solution. Evolution of carbon dioxide is detected by a color change in the indicator long before active fermentation becomes evident.

In some instances (e.g., detection of osmotolerant yeasts in the presence of osmosensitive strains) it may be desirable to dilute the sample in phosphate dissolved in a 45° Brix sugar syrup in order not to reduce excessively the osmolarity of the fruit juice. Under such conditions the rate of fermentation is slower but appropriate incubation periods can be derived for various contamination levels.

Although not strictly quantitative, the method provides an indication of very low levels of contamination in a shorter time than could be achieved by many other methods. Mossel (1951) previously recommended the use of Einhorn tubes for a test which permitted detection of both aerobic and anaerobic carbohydrate metabolism.

The Ingram (1960) method can be used to assess preservative inhibition of realistically low numbers of yeast cells suspended in liquid substrates. If required, the method can be made more quantitative by the use of micromanometers to assess the rate of evolution of carbon dioxide.

Instrumental Methods.—The growing requirement for rapid microbiological methods of food and beverage analysis has led to the development of specialized analytical instrumentation. Although not developed specifically for estimation of fungi, some of the techniques may find future application in food and beverage mycology.

Electrical Impedance Measurements.—Growth of microorganisms results in minute changes in the electrical conductivity of culture media as a consequence of metabolism. Apparatus to detect changes in conductivity have been developed commercially; the test culture is held in specially constructed cells fitted with metal electrodes (Ur and Brown 1974; Wheeler and Goldschmidt 1975). Although the method gives an indication of microbial activity which can be related to cell numbers, it lacks sensitivity and many food constituents cause interference.

Determination of Microbial ATP.—The Luminescence Biometer (DuPont Instruments) provides a simple method for rapid detection of picogram quantities of adenosine triphosphate (ATP) based on the luciferin-luciferase photochemical reaction. Details of the method are given by Sharpe (1973). Unfortunately, the intrinsic levels of ATP in many foods exceed the levels present in food-borne microbial cells, so that the future application of the method for food analysis is limited (Sharpe et al. 1970; Williams 1971; Sharpe 1973).

Particle Counting.—Several electronic particle counters are available commercially and have been used for enumeration of yeast cells in mixed culture (see, for instance, Drake and Tsuchiya 1973). Work in the author's laboratory has been directed toward use of the Coulter Counter (Coulter Electronics) for viable counts of microcolonies, based on the earlier work of Surhen (1975). While the method has proved successful for bacteria, only limited success has been achieved with yeasts which do not readily maintain microcolony status after chemical fixing.

Other Approaches.—The prognosis for quantitative instrumental analysis of microorganisms would be improved if cells could be separated from food particles, "cleaned," and concentrated before analysis. Wood *et al.* (1976) showed that it is possible to separate microorganisms from food homogenates using short columns of ion exchange resin. Yeasts and mold spores are readily adsorbed to the resins and can be eluted almost quantitatively by appropriate changes in eluant. Good recoveries of yeasts from wines and other beverages have been achieved, but to date the quantitative recovery of yeasts and molds from foods has not been resolved.

KEEPING-QUALITY TESTS

Assessment of the potential shelf-life of nonsterile preserved foods (e.g., sauces, pickles, intermediate moisture foods, etc.) can be achieved by monitoring the levels of particular groups of microorganisms both before and after various periods of storage (Tuynenburg-Muys 1965). The counts obtained following storage should not be significantly greater than those determined initially if the food is adequately preserved. Mossel (1975) has interpreted "significantly" as meaning "not greater than three times the initial count," but such an increase may not allow adequately for the vagaries of viable estimation methods for molds, or for uneven distribution of organisms in foods. If required, stability tests can be accelerated to some degree by increasing the temperature of incubation above ambient, but care must then be used in interpretation of results, since different populations of organisms may grow at elevated temperatures.

It has been suggested that such procedures might be used to establish stability specifications for food commodities marketed between countries in the European Economic Community. The approach has been used also for challenge testing preserved foods with laboratory-prepared inocula. In challenge testing it is essential to ensure (1) an appropriate choice of microorganisms, (2) a uniform inoculum suspension, (3) even dispersal of the inoculum on, or in, the food, (4) multireplicate analyses at all stages of testing, and (5) uniformity of analytical procedures, especially between different collaborating laboratories.

INTERPRETATION OF MYCOLOGICAL RESULTS

The interpretation of laboratory results depends in part on the tests undertaken, which in turn depend on the nature of the sample, the conditions under which it has been stored, and the purposes for which it is to be used. Microscopic examination permits a rapid estimate of the mycological quality of a commodity but does not differentiate between live and dead organisms. However, it provides a historical appraisal of the sample since high numbers of fungal cells will generally indicate heavy growth at some stage prior to analysis. It is particularly useful for foods which have been subjected to thermal processing, or those having biocidal preservative properties. Cultural tests may be selected to detect general fungal contamination, or may be directed toward particular groups of organisms. The significance of different fungi in foods and beverages has been considered in earlier chapters. The critical question of interpretation concerns levels of organisms in foods.

Colony counts of yeasts presuppose that individual colonies arise from single cells, clumps, or aggregates of cells. Organoleptic spoilage of many foods by yeasts is associated with counts of about 10^6 colony-forming units (cfu) per g, although spoiled fermented beverages can often have counts of only 10^4 cfu per ml. The difference in these levels undoubtedly relates to different forms of metabolism. The lower levels of cells (relative to bacteria) which cause organoleptic spoilage probably relate to the greater cell size and metabolic activity of yeasts.

Colony (or more correctly, propagule) counts of molds are more difficult to interpret. In samples where mold growth is not active, the count of propagules will indicate the level of contamination by viable mold conidia (or other spore forms) plus hyphal fragments produced by homogenization of preformed mycelium. Such counts can be useful for assessing the quality of dry food ingredients and post-process contamination of cooked products. With highly selective conditions of incubation, counts can also be of value in indicating the number of propagules likely to grow in a food of specific composition (e.g., the water phase of fat-water emulsions; Tuynenburg-Muys 1965).

However, when active mold growth has occurred, high propagule counts will result not only from the presence of viable spores, but also from fragments of mycelium and disrupted spore structures. The degree to which fragmentation will occur is dependent upon the conditions and type of homogenization used; while still disrupting spore structures, processes such as stomaching have less effect on mycelium than mechanical maceration. More objective methods of analysis are therefore desirable when active mold growth has occurred.

Overt mold growth is normally taken as *prima facie* evidence that the food is unfit for consumption, although many consumers and some manufacturers are still prepared to scrape or strain mold growth from the surface of foods or beverages. The potential hazards of such practices have been discussed elsewhere (Jarvis 1975); it is essential nowadays to consider fungi not only as spoilage organisms, but also as potential health hazards.

Acknowledgments

I am grateful to colleagues in the Microbiology section and the Information section at the Leatherhead Food R.A. for their assistance and for helpful comments and suggestions during the preparation of this chapter. In particular I should like to acknowledge the continuing assistance of my wife, Marjorie, in searching the literature and for checking the bibliography.

REFERENCES

ACHA, I. G., and VILLANUEVA, J. R. 1961. A selective medium for the formation of ascospores by *Aspergillus nidulans*. Nature (London) *189*, 328.

AM. PUBLIC HEALTH ASSOC. 1966. Recommended Methods for the Microbiological Examination of Foods. Am. Public Health Assoc., New York.

AM. PUBLIC HEALTH ASSOC. 1967. Standard Methods for the Examination of Dairy Products. Am. Public Health Assoc., New York.

ANAGNOSTAKIS, S. L., and HANKIN, L. 1975. Use of selective media to detect enzyme production by microorganisms in food products. J. Milk Food Technol. *38*, 570-572.

ANGELOTTI, R., WILSON, J. L., LITSKY, W., and WALTER, W. G. 1964. Comparative evaluation of the cotton wool swab and Rodac methods for the recovery of *Bacillus subtilis* spore contamination from stainless steel surfaces. Health Lab. Sci. *1*, 289-296.

ASSOC. OFF. ANAL. CHEM. 1975. Official Methods of Analysis of the AOAC, 12th Edition. Assoc. Off. Anal. Chem., Washington, D.C.

AUSTWICK, P. K. C. 1976. Personal communication. London.

BARTNICKI-GARCIA, S. 1968. Cell wall chemistry. 1. Morphogenesis and taxonomy of fungi. Annu. Rev. Microbiol. *22*, 87-108.

BEECH, F. W., and CARR, J. G. 1955. A survey of inhibitory compounds for the separation of yeast and bacteria in apple juices and ciders. J. Gen. Microbiol. *12*, 85-94.

BEECH, F. W., and DAVENPORT, R. R. 1971. Isolation, purification and maintenance of yeasts. *In* Methods in Microbiology, Vol. 4. C. Booth (Editor). Academic Press, London.

BELL, D. K., and CRAWFORD, J. L. 1967. A botran-amended medium for isolating *Aspergillus flavus* from peanuts and soil. Phytopathology *57*, 939-941.

BLOOD, R. M. 1976. Personal communication. Leatherhead, England.

BOARD, R. G., and LOVELOCK, D. W. 1973. Sampling—Microbiological Monitoring of Environments. Academic Press, London.

BOOTH, C. 1971A. Introduction to general methods. *In* Methods in Microbiology, Vol. 4. C. Booth (Editor). Academic Press, London.

BOOTH, C. 1971B. Fungal culture media. *In* Methods in Microbiology, Vol. 4. C. Booth (Editor). Academic Press, London.

BOTHAST, R. J., and FENNELL, D. I. 1974. A medium for rapid identification and enumeration of *Aspergillus flavus* and related organisms. Mycologia *66*, 365-369.

BOURS, J., and MOSSEL, D. A. A. 1973. A comparison of methods for the determination of lipolytic properties of yeasts, mainly isolated from margarine, moulds and bacteria. Arch. Lebensmittelhyg. *24*, 197-203.

BOWEN, J. F., and BEECH, F. W. 1967. Yeast flora of cider factories. J. Appl. Bacteriol. 30, 475-483.

CARR, J. G. 1956. The occurrence and role of lactic acid bacteria in cider making. Ph.D. Thesis. University of Bristol, England.

CATE, L. TEN 1965. A note on a simple method of bacteriological sampling by means of agar sausages. J. Appl. Bacteriol. 28, 221-223.

CHRISTENSEN, C. M. 1946. The quantitative determination of molds in flour. Cereal Chem. 23, 322-329.

CICHOWICZ, S. M., and EISENBERG, W. V. 1974. Collaborative study of the determination of Geotrichum mold in selected canned fruits and vegetables. J. Assoc. Off. Anal. Chem. 57, 957-960.

CORRY, J. E. L. 1976. The effect of sugars and polyols on the heat resistance and microscopic morphology of osmophilic yeasts. J. Appl. Bacteriol. 40, 269-276.

CRANSTON, P. M., and CALVER, J. H. 1974. Quantitative fluorescent microscopy of yeasts in beverages. Food Technol. Aust. 26, 15-17.

DAKIN, J. C. 1964. The validity of the Howard Mould Count as a means of assessing the quality of tomato puree—a review. Food Trade Rev. 34, 41-44, 69.

DAVENPORT, R. R. 1967. The microflora of cider apple fruit buds. Rep. Agric. Hortic. Res. Stn., Univ. Bristol for 1966, 246-248.

DAVENPORT, R. R. 1973. Vineyard yeasts, an environmental study. In Sampling—Microbiological Monitoring of Environments. R. G. Board and D. W. Lovelock (Editors). Academic Press, London.

DAVIES, R. R. 1971. Air sampling for fungi, pollens and bacteria. In Methods in Microbiology, Vol. 4. C. Booth (Editor). Academic Press, London.

DAVIS, J. G. 1931. Standardization of media in the acid ranges with special reference to the use of citric acid and buffer mixtures for yeast and mold media. J. Dairy Res. 3, 133-141.

DAVIS, J. G. 1958. A convenient semi-synthetic medium for yeast and mould counts. Lab. Pract. 7, 30.

DENIZEL, T. 1974. Factors affecting aflatoxin formation in Turkish pistachio nuts. Ph.D. Thesis. Univ. of Reading, England.

DENIZEL, T., ROLFE, E. J., and JARVIS, B. 1976. Moisture-equilibrium relative humidity relationships in pistachio nuts with particular regard to control of aflatoxin formation. J. Sci. Food Agric. 27, 1027-1034.

DEVILLERS, P. 1957. Enumeration of osmophilic yeasts and molds in sucrose and syrups. Ind. Aliment. Agric. 74, 269-271. (French)

DRAKE, J. F., and TSUCHIYA, H. M. 1973. Differential counting in mixed cultures with Coulter counters. Appl. Microbiol. 26, 9-13.

EISENBERG, W. V. 1968. Mold counts of tomato products as influenced by different degrees of product comminution. Q. Bull. Assoc. Food Drug. Off., U.S. 32, 173-179.

ENDO, R. M. 1966. A cellophane tape-cover glass technique for preparing microscopic slide mounts of fungi. Mycologia 58, 655-659.

ETCHELLS, J. L., and JONES, I. D. 1946. Procedure for bacteriological examination of brined, salted and pickled vegetables and vegetable products. Am. J. Public Health 36, 1112-1123.

FLANNIGAN, B. 1974. The use of acidified media for the enumeration of yeasts and moulds. Lab. Pract. 23, 633-634.

FOSTER, W. D. 1960. Environmental staphylococcal contamination—a study by a new method. Lancet, 670-673.

FOX, J. G., and GOULD, W. A. 1976. Geotrichum candidum: The new FDA indicator of plant sanitation for the food processing industry. Res. Prog. Rep. Hortic. Ser. No. 437, 41-45. Ohio State Univ., Columbus.

GILCHRIST, J. E., CAMPBELL, J. E., DONELLY, C. B., PELLER, J. T., and DELANEY, J. M. 1973. Spiral plate method for bacterial determination. Appl. Microbiol. 25, 244-252.

GOLDBERG, H. S. 1959. Antibiotics. Van Nostrand, New York.

GREEN, S. R., and GRAY, P. P. 1950. A differential procedure applicable to bacteriological investigation in brewing. Wallerstein Lab. Commun. 13, 357-366.

GWYNNE-VAUGHAN, H. C. I., and BARNES, B. 1927. The Structure and Development of the Fungi. Cambridge Univ. Press, Cambridge, England.

HALDEN, H. E., LEETHAM, D. D., and EIS, F. G. 1960. Determination of microorganisms in sugar products by the Millipore method. J. Am. Soc. Sugar Beet Technol. 11, 137-142.

HANKIN, L., and ANAGNOSTAKIS, S. L. 1975. The use of solid media for detection of enzyme production by fungi. Mycologia 67, 597-607.

HARA, S., FENNELL, D. I., and HESSELTINE, C. W. 1974. Aflatoxin-producing strains of Aspergillus flavus detected by fluorescence of agar medium under ultraviolet light. Appl. Microbiol. 27, 1118-1123.

HOWARD, B. J. 1911. Tomato ketchup under the microscope with practical suggestions to insure a cleanly product. U.S. Dep. Agric., Bur. Chem. Circ. No. 68.

HUP, G., and STADHOUDERS, J. 1972. Comparison of media for the enumeration of yeasts and moulds in dairy products. Neth. Milk Dairy J. 26, 131-140.

ICMSF. 1974. Micro-organisms in Foods, Vol. 2. Sampling for Microbiological Analysis. Principles and Specific Applications. Univ. of Toronto Press, Toronto.

INGRAM, M. 1960. Fermentation tests to detect yeasts in fruit juices and similar products. Ann. Inst. Pasteur Lille 11, 203-208.

JARVIS, B. 1973. Comparison of an improved rose bengal chlortetracycline agar with other media for the selective isolation and enumeration of moulds and yeasts in foods. J. Appl. Bacteriol. 36, 723-727.

JARVIS, B. 1975. Mycotoxins in foods. In Microbiological Trends in Agriculture, Fisheries and Food. F. A. Skinner and J. G. Carr (Editors). Academic Press, London.

JARVIS, B. 1976. Unpublished data. Leatherhead, England.

JARVIS, B., and WILLIAMS, A. P. 1975. Unpublished data. Leatherhead, England.

KEYWORTH, W. G. 1951. A petri-dish moist chamber. Trans. Br. Mycol. Soc. 34, 291-292.

KOBURGER, J. A. 1970. Fungi in foods. I. Effect of inhibitor and incubation temperature on enumeration. J. Milk Food Technol. 33, 433-434.

KOBURGER, J. A. 1971. Fungi in foods. II. Some observations on acidulants used to adjust media pH for yeast and mold counts. J. Milk Food Technol. 34, 475-477.

KOBURGER, J. A. 1972A. Fungi in foods. III. The enumeration of lipolytic and proteolytic organisms. J. Milk Food Technol. 35, 117-118.

KOBURGER, J. A. 1972B. Fungi in foods. IV. Effect of plating medium pH on counts. J. Milk Food Technol. 35, 659-660.

KOBURGER, J. A. 1973. Fungi in foods. V. Response of natural populations to incubation temperatures between 12° and 32°C. J. Milk Food Technol. 36, 434-435.

KOBURGER, J. A., and FARHAT, B. Y. 1975. Fungi in foods. VI. A comparison of media to enumerate yeasts and molds. J. Milk Food Technol. 38, 466-468.

KOBURGER, J. A., and NORDEN, A. R. 1975. Fungi in foods. VII. A comparison of the surface, pour plate and most probable number methods for enumeration of yeasts and molds. J. Milk Food Technol. 38, 745-746.

LADIGES, W. C., FOSTER, J. F., and JORGENSEN, J. J. 1974. Comparison of media for enumerating fungi in precooked frozen convenience foods. J. Milk Food Technol. 37, 302-304.

LASZILO, E., and NYIREDY, I. 1974. The effect of shaking on the microbiological evaluation of feeds at the base dilution. Magy. Allatorv. Lapja 29, 612-616. (Hungarian)

LEACH, C. M. 1971. A practical guide to the effects of visible and ultra-violet light on fungi. In Methods in Microbiology, Vol. 4. C. Booth (Editor). Academic Press, London.

LeBARS, J. 1972. On obtaining a defined series of dilutions from a suspension of fungal germinal elements. Ann. Rech. Vet. 3, 435-447. (French)

LIN, Y. 1973. Detection of wild yeasts in the brewery: a new anterior. Brew. Dig. 48, 60-69.

LLOYD, A. C. 1975. Osmophilic yeasts in preserved ginger products. J. Food Technol. 10, 575-581.

LODDER, J. 1970, The Yeasts—A Taxonomic Study, 2nd Edition. North-Holland Publishing Co., Amsterdam.

LUND, A. 1956. Yeasts in nature. Wallerstein Lab. Commun. 19, 221-236.

MARSHALL, C. R., and WALKLEY, V. T. 1951. Some aspects of microbiology applied to commercial apple juice production. I. Distribution of microorganisms on the fruit. Food Res. 16, 448-458.

MARTIN, J. P. 1950. Use of acid, rose bengal and streptomycin in the plate method for estimating soil fungi. Soil Sci. 69, 215-232.

MILES, A. A., and MISRA, S. S. 1938. Estimation of the bactericidal power of blood. J. Hyg. 38, 732-749.

MILLER, E. J., and WEBB, N. S. 1954. Isolation of yeasts from soil with the aid of acid, rose bengal and oxgall. Soil Sci. 77, 197-204.

MILLIPORE CORP. 1967. Techniques for microbiological analysis. Bull. No. ADM 40, Millipore Corp., Bedford, Mass.

MILLIPORE CORP. 1973. Microbiological analysis of beverages. Bull. No. AB 601, Millipore Corp., Bedford, Mass.

MILLIPORE CORP. 1974. Samplers for monitoring microorganisms in liquids. Bull. No. PB 407, Millipore Corp., Bedford, Mass.

MILLIPORE CORP. 1975. Rapid Yeast Detection. Bull. No. AB 807, Millipore Corp., Bedford, Mass.

MINISTRY OF AGRICULTURE, FISHERIES, AND FOOD. 1968. Bacteriological Techniques for Dairy Purposes. Minist. Agric., Fisheries, Food Tech. Bull. 17, H. M. Stationery Office, London.

MOSSEL, D. A. A. 1951. Investigation of a case of fermentation in fruit juice products, rich in sugars. Antonie van Leeuwenhoek J. Microbiol. Serol. 17, 146-152.

MOSSEL, D. A. A. 1975. Occurrence, prevention and monitoring of microbial quality loss of foods and dairy products. CRC Crit. Rev. Environ. Control 5, 1-139.

MOSSEL, D. A. A., HARREWIJN, G. A., and VAN SPRANG, F. J. 1974. Microbiological quality assurance for weaning formulae. In The Microbiological Safety of Food. B. C. Hobbs and J. H. B. Christian (Editors). Academic Press, London.

MOSSEL, D. A. A., KLEYNEN-SEMMELING, A. M. C., and VINCENTIE, H. M. 1970. Oxytetracycline-glucose-yeast extract agar for selective enumeration of moulds and yeasts in foods and clinical material. J. Appl. Bacteriol. 33. 454-457.

MOSSEL, D. A. A., VEGA, C. L., and PUT, H. M. C. 1975. Further studies on the suitability of various media containing antibacterial antibiotics for the enumeration of moulds in food and food environments. J. Appl. Bacteriol. 39, 15-22.

MOSSEL, D. A. A., VISSER, M., and MENGERINK, W. H. J. 1962. A comparison of media for the enumeration of moulds and yeasts in foods and beverages. Lab. Pract. 11, 109-112.

OVERCAST, W. W., and WEAKLEY, D. J. 1969. An aureomycin rose bengal agar for enumeration of yeasts and molds in cottage cheese. J. Milk Food Technol. 32, 442-445.

PITT, J. I. 1975. Xerophilic fungi and the spoilage of foods of plant origin. In Water Relations of Foods. R. B. Duckworth (Editor). Academic Press, London.

PREECE, T. F. 1971A. Fluorescent techniques in mycology. In Methods in Microbiology; Vol. 4. C. Booth (Editor). Academic Press, London.

PREECE, T. F. 1971B. Immunological techniques in mycology. In Methods in Microbiology, Vol. 4. C. Booth (Editor). Academic Press, London.

PREECE, T. F., and COOPER, D. J. 1969. The preparation and use of a fluorescent antibody reagent for Botrytis cinerea grown on glass slides. Trans. Br. Mycol. Soc. 52, 99-104.

PUT, H. M. C., and CONWAY, C. C. 1974. The limitations of oxytetracycline for the enumeration of fungi in soil, feeds and foods in comparison with the selectivity obtained by globenicol (chloramphenicol). Arch. Lebensmittelhyg. 25, 73-83.

PUT, H. M. C., VAN REEVEN-BOUWHUIS, A., and KRUISWIJK, J. TH. 1968. Microbiological spoilage of pasteurized strawberries in syrup, due to Byssochlamys. Ann. Inst. Pasteur Lille 19, 171-189. (French)

REISS, J. 1972. A selective culture medium for the detection of Aspergillus flavus in mouldy bread. Zentralbl. Bakteriol. Parasitenkd. Infektionskr. Hyg. Abt. 1: Orig. Reihe A 220, 564-566. (German)

RIDE, J. P., and DRYSDALE, R. B. 1971. A chemical method for estimating Fusarium oxysporum f. lycopersici in infected tomato plants. Physiol. Plant Pathol. 1, 409-420.

RIDE, J. P., and DRYSDALE, R. B. 1972. A rapid method for the chemical estimation of filamentous fungi in plant tissue. Physiol. Plant Pathol. 2, 7-15.

SCARR, M. P. 1959. Selective media used in the microbiological examination of sugar products. J. Sci. Food Agric. 10, 678-681.

SEIDEL, H. 1973. Differentiation between brewery culture yeasts and "wild" yeasts. Part II. Experiences in the detection of "wild" yeasts by SDM (Schwarz Differential Medium). Brauwissenschaft 26, 179-183. (German)

SHARPE, A. N. 1973. Automation and instrumentation developments for the bacteriology laboratory. In Sampling—Microbiological Monitoring of Environments. R. G. Board and D. W. Lovelock (Editors). Academic Press, London.

SHARPE, A. N., WOODROW, M. N., and JACKSON, A. K. 1970. Adenosinetriphosphate (ATP) levels in foods contaminated with bacteria. J. Appl. Bacteriol. 33, 758-767.

SLOAN, B. J., ROUTIERI, J. B., and MILLER, V. P. 1961. Increased sporulation in fungi. Mycologia 52, 47-63.

SPLITTSTOESSER, D. F., KUSS, F. R., and HARRISON, W. 1970. Enumeration of Byssochlamys and other heat resistant molds. Appl. Microbiol. 20, 393-397.

SURHEN, G. 1975. Computer image analysis and electronic microcolony counting for the determination of microorganisms in milk. Arch. Lebensmittelhyg. 26, 10. (German)

SWIFT, M. J. 1973. The estimation of mycelial biomass by determination of hexosamine content of wood tissue decayed by fungi. Soil Biol. Biochem. 5, 321-332.

THATCHER, F. S., and CLARK, D. S. 1968. Microorganisms in Foods—Their Significance and Methods of Enumeration. Univ. of Toronto Press, Toronto.

TILBURY, R. H. 1976. The microbial stability of intermediate moisture foods with respect to yeasts. In Intermediate Moisture Foods. R. Davies, G. G. Birch, and K. J. Parker (Editors). Applied Science Publishers, London.

TROY, V. S. 1968. Mold Counting of Tomato Products, 3rd Edition. Continental Can Co., Chicago.

TUTTLEBEE, J. W. 1975. The Stomacher—its use for homogenization in food microbiology. J. Food Technol. 10, 113-122.

TUYNENBURG-MUYS, G. 1965. Microbiological quality of edible emulsions during manufacture and storage. Chem. Ind. (London) 1965, 1245-1250.

UR, A., and BROWN, D. F. J. 1974. Rapid detection of bacterial activity using impedance measurements. Biomed. Eng. 9, 18-20.

VAN DER WALT, J. P. 1962. Utilization of ethylamine by yeasts. Antonie van Leeuwenhoek J. Microbiol. Serol. 28, 91-96.

VAN DER WALT, J. P. 1963. Nitrite utilization by Brettanomyces. Antonie van Leeuwenhoek J. Microbiol. Serol. 29, 52-56.

VAN DER WALT, J. P. 1970. Criteria and methods used in classification. In The Yeasts —A Taxonomic Study, 2nd Edition. J. Lodder (Editor). North-Holland Publishing Co., Amsterdam.

VAN DER WALT, J. P., and VAN KERKEN, A. E. 1961. The wine yeasts of the Cape. Part IV. Ascospore formation in the genus Brettanomyces. Antonie van Leeuwenhoek J. Microbiol. Serol. 26, 292-296.

VAS, K., FABRI, I., KUTZ, N., LANG, A., ORBANYI, T., and SZABO, G. 1959. Factors involved in the interpretation of mold counts of tomato products. Food Technol. 13, 318-322.

WALTER, A. H. 1967. Hard surface disinfection and its evaluation. J. Appl. Bacteriol. 30, 56-65.

WALTERS, L. S., and THISTLETON, M. R. 1953. Utilization of lysine by yeasts. J. Inst. Brew. London 59, 401-404.

WARNOCK, D. W. 1973. Use of immunofluorescence to detect mycelium of Alternaria, Aspergillus and Penicillium in barley grains. Trans. Br. Mycol. Soc. 61, 547-552.

WHEELER, T. G., and GOLDSCHMIDT, M. C. 1975. Determination of bacterial cell concentration by electrical measurement. J. Clin. Microbiol. 1, 25-29.

WICKERHAM, L. J. 1957. Presence of nitrite-assimilating species of Debaryomyces in lunch meats. J. Bacteriol. 74, 832-833.

WILLIAMS, M. L. B. 1971. The limitations of the DuPont Biometer in the microbiological analysis of food. Can. Inst. Food Technol. J. 4, 187-189.

WOOD, J. M., JARVIS, B., and WISEMAN, A. 1976. The separation of microorganisms from foods. J. Sci. Food Agric. 27, 783-784.

Appendix

Regulatory Action Levels for Mold Defects in Foods

The United States Food and Drug Administration maintains a compilation of current levels for natural or unavoidable defects in food for human use that present no health hazard. Products that may be harmful to consumers are acted against by the Food and Drug Administration on the basis of their hazard to health, whether or not they exceed defect action levels. Poor manufacturing practices may also result in regulatory action, whether foods produced by the manufacturer are above or below the defect level.

An array of types of defects are under constant review by the Food and Drug Administration. As technology improves, attempts are made to lower the action levels. Defects listed in current levels include the presence of rodent hairs, insects, insect fragments, insect infestation and/or damage, worms, excreta, pit or shell fragments, ash, and/or mold. Action levels for mold defects are usually based on a percentage, either by weight or count, of the food containing mold or a combination of mold and other defects. Table A.1 lists current action levels for natural or unavoidable mold defects in food for human use that present no health hazard. A complete list of action levels for other defects can be obtained by writing to the Department of Health, Education, and Welfare, Public Health Service, Food and Drug Administration, 200 C Street, S. W., Washington, D. C. 20204. Procedures for determining defect levels in foods are given in *Official Methods of Analysis of the Association of Analytical Chemists, 12th Edition* (1975), published by the Association of Official Analytical Chemists, P.O. Box 540, Benjamin Franklin Station, Washington, D.C. 20044.

TABLE A.1

CURRENT LEVELS FOR NATURAL AND UNAVOIDABLE DEFECTS DUE TO MOLDS IN FOOD FOR HUMAN USE THAT PRESENT NO HEALTH HAZARD

Product	Defect Action Level
Chips and specialty items	
Potato chips	6% of the chips by weight contain rot
Chocolate and chocolate products	
Cacoa beans	4% by count show mold or 4% by count insect infested or damaged; or total of 6% by count show mold and are insect infested
Coffee beans	10% by count are insect infested, insect damaged, or show mold
Fruits	
Drupelet berries (blackberries, raspberries, etc.; canned or frozen)	Microscopic mold count average of 60%
Cherries (fresh, canned, or frozen)	Average of 7% rejects due to rot
Citrus fruit juices (canned)	Microscopic mold count average of 10%
Dates (whole or pitted)	5% by count rejects (moldy, dead insects, insect excreta, sour, dirty, and/or worthless) determined by macroscopic sequential examination
Figs	Average of 10% by count insect infested and/or show mold and/or dirty pieces of fruit
Olives (salt-cured)	Average of 25% by count showing mold
Peaches (canned)	Average of 5% wormy or moldy fruit by count or 4% if a whole larva or equivalent is found in 20% of the cans
Pineapple (canned, crushed)	Microscopic mold count average of 30%
Plums (canned)	Average of 5% by count with rot spots larger than the area of a circle 12 mm in diam
Prunes (dried and dehydrated, low moisture)	Average of 5% by count insect infested and/or showing mold and/or dirty fruits or pieces of fruit
Raisins	Average of 5% by count of natural raisins showing mold
Strawberries (frozen; whole or sliced)	Microscopic mold count average of 45% and mold count of 55% in one-half of the subsamples

TABLE A.1 *(Continued)*

Product	Defect Action Level
Jams, jellies, and fruit butters	
Apple butter	Microscopic mold count average of 12%
Black currant jam	Microscopic mold count average of 75%
Cherry jam	Microscopic mold count average of 30%
Nuts	
Tree nuts	Reject nuts (insect infested, rancid, moldy, gummy and shriveled, or empty shells) determined by macroscopic examination of the following limits:

Unshelled		*Shelled*	
Almonds	5%	Almonds	5%
Brazils	10%	Brazils	5%
Green chestnuts	15%	Cashews	5%
Baked chestnuts	10%	Dried chestnuts	5%
Filberts	10%	Filberts	5%
Pecans	10%	Pecans	5%
Pistachios	10%	Pistachios	5%
Walnuts	10%	Walnuts	5%
Lichee nuts	15%	Pili nuts	10%
Pili nuts	15%		

Mixed nuts in shell

The above percentage of reject nuts for any one type of nut applies to that type of nut in a mixture. The above limits apply for orchard type insect infestation.

Product	Defect Action Level
Peanuts	*Unshelled*—Average of 10% insect infested; moldy; otherwise decomposed; blanks and shriveled nuts
	Shelled—Average of 5% insect infested; moldy; otherwise decomposed; shriveled nuts
Spices	
Allspice	Average of 5% moldy berries by weight
Bay (laurel) leaves	Average of 5% moldy pieces by weight
Capsicum	Average of 3% insect infested and/or moldy pods by weight; microscopic mold count average of 20% for powder

TABLE A.1 *(Continued)*

Product	Defect Action Level
Cassia or cinnamon (whole)	Average of 5% moldy pieces by weight
Ginger (whole)	Average of 3% moldy and/or insect infested pieces by weight
Leafy spices, other than bay leaves	Average of 5% insect infested and/or moldy pieces by weight
Mace	Average of 3% insect infested and/or moldy pieces by weight
Nutmegs	Average of 10% insect infested and/or pieces showing mold by count
Pepper	Average of 1% insect infested and/or moldy pieces by weight
Sesame seeds	Average of 5% insect infested or decomposed seeds by weight
Tomatoes and tomato products	
Canned tomatoes packed in tomato puree	Microscopic mold count average of the drained packing media is 25%
Canned tomatoes, with or without added tomato juice	Microscopic mold count average of the drained juice is 12%
Pizza sauce (based on 6% total tomato solids after pulping)	Microscopic mold count average of 30%
Tomato catsup	Microscopic mold count average of 30%
Tomato juice	Microscopic mold count average of 20%
Tomato paste or puree	Microscopic mold count average of 40%
Tomato sauce (undiluted)	Microscopic mold count average of 40%
Tomato soup and other tomato products	Microscopic mold count average of 40%
Vegetables	
Beets (canned)	Average of 5% by weight of pieces with dry rot
Greens (canned)	Average of 10% of leaves by count or weight showing mildew 1.27 cm in diam
Miscellaneous	
Corn husks (for tamales)	5% by weight of the corn husks examined are insect infested (including insect damaged) or moldy

Index

Other AVI Books

BASIC FOOD CHEMISTRY
  Lee
BEVERAGES: CARBONATED AND NONCARBONATED
  Woodroof and Phillips
CARBOHYDRATES AND HEALTH
  Hood, Wardrip and Bollenback
CITRUS SCIENCE AND TECHNOLOGY
  Vol. 1 and 2 *Nagy, Shaw and Veldhuis*
DAIRY TECHNOLOGY AND ENGINEERING
  Harper and Hall
EVALUATION OF PROTEINS FOR HUMANS
  Bodwell
FOOD COLLOIDS
  Graham
FOOD MICROBIOLOGY: PUBLIC HEALTH AND
SPOILAGE ASPECTS
  deFigueiredo and Splittstoesser
FOOD PRODUCTS FORMULARY
  Vol. 1 *Komarik, Tressler and Long*
  Vol. 2 *Tressler and Sultan*
  Vol. 3 *Tressler and Woodroof*
FOOD PROTEINS
  Whitaker and Tannenbaum
FOOD QUALITY ASSURANCE
  Gould
IMMUNOLOGICAL ASPECTS OF FOODS
  Catsimpoolas
INTRODUCTORY FOOD CHEMISTRY
  Garard
MICROBIOLOGY OF FOOD FERMENTATIONS
  Pederson
NUTRITIONAL EVALUATION OF FOOD PROCESSING
  2nd Edition *Harris and Karmas*
POSTHARVEST BIOLOGY AND HANDLING OF FRUITS
AND VEGETABLES
  Haard and Salunkhe
PRACTICAL FOOD MICROBIOLOGY AND TECHNOLOGY
  2nd Edition *Weiser, Mountney and Gould*
PRINCIPLES OF FOOD CHEMISTRY
  deMan
RHEOLOGY AND TEXTURE IN FOOD QUALITY
  deMan, Voisey, Rasper and Stanley
SYSTEMS ANALYSIS FOR THE FOOD INDUSTRY
  Bender, Kramer and Kahan
THE TECHNOLOGY OF FOOD PRESERVATION
  4th Edition *Desrosier and Desrosier*